Fourth Edition

Essential Mathematics

Margaret L. Lial
American River College

Stanley A. Salzman
American River College

With the assistance of
Jill K. Owens, Diablo Valley College

PEARSON

Boston Columbus Indianapolis New York San Francisco Upper Saddle River
Amsterdam Cape Town Dubai London Madrid Milan Munich Paris Montréal Toronto
Delhi Mexico City São Paulo Sydney Hong Kong Seoul Singapore Taipei Tokyo

Editorial Director	Christine Hoag
Editor in Chief	Maureen O'Connor
Executive Content Editor	Kari Heen
Content Editor	Christine Whitlock
Senior Content Editor	Lauren Morse
Assistant Editor	Rachel Haskell
Senior Managing Editor	Karen Wernholm
Senior Production Project Manager	Kathleen A. Manley
Digital Assets Manager	Marianne Groth
Supplements Production Coordinator	Kerri Consalvo
Media Producer	Stephanie Green
Software Development	Rebecca Williams, MathXL; Mary Durnwald, TestGen
Marketing Manager	Rachel Ross
Senior Author Support/Technology Specialist	Joe Vetere
Rights and Permissions Advisor	Cheryl Besenjak
Image Manager	Rachel Youdelman
Procurement Manager	Evelyn Beaton
Procurement Specialist	Debbie Rossi
Media Procurement Specialist	Ginny Michaud
Associate Director of Design	Andrea Nix
Senior Designer	Barbara Atkinson
Text Design, Production Coordination, Composition, and Illustrations	Cenveo Publisher Services/ Nesbitt Graphics, Inc.
Cover Image	*Between Seasons* © Lorraine Cota Manley

For permission to use copyrighted material, grateful acknowledgment is made to the copyright holders on page I-4, which is hereby made part of this copyright page.

Library of Congress Cataloging-in-Publication Data

Lial, Margaret L.

Essential mathematics. / Margaret L. Lial, Stanley A. Salzman. —4th ed.

p. cm.

Includes index.

ISBN: 978-0-321-84505-4

1. Arithmetic—Textbooks. I. Salzman, Stanley A. II. Title.

QA107.2.L53 2013

513—dc23

2012013807

2 3 4 5 6 7 8 9 10—V011—16 15 14 13

www.pearsonhighered.com

ISBN 13: 978-0-321-84505-4
ISBN 10: 0-321-84505-6

This book is dedicated to Margaret L. Lial

Always passionate about mathematics and teaching,

Always a valued colleague, a mentor, and a friend,

Always in our memory.

In appreciation of your lasting support and never-ending enthusiasm: family, colleagues, and more than a generation of motivated students.

Stan Salzman

Contents

Preface

The fourth edition of *Essential Mathematics* continues our ongoing commitment to provide the best possible text and supplements package that will help instructors teach and students succeed. To that end, we have addressed the diverse needs of today's students by creating a tightly coordinated text and technology package that includes integrated activities to help students improve their study skills, an attractive design, updated applications and graphs, helpful features, and careful explanations of concepts. We've also expanded the supplements and study aids. We've revamped the video series into a complete Lial Video Library with expanded video coverage and new, easier navigation. And we've added the new Lial MyWorkBook. We have also responded to the suggestions of users and reviewers and have added many new examples and exercises based on their feedback.

The program is designed to help students achieve success in a developmental mathematics program. It provides the necessary review and coverage of whole numbers, fractions, decimals, ratio and proportion, percent, and measurement, as well as an introduction to algebra and geometry and a preview of statistics. It is part of a series that also includes the following books:

- *Basic College Mathematics*, Ninth Edition, by Lial, Salzman, and Hestwood
- *Prealgebra,* Fifth Edition, by Lial and Hestwood
- *Introductory Algebra*, Tenth Edition, by Lial, Hornsby, and McGinnis
- *Intermediate Algebra,* Tenth Edition, by Lial, Hornsby, and McGinnis
- *Introductory and Intermediate Algebra,* Fifth Edition, by Lial, Hornsby, and McGinnis
- *Prealgebra and Introductory Algebra,* Fourth Edition, by Lial, Hestwood, Hornsby, and McGinnis
- *Developmental Mathematics: Basic Mathematics and Algebra,* Third Edition, by Lial, Hornsby, McGinnis, Salzman, and Hestwood

WHAT'S NEW IN THIS EDITION

The scope and sequence of topics in *Essential Mathematics* has stood the test of time and rates highly with our reviewers. Therefore, you will find the table of contents intact, making the transition to the new edition easier.

▶ ***Examples and Exercises*** Throughout the text, examples and exercises have been adjusted or replaced to reflect current data and practices. Applications have been updated and cover a wider variety of topics, such as the fields of technology, ecology, and health sciences.

▶ ***Guided Solutions*** Selected exercises in the margins and in the exercise sets, marked with a GS icon, now show the first few solution steps. This provides guidance to students as they start learning a new concept or procedure and gets them off to a successful start. (See p. 301 margin, and p. 356, Exercises 3-4.)

▶ ***Concept Checks*** The Concept Checks at the beginning of each exercise set assure students that they have necessary background skills or information to proceed. Concept Checks later in the exercise sets help students summarize and consolidate their learning by writing explanations, correcting common errors, and practicing mathematical processes. (See pp. 285-286.)

▶ ***Vocabulary Tips*** Many students at this level of mathematics do not have strong reading skills. The vocabulary tips included in the margins throughout the book help them to learn the meaning of root words and prefixes commonly used in mathematics vocabulary (for example, *equ-*, *centi*, *tri-*), distinguish the mathematical meaning from the common usage of particular words (such as *volume*, *average*), and provide tips for remembering the difference between often-confused terms (such as LCM and LCD on p. 215).

- ***Teaching Tips*** Although the mathematical content in this text is familiar to instructors, they may not have experience in teaching the material to adult students. The Teaching Tips, printed in the margins of the Annotated Instructor's Edition, provide helpful comments from colleagues with successful experience at this level. Common trouble spots are noted, with suggestions for improving student understanding. (See p. 276.)

- ***Lial Video Library*** The Lial Video Library, available in MyMathLab and on the Video Resources DVD, provides students with a wealth of video resources to help them navigate the road to success. All video resources in the library include optional subtitles in English. The Lial Video Library includes Section Lecture Videos, Solutions Clips, Quick Review Lectures, and Chapter Test Prep Videos. The Chapter Test Prep Videos are also available on YouTube (searchable using author name and book title), or by scanning the QR code on the inside back cover for easy access.

- ***MyWorkBook*** This new workbook provides Guided Examples and corresponding Now Try Exercises for each text objective. The extra practice exercises for every section of the text, with ample space for students to show their work, are correlated to Examples, Lecture Videos, and Exercise Solution Clips, to give students the help they need to successfully complete problems. Additionally, MyWorkBook lists the learning objectives and key vocabulary terms for every text section, along with vocabulary practice problems.

- ***Math in the Media*** Each one-page activity presents a relevant look at how mathematics is used in the media. Designed to help instructors answer the often-asked question, "When will I ever use this stuff?," these activities ask students to read and interpret data from newspaper articles, the Internet, and other familiar, real-world sources. (See p. 196.) The activities are well-suited to collaborative work or they can be completed by individuals or used for open-ended class discussions.

- ***Study Skills*** Thirteen carefully designed study skills activities provide opportunities for students to practice proven strategies for learning mathematics. Poor study skills and behaviors are major factors in low success rates in mathematics courses. Research shows that a few generic tips sprinkled here and there are not enough to help students change their study behaviors. Because students need specific instruction in study skills, we have contextualized them, integrating them into the text material. Topics include note taking, homework, study cards, math anxiety, test preparation, test taking, preparing for a final exam, and more. (See pp. 138–139, 180–181, and 182–183.) Most are located within the first few chapters so that students can use the skills throughout the course. (See the Contents for titles and locations.) The first activity, "Your Brain *Can* Learn Mathematics," explains how the brain actually learns and remembers so that students understand why the study skills will help them succeed in the course.

HALLMARK FEATURES

We believe students and instructors will welcome these familiar hallmark features.

- ***Chapter Openers*** The new and engaging Chapter Openers portray real life situations making math relevant for students. (See Chapter 2, p. 113.)

- ***Real-Life Applications*** We are always on the lookout for interesting data to use in real-life applications. As a result, we have included many new or updated examples and exercises throughout the text that focus on real-life applications of mathematics. Students are often asked to find data in a table, chart, graph, or advertisement. (See pp. 156 and 157.) These applied problems provide an up-to-date flavor that will appeal to and motivate students.

- ***Figures and Photos*** Today's students are more visually oriented than ever. Thus, we have made a concerted effort to include mathematical figures, diagrams, tables, and graphs whenever possible. (See p. 157.) Many of the graphs use a style similar to that seen by students in today's print and electronic media. Photos have been incorporated to enhance applications in examples and exercises. (See p. 158.)

- ***Emphasis on Problem Solving*** Introduced at the end of Chapter 1, our six-step process for solving application problems is integrated throughout the text. The six steps, *Read, Plan, Estimate, Solve, State the Answer,* and *Check,* are emphasized in boldface type and repeated in specific problem-solving examples in Chapters 1, 2, 3, 5, and 6. (See p. 162.)

- ***Learning Objectives*** Each section begins with clearly stated, numbered objectives, and the material within sections is keyed to these objectives so that students know exactly what concepts are covered. (See p. 114.)

- ***Pointers*** More pointers have been added to examples to provide students with important on-the-spot reminders and warnings about common pitfalls. (See pp. 132 and 162.)

- ***Cautions and Notes*** These color-coded and boxed comments, one of the most popular features of previous editions, warn students about common errors and emphasize important ideas throughout the exposition. (See pp. 144–145.) Cautions are highlighted in yellow and Notes are highlighted with blue tabs.

- ***Calculator Tips*** These optional tips, marked with a red calculator icon, offer helpful information and instruction for students using calculators in the course. (See p. 268.)

- ***Margin Problems*** Margin problems, with answers immediately available on the bottom of the page, are found in every section of the text. (See pp. 121–122.) This key feature allows students to immediately practice the material covered in the examples in preparation for the exercise sets.

- ***Ample and Varied Exercise Sets*** The text contains a wealth of exercises to provide students with opportunities to practice, apply, connect, and extend the skills they are learning. Numerous illustrations, tables, graphs, and photos help students visualize the problems they are solving. Problem types include skill building, writing, estimation, and calculator exercises, as well as applications and correct-the-error problems. In the Annotated Instructor's Edition of the text, the writing exercises are marked with an icon so that instructors may assign these problems at their discretion. Exercises suitable for calculator work are marked in both the student and instructor editions with a calculator icon. (See pp. 175–179.) Students can watch an instructor work through the complete solution for all exercises marked with a Play Button icon on the Videos on DVD or in MyMathLab.

- ***Relating Concepts Exercises*** These help students tie concepts together and develop higher level problem-solving skills as they compare and contrast ideas, identify and describe patterns, and extend concepts to new situations. (See pp. 151, 294, and 306.) These exercises make great collaborative activities for pairs or small groups of students.

- ***Solutions*** Solutions to selected section exercises are included in the back of the book (following the Answers section). This provides students with easily accessible step-by-step help in solving the exercises that are most commonly missed. Solutions are provided for the exercises marked with a square of blue color around the exercise number, for example, **15.**

- ***Summary Exercises*** All chapters now include this helpful mid-chapter review. These exercises provide students with the all-important *mixed* practice they need at these critical points in their skill development. (See pp. 140–141.)

- ***Ample Opportunity for Review*** Each chapter ends with a Chapter Summary featuring: Key Terms with definitions and helpful graphics, New Formulas, New Symbols, Test Your Word Power, and a Quick Review of each section's content with additional examples. Also included is a comprehensive set of Chapter Review Exercises keyed to individual sections, a set of Mixed Review Exercises, and a Chapter Test. Students can watch an instructor work out the full solutions to the Chapter Test problems in the Chapter Test Prep Videos. Beginning with Chapter 2, each chapter concludes with a set of Cumulative Review Exercises. (See pp. 184–195.)

- ***Test Your Word Power*** This feature, incorporated into each Chapter Summary, helps students understand and master mathematical vocabulary. Key terms from the chapter are presented along with three possible definitions in a multiple-choice format. Answers and examples illustrating each term are provided. (See p. 185.)

STUDENT SUPPLEMENTS

Student's Solutions Manual

- By Jeffery A. Cole, Anoka-Ramsey Community College
- Provides detailed solutions to the odd-numbered section-level exercises and to all margin, Relating Concepts, Summary, Chapter Review, Chapter Test, and Cumulative Review Exercises

ISBNs: 0-321-83660-X, 978-0-321-83660-1

NEW MyWorkBook

- Provides Guided Examples and corresponding Now Try Exercises for each text objective
- Refers students to correlated Examples, Lecture Videos, and Exercise Solution Clips
- Includes extra practice exercises for every section of the text with ample space for students to show their work
- Lists the learning objectives and key vocabulary terms for every text section, along with vocabulary practice problems

ISBNs: 0-321-83682-0, 978-0-321-83682-3

NEW Lial Video Library

The Lial Video Library, available in MyMathLab and on the Video Resources DVD, provides students with a wealth of video resources to help them navigate the road to success! All video resources in the library include optional subtitles in English. The Lial Video Library includes the following resources:

- **Section Lecture Videos** offer a new navigation menu that allows students to easily focus on the key examples and exercises that they need to review in each section. Optional Spanish subtitles are available.
- **Solutions Clips** show an instructor working through the complete solutions to selected exercises from the text. Exercises with a solution clip are marked in the text and e-book with a Play Button icon ▶.
- **Quick Review Lectures** provide a short summary lecture of each key concept from the Quick Reviews at the end of every chapter in the text.
- **The Chapter Test Prep Videos** provide step-by-step solutions to all exercises from the Chapter Tests. These videos provide guidance and support when students need it the most: the night before an exam. The Chapter Test Prep Videos are also available on YouTube (searchable using author name and book title), or by scanning the QR code on the inside back cover for easy access.

INSTRUCTOR SUPPLEMENTS

Annotated Instructor's Edition

- Provides answers to all text exercises in color next to the corresponding problems
- Icons identify writing and calculator exercises

ISBNs: 0-321-82631-0, 978-0-321-82631-2

Instructor's Solutions Manual (Download only)

- By Jeffery A. Cole, Anoka-Ramsey Community College
- Provides complete solutions to all exercises in the text
- Available for download at www.pearsonhighered.com

ISBNs: 0-321-82568-3, 978-0-321-82568-1

Instructor's Resource Manual with Tests and Mini-Lectures (Download only)

- Contains a test bank with two diagnostic pretests, six free-response and two multiple-choice test forms per chapter, and two final exams
- Contains a mini-lecture for each section of the text with objectives, key examples, and teaching tips
- Includes a correlation guide from the eighth to the ninth edition and phonetic spellings for all key terms in the text
- Includes resources to help both new and adjunct faculty with course preparation and classroom management, by offering helpful teaching tips correlated to the sections of the text
- Available for download at www.pearsonhighered.com

ISBNs: 0-321-57461-3, 978-0-321-57461-9

ADDITIONAL MEDIA SUPPLEMENTS

MyMathLab®

MyMathLab® Online Course (access code required)

MyMathLab delivers **proven results** in helping individual students succeed. It provides **engaging experiences** that personalize, stimulate, and measure learning for each student. And, it comes from a **trusted partner** with educational expertise and an eye on the future.

To learn more about how MyMathLab combines proven learning applications with powerful assessment, visit **www.mymathlab.com** or contact your Pearson representative.

MyMathLab® Ready to Go Course (access code required)
These new Ready to Go courses provide students with all the same great MyMathLab features that they are used to but make it easier for instructors to get started. Each course includes pre-assigned homework and quizzes to make creating a course even simpler.

Ask your Pearson representative about the details for this particular course or to see a copy of this course.

MyMathLab® Plus/MyStatLab™ Plus
MyLabsPlus combines proven results and engaging experiences from MyMathLab® and MyStatLab™ with convenient management tools and a dedicated services team. Designed to support growing math and statistics programs, it includes additional features such as:

- **Batch Enrollment:** Schools can create the login name and password for every student and instructor, so everyone can be ready to start class on the first day. Automation of this process is also possible through integration with the school's Student Information System.
- **Login from your campus portal:** Instructors and their students can link directly from their campus portal into the MyLabsPlus courses. A Pearson service team works with the institution to create a single sign-on experience for instructors and students.
- **Advanced Reporting:** MyLabsPlus's advanced reporting enables instructors to review and analyze students' strengths and weaknesses by tracking their performance on tests, assignments, and tutorials. Administrators can review grades and assignments across all courses on a MyLabsPlus campus for a broad overview of program performance.
- **24/7 Support:** Students and instructors receive 24/7 support, 365 days a year, by email or online chat.

MyLabsPlus is available to qualified adopters. For more information, visit our website at *www.mylabsplus.com* or contact your Pearson representative.

MathXL®

MathXL® Online Course (access code required)
MathXL® is the homework and assessment engine that runs MyMathLab. (MyMathLab is MathXL plus a learning management system.)

With MathXL, instructors can:

- Create, edit, and assign online homework and tests using algorithmically-generated exercises correlated at the objective level to the textbook.
- Create and assign their own online exercises and import TestGen tests for added flexibility.
- Maintain records of all student work tracked in MathXL's online gradebook.

With MathXL, students can:

- Take chapter tests in MathXL and receive personalized study plans and/or personalized homework assignments based on their test results.
- Use the study plan and/or the homework to link directly to tutorial exercises for the objectives they need to study.
- Access supplemental animations and video clips directly from selected exercises.

MathXL is available to qualified adopters. For more information, visit our website at *www.mathxl.com*, or contact your Pearson representative.

TestGen®
TestGen® (*www.pearsoned.com/testgen*) enables instructors to build, edit, print, and administer tests using a computerized bank of questions developed to cover all the objectives of the text. TestGen is algorithmically based, allowing instructors to create multiple but equivalent versions of the same question or test with the click of a button. Instructors can also modify test bank questions or add new questions. The software and testbank are available for download from Pearson Education's online catalog.

PowerPoint® Lecture Slides

- Present key concepts and definitions from the text
- Available for download at *www.pearsonhighered.com* or in MyMathLab

ACKNOWLEDGMENTS

The comments, criticisms, and suggestions of users, nonusers, instructors, and students have positively shaped this textbook over the years, and we are most grateful for the many responses we have received. The feedback gathered for this revision of the text was particularly helpful, and we especially wish to thank the following individuals who provided invaluable suggestions for this and the previous edition:

George Alexander, *University of Wisconsin*
Sonya Armstrong, *West Virginia State University*
Vernon Bridges , *Durham Technical Community College*
Solveig R. Bender, *William Rainey Harper College*
Barbara Brown, *Anoka-Ramsey Community College*
Ernie Chavez, *Gateway Community College*
Terry Joe Collins, *Hinds Community College*
Martha Daniels, *Central Oregon Community College*
Cheryl Eichenseer, *St. Charles Community College*
Matthew Flacche, *Camden Community College*
James V. Figliolia, *Cape Fear Community College*
Donna Foster, *Piedmont Technical College*
Lindsay A. Gold, *Sinclair Community College*
Mark Gollwitzer, *Greenville Technical College*
Lourdes Gonzalez, *Miami-Dade Community College*
Kimberly Gregor, *Delaware Technical & Community College*
Lance Hemlow, *Raritan Valley Community College*
Joe Howe, *St. Charles County Community College*
Matthew Hudock, *St Philip's College*
Rose Kaniper, *Burlington County College*
Cameron S. Kishel, *Columbus State Community College*
Douglas Lewis, *Yakima Valley Community College*
Yixia Lu, *South Suburban College*
Valerie Maley, *Cape Fear Community College*
Connie L. McLean, *Black Hawk College*
Judy Mee, *Oklahoma City Community College*
Wayne Miller, *Lee College*
John Notini, *William Rainey Harper College*
Kathy Peay, *Sampson Community College*
Thea Philliou, *College of Santa Fe*
Jane Roads, *Moberly Area Community College*
Richard D. Rupp, *Del Mar College*
Ellen Sawyer, *College of DuPage*
Lois Schuppig, *College of Mount St. Joseph*
Mary Lee Seitz, *Erie Community College—City Campus*
Kathryn Taylor, *Santa Ana College*
Mike Tieleman-Ward, *Anoka Technical College*
Sven Trenholm, *North Country Community College*
Bettie A. Truitt, *Black Hawk College*
Jackie Wing, *Angelina College*

Our sincere thanks go to the dedicated individuals at Pearson who have worked hard to make this revision a success: Maureen O'Connor, Kathy Manley, Barbara Atkinson, Michelle Renda, Rachel Ross, Kari Heen, Christine Whitlock, Lauren Morse, Stephanie Green, and Rachel Haskell. We are also grateful to Marilyn Dwyer and Carol Merrigan of Cenveo/Nesbitt Graphics for their excellent production work; Bonnie Boehme, for supplying her copyediting expertise; Beth Anderson, for her fine photo research; Lucie Haskins, for producing a useful Index; Jeff Cole, for writing the Solutions Manuals; Perian Herring, for writing many of the Vocabulary Tips and accuracy checking the manuscript; Sam Tinsley, for his accuracy checking; and Judy Martinez, for her excellent typing skills.

Special thanks go to Jill Owens, an instructor at Diablo Valley College, who assisted with manuscript development and provided accuracy checking throughout this edition, and to Janis M. Cimperman, Associate Professor of Mathematics at St. Cloud State University, for her careful review of the new material in selected chapters.

Linda Russell, who developed and wrote the Study Skills activities that appear throughout this text, also worked to make the text more readable for developmental-level students, wrote many of the vocabulary tips, and provided much-needed help during the production phase. Her many years of teaching and working with students at this level was invaluable.

The ultimate measure of this textbook's success is whether it helps students master basic skills, develop problem-solving techniques, and increase their confidence in learning and using mathematics. In order for us to know what to keep and what to improve for the next edition, we need to hear from you, the instructor, and you, the student. Please tell us what you like and where you need additional help by sending an e-mail to *math@pearson.com*. We appreciate your feedback!

Stanley A. Salzman

1 Whole Numbers

The Golden Gate Bridge, connecting San Francisco and Marin Counties, was completed in 1937 and is 8981 feet long. The bridge is supported by two main towers which are constructed with 1,200,000 rivets, rises 746 feet above the water, and has had more than 2,153,896,448 cars and trucks cross over it. In this chapter we discuss whole numbers, which are used daily in our lives.

1.1 Reading and Writing Whole Numbers

OBJECTIVES

1. Identify whole numbers.
2. Give the place value of a digit.
3. Write a number in words or digits.
4. Read a table.

Knowing how to read and write numbers is important in mathematics.

OBJECTIVE 1 Identify whole numbers. The **decimal system** of writing numbers uses the ten digits

$$0, 1, 2, 3, 4, 5, 6, 7, 8, 9$$

to write any number. These digits can be used to write **the whole numbers:**

$$0, 1, 2, 3, 4, 5, 6, 7, 8, 9, 10, 11, 12, 13\ldots$$

The three dots indicate that the list goes on forever.

OBJECTIVE 2 Give the place value of a digit. Each digit in a whole number has a **place value,** depending on its position in the whole number. The following place value chart shows the names of the different places used most often and has the whole number 402,759,780 entered.

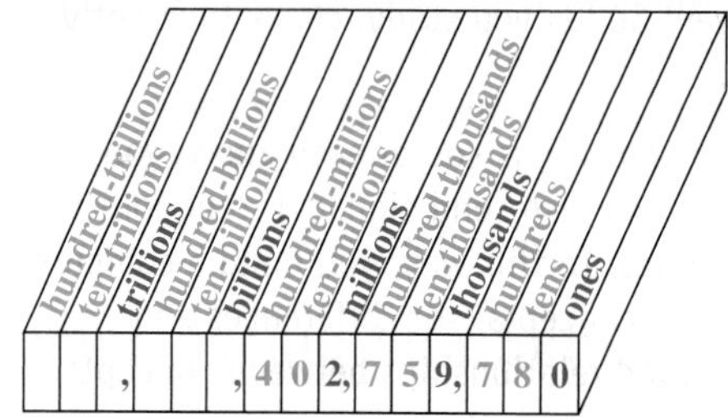

The United States is the leading consumer of coffee in the world. Each day we drink 402,759,780 cups of coffee. Each of the 7s in 402,759,780 represents a different amount because of its position, or *place value*, within the number. The *place value* of the 7 on the left is 7 hundred-thousands (700,000). The *place value* of the 7 on the right is 7 hundreds (700).

EXAMPLE 1 Identifying Place Values

Identify the place value of 8 in each whole number.

Each "8" has a different value.

(a) 28 — 8 ones

(b) 85 — 8 tens

(c) 869 — 8 hundreds

Notice that the value of 8 in each number is different, depending on its location (place) in the number.

◀ Work Problem 1 at the Side.

EXAMPLE 2 Identifying Place Values

Identify the place value of each digit in the number 725,283.

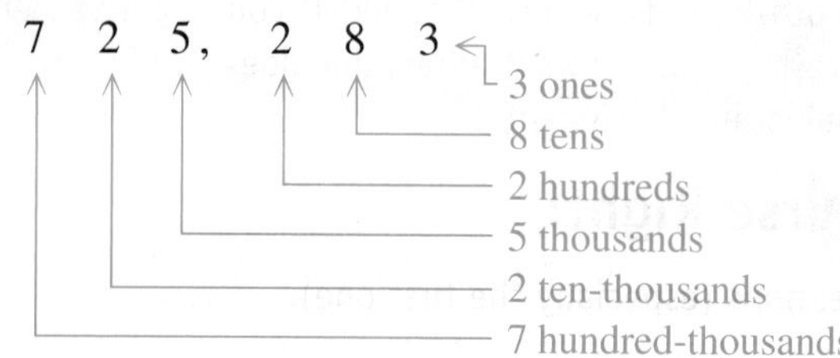

Continued on Next Page

VOCABULARY TIP

Place value In our decimal number system, each place has a value of 10 times the place to its right.

1 Identify the place value of the 4 in each whole number.

GS **(a)** 342

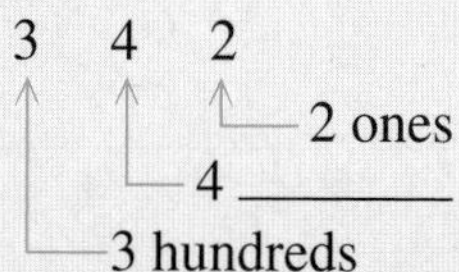

(b) 714

(c) 479

Answers

1. (a) tens (b) ones (c) hundreds

Notice the comma between the hundreds and thousands position in the number 725,283 in **Example 2.**

Work Problem 2 at the Side. ▶

Using Commas

Commas are used to separate each group of three digits, starting from the right. This makes numbers easier to read. (An exception: Commas are frequently omitted in four-digit numbers such as 9748 or 1329.) Each three-digit group is called a **period.** Some instructors prefer to just call them **groups.**

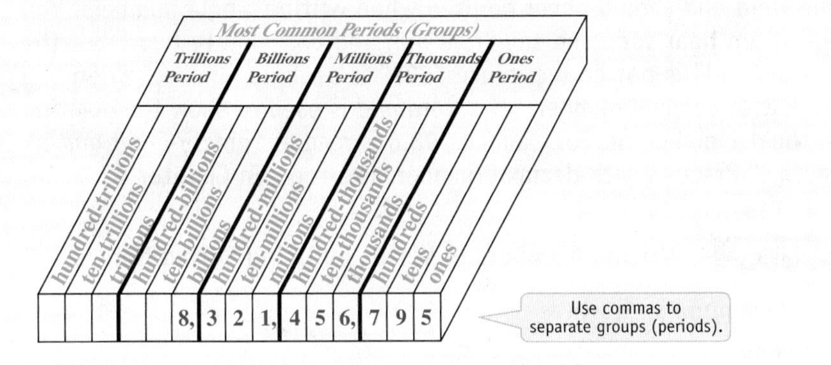

EXAMPLE 3 Knowing the Period or Group Names

Write the digits in each period of 8,321,456,795.

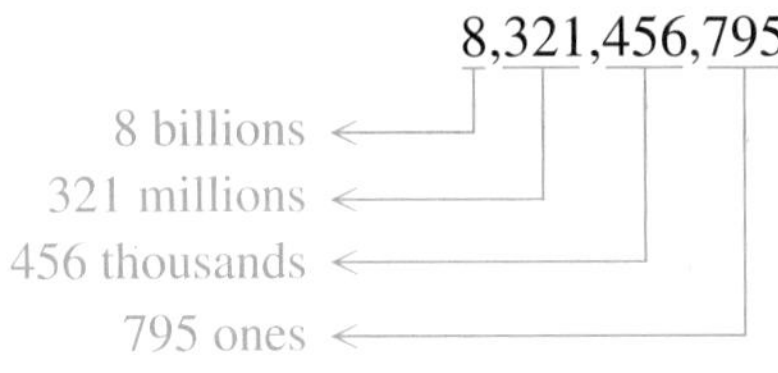

Work Problem 3 at the Side. ▶

Use the following rule to read a number with more than three digits.

Writing Numbers in Words

Start at the left when writing a number in words or saying it aloud. Write or say the digit names in each period (group), followed by the name of the period, except for the period name "ones," which is *not* used.

OBJECTIVE 3 Write a number in words or digits. The following examples show how to write names for whole numbers.

EXAMPLE 4 Writing Numbers in Words

Write each number in words.

(a) 57

This number means 5 tens and 7 ones, or 50 ones and 7 ones. Write the number as

fifty-seven.

Continued on Next Page

2 Identify the place value of each digit.

GS **(a)** 14,218

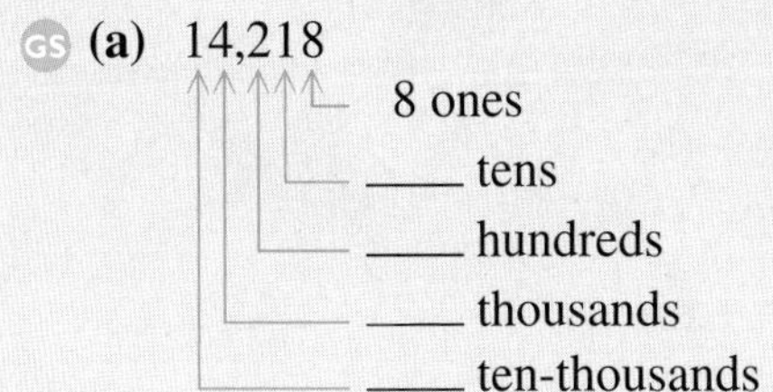

(b) 460,329

3 In the number 3,251,609,328 identify the digits in each period (group).

(a) billions period

(b) millions period

(c) thousands period

(d) ones period

Answers

2. (a) 1 : ten-thousands
4 : thousands
2 : hundreds
1 : tens
8 : ones
(b) 4 : hundred-thousands
6 : ten-thousands
0 : thousands
3 : hundreds
2 : tens
9 : ones

3. (a) 3 **(b)** 251 **(c)** 609 **(d)** 328

4 Write each number in words.

(a) 18

(b) 36

(c) 418

(d) 902

5 Write each number in words.

GS (a) 3104

three __________,

one __________ four

GS (b) 95,372

ninety-five __________,
three hundred seventy-two

(c) 100,075,002

(d) 11,022,040,000

6 Rewrite each number using digits.

(a) one thousand, four hundred thirty-seven

(b) nine hundred seventy-one thousand, six

(c) eighty-two million, three hundred twenty-five

Answers

4. (a) eighteen
(b) thirty-six
(c) four hundred eighteen
(d) nine hundred two

5. (a) three thousand, one hundred four
(b) ninety-five thousand, three hundred seventy-two
(c) one hundred million, seventy-five thousand, two
(d) eleven billion, twenty-two million, forty thousand

6. (a) 1437 (b) 971,006 (c) 82,000,325

(b) 94

ninety-four

Remember: Start at the *left* to read a number.

(c) 874

eight hundred seventy-four

(d) 601

six hundred one

◀ **Work Problem 4 at the Side.**

CAUTION

The word *and* should never be used when writing whole numbers. You will often hear someone say "five hundred *and* twenty-two," but the use of "and" is not correct since "522" is a whole number. When you work with decimal numbers, the word *and* is used to show the position of the decimal point. For example, 98.6 is read as "ninety-eight *and* six tenths." Practice with decimal numbers is a topic in **Chapter 4**.

EXAMPLE 5 Writing Numbers in Words by Using Period Names

Write each number in words.

(a) 725,283

seven hundred twenty-five **thousand,** two hundred eighty-three

(seven hundred twenty-five: Number in period; thousand: Name of period; two hundred eighty-three: Number in period (not necessary to write "ones"))

(b) 7252

Careful: *Do not* use "and" when reading a whole number.

seven **thousand,** two hundred fifty-two

(thousand: Name of period; No period name needed)

(c) 111,356,075

one hundred eleven **million,** three hundred fifty-six **thousand,** seventy-five

The period name is not used for the ones period.

(d) 17,000,017,000

seventeen **billion,** seventeen **thousand**

◀ **Work Problem 5 at the Side.**

EXAMPLE 6 Writing Numbers in Digits

Rewrite each number using digits.

(a) six **thousand,** twenty-two

6022

With 4 digits or fewer, *no comma* is needed.

(b) two hundred fifty-six **thousand,** six hundred twelve

256,612

(c) nine **million,** five hundred fifty-nine

9,000,559

Zeros indicate there are no thousands.

◀ **Work Problem 6 at the Side.**

Calculator Tip

Does your calculator show a comma between each group of three digits? Probably not, but try entering a long number such as 34,629,075. Notice that there is no key with a comma on it, so you do not enter commas. A few calculators may show the position of the commas *above* the digits, like this

34'629'075

Most of the time you will have to write in the commas where needed.

OBJECTIVE 4 **Read a table.** A common way of showing number values is by using a **table.** Tables organize and display facts so that they are more easily understood. The following table shows some past facts and future predictions for the United States. These numbers give us a glimpse of what we can expect in the 21st century.

NUMBERS FOR THE 21ST CENTURY

Year	1990	2010	2020*
U.S. population	261 million	309 million	338 million
Household income	$42,936	$46,326	$55,735
Average yearly salary	$21,129	$28,834	$32,080

*Estimated figures

Source: *Family Circle* magazine; U.S. Census Bureau.

If you read from left to right along the row labeled "U.S. population," you find that the population in 1990 was 261 million, then the population in 2010 was 309 million, and the estimated population for 2020 is 338 million.

EXAMPLE 7 Reading a Table

Use the table to find each number, and write the number in words.

(a) The estimated household income in the year 2020

Read from left to right along the row labeled "Household income" until you reach the 2020 column and find $55,735.

Fifty-five thousand, seven hundred thirty-five dollars

(b) The average yearly salary in 1990

Read from left to right along the row labeled "Average yearly salary." In the 1990 column you find $21,129.

Remember: Use hyphens when necessary.

Twenty-one thousand, one hundred twenty-nine dollars

Work Problem 7 at the Side. ▶

Note

Notice in **Example 7** that hyphens are used when writing numbers in words. A hyphen is used when writing the numbers 21 through 99 (twenty-one through ninety-nine), except for numbers ending in zero (20, 30, 40, ... 90).

7 Use the table to find each number, and write the number in digits when given in words, or write the number in words when given in digits.

GS **(a)** The population in 2010

The U.S. population in the 2010 column is 309 million and is written in digits as 3__ __,000,000

(b) The estimated population in 2020

(c) Household income in 1990

(d) The estimated average yearly salary in 2020

Answers

7. **(a)** 0; 9; 309,000,000
(b) 338,000,000
(c) forty-two thousand, nine hundred thirty-six dollars
(d) thirty-two thousand, eighty dollars

1.1 Exercises

FOR EXTRA HELP

Download the MyDashBoard App

MyMathLab®

CONCEPT CHECK *Choose the letter of the correct response.*

1. The digit in the hundreds place in the whole number 3065 is

(a) 5 **(b)** 3 **(c)** 0 **(d)** 6

2. The digit in the ten-thousands place in the whole number 134,681 is

(a) 6 **(b)** 3 **(c)** 8 **(d)** 1

Write the digit for the given **place value** *in each whole number.* **See Examples 1 and 2.**

3. 18,015
ten-thousands
hundreds

4. 86,332
ten-thousands
ones

5. 7,628,592,183
millions
thousands

6. 1,700,225,016
billions
millions

CONCEPT CHECK *Identify the correct period.*

7. Write the digits in the thousands period in the whole number 552,687,318.

8. Write the digits in the millions period in the whole number 947,321,876,528.

Write the digits for the given **period** *(group) in each whole number.* **See Example 3.**

9. 3,561,435
millions
thousands
ones

10. 100,258,100,006
billions
millions
thousands
ones

11. Do you think the fact that humans have four fingers and a thumb on each hand explains why we use a number system based on ten digits? Explain.

12. The decimal system uses ten digits. Fingers and toes are often referred to as digits. In your opinion, is there a relationship here? Explain.

CONCEPT CHECK *Answer* true *or* false *for each statement.*

13. The number 23,115 is written in words as twenty-three thousand and one hundred and fifteen.

14. The number 37,886 is written in words as thirty-seven thousand, eight hundred eighty-six.

Write each number in words. **See Examples 4 and 5.**

15. 346,009
GS three hundred forty six _______, _______

16. 218,033
GS two hundred eighteen _______, thirty-_______

17. 25,756,665

18. 999,993,000

Write each number using digits. **See Example 6.**

19. sixty-three thousand, one hundred sixty-three
GS 6 __, 1 __ __

20. ninety-five thousand, one hundred eleven
GS 9 __, __ __ 1

21. ten million, two hundred twenty-three

22. one hundred million, two hundred

Write the numbers from each sentence using digits. ***See Example 6.***

23. There are three million, two hundred thousand parachute jumps in the United States each year. (*Source:* History Channel.)

24. A full-grown caterpillar is 27,000 times its birth size. A 9-pound human baby growing at the same rate would weigh two hundred forty-three thousand pounds by college graduation. (*Source: Spirit Magazine.*)

25. The number of cans of Pepsi Cola sold each day is fifty million, fifty-one thousand, five hundred seven. (*Source:* Andy Rooney, *60 Minutes.*)

26. In the United States we drink one hundred forty-six billion, three hundred eighty-five million cups of coffee every year. (*Source: Spirit Magazine.*)

27. There are fifty-four million, seven hundred fifty thousand Hot Wheels sold each year. (*Source:* Andy Rooney, *60 Minutes.*)

28. A middle-income family will typically spend two hundred twenty-one thousand dollars to raise a child to the age of eighteen. (*Source: Los Angeles Times.*)

29. Rewrite eight hundred trillion, six hundred twenty-one million, twenty thousand, two hundred fifteen by using digits.

30. Rewrite 2,153,896,448, the number of vehicles that have crossed the Gloden Gate Bridge, in words.

The table at the right shows various ways people get to work. Use the table to answer Exercises 31–34. ***See Example 7.***

31. Which method of transportation is least used? Write the number in words.

32. Which method of transportation is most used? Write the number in words.

33. Find the number of people who walk to work or work at home, and write it in words.

34. Find the number of people who carpool, and write it in words.

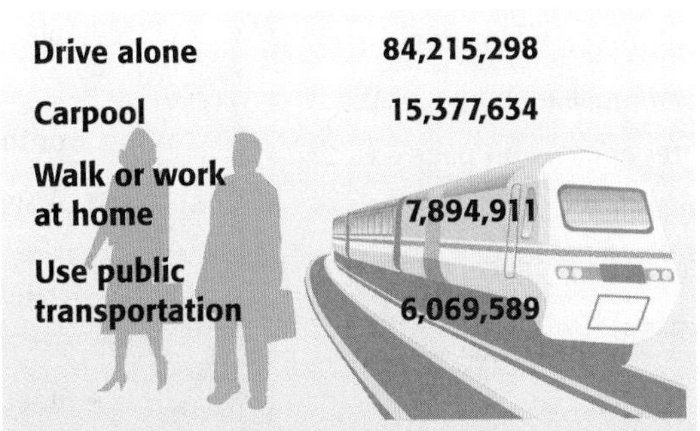

GETTING TO WORK

How workers 16 and over get to work:

Drive alone	**84,215,298**
Carpool	**15,377,634**
Walk or work at home	**7,894,911**
Use public transportation	**6,069,589**

Source: U.S. Census Bureau.

Study Skills
USING YOUR TEXTBOOK

OBJECTIVES

1. Explain the meaning of text features such as section numbering, objectives, margin exercises.
2. Locate the Index, Answers, and Solutions sections.

Be sure to read Your Brain* Can *Learn Mathematics before this activity. It is at the beginning of this chapter. You'll find out how your brain learns and remembers.

Your textbook can be very helpful. Find out what it has to offer. First, let's look at some general features that will help in all chapters.

Table of Contents

Look at page v in the very front for the Table of Contents. Before you start Chapter 1, you should look at the Preface, which begins on page vii. On page x is a list of Supplementary Resources for students. If you are interested in any of these, ask your instructor if they are available.

Section Numbering

Each chapter is divided into sections, and each section has a number, such as 1.3 or 3.5. Your instructor will use these numbers to assign readings and homework.

Chapter 3 ⟶ **3.5** ⟵ Section 5 within Chapter 3

Chapter Features

There are four features to pay special attention to as you work in your book.

- **Objectives.** Each section lists the objectives in the upper corner of the first page. The objectives are listed again as each one is introduced. An objective tells you *what you will be able to do after you complete the section.* An excellent way to check your learning is to go back to the list of objectives when you are finished with a section and ask yourself if you can do them all.
- **Margin Exercises.** The exercises in the blue shaded margins of the pages in your textbook give you immediate practice. **This is a perfect way to get your dendrites growing right away!** The answers are given at the bottom, so you can check yourself easily.
- **Cautions, Pointers, Notes, and Calculator Tips.**
  - **Caution!** A yellow box is a comment about a common error that students make or a common trouble spot you will want to avoid.
  - **Pointers** are little "clouds" next to worked examples. They point to specific places where common mistakes are made and give on-the-spot reminders.
  - Look for the specially marked **Note** boxes. They contain hints, explanations, or interesting side comments about a topic.
  - A small picture of a red calculator appears several places. In the main part of the chapter, the icon means that there is a **Calculator Tip,** which helps you learn more about using your calculator. A calculator in an Exercise section is a recommendation to use your calculator to work that exercise.
  - **Vocabulary Tips** appear in the margins to help you learn the language of mathematics. Use them to create study cards for terminology that is new to you.

List a page number from Chapter 1 for each of these features:

A *Caution* appears on page ____.

A *Pointer* appears on page ____.

A *Note* appears on page ____.

A *Calculator Tip* appears on page ____.

End of Chapter Features

Go back to the Table of Contents again. What is listed at the end of each chapter?

- **Chapter Summary.** Turn to page 98 to find the Summary for Chapter 1. It lists the chapter's **Key Terms** (arranged in the order that they appear in the chapter) and **New Symbols** and/or **New Formulas.** Then, **Test Your Word Power** checks your understanding of the math vocabulary. Next is a **Quick Review section.** It lists each topic in the chapter and shows a worked example, with tips.
- **Review Exercises** Use these exercises as a way to check your understanding of all the concepts in the chapter. You can practice every type of problem. If you get stuck, the numbers inside the dark blue rectangles tell you which section of the chapter to go back to for more explanations. Make sure you do the **Mixed Review Exercises** to practice for tests.
- **Chapter Test.** Plan to take the test as a practice exam. That way you can be sure you really know how to work all types of problems without looking back at the chapter.
- **Cumulative Review (starting with Chapter 2)** These exercises help you maintain the skills you've learned in all previous chapters. Working on previous skills throughout the course will be a big help on the final exam.

How will you make good use of the features at the end of each chapter?

____

____

____

____

Answers

How do you find out if you've worked the exercises correctly? Your textbook provides many of the answers. Throughout each chapter you should work the sample problems in the **margins.** The answers for those are at the **bottom of each page** in the margin area.

For homework, you can find the answers to all of the **odd-numbered section exercises** in the **Answers to Selected Exercises** section near the end of your textbook. Also, *all* of the answers are given for the Chapter Review Exercises, Chapter Tests, and Cumulative Reviews. Check your textbook now, and find the page on which the Answers section begins.

Flag the Answers section with a sticky note or other device, so that you can turn to it quickly.

Solutions

The **Solutions** section near the end of the book shows how to solve some of the harder odd-numbered exercises step by step. In the exercise sets, look for the exercise numbers with a square of blue shading around them. These are the ones that have a solution in the back of the book.

Index

The **Index** is the last thing in your book. It lists all the topics, vocabulary, and concepts in alphabetical order. For example, look up the words below. There may be several subheadings listed under the main word, or several page numbers listed. Notice that the page number printed in bold type is where the word is introduced and defined. Write down the boldface page number for each word. Then go to that page and find the word.

Commutative property of multiplication is defined on page ____.
Factors of numbers are defined on page ____.
Rounding of mixed numbers is explained on page ____.

Why Are These Features Brain Friendly?

The textbook authors included text features that make it easier for you to understand the mathematics. **Your brain naturally seeks organization and predictability.** When you pay attention to the regular features of your textbook, you are allowing your brain to get familiar with all of the helpful tips, suggestions, and explanations that your book has to offer. You will make the best possible use of your textbook.

1.2 Adding Whole Numbers

OBJECTIVES

1. Add two single-digit numbers.
2. Add more than two numbers.
3. Add when regrouping (carrying) is not required.
4. Add with regrouping (carrying).
5. Use addition to solve application problems.
6. Check the answer in addition.

There are four baseballs at the left and two at the right. In all, there are six.

The process of finding the total is called **addition.** Here 4 and 2 were added to get 6. Addition is written with a + sign, so that

$$4 + 2 = 6.$$

OBJECTIVE 1 **Add two single-digit numbers.** In addition, the numbers being added are called **addends,** and the resulting answer is called the **sum** or **total.**

$$\begin{array}{rl} 4 & \leftarrow \text{Addend} \\ +\ 2 & \leftarrow \text{Addend} \\ \hline 6 & \leftarrow \text{Sum (total)} \end{array}$$

Addition problems can also be written horizontally, as follows.

$$\underset{\text{Addend}}{\underset{\uparrow}{4}} \quad + \quad \underset{\text{Addend}}{\underset{\uparrow}{2}} \quad = \quad \underset{\text{Sum}}{\underset{\uparrow}{6}}$$

Commutative Property of Addition

By the **commutative property of addition,** changing the order of the addends in an addition problem does not change the sum.

For example, the sum of $4 + 2$ is the same as the sum of $2 + 4$. This allows the addition of the same numbers in a different order.

EXAMPLE 1 **Adding Two Single-Digit Numbers**

Add, and then change the order of numbers to write another addition problem.

(a) $5 + 3 = 8$ and $3 + 5 = 8$

(b) $7 + 8 = 15$ and $8 + 7 = 15$

(c) $8 + 3 = 11$ and $3 + 8 = 11$

(d) $8 + 8 = 16$

Changing the order in addition does not change the sum.

◀ **Work Problem 1 at the Side.**

Associative Property of Addition

By the **associative property of addition,** changing the grouping of the addends in an addition problem does not change the sum.

For example, the sum of $3 + 5 + 6$ may be found as follows.

$$(3 + 5) + 6 = 8 + 6 = 14 \qquad \text{Parentheses tell us to add } 3 + 5 \text{ first.}$$

Another way to add the same numbers is

Changing the grouping of addends does not change the sum.

$$3 + (5 + 6) = 3 + 11 = 14. \qquad \text{Parentheses tell us to add } 5 + 6 \text{ first.}$$

Either grouping gives a sum of 14 because of the associative property of addition.

VOCABULARY TIP

Commutative property of addition You can remember the commutative property by thinking of the numbers "commuting," or changing places.

1 Add, and then change the order of numbers to write another addition problem.

GS **(a)** $2 + 6 \quad = \_\_\_\_$

$6 + \_\_\_\_ = \_\_\_\_$

(b) $9 + 5$

(c) $4 + 7$

(d) $6 + 9$

VOCABULARY TIP

Associative property of addition You can remember the associative property by thinking of two numbers associating with each other, and then one leaves to associate with another number.

Answers

1. **(a)** 8; $6 + 2 = 8$ **(b)** 14; $5 + 9 = 14$ **(c)** 11; $7 + 4 = 11$ **(d)** 15; $9 + 6 = 15$

OBJECTIVE 2 Add more than two numbers. To add several numbers, first write them in a column. Add the first number to the second. Add this sum to the third number. Continue until all the numbers are used.

EXAMPLE 2 Adding More Than Two Numbers

Add 2, 5, 6, 1, and 4.

```
    2 ]
    5 ]— 2 + 5 = 7
   (6)—————— 7 + 6 = 13
   (1)—————————— 13 + 1 = 14
 + (4)—————————————— 14 + 4 = 18
   18
```

Work Problem 2 at the Side. ▶

Note

By the commutative and associative properties of addition, numbers may also be added starting at the bottom of a column. Adding from the top or adding from the bottom will give the same answer.

OBJECTIVE 3 Add when regrouping (carrying) is not required. If numbers have two or more digits, you must arrange the numbers in columns so that the ones digits are in the same column, tens are in the same column, hundreds are in the same column, and so on. Next, you add column by column starting at the right.

EXAMPLE 3 Adding without Regrouping

Add 511 + 23 + 154 + 10.

First line up the numbers in columns, with the ones column at the right.

```
      Hundreds in a column
      Tens in a column
      Ones in a column
   511
    23
   154      Ones digits at
 +  10      the right
```

Now start at the right and add the ones digits. Add the tens digits next, and finally, the hundreds digits.

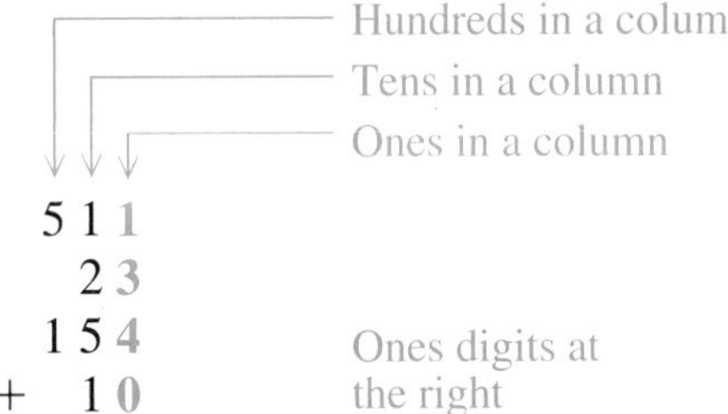

The sum of the four numbers is 698.

2 Add each column of numbers.

(a)
```
  3
  8
  5
  4
+ 6
```

(b)
```
  5
  6
  3
  2
+ 4
```

(c)
```
  9
  6
  8
  7
+ 3
```

(d)
```
  3
  8
  6
  4
+ 8
```

Answers

2. (a) 26 (b) 20 (c) 33 (d) 29

3 Add.

(a)
```
   26
 + 73
```

(b)
```
   534
 + 265
```

(c)
```
   42,305
 + 11,563
```

4 Add with regrouping

GS (a)
```
    1
   66
 + 27
   _3
```

(b)
```
   58
 + 33
```

(c)
```
   56
 + 37
```

(d)
```
   34
 + 49
```

Answers

3. (a) 99 (b) 799 (c) 53,868

4. (a) 9; 93 (b) 91 (c) 93 (d) 83

◀ Work Problem **3** at the Side.

OBJECTIVE 4 **Add with regrouping (carrying).** If the sum of the digits in any column is more than 9, use **regrouping** (sometimes called **carrying**).

EXAMPLE 4 Adding with Regrouping

Add 47 and 29.

Add ones.

```
   47
 + 29
```

7 ones and 9 ones = 16 ones

Regroup 16 ones as 1 ten and 6 ones. Write 6 ones in the ones column and write 1 ten in the tens column.

```
   1
   47      Write 1 ten in the tens column.
 + 29      7 ones and 9 ones = 16 ones
    6      Write 6 ones in the ones column.
```

Add the digits in the tens column, including the regrouped 1.

```
   1
   47
 + 29
   76
```

1 ten + 4 tens + 2 tens = 7 tens

◀ Work Problem **4** at the Side.

EXAMPLE 5 Adding with Regrouping

Add 324 + 7855 + 23 + 7 + 86.

Step 1 Add the digits in the ones column.

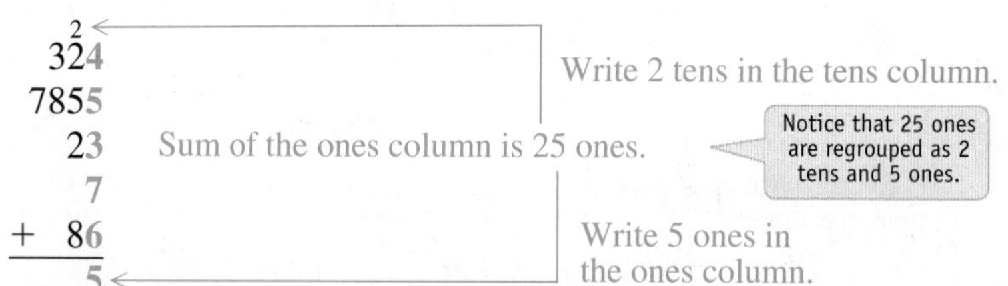

2
324
7855
23
7
+ 86
5
Write 2 tens in the tens column.
Sum of the ones column is 25 ones.
Notice that 25 ones are regrouped as 2 tens and 5 ones.
Write 5 ones in the ones column.

Step 2 Add the digits in the tens column, including the regrouped 2.

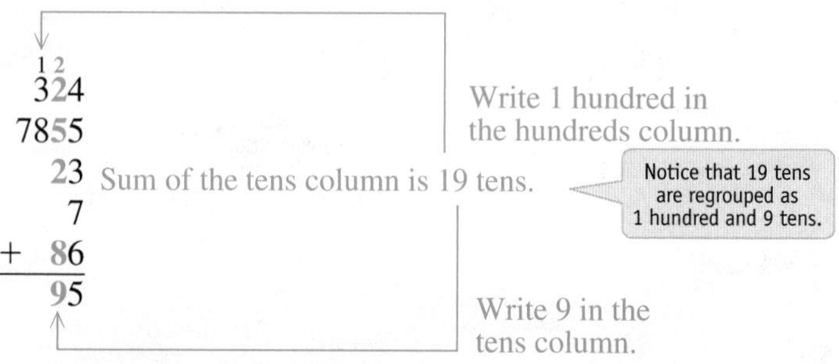

12
324
7855
23
7
+ 86
95
Write 1 hundred in the hundreds column.
Sum of the tens column is 19 tens.
Notice that 19 tens are regrouped as 1 hundred and 9 tens.
Write 9 in the tens column.

Continued on Next Page

Step 3 Add the hundreds column, including the regrouped 1.

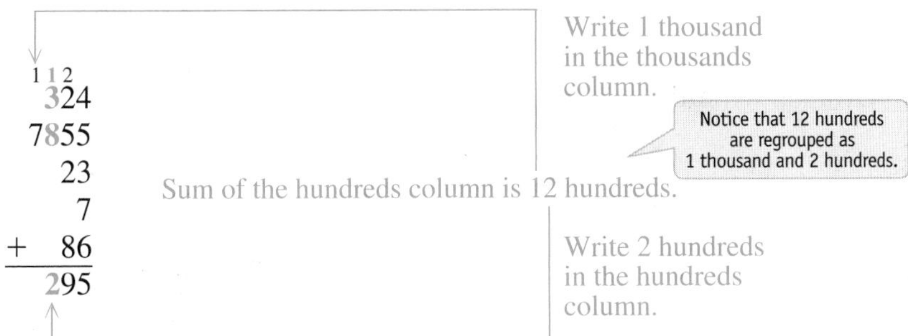

1 1 2
324
7855
23
7
+ 86
295
Write 1 thousand in the thousands column.
Notice that 12 hundreds are regrouped as 1 thousand and 2 hundreds.
Sum of the hundreds column is 12 hundreds.
Write 2 hundreds in the hundreds column.

Step 4 Add the thousands column, including the regrouped 1.

```
  1 1 2
   324
  7855
    23
     7
+   86
  8295
```

Sum of the thousands column is 8.

Finally, $324 + 7855 + 23 + 7 + 86 = 8295$.

Work Problem 5 at the Side. ▶

Note

For additional speed, you can try to regroup mentally. Do not write the regrouped number, but just remember it as you move to the top of the next column. Try this method, and use it if it works for you.

Work Problem 6 at the Side. ▶

OBJECTIVE 5 **Use addition to solve application problems.** In **Section 1.10** we will describe how to solve application problems in more detail. The next two examples are application problems that require adding.

EXAMPLE 6 **Applying Addition Skills**

On this map of the Walt Disney World area in Florida, the distance in miles from one location to another is written alongside the road. Find the shortest route from Altamonte Springs to Clear Lake.

Continued on Next Page

5 Add by regrouping as necessary.

(a)
```
  1    2 1
    1 6 2
  4 2 7 1
    3 7 2
+ 8 9 7 6
 1 3, _ _ 1
```

(b)
```
  7821
   435
    72
   305
+ 1693
```

6 Add by regrouping mentally.

(a)
```
   816
   363
    17
     2
     5
+ 7654
```

(b)
```
  15,829
     765
      78
      15
       9
       7
+ 13,179
```

Answers

5. **(a)** 7; 8; 13,781 **(b)** 10,326

6. **(a)** 8857 **(b)** 29,882

7 Use the map in **Example 6** to find the shortest route from Lake Buena Vista to Conway.

```
  4
  6
  3
+__
___
```

8 The road is closed between Orlando and Clear Lake, so this route cannot be used. Use the map in **Example 6** to find the next shortest route from Orlando to Clear Lake.

Approach Add the mileage along various routes to determine the distances from Altamonte Springs to Clear Lake. Then select the shortest route.

Solution One way from Altamonte Springs to Clear Lake is through Orlando. Add the mileage numbers along this route.

8	Altamonte Springs to Pine Hills
5	Pine Hills to Orlando
+ 8	Orlando to Clear Lake
21 →	miles from Altamonte Springs to Clear Lake, going through Orlando

Remember: Shortest distance is the fewest total miles.

Another way is through Casselberry, Bertha, and Winter Park. Add the mileage numbers along this route.

5	Altamonte Springs to Casselberry
6	Casselberry to Bertha
7	Bertha to Winter Park
+ 7	Winter Park to Clear Lake
25 →	miles from Altamonte Springs to Clear Lake through Bertha and Winter Park

The shortest route from Altamonte Springs to Clear Lake is 21 miles through Orlando.

◀ Work Problem **7** at the Side.

EXAMPLE 7 Finding a Total Distance

Use the map in **Example 6** to find the total distance from Shadow Hills to Casselberry to Orlando and back to Shadow Hills.

Approach Add the mileage from Shadow Hills to Casselberry to Orlando and back to Shadow Hills to find the total distance.

Solution Use the numbers from the map.

9	Shadow Hills to Bertha
6	Bertha to Casselberry
5	Casselberry to Altamonte Springs
8	Altamonte Springs to Pine Hills
5	Pine Hills to Orlando
8	Orlando to Clear Lake
+ 11	Clear Lake to Shadow Hills
52 →	miles from Shadow Hills to Casselberry to Orlando and back to Shadow Hills

◀ Work Problem **8** at the Side.

EXAMPLE 8 Finding a Perimeter

Find the number of feet of hedges needed to enclose the Civil War Veterans Park shown.

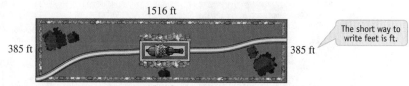

1516 ft

385 ft   385 ft

1516 ft

The short way to write feet is ft.

Approach Find the **perimeter,** or total distance around the park, by adding the lengths of all the sides.

Continued on Next Page

Answers

7. 6; 19 miles

8.
5	Orlando to Pine Hills
8	Pine Hills to Altamonte Springs
5	Altamonte Springs to Casselberry
6	Casselberry to Bertha
7	Bertha to Winter Park
+ 7	Winter Park to Clear Lake
38	miles

Solution Use the lengths shown.

$$\begin{array}{r} 1516 \\ 385 \\ 1516 \\ +\ 385 \\ \hline 3802 \text{ ft} \end{array}$$

The amount of hedge needed is 3802 ft, which is the perimeter of (distance around) the park.

Work Problem 9 at the Side. ▶

OBJECTIVE 6 Check the answer in addition. Checking the answer is an important part of problem solving. A common method for checking addition is to re-add from bottom to top. This is an application of the commutative and associative properties of addition.

EXAMPLE 9 Checking Addition

Check the following addition.

$$\begin{array}{r} 1428 \\ 738 \\ 63 \\ 125 \\ 17 \\ +\ 485 \\ \hline 1428 \end{array}$$

Add down. Adding down and adding up should give the same answer. To check, add up.

Add from bottom to top to check addition.

Here the answers agree, so the sum is probably correct.

EXAMPLE 10 Checking Addition

Check the following additions. Are they correct?

(a)

$$\begin{array}{r} 785 \\ 63 \\ +\ 185 \\ \hline 1033 \end{array} \qquad \begin{array}{r} 1033 \\ 785 \\ 63 \\ +\ 185 \\ \hline 1033 \end{array}$$

Correct, because both answers are the same. To check, add up.

(b)

$$\begin{array}{r} 635 \\ 73 \\ 831 \\ +\ 915 \\ \hline 2444 \end{array} \qquad \begin{array}{r} 2454 \\ 635 \\ 73 \\ 831 \\ +\ 915 \\ \hline 2444 \end{array}$$

Error, because the answers are different. To check, add up.

Avoid wrong answers by checking your work.

Re-add to find that the correct sum is 2454.

Work Problem 10 at the Side. ▶

9 Solve the problem. Find the number of feet of fencing needed to enclose the solar electricity generating project shown.

10 Check the following additions. If an answer is incorrect, find the correct answer.

(a)
$$\begin{array}{r} 63 \\ 4 \\ 9 \\ +\ 28 \\ \hline 104 \end{array}$$

(b)
$$\begin{array}{r} 927 \\ 395 \\ 64 \\ +\ 251 \\ \hline 1637 \end{array}$$

(c)
$$\begin{array}{r} 79 \\ 218 \\ 7 \\ +\ 639 \\ \hline 953 \end{array}$$

(d)
$$\begin{array}{r} 21{,}892 \\ 11{,}746 \\ +\ 43{,}925 \\ \hline 79{,}563 \end{array}$$

Answers

9. 1646 ft
10. **(a)** correct **(b)** correct
(c) incorrect; should be 943
(d) incorrect; should be 77,563

1.2 Exercises

FOR EXTRA HELP

Download the MyDashBoard App

MyMathLab®

Add. **See Examples 1–3.**

1. $\begin{array}{r} 43 \\ +\ 54 \\ \hline \end{array}$ **2.** $\begin{array}{r} 18 \\ +\ 11 \\ \hline \end{array}$ **3.** $\begin{array}{r} 56 \\ +\ 33 \\ \hline \end{array}$ **4.** $\begin{array}{r} 83 \\ +\ 15 \\ \hline \end{array}$ **5.** $\begin{array}{r} 317 \\ +\ 572 \\ \hline \end{array}$

6. $\begin{array}{r} 574 \\ +\ 325 \\ \hline \end{array}$ **7.** $\begin{array}{r} 318 \\ 151 \\ +\ 420 \\ \hline \end{array}$ **8.** $\begin{array}{r} 135 \\ 253 \\ +\ 410 \\ \hline \end{array}$ **9.** $\begin{array}{r} 6310 \\ 252 \\ +\ 1223 \\ \hline \end{array}$ **10.** $\begin{array}{r} 121 \\ 5705 \\ +\ 3163 \\ \hline \end{array}$

CONCEPT CHECK *Determine whether the following additions are* correct *or* incorrect.

11. $932 + 44 + 613 = 1589$

12. $517 + 131 + 250 = 1098$

13. $1251 + 4311 + 2114 = 7686$

14. $3241 + 1513 + 2014 = 6768$

Add. **See Examples 1–3.**

15. $12{,}142 + 43{,}201 + 23{,}103$

16. $41{,}124 + 12{,}302 + 23{,}500$

17. $3213 + 5715$

18. $6344 + 1655$

19. $38{,}204 + 21{,}020$ GS $\begin{array}{r} 38{,}2\ 0\ 4 \\ +\ 21{,}0\ 2\ 0 \\ \hline 59{,}\_\ \_\ 4 \end{array}$

20. $63{,}251 + 36{,}305$ GS $\begin{array}{r} 6\ 3{,}\ 2\ 5\ 1 \\ +\ 3\ 6{,}\ 3\ 0\ 5 \\ \hline 9\ \_{,}\ \_\ 5\ \_ \end{array}$

CONCEPT CHECK *Determine which answers are* correct *or* incorrect.

21. $\begin{array}{r} 87 \\ +\ 63 \\ \hline 150 \end{array}$ **22.** $\begin{array}{r} 19 \\ +\ 92 \\ \hline 101 \end{array}$ **23.** $\begin{array}{r} 86 \\ +\ 69 \\ \hline 155 \end{array}$ **24.** $\begin{array}{r} 37 \\ +\ 85 \\ \hline 132 \end{array}$ **25.** $\begin{array}{r} 47 \\ +\ 74 \\ \hline 111 \end{array}$

Add, regrouping as necessary. **See Examples 4 and 5.**

26. $\begin{array}{r} 97 \\ +\ 79 \\ \hline \end{array}$ **27.** $\begin{array}{r} 67 \\ +\ 78 \\ \hline \end{array}$ **28.** $\begin{array}{r} 96 \\ +\ 47 \\ \hline \end{array}$ **29.** $\begin{array}{r} 73 \\ +\ 29 \\ \hline \end{array}$ **30.** $\begin{array}{r} 68 \\ +\ 37 \\ \hline \end{array}$

31. $\begin{array}{r} 746 \\ +\ 905 \\ \hline \end{array}$ **32.** $\begin{array}{r} 621 \\ +\ 359 \\ \hline \end{array}$ **33.** $\begin{array}{r} 306 \\ +\ 848 \\ \hline \end{array}$ **34.** $\begin{array}{r} 798 \\ +\ 206 \\ \hline \end{array}$ **35.** $\begin{array}{r} 278 \\ +\ 135 \\ \hline \end{array}$

36. $\begin{array}{r} 172 \\ +\ 156 \\ \hline \end{array}$ **37.** $\begin{array}{r} 928 \\ +\ 843 \\ \hline \end{array}$ **38.** $\begin{array}{r} 686 \\ +\ 726 \\ \hline \end{array}$ **39.** $\begin{array}{r} 526 \\ +\ 884 \\ \hline \end{array}$ **40.** $\begin{array}{r} 116 \\ +\ 897 \\ \hline \end{array}$

41. $\begin{array}{r} 3574 \\ +\ 2817 \\ \hline \end{array}$

42. $\begin{array}{r} 6871 \\ +\ 7528 \\ \hline \end{array}$

43. $\begin{array}{r} 7896 \\ +\ 3728 \\ \hline \end{array}$

44. $\begin{array}{r} 9382 \\ +\ 7586 \\ \hline \end{array}$

45. $\begin{array}{r} 9625 \\ +\ 7986 \\ \hline \end{array}$

46. $\begin{array}{r} 5718 \\ 5623 \\ +\ 7436 \\ \hline \end{array}$

47. $\begin{array}{r} 9056 \\ 78 \\ 6089 \\ +\ 731 \\ \hline \end{array}$

48. $\begin{array}{r} 4022 \\ 709 \\ 8621 \\ +\ 37 \\ \hline \end{array}$

49. $\begin{array}{r} 18 \\ 708 \\ 9286 \\ +\ 636 \\ \hline \end{array}$

50. $\begin{array}{r} 1708 \\ 321 \\ 61 \\ +\ 8926 \\ \hline \end{array}$

51. $\begin{array}{r} 422 \\ 6074 \\ 435 \\ +\ 8663 \\ \hline \end{array}$

52. $\begin{array}{r} 6505 \\ 173 \\ 7044 \\ +\ 168 \\ \hline \end{array}$

53. $\begin{array}{r} 321 \\ 9603 \\ 8 \\ 21 \\ +\ 1604 \\ \hline \end{array}$

54. $\begin{array}{r} 7631 \\ 5983 \\ 7 \\ 36 \\ +\ 505 \\ \hline \end{array}$

55. $\begin{array}{r} 2109 \\ 63 \\ 16 \\ 3 \\ +\ 9887 \\ \hline \end{array}$

56. $\begin{array}{r} 322 \\ 6508 \\ 93 \\ 745 \\ 18 \\ +\ 2005 \\ \hline \end{array}$

57. $\begin{array}{r} 553 \\ 97 \\ 2772 \\ 437 \\ 63 \\ +\ 328 \\ \hline \end{array}$

58. $\begin{array}{r} 3187 \\ 810 \\ 527 \\ 76 \\ 2665 \\ +\ 317 \\ \hline \end{array}$

59. $\begin{array}{r} 413 \\ 85 \\ 9919 \\ 602 \\ 31 \\ +\ 1218 \\ \hline \end{array}$

60. $\begin{array}{r} 576 \\ 7934 \\ 60 \\ 781 \\ 5968 \\ +\ 371 \\ \hline \end{array}$

Check each addition. If an answer is incorrect, find the correct answer.
See Examples 9 and 10.

61. GS $\begin{array}{r} \underline{2\ \ 1} \\ 832 \\ 468 \\ +\ 791 \\ \hline 2091 \end{array}$

62. $\begin{array}{r} \underline{\quad\quad} \\ 326 \\ 852 \\ +\ 679 \\ \hline 1857 \end{array}$

63. GS $\begin{array}{r} \underline{7\ 9} \\ 179 \\ 214 \\ +\ 376 \\ \hline 759 \end{array}$

64. $\begin{array}{r} \underline{\quad\quad} \\ 17 \\ 296 \\ 713 \\ +\ 94 \\ \hline 1220 \end{array}$

65. $\begin{array}{r} \underline{\quad\quad} \\ 4713 \\ 28 \\ 615 \\ +\ 64 \\ \hline 5420 \end{array}$

66. $\begin{array}{r} \underline{\quad\quad} \\ 3\,628 \\ 72 \\ 564 \\ +\ 7\,319 \\ \hline 11{,}583 \end{array}$

67. $\begin{array}{r} \underline{\quad\quad} \\ 678 \\ 7\,952 \\ 56 \\ 718 \\ +\ 2\,173 \\ \hline 11{,}377 \end{array}$

68. $\begin{array}{r} \underline{\quad\quad} \\ 516 \\ 8\,760 \\ 24 \\ 189 \\ +\ 1\,723 \\ \hline 11{,}212 \end{array}$

69. $\begin{array}{r} \underline{\quad\quad} \\ 4\,714 \\ 27 \\ 77 \\ 8\,878 \\ +\ 636 \\ \hline 14{,}332 \end{array}$

70. $\begin{array}{r} \underline{\quad\quad} \\ 6\,715 \\ 283 \\ 9\,617 \\ 13 \\ +\ 81 \\ \hline 16{,}719 \end{array}$

71. Explain the commutative property of addition in your own words. How is this used when checking an addition problem?

72. Explain the associative property of addition. How can this be used when adding columns of numbers?

For Exercises 73–76, use the map to find the shortest route between each pair of cities. ***See Examples 6 and 7.***

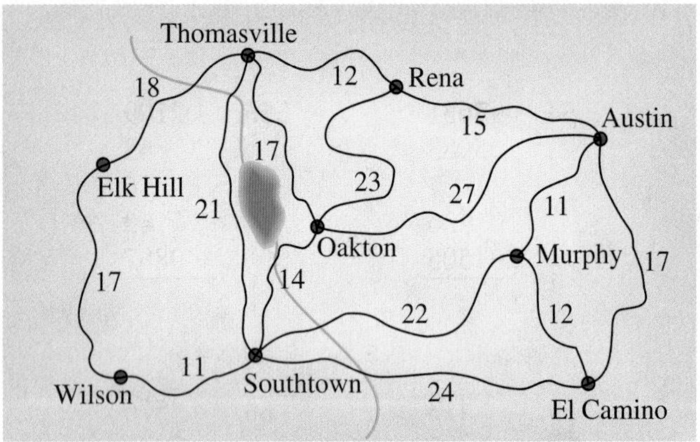

73. Southtown and Rena

Southtown to Thomasville	21 miles
Thomasville to Rena	+ 12 miles

74. Elk Hill and Oakton

Elk Hill to Thomasville	18 miles
Thomasville to Oakton	+ 17 miles

75. Thomasville and Murphy

76. Murphy and Thomasville

Solve each application problem.

77. The Twin Lakes Food Bank raised $3482 at a flea market and $12,860 at their annual auction. Find the total amount raised at these two events.

78. A ballpark vendor sold 185 hot dogs and 129 hamburgers. What was the total number of items sold?

79. There are 413 women and 286 men on the sales staff. How many people are on the sales staff?

80. One department in an office building has 283 employees while another department has 218 employees. How many employees are in the two departments?

81. This semester there are 13,786 students enrolled in on-campus day classes, 3497 students enrolled in night classes, and 2874 student's enrolled in on-line classes. Find the total number of students enrolled.

82. The number of tornadoes in each of the last 7 years has been 887, 223, 465, 683, 597, 214 and 1817. Find the total number of tornadoes in this 7-year period. (*Source:* Storm Prediction Center.)

Solve each problem involving perimeter. ***See Example 8.***

83. Find the total distance around a lot that has been developed as a go-cart track.

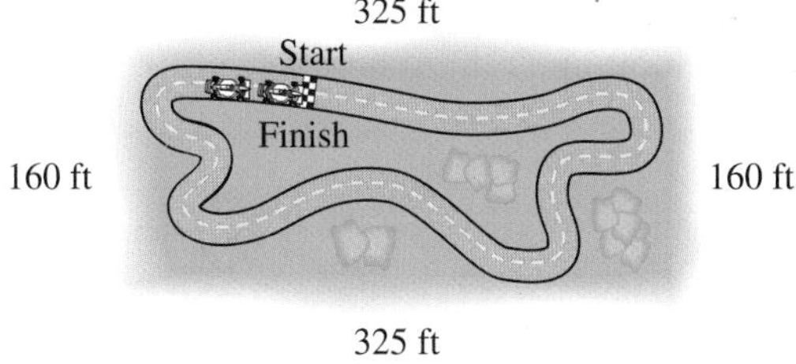

84. Because of heavy snowfall this winter, Maria needs to put new rain gutters around her entire roof. How many feet of gutters will she need?

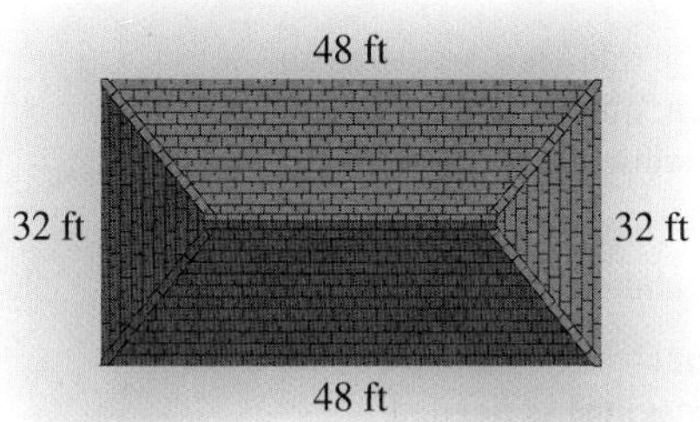

85. Martin plans to frame his back patio with redwood lumber. How many feet of lumber will he need?

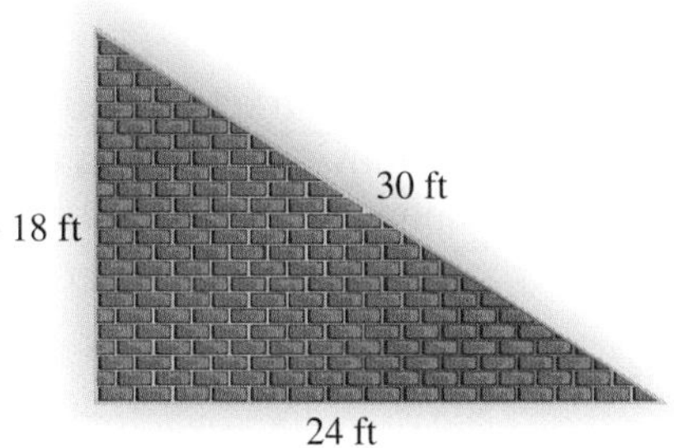

86. The university agriculture department is expanding the size of its experimental farm and wants to fence the area around the farm. How many meters of fencing will be needed?

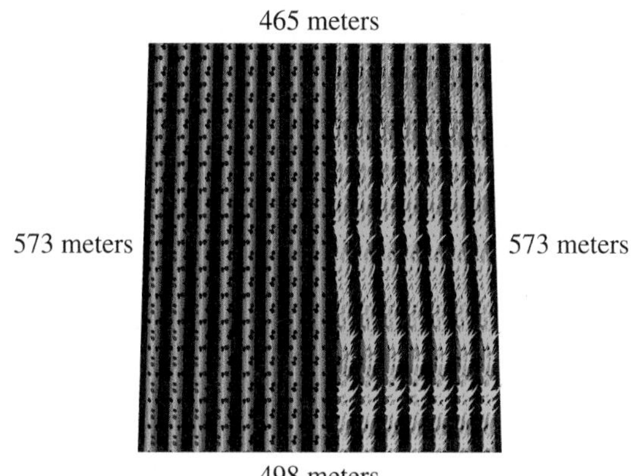

Relating Concepts (Exercises 87–94) For Individual or Group Work

Recall the place values of digits discussed in ***Section 1.1*** *and* ***work Exercises 87–94 in order.***

87. Write the largest four-digit number possible using the digits 4, 1, 9, and 2. Use each digit once.

88. Using the digits 4, 1, 9, and 2, write the smallest four-digit number possible. Use each digit once.

89. Write the largest five-digit number possible using the digits 6, 2, and 7. Use each digit at least once.

90. Using the digits 6, 2, and 7, write the smallest five-digit number possible. Use each digit at least once.

91. Write the largest seven-digit number possible using the digits 4, 3, and 9. Use each digit at least twice.

92. Using the digits 4, 3, and 9, write the smallest seven-digit number possible. Use each digit at least twice.

93. Explain your rule or procedure for writing the largest number in **Exercise 91.**

94. Explain your rule or procedure for writing the smallest number in **Exercise 92.**

1.3 Subtracting Whole Numbers

OBJECTIVES

1. Change addition problems to subtraction and subtraction problems to addition.
2. Identify the minuend, subtrahend, and difference.
3. Subtract when no regrouping (borrowing) is needed.
4. Check subtraction answers by adding.
5. Subtract with regrouping (borrowing).
6. Solve application problems with subtraction.

Suppose you have \$9, and you spend \$2 for parking. You then have \$7 left. There are two different ways of looking at these numbers.

As an addition problem:

\$2	+	\$7	=	\$9
↑		↑		↑
Amount spent		Amount left		Original amount

As a subtraction problem:

\$9	–	\$2	=	\$7
↑	↑	↑		↑
Original amount	Subtraction symbol	Amount spent		Amount left

OBJECTIVE 1 Change addition problems to subtraction and subtraction problems to addition. As shown in the box above, an addition problem can be changed to a subtraction problem and a subtraction problem can be changed to an addition problem.

EXAMPLE 1 Changing Addition Problems to Subtraction

Change each addition problem to a subtraction problem.

(a) $4 + 1 = 5$
Two subtraction problems are possible:

$$5 - 1 = 4 \quad \text{or} \quad 5 - 4 = 1$$

These figures show each subtraction problem.

(b) $8 + 7 = 15$

$$15 - 7 = 8 \quad \text{or} \quad 15 - 8 = 7$$

◄ Work Problem **1** at the Side.

EXAMPLE 2 Changing Subtraction Problems to Addition

Change each subtraction problem to an addition problem.

(a) $8 - 3 = 5$
↓ ↓ ↓
$8 = 3 + 5$

It is also correct to write $8 = 5 + 3$. (Recall that changing the order of addends does not change the sum.)

Continued on Next Page

1 Write two subtraction problems for each addition problem.

GS **(a)** $8 + 2 = 10$

$10 - \_\_\_\_ = 8$

$10 - \_\_\_\_ = \_\_\_\_$

(b) $7 + 4 = 11$

(c) $15 + 22 = 37$

(d) $23 + 55 = 78$

Answers

1. (a) 2; $10 - 2 = 8$ or 8; 2; $10 - 8 = 2$
(b) $11 - 4 = 7$ or $11 - 7 = 4$
(c) $37 - 22 = 15$ or $37 - 15 = 22$
(d) $78 - 55 = 23$ or $78 - 23 = 55$

(b) $18 - 13 = 5$

$18 = 13 + 5$ or $18 = 5 + 13$

(c) $29 - 13 = 16$

$29 = 13 + 16$ or $29 = 16 + 13$

Work Problem 2 at the Side. ▶

OBJECTIVE 2 **Identify the minuend, subtrahend, and difference.** In subtraction, as in addition, the numbers in a problem have names. For example, in the problem $8 - 5 = 3$, the number 8 is the **minuend,** 5 is the **subtrahend,** and 3 is the **difference** or answer.

$8 - 5 = 3$ ← Difference
(8: Minuend; 5: Subtrahend)

The answer in subtraction is the difference.

```
  8 ← Minuend
- 5 ← Subtrahend
  3 ← Difference
```

OBJECTIVE 3 **Subtract when no regrouping (borrowing) is needed.** Subtract two numbers by lining up the numbers in columns so the digits in the ones place are in the same column, the tens digits are in the same column, the hundreds digits are in the same column, and so on. Next, subtract by columns, starting at the right with the ones column.

EXAMPLE 3 **Subtracting Two Numbers**

Subtract.

Tens digits are lined up in the same column.
Ones digits are lined up in the same column.

(a)
```
  53
- 21
  32
```
3 ones − 1 one = 2 ones
5 tens − 2 tens = 3 tens

Ones digits are lined up.

(b)
```
  385
- 165
  220
```
Subtract from right to left.

5 ones − 5 ones = 0 ones
8 tens − 6 tens = 2 tens
3 hundreds − 1 hundred = 2 hundreds

(c)
```
  9437
-  210
  9227
```
7 ones − 0 ones = 7 ones
3 tens − 1 ten = 2 tens
4 hundreds − 2 hundreds = 2 hundreds
9 thousands − 0 thousands = 9 thousands

Work Problem 3 at the Side. ▶

OBJECTIVE 4 **Check subtraction answers by adding.** Use addition to check your answer to a subtraction problem. For example, check $8 - 3 = 5$ by *adding* 3 and 5.

$3 + 5 = 8$, so $8 - 3 = 5$ is correct.

2 Write an addition problem for each subtraction problem.

GS **(a)** $7 - 5 = 2$

$7 = 5 + \_\_\_\_$

(b) $9 - 4 = 5$

(c) $21 - 15 = 6$

(d) $58 - 42 = 16$

VOCABULARY TIP

Difference suggests comparing two things. In mathematics, this comparison is done by subtracting two numbers, the answer being the difference between them.

3 Subtract.

(a)
```
  74
- 43
```

(b)
```
  68
- 24
```

(c)
```
  429
- 318
```

(d)
```
  3927
- 2614
```

Answers

2. **(a)** 2; $7 = 5 + 2$ or $7 = 2 + 5$
(b) $9 = 4 + 5$ or $9 = 5 + 4$
(c) $21 = 15 + 6$ or $21 = 6 + 15$
(d) $58 = 42 + 16$ or $58 = 16 + 42$

3. **(a)** 31 **(b)** 44 **(c)** 111 **(d)** 1313

4 Use addition to determine whether each answer is correct. If incorrect, what should it be?

GS (a)
$$\begin{array}{r} 76 \\ -\ 45 \\ \hline 31 \end{array} \qquad \begin{array}{r} 45 \\ +\ \_\_ \\ \hline \_\_ \end{array}$$

(b)
$$\begin{array}{r} 53 \\ -\ 22 \\ \hline 21 \end{array}$$

(c)
$$\begin{array}{r} 374 \\ -\ 251 \\ \hline 113 \end{array}$$

(d)
$$\begin{array}{r} 7531 \\ -\ 4301 \\ \hline 3230 \end{array}$$

VOCABULARY TIP

Regrouping refers to both "carrying" in addition and "borrowing" in subtraction.

5 Subtract.

(a)
$$\begin{array}{r} 58 \\ -\ 19 \\ \hline \end{array}$$

(b)
$$\begin{array}{r} 86 \\ -\ 38 \\ \hline \end{array}$$

(c)
$$\begin{array}{r} 41 \\ -\ 27 \\ \hline \end{array}$$

(d)
$$\begin{array}{r} 863 \\ -\ 47 \\ \hline \end{array}$$

(e)
$$\begin{array}{r} 762 \\ -\ 157 \\ \hline \end{array}$$

Answers

4. **(a)** 31; 76; correct **(b)** incorrect; should be 31 **(c)** incorrect; should be 123 **(d)** correct

5. **(a)** 39 **(b)** 48 **(c)** 14 **(d)** 816 **(e)** 605

EXAMPLE 4 Checking Subtraction by Using Addition

Use addition to check each answer. If the answer is incorrect, find the correct answer.

(a)
$$\begin{array}{r} 89 \\ -\ 47 \\ \hline 42 \end{array}$$

Rewrite as an addition problem, as shown in **Example 2**.

Subtraction problem
$$\begin{array}{r} 89 \\ -\ 47 \\ \hline 42 \\ \hline 89 \end{array}$$
Addition problem
$$\begin{array}{r} \mathbf{47} \\ +\ \mathbf{42} \\ \hline \mathbf{89} \end{array}$$

Because $47 + 42 = 89$, the subtraction was done correctly.

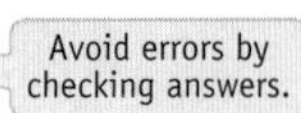

(b) $72 - 41 = 21$

Rewrite as an addition problem.

$$72 = 41 + 21$$

But, $41 + 21 = 62$, **not** 72, so the subtraction was done **incorrectly.** Rework the original subtraction to get the correct answer, 31. Then, $41 + 31 = 72$.

(c)
$$\begin{array}{r} 374 \\ -\ 141 \\ \hline 233 \end{array}$$
Match: $141 + 233 = 374$

The answer checks.

◀ **Work Problem 4 at the Side.**

OBJECTIVE 5 Subtract with regrouping (borrowing). When a digit in the minuend is less than the one directly below it, **regrouping** is necessary (also called **borrowing**).

EXAMPLE 5 Subtracting with Regrouping

Subtract 19 from 57.

Write the problem vertically.

$$\begin{array}{r} 57 \\ -\ 19 \\ \hline \end{array}$$

In the ones column, 7 is **less** than 9, so in order to subtract, we must regroup 1 ten as 10 ones.

5 tens − 1 ten = 4 tens → 4 17 ← 1 ten = 10 ones, and 10 ones + 7 ones = 17 ones

$$\begin{array}{r} {\scriptstyle 4\ 17} \\ \not{5}\,\not{7} \\ -\ 1\,9 \\ \hline \end{array}$$

Now subtract 9 ones from 17 ones in the o nes column. Then subtract 1 ten from 4 tens in the tens column.

$$\begin{array}{r} {\scriptstyle 4\ 17} \\ \not{5}\,\not{7} \\ -\ 1\,9 \\ \hline 3\,8 \end{array}$$ Difference

Finally, $57 - 19 = 38$. Check by adding 19 and 38; you should get 57.

◀ **Work Problem 5 at the Side.**

EXAMPLE 6 Subtracting with Regrouping

Subtract by regrouping when necessary.

(a) $\begin{array}{r} 7856 \\ -\ 137 \\ \hline \end{array}$

Regroup 1 ten as 10 ones. 10 ones + 6 ones = 16 ones

$$\begin{array}{r} 4\ 16 \\ 7\,8\,\cancel{5}\,\cancel{6} \\ -\ \ 1\,3\,7 \\ \hline 7\,7\,1\,9 \end{array}$$ Difference

(b) $\begin{array}{r} 635 \\ -\ 546 \\ \hline \end{array}$

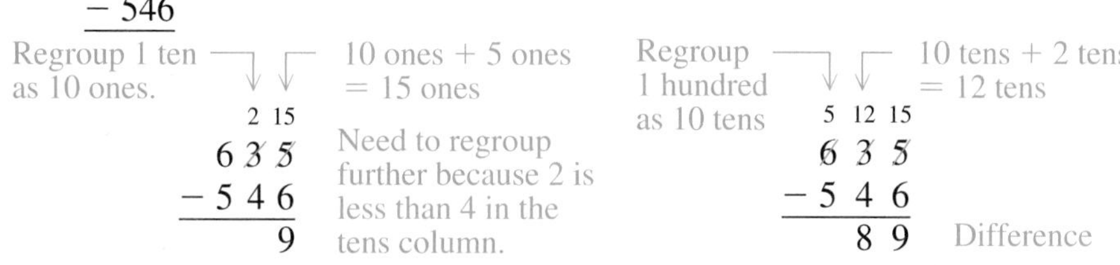

(c) $\begin{array}{r} 647 \\ -\ 489 \\ \hline \end{array}$

$$\begin{array}{r} 3\ 17 \\ 6\,\cancel{4}\,\cancel{7} \\ -\,4\,8\,9 \\ \hline 8 \end{array}$$

Need to regroup further because 3 is less than 8 in the tens column.

$$\begin{array}{r} 5\ 13\ 17 \\ \cancel{6}\,\cancel{4}\,\cancel{7} \\ -\,4\,8\,9 \\ \hline 1\,5\,8 \end{array}$$ Difference

Work Problem 6 at the Side. ▶

Sometimes a minuend has zeros in some of the positions. In such cases, regrouping may be a little more complicated than what we have shown so far.

EXAMPLE 7 Regrouping with Zeros

Subtract.

$$\begin{array}{r} 4607 \\ -\ 3168 \\ \hline \end{array}$$

There are no tens that can be regrouped into ones. So you must first regroup 1 hundred as 10 tens.

Regroup 1 hundred as 10 tens. Write 10 tens.

$$\begin{array}{r} 5\ 10\ \ \\ 4\,\cancel{6}\,\cancel{0}\,7 \\ -\,3\,1\,6\,8 \\ \hline \end{array}$$

Now we may regroup from the tens position.

Regroup 1 ten as 10 ones. 10 tens − 1 ten = 9 tens.

10 ones + 7 ones = 17 ones

$$\begin{array}{r} 9\ \ \\ 5\,10\,17 \\ 4\,\cancel{6}\,\cancel{0}\,\cancel{7} \\ -\,3\,1\,6\,8 \\ \hline 9 \end{array}$$

Continued on Next Page

6 Subtract.

(a) $\begin{array}{r} 8\ 12\ \ \\ \cancel{9}\ \cancel{2}\ 7 \\ -\quad 4\ 3 \\ \hline \_\ \_\ 4 \end{array}$

(b) $\begin{array}{r} 675 \\ -\ 86 \\ \hline \end{array}$

(c) $\begin{array}{r} 477 \\ -\ 389 \\ \hline \end{array}$

(d) $\begin{array}{r} 1417 \\ -\ 988 \\ \hline \end{array}$

(e) $\begin{array}{r} 8739 \\ -\ 3892 \\ \hline \end{array}$

Answers

6. **(a)** 8; 8; 884 **(b)** 589 **(c)** 88 **(d)** 429 **(e)** 4847

7 Subtract.

(a) $\begin{array}{r} 206 \\ -\,177 \\ \hline \end{array}$

(b) $\begin{array}{r} 703 \\ -\,415 \\ \hline \end{array}$

(c) $\begin{array}{r} 7024 \\ -\,2632 \\ \hline \end{array}$

8 Subtract.

GS (a) $\begin{array}{r} \overset{2}{\cancel{3}}\,\overset{\overset{9}{10}}{\cancel{0}}\,\overset{18}{\cancel{8}} \\ -\,1\,5\,9 \\ \hline \_\,\_\,9 \end{array}$

(b) $\begin{array}{r} 570 \\ -\,368 \\ \hline \end{array}$

(c) $\begin{array}{r} 1570 \\ -\,983 \\ \hline \end{array}$

(d) $\begin{array}{r} 7001 \\ -\,5193 \\ \hline \end{array}$

(e) $\begin{array}{r} 4000 \\ -\,1782 \\ \hline \end{array}$

Answers

7. (a) 29 (b) 288 (c) 4392

8. (a) 1; 4; 149 (b) 202 (c) 587 (d) 1808 (e) 2218

Complete the problem.

$$\begin{array}{r} 4\,\overset{5}{\cancel{6}}\,\overset{\overset{9}{10}}{\cancel{0}}\,\overset{17}{\cancel{7}} \\ -\,3\,1\,6\,8 \\ \hline 1\,4\,3\,9 \end{array}$$ Difference

Check by adding 1439 and 3168; you should get 4607.

◀ **Work Problem 7 at the Side.**

EXAMPLE 8 Regrouping with Zeros

Subtract.

(a) $\begin{array}{r} 708 \\ -\,149 \\ \hline \end{array}$

Write 10 tens.

Regroup 1 ten as 10 ones.

Regroup 1 hundred as 10 tens.

10 ones + 8 ones = 18 ones

$$\begin{array}{r} \overset{6}{\cancel{7}}\,\overset{\overset{9}{10}}{\cancel{0}}\,\overset{18}{\cancel{8}} \\ -\,1\,4\,9 \\ \hline 5\,5\,9 \end{array}$$

Remember to work from right to left.

(b) $\begin{array}{r} 380 \\ -\,276 \\ \hline \end{array}$

Regroup 1 ten as 10 ones.

Write 10 ones.

$$\begin{array}{r} 3\,\overset{7}{\cancel{8}}\,\overset{10}{\cancel{0}} \\ -\,2\,7\,6 \\ \hline 1\,0\,4 \end{array}$$

(c) $\begin{array}{r} 9000 \\ -\,6999 \\ \hline \end{array}$

$$\begin{array}{r} \overset{8}{\cancel{9}}\,\overset{\overset{9}{10}}{\cancel{0}}\,\overset{\overset{9}{10}}{\cancel{0}}\,\overset{10}{\cancel{0}} \\ -\,6\,9\,9\,9 \\ \hline 2\,0\,0\,1 \end{array}$$

Be extra careful when zeros are involved.

◀ **Work Problem 8 at the Side.**

As we have seen, an answer to a subtraction problem can be checked by adding.

EXAMPLE 9 Checking Subtraction by Using Addition

Use addition to check each answer.

(a) $\begin{array}{r} 613 \\ -\,275 \\ \hline 338 \end{array}$ Match **CHECK** $\begin{array}{r} 275 \\ +\,338 \\ \hline 613 \end{array}$ ✓ Correct

Continued on Next Page

(b)

```
  1915          CHECK  1635
- 1635                + 280
   280    Match        1915 ✓   Correct
```

(c)

```
  15,803         CHECK  7 325
-  7 325                + 8 578
   8 578   No Match     15,903   Error
```

It's always a good idea to check your work.

Rework the original problem to get the correct answer, 8478. Then, 7325 + 8478 **does** give 15,803.

Work Problem 9 at the Side. ▶

OBJECTIVE 6 Solve application problems with subtraction. As shown in the next example, subtraction can be used to solve an application problem.

EXAMPLE 10 Applying Subtraction Skills

Use the table to find how much more, on average, a person with an Associate of Arts degree earns each year than a high school graduate.

EDUCATION PAYS

The more education adults get, the higher their annual earnings.

Education Level	Average Earnings
Not a high school graduate	$33,435
High school graduate	$43,165
Some college, no degree	$50,359
Associate of Arts degree	$54,861
Bachelor's degree	$82,197
Master's degree	$99,516
Doctoral degree	$129,773
Professional degree	$166,065

Note: Average annual earnings for workers between ages 25 and 64.

Source: U.S. Census Bureau and Pearson Education, Inc.

Approach The average earnings for a person with an Associate of Arts degree is $54,861 each year. The average for a high school graduate is $43,165. Find how much more a college graduate earns by subtracting $43,165 from $54,861.

Solution

```
  $54,861 ← Associate of Arts degree
- $43,165 ← High school graduate
  $11,696 ← More earnings
```

Education pays.

On average, a person with an Associate of Arts degree earns $11,696 more each year than a high school graduate.

Work Problem 10 at the Side. ▶

---

9 Use addition to check each answer. If the answer is incorrect, find the correct answer.

(a)

```
  357        Check  1 6 8
- 168               + _ _ _
  189               3 5 7
```

(b)

```
  570
- 328
  252
```

(c)

```
  14,726
-  8 839
   5 887
```

10 Use the table from **Example 10** to find, on average,

(a) how much more a person with an Associate of Arts degree earns each year than a person who is not a high school graduate.

(b) how much more a person with a Bachelor's degree earns each year than a person with an Associate of Arts degree.

Answers

9. **(a)** 1; 8; 9; correct
(b) incorrect; should be 242
(c) correct

10. **(a)** $21,426 **(b)** $27,336

1.3 Exercises

FOR EXTRA HELP

MyMathLab®

CONCEPT CHECK *Write in the number needed to complete the check.*

1. $\begin{array}{r} 48 \\ -\ 32 \\ \hline 16 \end{array}$ $\begin{array}{r} 16 \\ +\ \_\_ \\ \hline 48 \end{array}$

2. $\begin{array}{r} 17 \\ -\ 13 \\ \hline 4 \end{array}$ $\begin{array}{r} \_\_ \\ +\ 13 \\ \hline 17 \end{array}$

3. $\begin{array}{r} 86 \\ -\ 53 \\ \hline 33 \end{array}$ $\begin{array}{r} 53 \\ +\ 33 \\ \hline \_\_ \end{array}$

4. $\begin{array}{r} 78 \\ -\ 35 \\ \hline 43 \end{array}$ $\begin{array}{r} \_\_ \\ +\ 43 \\ \hline 78 \end{array}$

5. $\begin{array}{r} 77 \\ -\ 60 \\ \hline 17 \end{array}$ $\begin{array}{r} 60 \\ +\ \_\_ \\ \hline 77 \end{array}$

Work each subtraction problem. Use addition to check each answer. ***See Examples 3 and 4.***

6. $\begin{array}{r} 87 \\ -\ 63 \\ \hline \end{array}$

7. $\begin{array}{r} 335 \\ -\ 122 \\ \hline \end{array}$

8. $\begin{array}{r} 602 \\ -\ 301 \\ \hline \end{array}$

9. $\begin{array}{r} 552 \\ -\ 451 \\ \hline \end{array}$

10. $\begin{array}{r} 888 \\ -\ 215 \\ \hline \end{array}$

11. $\begin{array}{r} 7352 \\ -\ 241 \\ \hline \end{array}$

12. $\begin{array}{r} 4420 \\ -\ 310 \\ \hline \end{array}$

13. $\begin{array}{r} 5546 \\ -\ 2134 \\ \hline \end{array}$

14. $\begin{array}{r} 1875 \\ -\ 1362 \\ \hline \end{array}$

15. $\begin{array}{r} 6259 \\ -\ 4148 \\ \hline \end{array}$

16. $\begin{array}{r} 9654 \\ -\ 4323 \\ \hline \end{array}$

17. $\begin{array}{r} 24{,}392 \\ -\ 11{,}232 \\ \hline \end{array}$

18. $\begin{array}{r} 57{,}921 \\ -\ 34{,}801 \\ \hline \end{array}$

19. $\begin{array}{r} 46{,}253 \\ -\ 5\ 143 \\ \hline \end{array}$

20. $\begin{array}{r} 75{,}904 \\ -\ 3\ 702 \\ \hline \end{array}$

Use addition to check each subtraction problem. If an answer is not correct, find the correct answer. ***See Example 4.***

21. GS $\begin{array}{r} 54 \\ -\ 42 \\ \hline 12 \end{array}$ $\begin{array}{r} 42 \\ +\ \_\_ \\ \hline 54 \end{array}$

22. GS $\begin{array}{r} 87 \\ -\ 43 \\ \hline 44 \end{array}$ $\begin{array}{r} 43 \\ +\ \_\_ \\ \hline 87 \end{array}$

23. $\begin{array}{r} 89 \\ -\ 27 \\ \hline 63 \end{array}$

24. $\begin{array}{r} 47 \\ -\ 35 \\ \hline 13 \end{array}$

25. $\begin{array}{r} 382 \\ -\ 261 \\ \hline 131 \end{array}$

26. $\begin{array}{r} 754 \\ -\ 342 \\ \hline 412 \end{array}$

27. $\begin{array}{r} 4683 \\ -\ 3542 \\ \hline 1141 \end{array}$

28. $\begin{array}{r} 5217 \\ -\ 4105 \\ \hline 1132 \end{array}$

29. $\begin{array}{r} 8643 \\ -\ 1421 \\ \hline 7212 \end{array}$

30. $\begin{array}{r} 9428 \\ -\ 3124 \\ \hline 6324 \end{array}$

31. CONCEPT CHECK *Underline the correct answer.*

In subtraction, regrouping is necessary when the digit in the (*minuend/subtrahend*) is less value than the digit in the subtrahend which is directly (*above/below*) it.

32. CONCEPT CHECK *Which of the subtraction problems will require regrouping?*

(a) $\begin{array}{r} 64 \\ -\ 51 \\ \hline \end{array}$ **(b)** $\begin{array}{r} 763 \\ -\ 473 \\ \hline \end{array}$

(c) $\begin{array}{r} 43{,}708 \\ -\ 22{,}607 \\ \hline \end{array}$ **(d)** $\begin{array}{r} 6208 \\ -\ 5126 \\ \hline \end{array}$

Subtract, regrouping when necessary. ***See Examples 5–8.***

33. $\begin{array}{r} 75 \\ -\ 37 \\ \hline \end{array}$

34. $\begin{array}{r} 86 \\ -\ 28 \\ \hline \end{array}$

35. $\begin{array}{r} 94 \\ -\ 49 \\ \hline \end{array}$

36. $\begin{array}{r} 68 \\ -\ 39 \\ \hline \end{array}$

37. $\begin{array}{r} 57 \\ -\ 38 \\ \hline \end{array}$

38. $\begin{array}{r} 47 \\ -\ 29 \\ \hline \end{array}$

39. $\begin{array}{r} 828 \\ -\ 547 \\ \hline \end{array}$

40. $\begin{array}{r} 916 \\ -\ 618 \\ \hline \end{array}$

41. $\begin{array}{r} 771 \\ -\ 252 \\ \hline \end{array}$

42. $\begin{array}{r} 973 \\ -\ 788 \\ \hline \end{array}$

43. $\begin{array}{r} 7538 \\ -\ 479 \\ \hline \end{array}$ **44.** $\begin{array}{r} 5863 \\ -\ 1295 \\ \hline \end{array}$ **45.** $\begin{array}{r} 9988 \\ -\ 2399 \\ \hline \end{array}$ **46.** $\begin{array}{r} 3576 \\ -\ 1658 \\ \hline \end{array}$ **47.** $\begin{array}{r} 80 \\ -\ 73 \\ \hline \end{array}$

48. $\begin{array}{r} 60 \\ -\ 37 \\ \hline \end{array}$ **49.** $\begin{array}{r} 308 \\ -\ 289 \\ \hline \end{array}$ **50.** $\begin{array}{r} 600 \\ -\ 599 \\ \hline \end{array}$ **51.** $\begin{array}{r} 4041 \\ -\ 1208 \\ \hline \end{array}$ **52.** $\begin{array}{r} 4602 \\ -\ 2063 \\ \hline \end{array}$

53. $\begin{array}{r} 9305 \\ -\ 1530 \\ \hline \end{array}$ **54.** $\begin{array}{r} 7120 \\ -\ 6033 \\ \hline \end{array}$ **55.** $\begin{array}{r} 1580 \\ -\ 1077 \\ \hline \end{array}$ **56.** $\begin{array}{r} 3068 \\ -\ 2105 \\ \hline \end{array}$ **57.** $\begin{array}{r} 2006 \\ -\ 1850 \\ \hline \end{array}$

58. $\begin{array}{r} 8203 \\ -\ 5365 \\ \hline \end{array}$ **59.** $\begin{array}{r} 8240 \\ -\ 6056 \\ \hline \end{array}$ **60.** $\begin{array}{r} 7050 \\ -\ 6045 \\ \hline \end{array}$ **61.** $\begin{array}{r} 8503 \\ -\ 2816 \\ \hline \end{array}$ **62.** $\begin{array}{r} 16{,}004 \\ -\ 5\ 087 \\ \hline \end{array}$

63. $\begin{array}{r} 80{,}705 \\ -\ 61{,}667 \\ \hline \end{array}$ **64.** $\begin{array}{r} 81{,}000 \\ -\ 55{,}456 \\ \hline \end{array}$ **65.** $\begin{array}{r} 66{,}000 \\ -\ 34{,}444 \\ \hline \end{array}$ **66.** $\begin{array}{r} 77{,}000 \\ -\ 65{,}308 \\ \hline \end{array}$ **67.** $\begin{array}{r} 20{,}080 \\ -\ 13{,}496 \\ \hline \end{array}$

CONCEPT CHECK *Fill in each blank with the correct response.*

68. To avoid errors when solving math problems, it's a good idea to ___________ your work.

69. When checking the accuracy of an answer to an addition problem, you can use ___________.

70. An answer to a subtraction problem may be checked using ___________.

Use addition to check each subtraction problem. If an answer is incorrect, find the correct answer. ***See Example 9.***

71. $\begin{array}{r} 9428 \\ -\ 4509 \\ \hline 4919 \end{array}$ **72.** $\begin{array}{r} 1671 \\ -\ 1325 \\ \hline 1346 \end{array}$ **73.** $\begin{array}{r} 2548 \\ -\ 2278 \\ \hline 270 \end{array}$ **74.** $\begin{array}{r} 5274 \\ -\ 1130 \\ \hline 4144 \end{array}$

75. $\begin{array}{r} 93{,}758 \\ -\ 52{,}869 \\ \hline 40{,}889 \end{array}$ **76.** $\begin{array}{r} 82{,}357 \\ -\ 14{,}396 \\ \hline 68{,}961 \end{array}$ **77.** $\begin{array}{r} 36{,}778 \\ -\ 17{,}405 \\ \hline 19{,}373 \end{array}$ **78.** $\begin{array}{r} 34{,}821 \\ -\ 17{,}735 \\ \hline 17{,}735 \end{array}$

79. An addition problem can be changed to a subtraction problem and a subtraction problem can be changed to an addition problem. Give two examples of each to demonstrate this.

80. Can you use the commutative and the associative properties in subtraction? Explain.

Solve each application problem. ***See Example 10.***

81. A man burns 187 calories during 60 minutes of sitting at a computer while a woman burns 140 calories at the same activity. How many fewer calories does a woman burn than a man in 60 minutes? (*Source:* www.cookinglight.com)

82. A woman burns 302 calories during an hour of walking, while a man burns 403 calories doing the same activity. How many more calories does a man burn than a woman during an hour of walking? (*Source:* www.cookinglight.com)

83. In April 2011, there were 612 tornadoes, shattering the old record of 543. How many more tornadoes were there than the old record number? (*Source:* National Oceanic and Atmospheres Administration.)

84. With an estimated 327 deaths, the tornado outbreak in April 2011 was the third deadliest on record, behind 747 deaths in April 1925 and 332 deaths in April 1932. How many more deaths were there in the deadliest tornado outbreak than in the tornado outbreak of April 2011? (*Source:* Accu Weather.)

85. GS The top of each main tower of the Golden Gate Bridge is 746 feet above the water and 500 feet above the roadway. How far above the water is the roadway? (*Source:* gocalifornia.about.com)

746 feet above water
−500 feet above roadway

86. In a recent three-month period there were 81,465 Ford Explorers and 70,449 Jeep Grand Cherokees sold. Which vehicle had greater sales? By how much? (*Source:* J. D. Power and Associates.)

87. Six years ago there were 6970 bridge and lock-tender jobs across the United States. Today there are 3700 that remain. How many of these jobs have been eliminated? (*Source:* Bureau of Labor Statistics.)

88. In 1964, its first year on the market, the Ford Mustang sold for $2500. In 2013, the Ford Mustang sold for $28,065. Find the increase in price. (*Source:* eBay.)

89. Patriot Flag Company manufactured 14,608 U.S. flags and sold 5069. How many flags remain unsold?

90. Eye exams have been given to 14,679 children in the school district. If there are 23,156 students in the school district, how many have not received eye exams?

91. The Jordanos now pay rent of $650 per month. If they buy a house, their housing expense will be $913 per month. How much more will they pay per month if they buy a house?

92. A retired couple who used to receive a Social Security payment of $1479 per month now receives $1568 per month. Find the amount of the monthly increase.

93. The distance from New York City to Buenos Aires, Argentina is 5299 miles, while the distance from Los Angeles to Dublin, Ireland is 5158 miles. How much further is one trip than the other? (*Source:* Map Crow Travel Calculator, mapcrow info)

94. In the year 2020 it is predicted that we will need 2,820,000 nurses in the United States, while only 1,810,000 nurses will be available. Find the shortage in the number of nurses. (*Source:* American Hospital Association.)

Solve each application problem. Add or subtract as necessary.

95. This year there were 264,311 hip replacement procedures in the United States. If 125,423 of the patients were 65 years of age or older, how many patients were under age 65? (*Source:* Federal Agency for Healthcare Research and Quality.)

96. This year there were 555,800 knee surgeries performed in the United States. The number of knee surgeries performed six years ago was 328,900. How many more of these surgeries were performed this year than six years ago? (*Source:* Agency for Healthcare Research and Quality.)

SUBWAY promotes healthy food choices by offering eight sandwiches that are low in fat. The nutritional information, printed on every SUBWAY napkin, appears below and includes information to answer Exercises 97–100. (Source: SUBWAY.*)*

SUBWAY

OUR 6" SANDWICHES:	CALORIES	FAT(g)
VEGGIE DELITE®	230	3
BLACK FOREST HAM	290	5
TURKEY BREAST	280	4
ROAST BEEF	320	5
SUBWAY CLUB	310	5
TURKEY BREAST & BLACK FOREST HAM	280	4
OVEN ROASTED CHICKEN	320	5
SWEET ONION CHICKEN TERIYAKI	380	5

SUBWAY® regular 6" subs include italian or wheat bread, veggies and meat. Addition of condiments or cheese alters nutrition content.

MUSTARD (2 tsp.)	5	0
CHEESE TRIANGLES (2)	40	4
OLIVE OIL (1 tsp.)	45	5
VERSUS:		
BIG MAC®	540	29
WHOPPER®	670	40

97. How many fewer calories and grams of fat are in a 6-inch Veggie Delite sandwich than a Big Mac?

98. How many fewer calories and grams of fat are in a 6-inch Turkey Breast and Black Forest Ham sandwich than a Whopper?

99. Find the total number of calories and grams of fat in an Oven Roasted Chicken sandwich with mustard and olive oil.

100. A customer ate two sandwiches, one with the least calories and one with the most calories. Find the total number of calories and grams of fat in the two sandwiches.

1.4 Multiplying Whole Numbers

OBJECTIVES

1. Identify the parts of a multiplication problem.
2. Do chain multiplication.
3. Multiply by single-digit numbers.
4. Use multiplication shortcuts for numbers ending in zeros.
5. Multiply by numbers having more than one digit.
6. Solve application problems with multiplication.

Suppose we want to know the total number of exercise bicycles available at the gym. The bicycles are arranged in four columns with three stations in each column. Adding the number 3 a total of 4 times gives 12.

$$3 + 3 + 3 + 3 = 12$$

This result can also be shown with a figure.

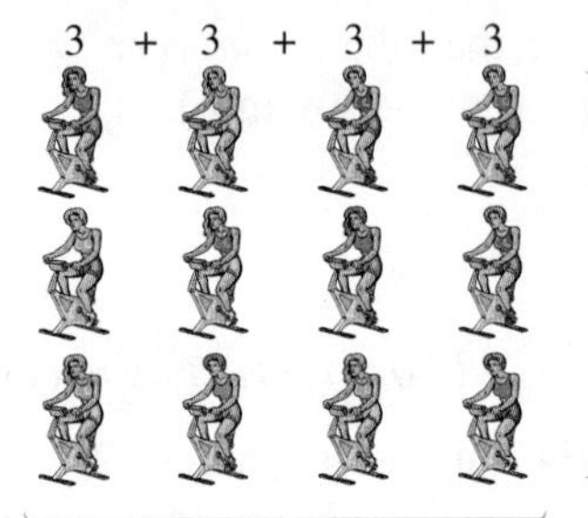

OBJECTIVE 1 Identify the parts of a multiplication problem. Multiplication is a shortcut for repeated addition. In the exercise bicycle example, instead of *adding* 3 + 3 + 3 + 3 to get 12, we can *multiply* 3 by 4 to get 12. The numbers being multiplied are called **factors.** The answer is called the **product.** For example, the product of 3 and 4 can be written with the symbol ×, a raised dot, or parentheses, as follows.

$$\begin{array}{r} 3 \\ \times\ 4 \\ \hline 12 \end{array}$$

3 ← Factor (also called *multiplicand*)
× 4 ← Factor (also called *multiplier*)
12 ← Product (answer)

$3 \times 4 = 12$ *or* $3 \cdot 4 = 12$ *or* $(3)(4) = 12$ *or* $3(4) = 12$

◀ Work Problem 1 at the Side.

Commutative Property of Multiplication

By the **commutative property of multiplication,** the product (answer) remains the same when the order of the factors is changed. For example,

$$3 \times 5 = 15 \quad \text{and} \quad 5 \times 3 = 15.$$

Multiply numbers in any order.

EXAMPLE 1 Multiplying Two Numbers

Multiply.

(a) $3 \times 4 = 12$

(b) $6 \cdot 0 = 0$ — Multiply any number by zero and the answer is always zero.

(c) $4(8) = 32$

Continued on Next Page

1 Identify the factors and the product in each multiplication problem.

(a) $8 \times 5 = 40$

(b) $6(4) = 24$

(c) $7 \cdot 6 = 42$

(d) $(3)(9) = 27$

Answers

1. (a) factors: 8, 5; product: 40
   (b) factors: 6, 4; product: 24
   (c) factors: 7, 6; product: 42
   (d) factors: 3, 9; product: 27

Learning the multiplication table will help you in later chapters.

Multiplication Table

×	1	2	3	4	5	6	7	8	9
1	1	2	3	4	5	6	7	8	9
2	2	4	6	8	10	12	14	16	18
3	3	6	9	12	15	18	21	24	27
4	4	8	12	16	20	24	28	32	36
5	5	10	15	20	25	30	35	40	45
6	6	12	18	24	30	36	42	48	54
7	7	14	21	28	35	42	49	56	63
8	8	16	24	32	40	48	56	64	72
9	9	18	27	36	45	54	63	72	81

Recall that any number multiplied by 1 is always the number itself.

Work Problem **2** at the Side. ▶

OBJECTIVE 2 **Do chain multiplication.** Some multiplications involve more than two factors.

Associative Property of Multiplication

By the **associative property of multiplication,** grouping the factors differently does not change the product.

EXAMPLE 2 Multiplying Three Numbers

Multiply $2 \times 3 \times 5$.

$$(2 \times 3) \times 5 \qquad \text{Parentheses show what to do first.}$$

$$6 \times 5 = 30$$

Also,

$$2 \times (3 \times 5)$$

$$2 \times 15 = 30$$

Either grouping results in the same product.

Work Problem **3** at the Side. ▶

Calculator Tip

The calculator approach to **Example 2** uses chain calculations.

2 (×) 3 (×) 5 (=) **30**

A problem with more than two factors, such as the one in **Example 2**, is called a **chain multiplication** problem.

VOCABULARY TIP

Chain multiplication Just like a chain is a connected series of objects—a "chain of events," for example—chain multiplication is a series of numbers connected by the operation of multiplication.

2 Multiply.

(a) 7×4

(b) 0×9

(c) $8(5)$

(d) $6 \cdot 5$

(e) $(1)(8)$

3 Multiply.

(a) $3 \times 2 \times 5$

(b) $4 \cdot 7 \cdot 1$

(c) $(8)(3)(0)$

Answers

2. **(a)** 28 **(b)** 0 **(c)** 40 **(d)** 30 **(e)** 8
3. **(a)** 30 **(b)** 28 **(c)** 0

OBJECTIVE 3 **Multiply by single-digit numbers.** Regrouping may be needed in multiplication problems with larger factors.

EXAMPLE 3 **Multiplying with Regrouping**

Multiply.

(a) $\begin{array}{r} 53 \\ \times\ 4 \\ \hline \end{array}$

Start by multiplying in the ones column.

$\begin{array}{r} {}^{1} \\ 53 \\ \times\ 4 \\ \hline 2 \end{array}$ $4 \times 3 = \mathbf{12}$ ones

Write 1 ten in the tens column.

Write 2 ones in the ones column.

Next, multiply 4 times 5 tens.

$\begin{array}{r} {}^{1} \\ 5\,3 \\ \times\ 4 \\ \hline 2 \end{array}$ 4×5 tens $=$ **20** tens

Add the 1 ten that was written at the top of the tens column.

$\begin{array}{r} {}^{1} \\ 5\,3 \\ \times\ 4 \\ \hline 212 \end{array}$ 20 tens + 1 ten = 21 tens

(b) $\begin{array}{r} 724 \\ \times\ 5 \\ \hline \end{array}$

Work as shown.

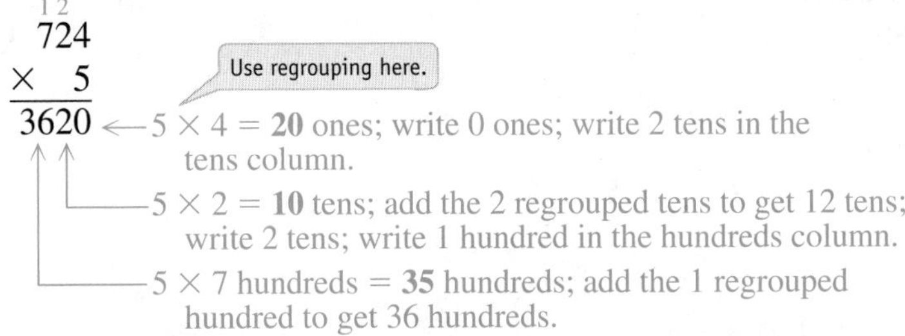

1 2
724
× 5
3620
Use regrouping here.
5 × 4 = 20 ones; write 0 ones; write 2 tens in the tens column.
5 × 2 = 10 tens; add the 2 regrouped tens to get 12 tens; write 2 tens; write 1 hundred in the hundreds column.
5 × 7 hundreds = 35 hundreds; add the 1 regrouped hundred to get 36 hundreds.

◀ Work Problem **4** at the Side.

4 Multiply.

GS **(a)** $\begin{array}{r} {}^{1} \\ 5\,3 \\ \times\ 5 \\ \hline \_\,\_\,5 \end{array}$

(b) $\begin{array}{r} 79 \\ \times\ 0 \\ \hline \end{array}$

(c) $\begin{array}{r} 758 \\ \times\ 8 \\ \hline \end{array}$

(d) $\begin{array}{r} 2831 \\ \times\ 7 \\ \hline \end{array}$

(e) $\begin{array}{r} 4714 \\ \times\ 8 \\ \hline \end{array}$

OBJECTIVE 4 **Use multiplication shortcuts for numbers ending in zeros.** The product of two whole number factors is also called a **multiple** of either factor. For example, since $4 \cdot 2 = 8$, the whole number 8 is a multiple of both 4 and 2. *Multiples of 10* are very useful when multiplying. A **multiple of 10** is a whole number that ends in 0, such as 10, 20, or 30; 100, 200, or 300; 1000, 2000, or 3000; and so on. There is a short way to multiply by these multiples of 10. Look at the following examples.

$$26 \times 1 = 26$$

$$26 \times 10 = 260$$

$$26 \times 100 = 2600$$

$$26 \times 1000 = 26{,}000$$

Do you see a pattern? These examples suggest the rule that follows.

Answers

4. **(a)** 2; 6; 265 **(b)** 0 **(c)** 6064 **(d)** 19,817 **(e)** 37,712

Multiplying by Multiples of 10

To multiply a whole number by 10, 100, or 1000, attach one, two, or three zeros, respectively, to the right of the whole number.

EXAMPLE 4 **Using Multiples of 10 to Multiply**

Multiply.

(a) $59 \times 10 = 590$ ← Attach 0.

(b) $74 \times 100 = 7400$ ← Attach 00.

(c) $803 \times 1000 = 803{,}000$ ← Attach 000.

Work Problem 5 at the Side. ▶

You can also find the product of other multiples of 10 by attaching zeros.

EXAMPLE 5 **Using Multiples of 10 to Multiply**

Multiply.

(a) 75×3000

Multiply 75 by 3, and then attach three zeros.

$75 \times 3000 = 225{,}000$

Use useful shortcuts.

$$\begin{array}{r} 75 \\ \times\ 3 \\ \hline 225 \end{array}$$

Attach 000.

(b) 150×70

Multiply 15 by 7, and then attach two zeros.

$150 \times 70 = 10{,}500$ ← Attach 00.

$$\begin{array}{r} 15 \\ \times\ 7 \\ \hline 105 \end{array}$$

Work Problem 6 at the Side. ▶

OBJECTIVE 5 **Multiply by numbers having more than one digit.** The next example shows multiplication when both factors have more than one digit.

EXAMPLE 6 **Multiplying with More Than One Digit**

Multiply 46 and 23.

First multiply 46 by 3.

$$\begin{array}{r} \overset{1}{4}6 \\ \times\ 3 \\ \hline 138 \end{array}$$

Regrouping is needed here.

138 ← $46 \times 3 = 138$

Continued on Next Page

5 Multiply.

GS **(a)** $63 \times 10 = 63$____

(b) 305×100

(c) 714×1000

6 Multiply.

GS **(a)** 17×50

$$\begin{array}{r} 17 \\ \times\ 5 \\ \hline 85\_ \end{array}$$

Attach one zero.

(b) 73×400

(c)
$$\begin{array}{r} 180 \\ \times\ 30 \\ \hline \end{array}$$

(d)
$$\begin{array}{r} 4200 \\ \times\ \ 80 \\ \hline \end{array}$$

(e)
$$\begin{array}{r} 800 \\ \times\ 600 \\ \hline \end{array}$$

Answers

5. **(a)** 0; 630 **(b)** 30,500 **(c)** 714,000

6. **(a)** 0; 850 **(b)** 29,200 **(c)** 5400 **(d)** 336,000 **(e)** 480,000

7 Complete each multiplication.

GS **(a)**
```
     2
     2
    35
  × 54
   140
  175
  __90
```

GS **(b)**
```
     2
     5
    76
  × 49
   684
  304
  ____
```

8 Multiply.

(a) 52×16

(b) 81×49

(c) 234×73

(d) 835×189

Answers

7. **(a)** 1; 8; 1890 **(b)** 3724
8. **(a)** 832 **(b)** 3969
(c) 17,082 **(d)** 157,815

Now multiply 46 by 20.

```
   1
   46
 × 20
  920  ←  46 × 20 = 920
```

Add the results.

```
    46
  × 23
   138  ←  46 × 3
 + 920  ←  46 × 20
  1058
   ↑
   └── Add.
```

Both 138 and 920 are called **partial products.** As a common practice and to save time, the 0 in 920 is usually not written.

```
    46
  × 23
   138
   92   ←  { 0 not written. Be very careful to
  1058       place the 2 in the tens column.
```

◀ **Work Problem 7 at the Side.**

EXAMPLE 7 **Using Partial Products**

Multiply.

(a)
```
     233
   × 132
     466
    699     (Tens lined up)
   233      (Hundreds lined up)
  30,756 ←  Product
```

Be certain to align numbers in columns.

(b)
```
   538
 ×  46
```

First multiply by 6.

```
   24  ←  Regrouping is needed here.
   538
 ×  46
  3228
```

Now multiply by 4, being careful to line up the tens.

```
   1 3
   2 4
   538
 ×  46
  3228 ┐
 2152  ┘── Finally, add the partial products.
 24,748
```

◀ **Work Problem 8 at the Side.**

When 0 appears in the multiplier, be sure to move the partial products to the left to account for the position held by the 0.

EXAMPLE 8 **Multiplying with Zeros**

Multiply.

(a)
```
    1 3 7
  × 3 0 6
  -------
    8 2 2
  0 0 0      (Tens lined up)
4 1 1        (Hundreds lined up)
---------
4 1,9 2 2
```

(b)
```
      1 4 0 6
   ×  2 0 0 1
   ----------
      1 4 0 6
    0 0 0 0      ← (Zeros to line up tens)
  0 0 0 0        ← (Zeros to line up hundreds)
2 8 1 2
-------------
2,8 1 3,4 0 6
```

Use extra caution when working with 0s.

```
      1 4 0 6
    × 2 0 0 1
    ---------
      1 4 0 6
2 8 1 2 0 0      ← Zeros are written so this partial product starts in the thousands column.
-------------
2,8 1 3,4 0 6
```

Note

In **Example 8(b)** in the alternative method on the right, zeros were inserted so that thousands were placed in the thousands column. This is a commonly used shortcut.

Work Problem 9 at the Side. ▶

OBJECTIVE 6 **Solve application problems with multiplication.** The next example shows how multiplication can be used to solve an application problem.

EXAMPLE 9 **Applying Multiplication Skills**

Find the total cost of 75 video games priced at $38 each.

Approach To find the cost of all the video games multiply the number of games (75) by the cost of one video game ($38).

Solution Multiply 75 by 38.

```
    7 5
  × 3 8
  -----
  6 0 0
2 2 5
-------
$2 8 5 0
```

The total cost of the video games is $2850.

Calculator Tip

If you are using a calculator for **Example 9**, you will do this calculation.

75 (×) 38 (=) 2850

Work Problem 10 at the Side. ▶

---

9 Multiply.

GS (a)
```
    2 8
  × 6 0
  -----
    _ _
1 6 8
-------
_ _ _ _
```

(b)
```
  728
×  50
```

(c)
```
  562
× 109
```

(d)
```
  3526
× 6002
```

10 Find the total cost of the following items.

GS (a) 314 garden sprayers at $14 per sprayer

(b) 64 tires priced at $139 each

(c) 12 delivery vans at $28,300 per van

Answers

9. (a) 0; 0; 1680 (b) 36,400 (c) 61,258 (d) 21,163,052

10. (a) $4396 (b) $8896 (c) $339,600

1.4 Exercises

FOR EXTRA HELP

Download the MyDashBoard App

MyMathLab®

CONCEPT CHECK *Fill in each blank with the correct response.*

1. In a chain multiplication, if you multiply by the largest number first, rather than the smallest number first, the product will always be _______.

2. The property described in **Exercise 1** is the ___________ property of multiplication.

3. When you multiply any number by zero, the answer is always _______.

4. When you multiply a whole number by 10, by 100 or by 1000, you can get the answer by attaching one, two, or three _______ to the _______ of the whole number.

Work each chain multiplication. ***See Example 2.***

5. $2 \times 6 \times 2$

6. $8 \times 6 \times 1$

7. $7 \cdot 8 \cdot 0$

8. $9 \cdot 0 \cdot 5$

9. $4 \cdot 1 \cdot 6$

10. $1 \cdot 5 \cdot 7$

11. $(4)(5)(2)$

12. $(4)(1)(9)$

13. Explain in your own words the commutative property of multiplication. How do the commutative properties of addition and multiplication compare to each other?

14. Explain in your own words the associative property of multiplication. How do the associative properties of addition and multiplication compare to each other?

Multiply. ***See Example 3.***

15. 35×6

16. 53×7

17. 34×7

18. 76×5

19. 642×5

20. 472×4

21. 624×3

22. 852×7

23. 2153×4 (carries: 2 1) GS
_ _12

24. 1137×3 (carries: 1 2) GS
_ _11

25. 2521×4

26. 2544×3

27. 2561×8

28. 7326×5

29. $36{,}921 \times 7$

30. $28{,}116 \times 4$

CONCEPT CHECK *Fill in each blank with the correct response.*

31. You can use multiples of 10 to multiply 86×200. First, multiply _______ $\times$ 2 to get _______. Then, attach _______ zeros to the right of this number for a final answer of _______.

32. You can use multiples of 10 to multiply 7800×450. First, multiply $78 \times$ _______ to get _______. Then, attach _______ zeros to the right of this number for a final answer of _______.

Multiply. ***See Examples 4 and 5.***

33. 80×6

34. 70×5

35. 740×3

36. 400×8

37. 600×6

38. 860×7

39. 125×30

40. 246×50

41. 1635×40

42. 7311×50

43. 900×300

44. 400×700

45. $43{,}000 \times 2\,000$

$43 \times 2 = 86$ Attach 000,000.

46. $11{,}000 \times 9\,000$

$11 \times 9 = 99$ Attach 000,000.

47. 970 • 50

48. 730 • 40

49. 800 • 900

50. 850 • 700

51. 9700 • 200

52. 10,050 • 300

Multiply. ***See Examples 6–8.***

53. 28×17

54. 16×34

55. 75×32

56. 82×32

57. 83×45

58. (75)(21)

59. (58)(41)

60. (82)(67)

61. (67)(92)

62. (26)(33)

63. (28)(564)

64. (58)(312)

65. (619)(35)

66. (681)(47)

67. (55)(286)

68. 286×574

69. 735×112

70. 621×415

71. 538×342

72. 3228×751

73. 9352×264

74. 528×106

75. 215×307

76. 218×106

77. 428×201

78. 3706×208

79. 6310×3078

80. 3533×5001

81. 2195×1038

82. 1502×2009

83. A classmate of yours is not clear on how to use a shortcut to multiply a whole number by 10, by 100, or by 1000. Write a short note explaining how this can be done.

84. Show two ways to multiply when a 0 is in the multiplier. Use the problem 291×307 to show this.

Solve each application problem. ***See Example 9.***

85. Carepanian Company, a health care supplier, purchased 300 cartons of Thera Bond Gym Balls. If there are 10 balls in each carton, find the total number of balls purchased.

86. A medical supply house has 30 bottles of vitamin C tablets, with each bottle containing 500 tablets. Find the total number of vitamin C tablets in the supply house.

87. The most expensive U.S. City for a hotel room is New York City, with an average cost of $194 a night. Find the cost of a 12-night stay. (*Source:* hotels.com)

88. Judge Judith Sheindlin, known as Judge Judy on court television, recently signed a four-year contract paying her $45 million each year. Find the total amount of her earnings on this contract. (*Source:* KSTE Radio News.)

89. The average amount of water used per person each day in the United States is 66 gallons. How much water does the average person use in one year? (1 year = 365 days). (*Source:* Oxfam.)

$365 \times 66 =$ ___ ___,090

90. Squid are being hauled out of the Santa Barbara Channel by the ton. They are then processed, renamed calamari, and exported. Last night 27 fishing boats each hauled out 40 tons of squid. What was the total catch for the night? (*Source: Santa Barbara News Press.*)

27×40
$27 \times 4 = 108$ Attach _______.

Find the total cost of the following items. ***See Examples 7–9.***

91. 75 first-aid kits at $8 per kit

92. 27 days of child care at $82 per day

93. 65 rebuilt alternators at $24 per alternator

94. 62 wheelchair cushions at $44 per cushion

95. 206 laptop computers at $548 per computer

96. 520 printers at $219 per printer

Multiply.

97. $21 \cdot 43 \cdot 56$

98. $(600)(8)(75)(40)$

Use addition, subtraction, or multiplication to solve each application problem.

99. In a forest-planting project, 450 trees are planted on each acre. Find the number of trees needed to plant 85 acres.

100. The largest living land mammal is the African elephant, and the largest mammal of all time is the blue whale. An African elephant weighs 15,225 pounds and a blue whale weighs 28 times that amount. Find the weight of the blue whale.

101. New York City has a population of 8,391,881, the largest in the country. Boston, in twenty-second place, has a population of 645,169. How many more people live in New York City than in Boston? (*Source:* U.S. Census Bureau.)

102. Los Angeles, the second largest city in the country, has a population of 3,849,378. Dallas, at ninth largest, has a population of 1,232,940. Find the difference in the population of these two cities. (*Source:* U.S. Census Bureau.)

103. A medical center purchased 12 laptop computers at \$970 each and 8 printers at \$315 each. Find the total cost of this equipment.

104. In the first 17 years of food drives, the Postal Workers have collected 954 million pounds of food. This year they collected 77 million pounds of food. Find the total food collection in 18 years. (*Source:* Stamp Out Hunger, U.S. Postal Service.)

Relating Concepts (Exercises 105–114) For Individual or Group Work

Work Exercises* 105–114 *in order.

105. Add.

(a) 189 + 263

(b) 263 + 189

106. Your answers to **Exercise 105(a) and (b)** should be the same. This shows that the order of numbers in an addition problem does not change the sum. This is known as the __________ property of addition.

107. Add. Recall that parentheses show you what to do first.

(a) (65 + 81) + 135

(b) 65 + (81 + 135)

108. Since the answers to **Exercise 107(a) and (b)** are the same, we see that grouping the numbers differently when adding does not change the sum. This is known as the __________ property of addition.

109. Multiply.

(a) 220×72

(b) 72×220

110. Since the answers to **Exercise 109(a) and (b)** are the same, we see that the product remains the same when the order of the factors is changed. This is known as the __________ property of multiplication.

111. Multiply. Recall that parentheses tell you what to do first.

(a) $(26 \times 18) \times 14$

(b) $26(18 \times 14)$

112. Since the answers to **Exercise 111(a) and (b)** are the same, we see that grouping the numbers differently when multiplying does not change the product. This is known as the __________ property of multiplication.

113. Do the commutative and associative properties apply to subtraction? Explain your answer using several examples.

114. Do you think that the commutative and associative properties will apply to division? Explain your answer using several examples.

1.5 Dividing Whole Numbers

OBJECTIVES

1. Write division problems in three ways.
2. Identify the parts of a division problem.
3. Divide 0 by a number.
4. Recognize that a number cannot be divided by 0.
5. Divide a number by itself.
6. Divide a number by 1.
7. Use short division.
8. Use multiplication to check the answer to a division problem.
9. Use tests for divisibility.

Suppose the total cost of lunch at a SUBWAY is $18 and is to be divided equally by three friends. Each person would pay $6, as shown here.

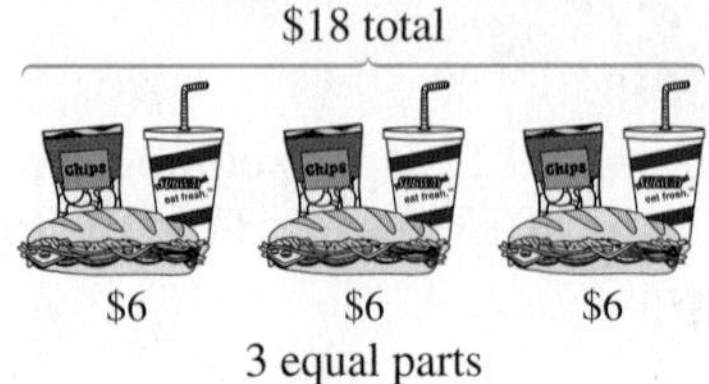

OBJECTIVE 1 Write division problems in three ways. Just as $3 \cdot 6$, 3×6, and (3)(6) are different ways of indicating the multiplication of 3 and 6, there are several ways to write 18 divided by 3.

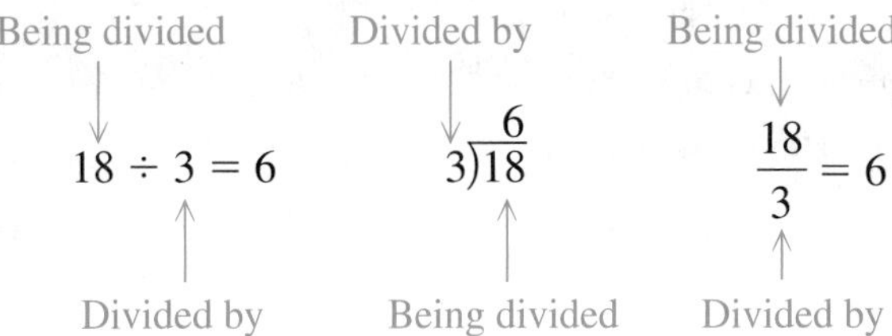

We will use all three division symbols, ÷, $\overline{)}$, and —. In courses such as algebra, a slash symbol, /, or a fraction bar, —, is most often used.

EXAMPLE 1 Using Division Symbols

Write each division problem using two other symbols.

(a) $18 \div 6 = 3$

This division can also be written as shown below.

$$6\overline{)18} \text{ with quotient } 3 \quad \text{or} \quad \frac{18}{6} = 3$$

Remember the three division symbols.

(b) $\frac{15}{5} = 3$ $\qquad 15 \div 5 = 3$ or $5\overline{)15}$ with quotient 3

(c) $5\overline{)20}$ with quotient 4 $\qquad 20 \div 5 = 4$ or $\frac{20}{5} = 4$

◀ Work Problem 1 at the Side.

OBJECTIVE 2 Identify the parts of a division problem. In division, the number being divided is the **dividend,** the number divided by is the **divisor,** and the answer is the **quotient.**

$$\text{dividend} \div \text{divisor} = \textbf{quotient}$$

$$\text{divisor}\overline{)\text{dividend}} \text{ with } \textbf{quotient} \text{ on top} \qquad \frac{\text{dividend}}{\text{divisor}} = \textbf{quotient}$$

---

1 Write each division problem using two other symbols.

GS **(a)** $24 \div 6 = 4$

$6\overline{)24}$ with quotient 4 $\qquad \frac{24}{\_\_\_} = 4$

(b) $9\overline{)36}$ with quotient 4

(c) $48 \div 6 = 8$

(d) $\frac{42}{6} = 7$

Answers

1. (a) $6\overline{)24}$ with quotient 4 and $\frac{24}{6} = 4$
(b) $36 \div 9 = 4$ and $\frac{36}{9} = 4$
(c) $6\overline{)48}$ with quotient 8 and $\frac{48}{6} = 8$
(d) $6\overline{)42}$ with quotient 7 and $42 \div 6 = 7$

EXAMPLE 2 Identifying the Parts of a Division Problem

Identify the dividend, divisor, and quotient.

(a) $35 \div 7 = 5$

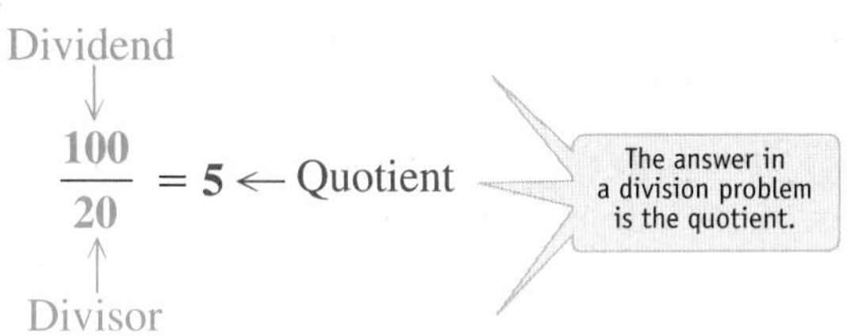

(b) $\frac{100}{20} = 5$

Dividend ↓ $\frac{100}{20} = 5$ ← Quotient; ↑ Divisor

The answer in a division problem is the quotient.

(c) $8\overline{)72}$ with quotient 9

9 ← Quotient
$8\overline{)72}$ ← Dividend
↑ Divisor

Work Problem **2** at the Side. ▶

OBJECTIVE 3 **Divide 0 by a number.** If no money, or \$0, is divided equally among five people, each person gets \$0. The general rule for dividing 0 follows.

Dividing 0 by a Number

The number **0** divided by any nonzero number is **0.**

EXAMPLE 3 Dividing 0 by a Number

Divide.

(a) $0 \div 12 = 0$

(b) $0 \div 1728 = 0$

(c) $\frac{0}{375} = 0$

Zero divided by any nonzero number is zero.

(d) $129\overline{)0}$ with quotient 0

Work Problem **3** at the Side. ▶

Just as a subtraction such as $8 - 3 = 5$ can be written as the addition $8 = 3 + 5$, any division can be written as a multiplication. For example, $12 \div 3 = 4$ can be written as

$$3 \times 4 = 12 \quad \text{or} \quad 4 \times 3 = 12.$$

2 Identify the dividend, divisor, and quotient.

(a) $15 \div 3 = 5$

(b) $18 \div 6 = 3$

(c) $\frac{28}{7} = 4$

(d) $9\overline{)27}$ with quotient 3

3 Divide.

(a) $0 \div 5$

(b) $\frac{0}{9}$

(c) $\frac{0}{24}$

(d) $37\overline{)0}$

Answers

2. **(a)** dividend: 15; divisor: 3; quotient: 5
(b) dividend: 18; divisor: 6; quotient: 3
(c) dividend: 28; divisor: 7; quotient: 4
(d) dividend: 27; divisor: 9; quotient: 3

3. all 0

4 Write each division problem as a multiplication problem.

GS **(a)** $5\overline{)15}$ with quotient 3

5 • 3 = ____

____ • 5 = ____

(b) $\frac{32}{4} = 8$

(c) $45 \div 9 = 5$

Answers

4. **(a)** 15; 5 • 3 = 15 or 3; 15; 3 • 5 = 15
(b) 4 • 8 = 32 or 8 • 4 = 32
(c) 9 • 5 = 45 or 5 • 9 = 45

EXAMPLE 4 Changing Division Problems to Multiplication

Change each division problem to a multiplication problem.

(a) $\frac{20}{4} = 5$ becomes $4 \cdot 5 = 20$ or $5 \cdot 4 = 20$.

(b) $8\overline{)48}$ with quotient 6 becomes $8 \cdot 6 = 48$ or $6 \cdot 8 = 48$.

(c) $72 \div 9 = 8$ becomes $9 \cdot 8 = 72$ or $8 \cdot 9 = 72$.

◀ Work Problem **4** at the Side.

OBJECTIVE 4 Recognize that a number cannot be divided by 0. Division of any number by 0 cannot be done. To see why, try to find

$$9 \div 0 = ?$$

As we have just seen, any division problem can be converted to a multiplication problem so that

divisor • quotient = dividend.

If you convert the preceding problem to its multiplication counterpart, it reads as follows.

$$0 \cdot ? = 9$$

You already know that 0 times any number must always be 0. Try any number you like to replace the "?" and you'll aways get 0 instead of 9. Therefore, the division problem 9 ÷ 0 cannot be done. Mathematicians say it is ***undefined*** and have agreed never to divide by 0. However, 0 ÷ 9 *can* be done. Check by rewriting it as a multiplication problem.

$$0 \div 9 = 0 \quad \text{because} \quad 9 \cdot 0 = 0 \text{ is true.}$$

Dividing a Number by 0

Since dividing any number by 0 cannot be done, we say that division by **0** is ***undefined.*** It is impossible to compute an answer.

EXAMPLE 5 Dividing Numbers by 0

All the following divisions are undefined.

(a) $\frac{6}{0}$ is undefined.

(b) $0\overline{)8}$ is undefined.

(c) $18 \div 0$ is undefined. (You **cannot** divide a number by zero.)

(d) $\frac{3}{0}$ is undefined.

Division Involving 0

$$0 \div \text{nonzero number} = 0 \quad \text{and} \quad \frac{0}{\text{nonzero number}} = 0$$

but

$$\text{nonzero number} \div 0 \quad \text{and} \quad \frac{\text{nonzero number}}{0} \quad \text{are } \textbf{undefined.}$$

CAUTION

When 0 is the divisor in a problem, you write "undefined" as the answer. Never divide by 0.

Work Problem **5** at the Side. ▶

Calculator Tip

Try these two problems on your calculator. Jot down your answers.

9 (÷) 0 (=) ______ 0 (÷) 9 (=) ______

When you try to divide by 0, the calculator cannot do it, so it shows the word "Error" or the letter "E" (for error) in the display. But, when you divide 0 by 9 the calculator displays 0, which is the correct answer.

OBJECTIVE 5 Divide a number by itself. What happens when a number is divided by itself? For example, what is 4 ÷ 4 or 97 ÷ 97?

Dividing a Number by Itself

Any ***nonzero*** number divided by itself is **1.**

EXAMPLE 6 Dividing a Nonzero Number by Itself

Divide.

(a) $16 \div 16 = 1$

(b) $32\overline{)32}$ with quotient 1

A nonzero number divided by itself is 1.

(c) $\frac{57}{57} = 1$

Work Problem **6** at the Side. ▶

OBJECTIVE 6 Divide a number by 1. What happens when a number is divided by 1? For example, what is 5 ÷ 1 or 86 ÷ 1?

Dividing a Number by 1

Any number divided by 1 is itself.

5 Divide. If the division is not possible, write "undefined."

(a) $\frac{4}{0}$

(b) $\frac{0}{4}$

(c) $0\overline{)36}$

(d) $36\overline{)0}$

(e) $100 \div 0$

(f) $0 \div 100$

6 Divide.

(a) $8 \div 8$

(b) $15\overline{)15}$

(c) $\frac{37}{37}$

Answers

5. **(a)** undefined **(b)** 0 **(c)** undefined **(d)** 0 **(e)** undefined **(f)** 0

6. all 1

EXAMPLE 7 Dividing Numbers by 1

Divide.

(a) $5 \div 1 = 5$

(b) $1\overline{)26}$ with quotient 26

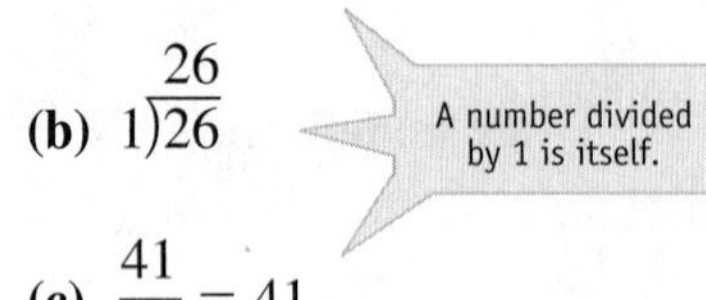
A number divided by 1 is itself.

(c) $\dfrac{41}{1} = 41$

◀ Work Problem 7 at the Side.

7 Divide.

(a) $9 \div 1$

(b) $1\overline{)18}$

(c) $\dfrac{43}{1}$

OBJECTIVE 7 **Use short division.** **Short division** is a method of dividing a number by a one-digit divisor.

EXAMPLE 8 Using Short Division

Divide using short division. $3\overline{)96}$

First, divide 9 by 3.

$$\begin{array}{r} 3 \\ 3\overline{)96} \end{array} \leftarrow \frac{9}{3} = 3$$

Next, divide 6 by 3.

$$\begin{array}{r} 32 \\ 3\overline{)96} \end{array} \leftarrow \frac{6}{3} = 2$$

◀ Work Problem 8 at the Side.

8 Divide using short division.

(a) $2\overline{)24}$

(b) $3\overline{)93}$

(c) $4\overline{)88}$

(d) $2\overline{)624}$

VOCABULARY TIP

Remainder In division, when the answer (quotient) is not a whole number, the portion left over (remains) is called the *remainder.*

When two numbers do not divide exactly, the leftover portion is called the **remainder.** The remainder must always be less than the divisor.

EXAMPLE 9 Using Short Division with a Remainder

Divide 147 by 4 using short division.
Rewrite the problem.

$$4\overline{)147}$$

Because 1 cannot be divided by 4, divide 14 by 4. Notice that the 3 is placed over the 4 in 14.

Since 14 is being divided by 4, the answer (3) is placed over the 4.

$$\begin{array}{r} 3 \\ 4\overline{)14^{2}7} \end{array} \qquad \frac{14}{4} = 3 \text{ with 2 left over}$$

Next, divide 27 by 4. The final number left over is the remainder. Use **R** to indicate the remainder, and write the remainder to the side.

$$\begin{array}{r} 3\;\;6\;\mathbf{R}3 \\ 4\overline{)14^{2}7} \end{array} \qquad \frac{27}{4} = 6 \text{ with 3 left over}$$

◀ Work Problem 9 at the Side.

9 Divide using short division.

(a) $2\overline{)125}$

(b) $3\overline{)215}$

(c) $4\overline{)538}$

(d) $\dfrac{819}{5}$

Answers

7. (a) 9 (b) 18 (c) 43
8. (a) 12 (b) 31 (c) 22 (d) 312
9. (a) 62 **R**1 (b) 71 **R**2 (c) 134 **R**2 (d) 163 **R**4

EXAMPLE 10 **Dividing with a Remainder**

Divide 1809 by 7.

Divide 18 by 7.

$$7\overline{)18^{4}09}\ \text{quotient } 2 \qquad \frac{18}{7} = 2 \text{ with 4 left over}$$

Divide 40 by 7.

$$7\overline{)18^{4}0^{5}9}\ \text{quotient } 2\ 5 \qquad \frac{40}{7} = 5 \text{ with 5 left over}$$

Divide 59 by 7.

$$7\overline{)18^{4}0^{5}9}\ \text{quotient } 2\ 5\ 8\ \mathbf{R}3 \qquad \frac{59}{7} = 8 \text{ with 3 left over}$$

The remainder must be less than the divisor.

Work Problem **10** at the Side. ▶

Note

Short division takes practice but is useful when the divisor is a one-digit number.

OBJECTIVE 8 **Use multiplication to check the answer to a division problem. Check** the answer to a division problem as follows.

Checking Division

(divisor × quotient) + remainder = dividend

Parentheses tell you what to do first: Multiply the divisor by the quotient, then add the remainder.

EXAMPLE 11 **Checking Division by Using Multiplication**

Check each answer.

(a) $5\overline{)458}$ quotient 91 **R**3

(divisor × quotient) + remainder = dividend

(5 × 91) + 3

455 + 3 = 458

Be careful! Always add the remainder when checking division.

Matches original dividend, so the division was done correctly.

Continued on Next Page

10 Divide.

(a) $4\overline{)5^{1}3^{1}0}$ quotient 1 3 _ **R** _

(b) $\frac{515}{7}$

(c) $3\overline{)1885}$

(d) $6\overline{)1415}$

Answers

10. **(a)** 2; 2; 132 **R**2 **(b)** 73 **R**4 **(c)** 628 **R**1 **(d)** 235 **R**5

11 Use multiplication to check each division. If an answer is incorrect, give the correct answer.

GS **(a)** $2\overline{)65}$ = 32 **R**1

$2 \times 32 = 64$

___ + ___ = 65

(b) $7\overline{)586}$ = 83 **R**4

(c) $3\overline{)1223}$ = 407 **R**2

(d) $5\overline{)2383}$ = 476 **R**3

Answers

11. **(a)** 64; 1; correct **(b)** incorrect; should be 83 **R**5 **(c)** correct **(d)** correct

(b) $6\overline{)1437}$ = 239 **R4**

(divisor × quotient) + remainder = dividend

$(6 \times 239) + 4$

$1434 + 4 = \mathbf{1438}$

Does not match original dividend.

The answer does **not** check. Rework the original problem to get the correct answer, 239 **R**3. Then, $(6 \times 239) + 3$ **does** give 1437.

CAUTION

A common error when checking division is to forget to add the remainder. Be sure to add any remainder when checking a division problem.

◀ Work Problem **11** at the Side.

OBJECTIVE 9 **Use tests for divisibility.** It is often important to know whether a number is *divisible* by another number. You will find this useful in **Chapter 2** when writing fractions in lowest terms.

Divisibility

One whole number is **divisible** by another if the remainder is 0.

Use the following tests to decide whether one number is divisible by another number.

Tests for Divisibility

A number is divisible by

2 if it ends in 0, 2, 4, 6, or 8. These are the even numbers.
3 if the sum of its digits is divisible by 3.
4 if the last two digits make a number that is divisible by 4.
5 if it ends in 0 or 5.
6 if it is divisible by both 2 and 3.
7 has no simple test.
8 if the last three digits make a number that is divisible by 8.
9 if the sum of its digits is divisible by 9.
10 if it ends in 0.

The most commonly used tests are those for 2, 3, 5, and 10.

Divisibility by 2

A number is divisible by **2** if the number ends in 0, 2, 4, 6, or 8. All even numbers are divisible by 2.

EXAMPLE 12 **Testing for Divisibility by 2**

Are the following numbers divisible by 2?

(a) 986

Ends in 6

All even numbers are divisible by 2.

Because the number ends in 6, which is an *even number,* the number 986 is divisible by 2.

(b) 3255 is not divisible by 2.

Ends in 5, and not in 0, 2, 4, 6, or 8

Work Problem 12 at the Side. ▶

Divisibility by 3

A number is divisible by **3** if the sum of its digits is divisible by **3.**

EXAMPLE 13 **Testing for Divisibility by 3**

Are the following numbers divisible by 3?

(a) 4251
Add the digits.

4 2 5 1

$4 + 2 + 5 + 1 = 12$

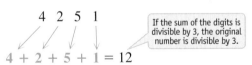

If the sum of the digits is divisible by 3, the original number is divisible by 3.

Because 12 is divisible by 3, the number 4251 is also divisible by 3.

(b) 29,806
Add the digits.

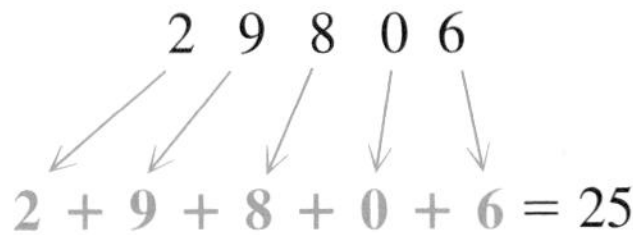

2 9 8 0 6

$2 + 9 + 8 + 0 + 6 = 25$

Because 25 is *not* divisible by 3, the number 29,806 is *not* divisible by 3.

CAUTION

Be careful when testing for divisibility by adding the digits. This method works only for the numbers 3 and 9.

Work Problem 13 at the Side. ▶

12 Which numbers are divisible by 2?

(a) 258

(b) 307

(c) 4216

(d) 73,000

13 Which numbers are divisible by 3?

GS **(a)** 743

$7 + 4 + 3 = 14$

14 is *not* divisible by 3.

So, 743 is *not* divisible by 3.

GS **(b)** 5325

$5 + 3 + 2 + 5 =$ ____

____ is divisible by 3.

So, 5325 is divisible by ____.

(c) 374,214

(d) 205,633

Answers

12. all but b

13. 15; 15; 3; b and c

14 Which numbers are divisible by 5?

(a) 180

(b) 635

(c) 8364

(d) 206,105

15 Which numbers are divisible by 10?

(a) 270

(b) 495

(c) 5030

(d) 14,380

Answers

14. all but (c)
15. all but (b)

Divisibility by 5 and by 10

A number is divisible by **5** if it ends in 0 or 5.
A number is divisible by **10** if it ends in 0.

EXAMPLE 14 Testing for Divisibility by 5

Are the following numbers divisible by 5?

(a) 12,900 ends in 0 and is divisible by 5.

(b) 4325 ends in 5 and is divisible by 5.

If the number ends in 0 or 5, it's divisible by 5.

(c) 392 ends in 2 and is *not* divisible by 5.

◀ Work Problem **14** at the Side.

EXAMPLE 15 Testing for Divisibility by 10

Are the following numbers divisible by 10?

(a) 700 and 9140 both end in 0 and are divisible by 10.

If the number ends in 0, it's divisible by 10.

(b) 355 and 18,743 do not end in 0 and are *not* divisible by 10.

◀ Work Problem **15** at the Side.

1.5 Exercises

FOR EXTRA HELP

Download the MyDashBoard App

MyMathLab®

1. CONCEPT CHECK Write the three common symbols used to show multiplication.

2. CONCEPT CHECK Write the three common symbols used to show division.

Write each division problem using two other symbols. ***See Example 1.***

3. $24 \div 4 = 6$

4. $36 \div 3 = 12$

5. $\frac{45}{9} = 5$

6. $\frac{56}{8} = 7$

7. $2\overline{)16}$ with quotient 8

8. $8\overline{)48}$ with quotient 6

9. CONCEPT CHECK When a number is divided by 1, the answer is always the ______ itself.

10. CONCEPT CHECK When zero is divided by a number, the answer is always ______.

Divide. If the division is not possible, write "undefined." ***See Examples 3–7.***

11. $9 \div 9$

12. $36 \div 9$

13. $\frac{14}{2}$

14. $\frac{10}{0}$

15. $22 \div 0$
GS When 0 is the divisor, write ________ as the answer.

16. $6 \div 6$
GS When a number is divided by itself, write ________ as the answer.

17. $\frac{24}{1}$

18. $\frac{12}{1}$

19. $15\overline{)0}$
GS When dividing 0 by a nonzero number, the answer is ________.

20. $\frac{0}{12}$

21. $0\overline{)43}$

22. $\frac{8}{0}$

CONCEPT CHECK *Use the tests for divisibility by 2, 3, 5, and 10. Circle the numbers that will divide evenly into the given number.*

23. 8670
2 3 5 10

24. 13,785
2 3 5 10

25. 9,221,784
2 3 5 10

26. 5,409,720
2 3 5 10

Divide by using short division. Use multiplication to check each answer.
See Examples 8–10.

27. $3\overline{)75}$

28. $5\overline{)85}$

29. $7\overline{)126}$

30. $6\overline{)168}$

31. $4\overline{)1216}$

32. $5\overline{)2305}$

33. $4\overline{)2509}$

34. $8\overline{)1335}$

35. $6\overline{)9137}$

36. $9\overline{)8371}$

37. $6\overline{)1854}$

38. $8\overline{)856}$

39. 12,020 ÷ 4

40. 8012 ÷ 4

41. 30,036 ÷ 6

42. 32,008 ÷ 8

43. 2434 ÷ 3

44. 5993 ÷ 7

45. 12,947 ÷ 5

46. 33,285 ÷ 9

47. $\frac{21{,}040}{8}$

48. $\frac{8199}{9}$

49. $\frac{74{,}751}{6}$

50. $\frac{72{,}543}{5}$

51. $\frac{71{,}776}{7}$

52. $\frac{77{,}621}{3}$

53. $\frac{128{,}645}{7}$

54. $\frac{172{,}255}{4}$

Use multiplication to check each answer. If an answer is incorrect, find the correct answer.
See Example 11.

55. $\begin{array}{r} 375 \ \textbf{R}2 \\ 5\overline{)1877} \end{array}$

GS **CHECK**

5 • 375 + 2 = ____

correct

56. $\begin{array}{r} 427 \ \textbf{R}1 \\ 3\overline{)1282} \end{array}$

GS **CHECK**

3 • 427 + 1 = ____

correct

57. $\begin{array}{r} 1908 \ \textbf{R}2 \\ 3\overline{)5725} \end{array}$

58. $\begin{array}{r} 432 \ \textbf{R}3 \\ 5\overline{)2158} \end{array}$

59. $\begin{array}{r} 650 \ \textbf{R}2 \\ 7\overline{)4692} \end{array}$

60. $\begin{array}{r} 663 \ \textbf{R}5 \\ 9\overline{)5974} \end{array}$

61. $\begin{array}{r} 3\ 568 \ \textbf{R}2 \\ 6\overline{)21{,}409} \end{array}$

62. $\begin{array}{r} 532 \\ 6\overline{)3192} \end{array}$

63. $\begin{array}{r} 2\ 002 \ \textbf{R}3 \\ 8\overline{)16{,}019} \end{array}$

64. $\begin{array}{r} 4\ 208 \\ 8\overline{)33{,}664} \end{array}$

65. $\begin{array}{r} 11{,}523 \ \textbf{R}2 \\ 6\overline{)69{,}140} \end{array}$

66. $\begin{array}{r} 27{,}532 \ \textbf{R}1 \\ 3\overline{)82{,}598} \end{array}$

67. $\begin{array}{r} 9\ 628 \ \textbf{R}7 \\ 9\overline{)86{,}655} \end{array}$

68. $\begin{array}{r} 7\ 258 \ \textbf{R}4 \\ 7\overline{)50{,}809} \end{array}$

69. $\begin{array}{r} 27{,}822 \\ 8\overline{)222{,}576} \end{array}$

70. $\begin{array}{r} 77{,}804 \\ 4\overline{)311{,}216} \end{array}$

71. Explain in your own words how to check a division problem using multiplication. Be sure to tell what must be done if the quotient includes a remainder.

72. Describe the three divisibility rules that you think will be most useful and tell why.

Solve each application problem.

73. The Carnival Cruise Line has 2624 linen napkins. If it takes eight napkins to set each table, find the number of tables that can be set. (*Source: USA Today.*)

74. A school district will distribute 1620 new science books equally among 12 schools. How many books will each school receive?

75. In one 8-hour day Dreyer's Edy's can produce 76,800 ice cream drumsticks. How many are produced each hour? (*Source:* History Channel, *Modern Marvels: Snack Food Tech.*)

76. Tootsie Roll Industries produces 415,000,000 Tootsie Rolls in a 5-day week. Find the number produced each day. (*Source:* History Channel, *Modern Marvels: Snack Food Tech.*)

77. Lottery winnings of \$436,500 are divided equally among nine Starbucks employees. Find the amount received by each employee.

78. How many 5-pound bags of organic whole wheat flour can be filled from a 17,175-pound bin of flour?

79. McDonald's Restaurants is hiring 660 new employees, from crew workers to managers, in one area. If 4 new employees are hired at each location, find the number of locations in this area. (*Source: Sacramento Bee.*)

80. The Wii remains the top-selling video game system. Nintendo sold 85 million Wiis in the first 5 years. Find the average number sold each year. (*Source: USA Today.*)

81. A class-action lawsuit settlement of $6,825,000 is divided evenly among six injured people. Find the amount received by each person.

82. A 12,000-square foot condominium at the edge of Central Park in Manhattan sold for a record $45,000,000. The buyer paid for the condominium in eight equal payments. Find the amount of each payment. (*Source: USA Today.*)

83. The record for picking blueberries in one day was set in the state of Maine and was 6900 pounds. Since there are 1000 blueberries in a pound, this amounted to 6,900,000 blueberries. If these berries were picked in 8 hours, find the average number of berries picked each hour. (*Source:* Discovery Channel, *Dirty Jobs.*)

84. A professional basketball player signed a 4-year contract for $21,937,500. How much is this each year?

Put a ✓ mark in the blank if the number at the left is divisible by the number at the top. Put an X in the blank if the number is not divisible by the number at the top.
See Examples 12–15.

	2	3	5	10
85. 60	____	____	____	____
87. 92	____	____	____	____
89. 445	____	____	____	____
91. 903	____	____	____	____
93. 5166	____	____	____	____
95. 21,763	____	____	____	____

	2	3	5	10
86. 35	____	____	____	____
88. 96	____	____	____	____
90. 897	____	____	____	____
92. 500	____	____	____	____
94. 8302	____	____	____	____
96. 32,472	____	____	____	____

1.6 Long Division

If the total cost of 42 Sony iPod Docking Systems is \$3066, we can find the cost of each docking system using **long division.** Long division is used to divide by a number with more than one digit.

OBJECTIVES

1. Do long division.
2. Divide numbers ending in 0 by numbers ending in 0.
3. Use multiplication to check division answers.

OBJECTIVE 1 Do long division. In long division, estimate the various numbers by using a **trial divisor** to get a **trial quotient.**

VOCABULARY TIP

Trial divisor A method of determining how many times the true divisor goes into the dividend by estimating, using the leading digit of the divisor.

VOCABULARY TIP

Trial quotient The quotient that results from dividing the dividend by the trial divisor.

EXAMPLE 1 Using a Trial Divisor and a Trial Quotient

Divide. $42\overline{)3066}$

Because 42 is closer to 40 than to 50, use the first digit of the divisor as a trial divisor.

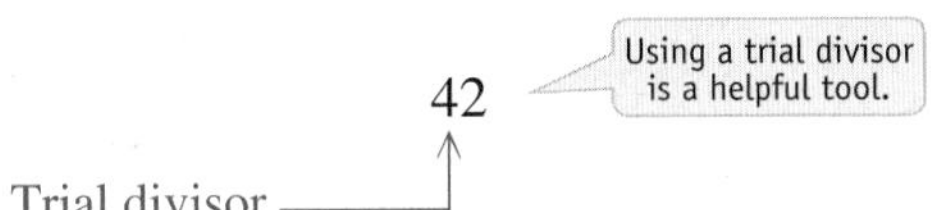
42
Using a trial divisor is a helpful tool.
Trial divisor

Try to divide the first digit of the dividend by 4. Since 3 cannot be divided by 4, use the first *two* digits, 30.

$$\frac{30}{4} = 7 \text{ with remainder } 2$$

$$\begin{array}{r} \mathbf{7} \\ 42\overline{)3066} \end{array} \leftarrow \text{Trial quotient}$$

7 goes over the 6, because $\frac{306}{42}$ is about 7.

Multiply 7 and 42 to get 294; next, subtract 294 from 306.

$$\begin{array}{r} 7 \\ 42\overline{)3066} \\ \underline{294} \\ 12 \end{array} \begin{array}{l} \\ \\ \leftarrow 7 \times 42 \\ \leftarrow 306 - 294 \end{array}$$

This number (12) must be smaller than 42, the divisor.

Bring down the 6 at the right.

$$\begin{array}{r} 7 \\ 42\overline{)3066} \\ \underline{294}\downarrow \\ 126 \end{array} \begin{array}{l} \\ \\ \\ \leftarrow 6 \text{ brought down} \end{array}$$

Use the trial divisor, 4.

First two digits of 126 $\rightarrow \frac{12}{4} = 3$

$$\begin{array}{r} 73 \\ 42\overline{)3066} \\ \underline{294} \\ 126 \\ \underline{126} \\ 0 \end{array} \begin{array}{l} \\ \\ \\ \\ \leftarrow 3 \times 42 = 126 \\ \\ \end{array}$$

The cost of each docking system is \$73.

Check the answer by multiplying 42 and 73. The product should be 3066.

1 Divide.

GS **(a)** $28\overline{)2296}$

$$\begin{array}{r} 8\_ \\ 28\overline{)2296} \\ \underline{224}\downarrow \\ 56 \\ \underline{56} \\ 0 \end{array}$$

(b) $16\overline{)1024}$

(c) $61\overline{)8784}$

(d) $\frac{2697}{93}$

2 Divide.

(a) $24\overline{)1344}$

(b) $72\overline{)4472}$

(c) $65\overline{)5416}$

(d) $89\overline{)6649}$

Answers

1. **(a)** 2; 82 **(b)** 64 **(c)** 144 **(d)** 29

2. **(a)** 56 **(b)** 62 **R**8 **(c)** 83 **R**21 **(d)** 74 **R**63

CAUTION

The ***first digit*** of the quotient in long division must be placed in the proper position over the dividend.

◀ **Work Problem 1 at the Side.**

EXAMPLE 2 Dividing to Find a Trial Quotient

Divide. $58\overline{)2730}$

Use 6 as a trial divisor, since 58 is closer to 60 than to 50.

First two digits of dividend → $\frac{27}{6} = 4$ with 3 left over

$$\begin{array}{r} 4 \\ 58\overline{)2730} \\ \underline{232} \\ 41 \end{array}$$

4 ← Trial quotient

232 ← $4 \times 58 = 232$

41 ← $273 - 232 = 41$ (smaller than 58, the divisor)

Bring down the 0.

$$\begin{array}{r} 4 \\ 58\overline{)2730} \\ \underline{232}\downarrow \\ 410 \end{array}$$

410 ← 0 brought down

First two digits of 410 → $\frac{41}{6} = 6$ with 5 left over

$$\begin{array}{r} 46 \\ 58\overline{)2730} \\ \underline{232} \\ 410 \\ \underline{348} \\ 62 \end{array}$$

46 ← Trial quotient

348 ← $6 \times 58 = 348$

62 ← Greater than 58

Do not leave a remainder that is **greater** than the divisor.

The remainder, 62, is greater than the divisor, 58, so 7 should be used instead of 6.

$$\begin{array}{r} 47\ \mathbf{R}4 \\ 58\overline{)2730} \\ \underline{232} \\ 410 \\ \underline{406} \\ 4 \end{array}$$

406 ← $7 \times 58 = 406$

4 ← $410 - 406$

Now the remainder, 4, is *less* than the divisor, 58.

◀ **Work Problem 2 at the Side.**

Sometimes it is necessary to write a 0 in the quotient.

EXAMPLE 3 Writing Zeros in the Quotient

Divide: $34\overline{)7068}$

Start as in **Examples 1 and 2.**

$$\begin{array}{r} 2 \\ 34\overline{)7068} \\ \underline{68} \leftarrow 2 \times 34 = 68 \\ 2 \leftarrow 70 - 68 = 2 \end{array}$$

Bring down the 6.

$$\begin{array}{r} 2 \\ 34\overline{)7068} \\ \underline{68}\downarrow \\ 26 \leftarrow \text{6 brought down} \end{array}$$

Since 26 cannot be divided by 34, write a 0 in the quotient as a placeholder.

$$\begin{array}{r} 20 \leftarrow \text{0 in quotient} \\ 34\overline{)7068} \\ \underline{68} \\ 26 \end{array}$$

Use a zero to hold a place in the quotient.

Bring down the final digit, the 8.

$$\begin{array}{r} 20 \\ 34\overline{)7068} \\ \underline{68}\downarrow \\ 268 \leftarrow \text{8 brought down} \end{array}$$

Complete the problem.

$$\begin{array}{r} 207 \ \mathbf{R}30 \\ 34\overline{)7068} \\ \underline{68} \\ 268 \\ \underline{238} \\ 30 \end{array}$$

The quotient is 207 **R**30.

CAUTION

There ***must be a digit*** in the quotient (answer) above every digit in the dividend once the answer has begun. Notice in **Example 3** that a **0** was used to ensure a digit in the quotient above every digit in the dividend.

Work Problem 3 at the Side. ▶

OBJECTIVE 2 Divide numbers ending in 0 by numbers ending in 0. When the divisor and dividend both contain zeros at the far right, recall that these numbers are multiples of 10. As with multiplication, there is a short way to divide these multiples of 10. Look at the following examples.

$$26{,}000 \div 1 = 26{,}000$$
$$26{,}000 \div 10 = 2600$$
$$26{,}000 \div 100 = 260$$
$$26{,}000 \div 1000 = 26$$

Do you see a pattern? These examples suggest the following rule.

3 Divide.

GS **(a)** $17\overline{)1823}$

$$\begin{array}{r} 1\_7\ \mathbf{R}\_ \\ 17\overline{)1823} \\ \underline{17}\downarrow \\ 12 \\ \underline{0}\downarrow \\ 123 \\ \underline{119} \\ \_ \end{array}$$

(b) $23\overline{)4791}$

(c) $39\overline{)15{,}933}$

(d) $78\overline{)23{,}462}$

Answers

3. **(a)** 0; 4; 4; 107 **R**4 **(b)** 208 **R**7 **(c)** 408 **R**21 **(d)** 300 **R**62

4 Divide.

(a) $70 \div 10$

(b) $2600 \div 100$

(c) $505{,}000 \div 1000$

5 Divide using the shortcut of dropping zeros.

GS **(a)** $50\overline{)6250}$ Drop one zero

Drop one zero: $5\overline{)6\_\_}$ with quotient $\_25$

```
     _25
  5)6__
    5
    12
    10
     25
     25
      0
```

(b) $130\overline{)131{,}040}$

(c) $3400\overline{)190{,}400}$

Answers

4. **(a)** 7 **(b)** 26 **(c)** 505
5. **(a)** 1; 2; 5; 125 **(b)** 1008 **(c)** 56

Dividing a Whole Number by 10, by 100, or by 1000

To divide a whole number by 10, by 100, or by 1000, drop the appropriate number of zeros from the whole number.

EXAMPLE 4 **Dividing by Multiples of 10**

Divide.

(a) $60 \div 10 = 6$ — One 0 in divisor; 0 dropped

(b) $3500 \div 100 = 35$ — Two zeros in divisor; 00 dropped

The same number of zeros must be dropped from the divisor and quotient.

(c) $915{,}000 \div 1000 = 915$ — Three zeros in divisor; 000 dropped

◀ **Work Problem 4 at the Side.**

Now we'll find the quotients for other multiples of 10 by dropping zeros.

EXAMPLE 5 **Dividing by Multiples of 10**

Divide:

(a) $40\overline{)11{,}000}$ Drop one zero from the divisor and the dividend.

```
     275
  4)1100
    8
    30
    28
     20
     20
      0
```

Since $1100 \div 4$ is 275, then $11{,}000 \div 40$ is also 275.

(b) $3500\overline{)31{,}500}$ Drop two zeros from the divisor and the dividend.

```
       9
  35)315
     315
       0
```

Since $315 \div 35$ is 9, then $31{,}500 \div 3500$ is also 9.

Note

Dropping zeros when dividing by multiples of 10 ***does not*** change the quotient (answer).

◀ **Work Problem 5 at the Side.**

OBJECTIVE 3 Use multiplication to check division answers. Answers in long division can be checked just as answers in short division were checked.

EXAMPLE 6 Checking Division by Using Multiplication

Check each answer.

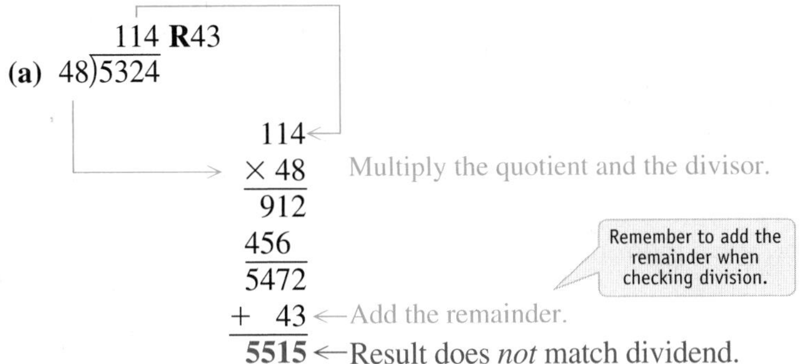

(a) 48)5324 = 114 R43
114
× 48
912
456
5472
+ 43
5515
Multiply the quotient and the divisor.
Remember to add the remainder when checking division.
Add the remainder.
Result does *not* match dividend.

The answer does ***not*** check. Rework the original problem to get 110 **R**44. Then (110 × 48) + 44 **does** give 5324.

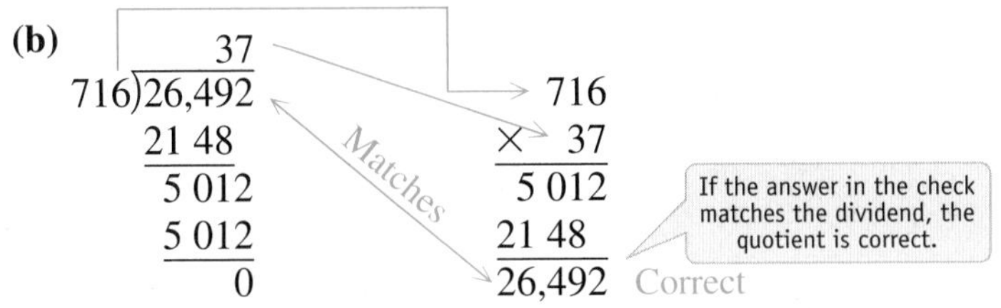

(b) 716)26,492 = 37
21 48
5 012
5 012
0
Matches
716
× 37
5 012
21 48
26,492
Correct
If the answer in the check matches the dividend, the quotient is correct.

Calculator Tip

To check the answer to **Example 6(a)**, don't forget to add the remainder.

CAUTION

When checking a division problem, first multiply the quotient and the divisor. Then be sure to ***add any remainder*** before checking it against the original dividend.

Work Problem 6 at the Side. ▶

6 Decide whether each answer is correct. If the answer is incorrect, find the correct answer.

(a) 16)608 = 38

```
     38
16)608
   48
   128
   128
     0
```

(b) 426)19,170 = 42 **R**178

```
        42 R178
426)19,170
    17 040
     1 130
       952
       178
```

(c) 514)29,316 = 57 R18

```
        57 R18
514)29,316
    25 700
     3 616
     3 598
        18
```

Answers

6. **(a)** correct **(b)** incorrect; should be 45 **(c)** correct

1.6 Exercises

FOR EXTRA HELP

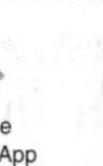

Download the MyDashBoard App

MyMathLab®

CONCEPT CHECK *Decide where the first digit in the quotient would be located. Then, without finishing the division, you can tell which of the three choices is the correct answer. Circle your choice.*

1. $50\overline{)2650}$
5 53 530

2. $14\overline{)476}$
3 34 304

3. $18\overline{)4500}$
2 25 250

4. $35\overline{)5600}$
16 160 1600

5. $86\overline{)10{,}327}$
12 120 **R**7 1200

6. $46\overline{)24{,}026}$
5 52 522 **R**14

7. $26\overline{)28{,}735}$
11 110 1105 **R**5

8. $12\overline{)116{,}953}$
974 **R**2 9746 **R**1 97,460

9. $21\overline{)149{,}826}$
71 713 7134 **R**12

10. $64\overline{)208{,}138}$
325 **R**2 3252 **R**10 32,521

11. $523\overline{)470{,}800}$
9 **R**100 90 **R**100 900 **R**100

12. $230\overline{)253{,}230}$
11 110 1101

Divide by using long division. Use multiplication to check each answer.
See Examples 1–3, 5, and 6.

13. $18\overline{)1319}$ →

$$\begin{array}{r} 73\ \textbf{R}\_ \\ 18\overline{)1319} \\ \underline{126} \\ 59 \\ \underline{54} \\ 5 \end{array}$$

CHECK
$73 \cdot 18 = 1314$
$1314 + \_\_\_ = 1319$

14. $58\overline{)3654}$ →

$$\begin{array}{r} 6\_ \\ 58\overline{)3654} \\ \underline{348} \\ 174 \\ \underline{174} \\ 0 \end{array}$$

CHECK
$\_\_\_ \cdot 58 = 3654$

15. $23\overline{)10{,}963}$

16. $83\overline{)39{,}692}$

17. $26\overline{)62{,}583}$

18. $28\overline{)84{,}249}$

19. $74\overline{)84{,}819}$

20. $238\overline{)186{,}948}$

21. $153\overline{)509{,}725}$

22. $308\overline{)26{,}796}$

23. $420\overline{)357{,}000}$

24. $900\overline{)153{,}000}$

Use multiplication to check each answer. If an answer is incorrect, find the correct answer. ***See Example 6.***

25. $\begin{array}{r}101\textbf{R}4\\35\overline{)3549}\end{array}$

26. $\begin{array}{r}42\textbf{R}26\\64\overline{)2712}\end{array}$

27. $\begin{array}{r}658\textbf{R}9\\28\overline{)18{,}424}\end{array}$

28. $\begin{array}{r}239\textbf{R}121\\145\overline{)34{,}776}\end{array}$

29. $\begin{array}{r}62\textbf{R}3\\614\overline{)38{,}068}\end{array}$

30. $\begin{array}{r}174\textbf{R}368\\557\overline{)97{,}286}\end{array}$

Solve each application problem by using addition, subtraction, multiplication, or division as needed. ***See Examples 3–5.***

31. The first female co-host of a game show, Vanna White, is "Television's Most Frequent Clapper." On the *Wheel of Fortune* she claps 720 times per episode, or 84,240 claps each season. Find the number of episodes filmed each season. (*Source: Guinness Book of World Records* and *Spirit Magazine*.)

32. The two main towers of the Golden Gate Bridge each used 600,000 rivets in their construction. To meet today's earthquake standards, each of these rivets is being replaced with high-strength bolts. If the bolts are shipped in containers holding 160 bolts, how many containers are used to complete the work on both towers? (*Source:* Consolidated Engineering Laboratories.)

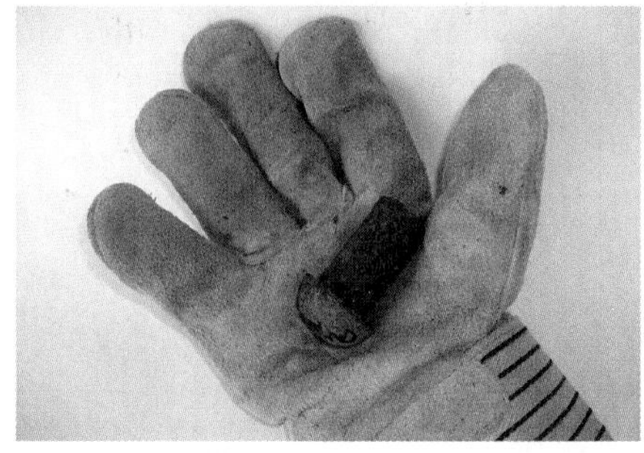

33. Don Gracey, the Mountain Timesmith, has serviced and repaired 636 clocks this year. He has worked on 272 wall clocks and 308 table clocks. The rest were standing floor clocks. Find the number of floor clocks he worked on this year.

total clocks − wall clocks − table clocks = floor clocks

636 − 272 − 308 = ________

34. There are 24,000,000 business enterprises in the United States. If 7000 of these are larger businesses (over 500 employees), find the number of businesses that are small to mid-size. (*Source:* U.S. Census Bureau.)

total business enterprises − larger businesses = small to mid-size businesses

24,000,000 − 7000 = ________

35. To complete her college education, Judy Martinez received education loans of $34,080 including interest. Find her monthly payment if the loan is to be paid off in 96 months (8 years).

36. A consultant charged $19,800 for evaluating a school's compliance with the Americans with Disabilities Act. If the consultant worked 225 hours, find the rate charged per hour.

37. Each minute there is one diamond ring sold on eBay's U.S. site. Find the number of diamond rings sold in 30 days. (*Source: Time Style and Design.*)

60 minutes • 24 hours • 30 days = ________

38. A retired milkman in Indianapolis has eaten a Twinkie every day for the last 60 years. How many Twinkies has he eaten over this time period?
Hint: 1 year = 365 days. (*Source:* History Channel, *Modern Marvels: Snack Food Tech.*)

60 years • 365 days = ___________

39. Don Gorske of Fond du Lac, Wisconsin has eaten 25,272 Big Macs over the last 39 years. (He is slim and claims a low cholestrol rate.) Find

(a) the average number of Big Macs he has eaten each year.

(b) whether Gorske has eaten more or less than 2 Big Macs each day.

Hint: 1 year = 365 days
(*Source: Guinness Book of World Records.*)

40. Major League baseball teams use more than 220,020 baseballs each season. If each of the 30 major league teams uses the same number of balls in a season, how many are used by each team? (*Source: Parade, The Sunday Newspaper Magazine.*)

Relating Concepts (Exercises 41–48) For Individual or Group Work

Knowing and using the rules of divisibility is necessary in problem solving.
Work Exercises 43–50 in order.

41. If you have $0 and you divide this amount among three people, how much will each receive?

42. When 0 is divided by any nonzero number, the result is ____.

43. Divide.
$8 \div 0$

44. We say that division by 0 is *undefined* because it is (*possible/impossible*) to compute the answer. Give an example involving cookies that will support your answer.

45. Divide.
(a) $14 \div 1$ **(b)** $1\overline{)17}$
(c) $\frac{38}{1}$

46. Any number divided by 1 is the number itself. Is this also true when multiplying by 1? Give three examples that support your answer.

47. Divide.
(a) $32{,}000 \div 10$
(b) $32{,}000 \div 100$
(c) $32{,}000 \div 1000$

48. Write a rule that explains the shortcut for doing divisions like the ones in **Exercise 47.**

Summary Exercises *Whole Numbers Computation*

CONCEPT CHECK *Write the digit for the given* ***place value*** *in each whole number.*

1. 631,548
ten-thousands ____
tens ____

2. 76,047,309
millions ____
hundred-thousands ____

3. 9,181,576,423
hundred-millions ____
thousands ____

Write each number in words.

4. 86,002

5. 425,208,733

Add or subtract as indicated.

6. $\begin{array}{r} 46 \\ +\ 51 \\ \hline \end{array}$

7. $166 + 739$

8. $\begin{array}{r} 82 \\ -\ 61 \\ \hline \end{array}$

9. $\begin{array}{r} 798 \\ -\ 389 \\ \hline \end{array}$

10. $6382 + 4062 + 7129$

11. $75 + 81{,}579 + 506 + 4$

12. $\begin{array}{r} 1704 \\ -\ 1027 \\ \hline \end{array}$

13. $\begin{array}{r} 55{,}000 \\ -\ 17{,}326 \\ \hline \end{array}$

14. $\begin{array}{r} 70{,}552 \\ -\ 34{,}663 \\ \hline \end{array}$

CONCEPT CHECK *Multiply using the shortcut for multiples of 10. Circle the correct answer.*

15. 56×10

56 560 5600

16. 140×40

5600 560 56,000

17. $\begin{array}{r} 500 \\ \times\ 700 \\ \hline \end{array}$

3,500,000 35,000 350,000

18. $\begin{array}{r} 3600 \\ \times\ 70 \\ \hline \end{array}$

25,200 252,000 2,520,000

Write the numbers in each sentence using digits.

19. There are one million, two hundred thirty-eight thousand, two hundred one nonprofit charitable organizations in the United States. (*Source:* The Giving USA Foundation.)

20. More than five million, five hundred forty-nine thousand, three hundred seventy-five households depend on the National Flood Insurance Program as their main source of protection against flooding, the most common natural disaster in the United States. (*Source:* National Association of Realators.)

Multiply or divide as indicated. If the division cannot be done, write "undefined."

21. $8 \div 8$

22. $0 \div 9$

23. $\frac{12}{0}$

24. $\frac{15}{1}$

25. 7×8

26. (6)(0)(5)

27. $8 \cdot 4 \cdot 3$

28. $\frac{608}{2}$

29. $3\overline{)8252}$

30. $4569 \div 6$

31. $\begin{array}{r} 65 \\ \times\ 52 \\ \hline \end{array}$

32. $\begin{array}{r} 507 \\ \times\ 435 \\ \hline \end{array}$

33. (28)(72)

34. (41)(36)

35. $25\overline{)1950}$

36. $18\overline{)3780}$

37. $\begin{array}{r} 3602 \\ \times\ 5008 \\ \hline \end{array}$

38. $62\overline{)31{,}400}$

39. $630\overline{)32{,}760}$

40. $351\overline{)424{,}011}$

41. $4587 \div 8$

42. $72\overline{)2952}$

43. $\begin{array}{r} 662 \\ \times\ 315 \\ \hline \end{array}$

44. $\begin{array}{r} 2186 \\ \times\ 504 \\ \hline \end{array}$

Built in the 1930s, the San Francisco Golden Gate Bridge is one of the most recognizable landmarks in the world. Some Golden Gate Bridge facts are listed in the table below. Use them to answer Exercises 45–50.

GOLDEN GATE BRIDGE FACTS

Description	Number
Total length, including approaches	8981 feet
Length of middle span	4200 feet
Clearance above water	220 feet
Total weight when built	894,500 tons
Total weight today (lighter road surface)	887,000 tons
Height of two main towers	746 feet
Number of rivets used in each tower	600,000 rivets
Number of strands of wire to make each of the two suspension cables	27,572 strands
Wire needed to make the two cables	80,000 miles
Average number of bridge crossings each year	40 million crossings
Total bridge crossings since construction	2154 million crossings
Number of full-time touch-up painters	38 painters

Source: gocalifornia.about.com

45. Write the total number of cars that have crossed the bridge, using digits (no words).

46. Write the average number of bridge crossings, each year, using digits.

47. If each touch-up painter works 8 hours a day, 5 days a week, how many total hours are spent on painting each week?

48. Find the difference in the weight of the bridge when it was built and the weight of the bridge today.

49. How many feet of wire are there in the two main bridge cables? Write your answer entirely in words. *Hint:* 1 mile = 5280 feet.

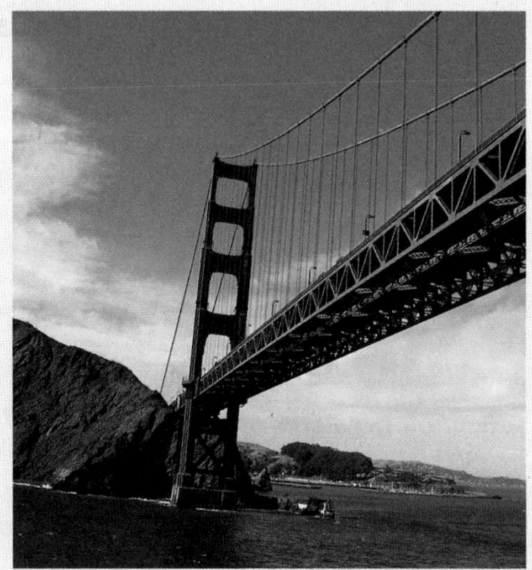

50. **(a)** Is the length of the bridge, including the approaches, more or less than a mile? How much more or less?

(b) Is the length of the middle span of the bridge more or less than a mile? How much more or less? *Hint:* 1 mile = 5280 feet.

1.7 Rounding Whole Numbers

OBJECTIVES

1. Locate the place to which a number is to be rounded.
2. Round numbers.
3. Round numbers to estimate an answer.
4. Use front end rounding to estimate an answer.

VOCABULARY TIP

Rounding Rounding is a useful tool for estimating the answer to a problem.

1 Locate and draw a line under the place to which each number is to be rounded. Then answer the question.

GS (a) 373 (nearest ten)

↑ Tens place

Is 373 closer to 370 or to 380?

____

(b) 1482 (nearest thousand)

Is 1482 closer to 1000 or to 2000?

____

(c) 89,512 (nearest hundred)

Is it closer to 89,500 or to 89,600?

____

(d) 546,325 (nearest ten-thousand)

Is it closer to 540,000 or to 550,000?

____

Answers

1. (a) 373 is closer to 370.
   (b) 1482 is closer to 1000.
   (c) 89,512 is closer to 89,500.
   (d) 546,325 is closer to 550,000.

One way to get a quick check on an answer is to *round* the numbers in the problem. **Rounding** a number means finding a number that is close to the original number, but easier to work with.

For example, the county planning commissioner might be discussing the need for more affordable housing. To demonstrate this, she probably would not need to say that the county is in need of 8235 more affordable housing units—she probably could say that the county needs 8200 or even 8000 housing units.

OBJECTIVE 1 Locate the place to which a number is to be rounded. The first step in rounding a number is to locate the *place to be rounded.*

EXAMPLE 1 Finding the Place to Which a Number Is to Be Rounded

Locate and draw a line under the place to which each number is to be rounded.

(a) Round 83 to the nearest ten. Is 83 closer to 80 or to 90?

83 is closer to 80. (83 is closer to 80 than to 90.)

Tens place

(b) Round 54,702 to the nearest thousand. Is it closer to 54,000 or to 55,000?

54,702 is closer to 55,000.

Thousands place

(c) Round 2,806,124 to the nearest hundred-thousand. Is it closer to 2,800,000 or to 2,900,000?

2,806,124 is closer to 2,800,000.

Hundred-thousands place

◀ Work Problem 1 at the Side.

OBJECTIVE 2 Round numbers. Use the rules for rounding whole numbers.

Rounding Whole Numbers

Step 1 Locate the *place* to which the number is to be rounded. Draw a line under that place.

Step 2(a) Look only at the next digit to the right of the one you underlined. If it is *5 or more, increase* the underlined digit by 1.

Step 2(b) If the next digit to the right is *4 or less, do not change* the digit in the underlined place.

Step 3 *Change* all digits to the right of the underlined place to zeros.

EXAMPLE 2 Using Rounding Rules for 4 or Less

Round 349 to the nearest hundred.

Step 1 Locate the place to which the number is being rounded. Draw a line under that place.

349

Hundreds place

Continued on Next Page

Step 2 Because the next digit to the right of the underlined place is 4, which is 4 or less, do *not* change the digit in the underlined place.

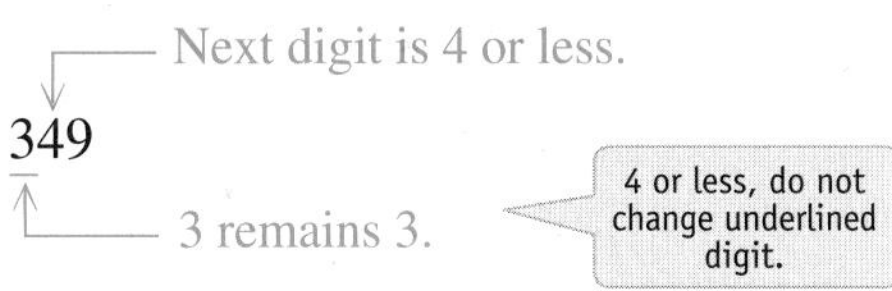
Next digit is 4 or less.
349
3 remains 3.
4 or less, do not change underlined digit.

Step 3 Change all digits to the right of the underlined place to zeros.

349 rounded to the nearest hundred is 300.

In other words, 349 is closer to 300 than to 400.

Work Problem 2 at the Side. ▶

EXAMPLE 3 Using Rounding Rules for 5 or More

Round 36,833 to the nearest thousand.

Step 1 Find the place to which the number is to be rounded and draw a line under that place.

36,833
Thousands

Step 2 Because the next digit to the right of the underlined place is 8, which is 5 or more, add 1 to the underlined place.

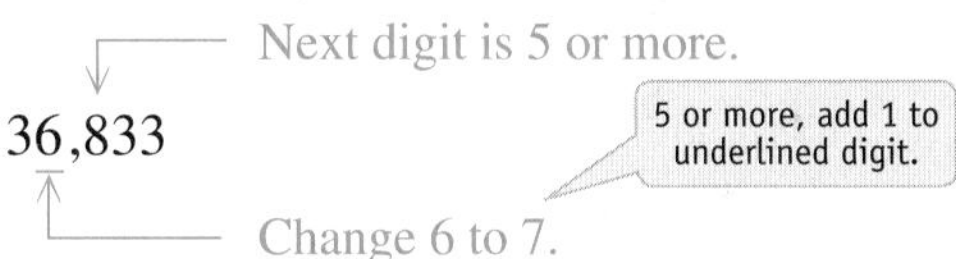
Next digit is 5 or more.
36,833
Change 6 to 7.
5 or more, add 1 to underlined digit.

Step 3 Change all digits to the right of the underlined place to zeros.

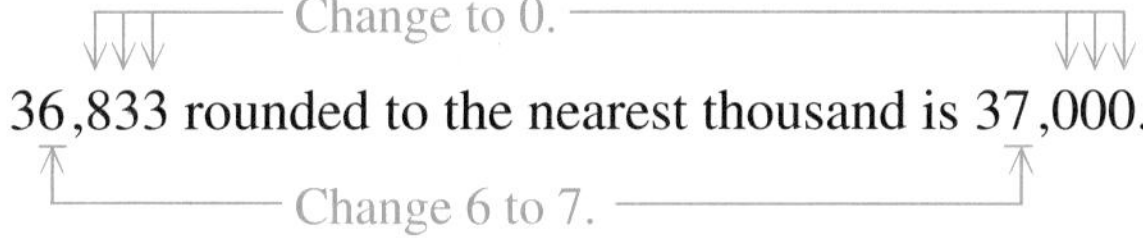
Change to 0.
36,833 rounded to the nearest thousand is 37,000.
Change 6 to 7.

In other words, 36,833 is closer to 37,000 than to 36,000.

Work Problem 3 at the Side. ▶

EXAMPLE 4 Using Rounding Rules

(a) Round 2382 to the nearest ten.

Step 1 2382
Tens place

Step 2 The next digit to the right is 2, which is 4 or less.

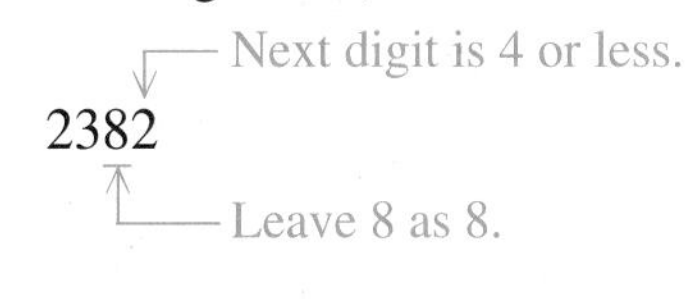
Next digit is 4 or less.
2382
Leave 8 as 8.

Step 3 2382 Change to 0.

2382 rounded to the nearest ten is 2380.

In other words, 2382 is closer to 2380 than to 2390.

Continued on Next Page

2 Round to the nearest ten.

GS **(a)** 62

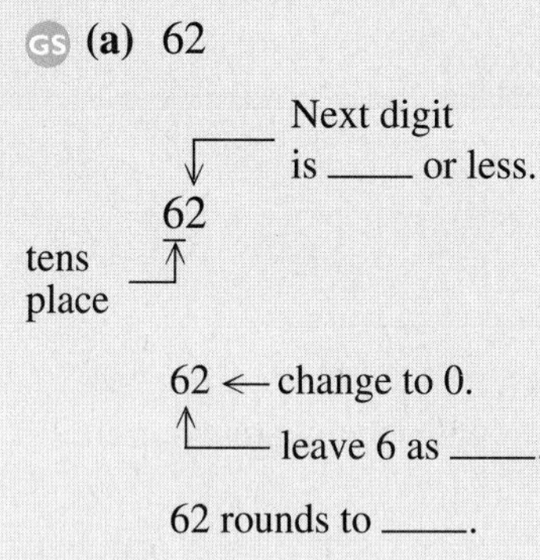
Next digit is ____ or less.
62
tens place

62 ← change to 0.
leave 6 as ____.

62 rounds to ____.

(b) 94

(c) 134

(d) 7543

3 Round to the nearest thousand.

(a) 3683

(b) 6502

(c) 84,621

(d) 55,960

Answers

2. **(a)** 4; 6; 60 **(b)** 90 **(c)** 130 **(d)** 7540

3. **(a)** 4000 **(b)** 7000 **(c)** 85,000 **(d)** 56,000

4 Round each number as indicated.

(a) 3458 to the nearest ten

(b) 6448 to the nearest hundred

(c) 73,077 to the nearest hundred

73,077 — Next digit is ____ or more.

Hundreds place rounds to 73, 1___ ___.

(d) 85,972 to the nearest hundred

5 Round each number as indicated.

(a) 14,598 to the nearest ten-thousand

(b) 724,518,715 to the nearest million

Answers

4. **(a)** 3460 **(b)** 6400 **(c)** 5; 0; 0; 73,100 **(d)** 86,000

5. **(a)** 10,000 **(b)** 725,000,000

(b) Round 13,961 to the nearest hundred.

Step 1 13,961 — Hundreds place

Step 2 The next digit to the right is 6.

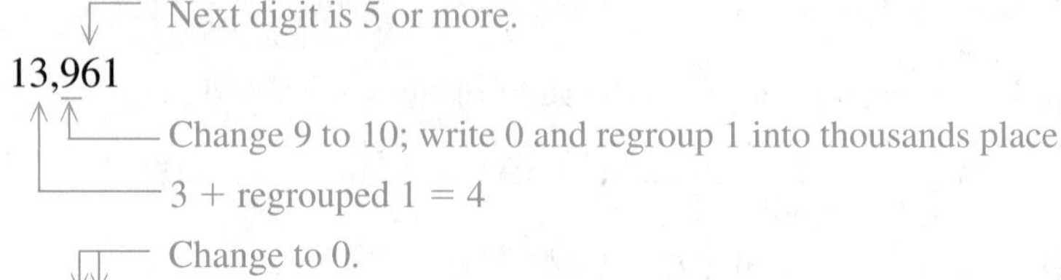

Next digit is 5 or more.
13,961
Change 9 to 10; write 0 and regroup 1 into thousands place.
3 + regrouped 1 = 4
Change to 0.

Step 3 14,061

13,961 rounded to the nearest hundred is 14,000.
In other words, 13,961 is closer to 14,000 than to 13,900.

Note

In *Step 2* of **Example 4(b),** notice that the first three digits increased from 139 to 140 when we added 1 to the hundreds place.

(13,9)61 rounded to (14,0)00

◀ **Work Problem 4 at the Side.**

EXAMPLE 5 Rounding Large Numbers

(a) Round 37,892 to the nearest ten-thousand.

Step 1 37,892 — Ten-thousands place

Remember to *underline* the place to which you are rounding.

Step 2 The next digit to the right is 7.

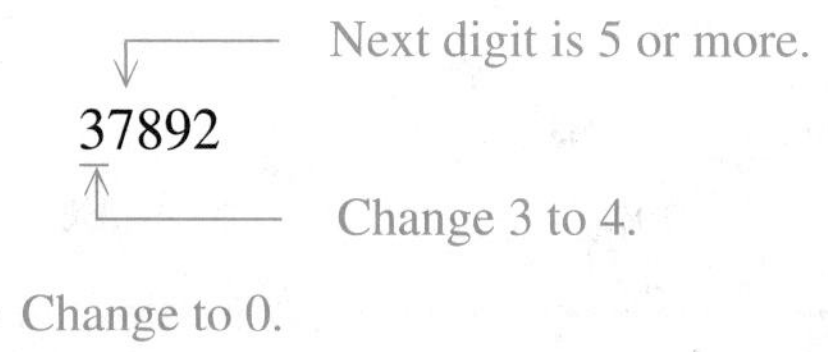

Next digit is 5 or more.
37892
Change 3 to 4.

Change to 0.

Step 3 47,892

37,892 rounded to the nearest ten-thousand is 40,000.

(b) Round 528,498,675 to the nearest million.

Step 1 528,498,675 — Millions place

Next digit is 4 or less.

Step 2 528,498,675 — Leave 8 as 8.

Change to 0.

Remember to change *everything* to the right of the place you have rounded to 0.

Step 3 528,498,675

528,498,675 rounded to the nearest million is 528,000,000.

◀ **Work Problem 5 at the Side.**

Sometimes a number must be rounded to different places.

EXAMPLE 6 Rounding to Different Places

Round 648 **(a)** to the nearest ten and **(b)** to the nearest hundred.

(a) to the nearest ten

Next digit is 5 or more.

648

Tens place (4 + 1 = 5)

648 rounded to the nearest ten is 650.

(b) to the nearest hundred

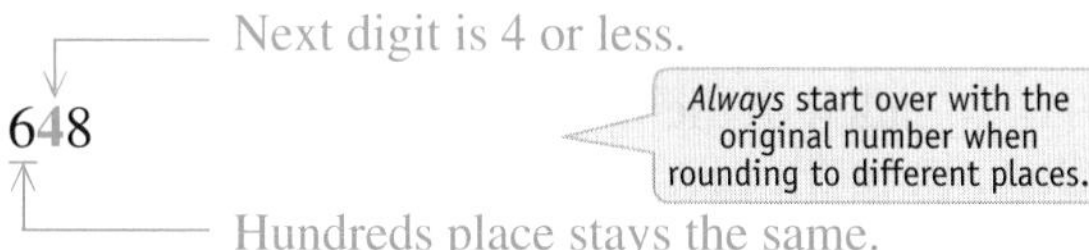

648 rounded to the nearest hundred is 600.

Notice that if 648 is rounded to the nearest ten (650), and then 650 is rounded to the nearest hundred, the result is 700. If, however, 648 is rounded directly to the nearest hundred, the result is 600 (not 700).

Work Problem 6 at the Side. ▶

CAUTION

Before rounding to a different place, always go back to the *original*, unrounded number.

EXAMPLE 7 Applying Rounding Rules

Round each number to the nearest ten, nearest hundred, and nearest thousand.

(a) 4358

First round 4358 to the nearest ten.

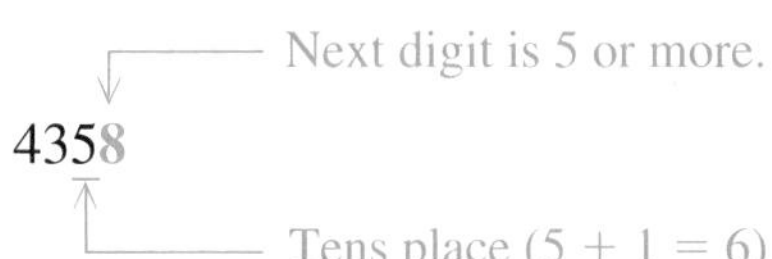

4358 rounded to the nearest ten is 4360.

Now go back to 4358, the *original* number, before rounding to the nearest hundred.

Next digit is 5 or more.

4358

Go back to the original number.

Hundreds place (3 + 1 = 4)

4358 rounded to the nearest hundred is 4400.

Again, go back to the *original* number before rounding to the nearest thousand.

4358 rounded to the nearest thousand is 4000.

Continued on Next Page

6 Round each number to the nearest ten and to the nearest hundred.

GS **(a)** 549

nearest ten: 5 __ 0

nearest hundred: __ __ 0

(b) 458

(c) 9308

Answers

6. **(a)** 5; 550; 5; 0; 500 **(b)** 460; 500 **(c)** 9310; 9300

7 Round each number to the nearest ten, nearest hundred, and nearest thousand.

(a) 4078

(b) 46,364

(c) 268,328

VOCABULARY TIP

Approximately equal to The symbol for "approximately equal to" is a wavy equal sign, ≈. It almost looks like an equal sign.

8 Estimate the answers by rounding each number to the nearest ten.

(a)

$$\begin{array}{r} 16 \\ 74 \\ 58 \\ +\ 31 \\ \hline \end{array}$$

GS **(b)**

$$\begin{array}{rcr} 53 & \rightarrow & 50 \\ -\ 19 & \rightarrow & -\ \_0 \\ \hline & & 30 \end{array}$$

GS **(c)**

$$\begin{array}{rcr} 46 & \rightarrow & \_0 \\ \times\ 74 & \rightarrow & \times\ 7\_ \\ \hline & & 35\_\_ \end{array}$$

Answers

7. **(a)** 4080; 4100; 4000
(b) 46,360; 46,400; 46,000
(c) 268,330; 268,300; 268,000

8. **(a)** 20 + 70 + 60 + 30 = 180
(b) 2; 50 − 20 = 30
(c) 5; 0; 0; 0; 70 × 50 = 3500

(b) 680,914

First, round to the nearest ten.

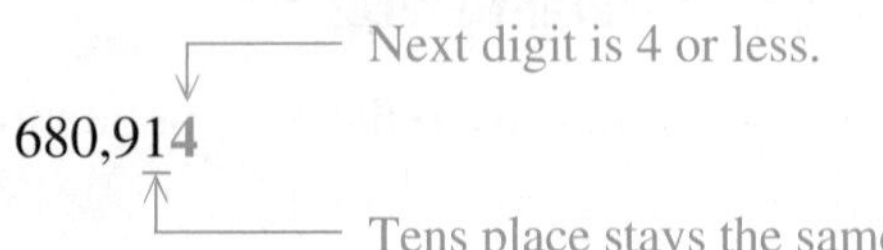

680,914 rounded to the nearest ten is 680,910.

Go back to 680,914, the *original* number, to round to the nearest hundred.

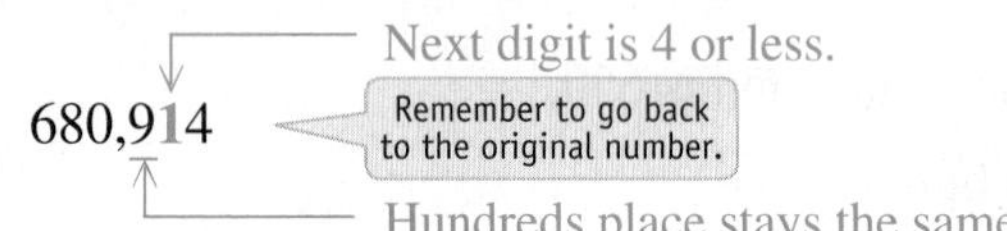

680,914 rounded to the nearest hundred is 680,900.

Go back to the *original* number to round to the nearest thousand.

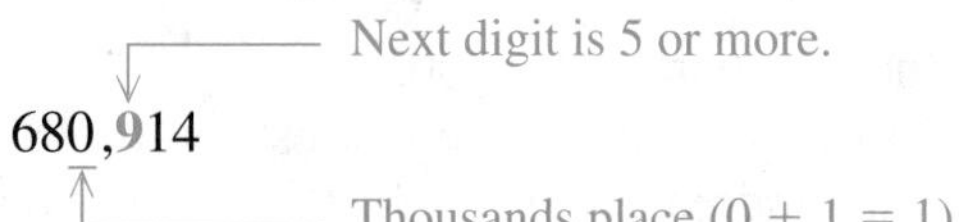

680,914 rounded to the nearest thousand is 681,000.

◀ **Work Problem 7 at the Side.**

OBJECTIVE 3 Round numbers to estimate an answer. Numbers may be rounded to **estimate** an answer. An estimated answer is one that is close to the exact answer and may be used as a check when the exact answer is found. The "≈" sign is often used to show that an answer has been rounded or estimated and is almost equal to the exact answer; ≈ means "approximately equal to."

EXAMPLE 8 Using Rounding to Estimate an Answer

Estimate each answer by rounding to the nearest ten.

(a)

$$\begin{array}{rcr} 76 & \longrightarrow & 80 \\ 53 & \longrightarrow & 50 \\ 38 & \longrightarrow & 40 \\ +\ 91 & \longrightarrow & +\ 90 \\ \hline & & 260 \end{array}$$

80, 50, 40, 90: Rounded to the nearest ten
260: Estimated answer

(b)

$$\begin{array}{rr} 27 & 30 \\ -\ 14 & -10 \\ \hline & 20 \end{array}$$

30, 10: Rounded to the nearest ten
20: Estimated answer

(c)

$$\begin{array}{rr} 16 & 20 \\ \times\ 21 & \times\ 20 \\ \hline & 400 \end{array}$$

20, 20: Rounded to the nearest ten
400: Estimated answer

◀ **Work Problem 8 at the Side.**

EXAMPLE 9 Using Rounding to Estimate an Answer

Estimate each answer by rounding to the nearest hundred.

(a)

252 →	300	Rounded to the nearest hundred
749 →	700	
576 →	600	
+ 819 →	+ 800	
	2400	Estimated answer

The hundreds position is 3 places to the left.

(b)

780	800	Rounded to the nearest hundred
− 536	− 500	
	300	Estimated answer

(c)

664	700	Rounded to the nearest hundred
× 834	× 800	
	560,000	Estimated answer

Work Problem 9 at the Side. ▶

OBJECTIVE 4 **Use front end rounding to estimate an answer.** A convenient way to estimate an answer is to use *front end rounding.* With **front end rounding,** we round to the highest possible place so that all the digits become 0 except the first one. For example, suppose you want to buy a big flat-screen television for $2449, a home theater system for $1759, and a reclining chair for $525. Using front end rounding, you can estimate the total cost of these purchases.

Television	$2449 →	2000
Home theater system	$1759 →	2000
Reclining chair	$525 →	+ 500
		$4500 ← Estimated total cost

Do not round the estimated answer.

EXAMPLE 10 Using Front End Rounding to Estimate an Answer

Estimate each answer using front end rounding.

(a)

3825	4000	All digits changed to 0 except first digit, which is rounded
72	70	
565	600	
+ 2389	+ 2000	
	6670	Estimated answer

(b)

6712	7000	First digit rounded and all others changed to 0
− 825	− 800	
	6200	Estimated answer

Notice: Front end rounding leaves *only* one nonzero digit.

(c)

725	700	
× 86	× 90	
	63,000	Estimated answer

Work Problem 10 at the Side. ▶

9 Estimate the answers by rounding each number to the nearest hundred.

(a) 358, 743, 822, + 978

(b) 842 − 475

(c) 723 × 478

VOCABULARY TIP

Front end rounding An estimation strategy that uses only the left-most digit of the number being rounded. This makes the number(s) the least accurate, but the easiest to work with.

10 Use front end rounding to estimate each answer.

GS **(a)**

36 →	4 0
3852 →	4 0 0 0
749 →	7 _ _
+ 5474 →	+ _ 0 0 0

(b) 2583 − 765

(c) 648 × 67

Answers

9. **(a)** 400 + 700 + 800 + 1000 = 2900
(b) 800 − 500 = 300
(c) 500 × 700 = 350,000

10. **(a)** 0; 0; 5; 40 + 4000 + 700 + 5000 = 9740
(b) 3000 − 800 = 2200
(c) 70 × 600 = 42,000

1.7 Exercises

FOR EXTRA HELP

Download the MyDashBoard App

MyMathLab®

CONCEPT CHECK *The following numbers have been rounded. Decide whether the number has been rounded to the nearest ten, nearest hundred, nearest thousand, or nearest ten-thousand, and fill in the blank.*

1. 624 to the nearest __________ is 620.

2. 509 to the nearest __________ is 510.

3. 86,813 to the nearest __________ is 86,800.

4. 17,211 to the nearest __________ is 17,200.

5. 78,499 to the nearest __________ is 78,000.

6. 14,314 to the nearest __________ is 14,000.

7. 12,987 to the nearest __________ is 10,000.

8. 6599 to the nearest __________ is 10,000.

Round each number as indicated. ***See Examples 1–5.***

9. 855 to the nearest ten

10. 946 to the nearest ten

11. 6771 to the nearest hundred

12. 5847 to the nearest hundred

13. 28,472 to the nearest hundred

GS 28,472
5 or more
hundreds place

14. 18,249 to the nearest hundred

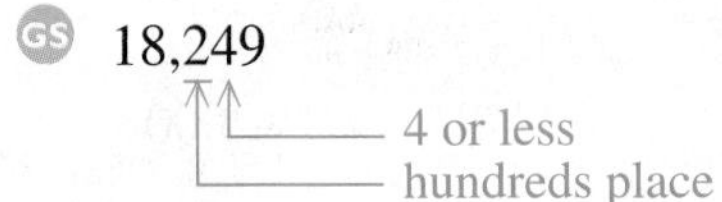

15. 5996 to the nearest hundred

16. 4452 to the nearest hundred

17. 15,758 to the nearest thousand

18. 28,465 to the nearest thousand

19. 7,760,058,721 to the nearest billion

20. 4,468,523,628 to the nearest billion

21. 595,008 to the nearest ten-thousand

22. 725,182 to the nearest ten-thousand

23. 4,860,220 to the nearest million

24. 13,713,409 to the nearest million

Round each number to the nearest ten, nearest hundred, and nearest thousand.
See Examples 6 and 7.

	Ten	Hundred	Thousand
25. 4476	_____	_____	_____
27. 3374	_____	_____	_____
29. 6048	_____	_____	_____

	Ten	Hundred	Thousand
26. 6483	_____	_____	_____
28. 7632	_____	_____	_____
30. 7065	_____	_____	_____

	Ten	Hundred	Thousand
31. 5343	______	______	______
32. 7456	______	______	______
33. 19,539	______	______	______
34. 59,806	______	______	______
35. 26,292	______	______	______
36. 78,519	______	______	______
37. 93,706	______	______	______
38. 84,639	______	______	______

39. Write in your own words the three steps that you would use to round a number when the digit to the right of the place to which you are rounding is *5 or more.*

40. Write in your own words the three steps that you would use to round a number when the digit to the right of the place to which you are rounding is *4 or less.*

Estimate the answer by rounding each number to the nearest ten. Then find the exact answer. ***See Example 8.***

41. *Estimate:* *Exact:*

Estimate		Exact
30	← Rounds to	25
60	←	63
50	←	47
+ 80	←	+ 84

42. *Estimate:* *Exact:*

Estimate		Exact
60	← Rounds to	56
20	←	24
90	←	85
+ 70	←	+ 71

43. *Estimate:* − ___ *Exact:* 78 − 43

44. *Estimate:* − ___ *Exact:* 57 − 24

45. *Estimate:* × ___ *Exact:* 67 × 34

46. *Estimate:* × ___ *Exact:* 53 × 75

Estimate the answer by rounding each number to the nearest hundred. Then find the exact answer. ***See Example 9.***

47. *Estimate:* *Exact:*

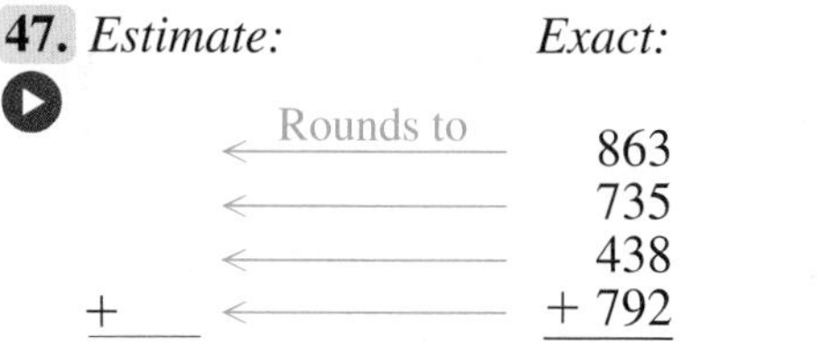

863
735
438
+ 792

48. *Estimate:* × ___ *Exact:*

623
362
189
+ 736

49. *Estimate:* *Exact:*

$$\begin{array}{r} \\ - \underline{\quad\quad} \end{array} \qquad \begin{array}{r} 883 \\ -\ 448 \\ \hline \end{array}$$

50. *Estimate:* *Exact:*

$$\begin{array}{r} \\ - \underline{\quad\quad} \end{array} \qquad \begin{array}{r} 614 \\ -\ 276 \\ \hline \end{array}$$

51. *Estimate:* *Exact:*

$$\begin{array}{r} \\ \times \underline{\quad\quad} \end{array} \qquad \begin{array}{r} 752 \\ \times\ 375 \\ \hline \end{array}$$

52. *Estimate:* *Exact:*

$$\begin{array}{r} \\ \times \underline{\quad\quad} \end{array} \qquad \begin{array}{r} 845 \\ \times\ 396 \\ \hline \end{array}$$

Estimate each answer using front end rounding. Then find the exact answer. **See Example 10.**

53. *Estimate:* *Exact:*

GS

$$\begin{array}{r} 8000 \\ 60 \\ 700 \\ +\ 4000 \\ \hline \end{array} \quad \xleftarrow{\text{Rounds to}} \quad \begin{array}{r} 8215 \\ 56 \\ 729 \\ +\ 3605 \\ \hline \end{array}$$

54. *Estimate:* *Exact:*

GS

$$\begin{array}{r} 3000 \\ 70 \\ 600 \\ +\ 7000 \\ \hline \end{array} \quad \xleftarrow{\text{Rounds to}} \quad \begin{array}{r} 2685 \\ 73 \\ 592 \\ +\ 7183 \\ \hline \end{array}$$

55. *Estimate:* *Exact:*

$$\begin{array}{r} \\ - \underline{\quad\quad} \end{array} \qquad \begin{array}{r} 687 \\ -\ 529 \\ \hline \end{array}$$

56. *Estimate:* *Exact:*

$$\begin{array}{r} \\ - \underline{\quad\quad} \end{array} \qquad \begin{array}{r} 543 \\ -\ 174 \\ \hline \end{array}$$

57. *Estimate:* *Exact:*

$$\begin{array}{r} \\ \times \underline{\quad\quad} \end{array} \qquad \begin{array}{r} 939 \\ \times\ \ 29 \\ \hline \end{array}$$

58. *Estimate:* *Exact:*

$$\begin{array}{r} \\ \times \underline{\quad\quad} \end{array} \qquad \begin{array}{r} 864 \\ \times\ \ 74 \\ \hline \end{array}$$

59. The number 3492 rounded to the nearest hundred is 3500, and 3500 rounded to the nearest thousand is 4000. But when 3492 is rounded directly to the nearest thousand it becomes 3000. Why is this true? Explain.

60. The use of rounding is helpful when estimating the answer to a problem. Why is this true? Give an example using either addition, subtraction, multiplication, or division to show how this works.

61. In 1900, the population of the United States was 76 million. Today it's 311 million. Round each of these numbers to the nearest ten-million. (*Source:* Reiman Publications and U.S. Census Bureau.)

62. Americans will eat more than 22,362,180 hot dogs in Major League ballparks this season. Round the number to the nearest ten-thousand and the nearest hundred-thousand. (*Source:* National Hot Dog and Sausage Council.)

63. There are 348,900 streets named Elm Street in the United States. Round this number to the nearest thousand and nearest ten-thousand. (*Source:* Expo Design Center.)

64. Of all the streets in the United States, the two most common candy-flavored street names are Peppermint and Chocolate. Ninety-five streets are named Peppermint and 27 are named Chocolate. Round each of these numbers to the nearest ten. (*Source:* Tel Atlas digital map database.)

65. There were 39,836,000 speeding tickets given in the United States last year. Round this number to the nearest ten-thousand, nearest hundred-thousand, and nearest million. (*Source:* Clark Howard Radio Show.)

66. Americans spend $57,463,625,000 on lottery tickets each year, or nine times as much as they spend on movie tickets. Round this number to the nearest ten-million, nearest hundred-million, and nearest billion. (*Source:* TLC-W, A Discovery Company.)

67. Ping-Pong is the fifth fastest growing sport in America. Today there are 19,265,780 players. Round this number to the nearest thousand, nearest ten-thousand, and the nearst hundred-thousand. (*Source: Parade, The Sunday Newspaper Magazine.*)

68. The highest annual production of Mitsubishi automobiles in Normal, Illinois was 221,543. Round this number to the nearest ten, nearest hundred, and nearest thousand. (*Source: Newsweek.*)

Relating Concepts (Exercises 69–75) For Individual or Group Work

To see how both rounding and front end rounding are used in solving problems, ***work Exercises 69–75 in order.***

69. A number rounded to the nearest thousand is 72,000. What is the *smallest* whole number this could have been before rounding?

70. A number rounded to the nearest thousand is 72,000. What is the *largest* whole number this could have been before rounding?

71. When front end rounding is used, a whole number rounds to 8000. What is the *smallest* possible original number?

72. When front end rounding is used, a whole number rounds to 8000. What is the *largest* possible original number?

The graph below shows the number of personal injuries in the United States each year for people participating in common activities.

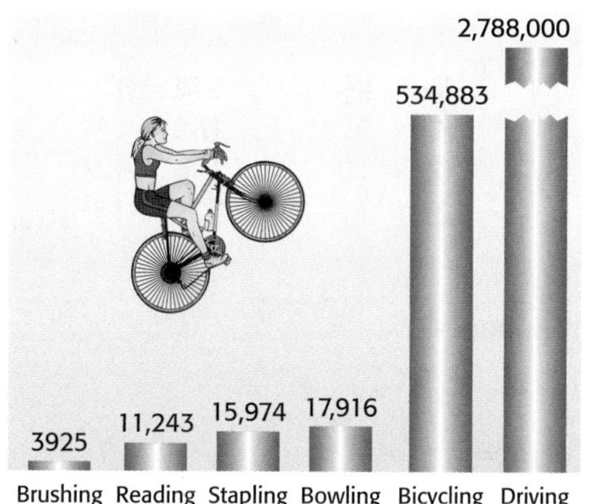

Source: AARP Magazine.

73. Round the number of accidents occurring in each activity to the nearest ten.

74. Use front end rounding to round the number of accidents in each activity.

75. **(a)** What is one advantage of using front end rounding instead of rounding to the nearest ten?

(b) What is one disadvantage?

1.8 Exponents, Roots, and Order of Operations

OBJECTIVES

1. Identify an exponent and a base.
2. Find the square root of a number.
3. Use the order of operations.

1 Identify the exponent and the base, and then simplify each expression.

(a) 4^2

(b) 5^3

(c) 3^4

(d) 2^6

VOCABULARY TIP

Perfect square It is helpful to have the first 12 or so perfect squares memorized so that they are easily recognizable.

2 Find each square root.

(a) $\sqrt{4}$

(b) $\sqrt{25}$

(c) $\sqrt{36}$

(d) $\sqrt{225}$

(e) $\sqrt{1}$

Answers

1. (a) 2; 4; 16 (b) 3; 5; 125
   (c) 4; 3; 81 (d) 6; 2; 64
2. (a) 2 (b) 5 (c) 6 (d) 15 (e) 1

OBJECTIVE 1 Identify an exponent and a base. The product 3 • 3 can be written as 3^2 (read as "3 squared"). The small raised number 2, called an **exponent,** says to use 2 factors of 3. The number 3 is called the **base.** Writing 3^2 as 9 is called *simplifying the expression.*

EXAMPLE 1 Simplifying Expressions

Identify the exponent and the base, and then simplify each expression.

(a) 4^3 Base $\longrightarrow 4^3 \longleftarrow$ Exponent $4^3 = 4 \times 4 \times 4 = 64$

The small raised number is the exponent.

(b. $2^5 = 2 \times 2 \times 2 \times 2 \times 2 = 32$
The base is 2 and the exponent is 5.

◄ Work Problem **1** at the Side.

OBJECTIVE 2 Find the square root of a number. Because $3^2 = 9$, the number 3 is called the **square root** of 9. The square root of a number is one of two identical factors of that number. Square roots of numbers are written with the symbol $\sqrt{\ }$.

Square Root

$$\sqrt{\textbf{number} \cdot \textbf{number}} = \sqrt{\textbf{number}^2} = \textbf{number}$$

For example: $\sqrt{36} = \sqrt{6 \cdot 6} = \sqrt{6^2} = 6$ The square root of 36 is **6**.

To find the square root of 64 ask, "What number can be multiplied by itself (that is, *squared*) to give 64?" The answer is 8, so

$$\sqrt{64} = \sqrt{8 \cdot 8} = \sqrt{8^2} = 8.$$

A **perfect square** is a number that is the square of a ***whole number.*** The first few perfect squares are listed here.

Perfect Squares Table

$\mathbf{0} = 0^2$	$\mathbf{16} = 4^2$	$\mathbf{64} = 8^2$	$\mathbf{144} = 12^2$
$\mathbf{1} = 1^2$	$\mathbf{25} = 5^2$	$\mathbf{81} = 9^2$	$\mathbf{169} = 13^2$
$\mathbf{4} = 2^2$	$\mathbf{36} = 6^2$	$\mathbf{100} = 10^2$	$\mathbf{196} = 14^2$
$\mathbf{9} = 3^2$	$\mathbf{49} = 7^2$	$\mathbf{121} = 11^2$	$\mathbf{225} = 15^2$

EXAMPLE 2 Using Perfect Squares

Find each square root.

Note that $\sqrt{16}$ is 4, not 4^2.

(a) $\sqrt{16}$ Because $4^2 = 16$, $\sqrt{16} = 4$. **(b)** $\sqrt{49} = 7$

(c) $\sqrt{0} = 0$ **(d)** $\sqrt{169} = 13$

◄ Work Problem **2** at the Side.

OBJECTIVE 3 Use the order of operations. Frequently problems may have parentheses, exponents, and square roots, and may involve more than one operation. Work these problems by following the **order of operations.**

Order of Operations

1. Do all operations inside ***parentheses*** or ***other grouping symbols.***
2. Simplify any expressions with ***exponents*** and find any ***square roots.***
3. ***Multiply*** or ***divide,*** proceeding from left to right.
4. ***Add*** or ***subtract,*** proceeding from left to right.

EXAMPLE 3 Understanding the Order of Operations

Use the order of operations to simplify each expression.

(a) $8^2 + 5 + 2$

$8^2 + 5 + 2$

$8 \cdot 8 + 5 + 2$ Evaluate exponent first; 8^2 is $8 \cdot 8$.

$64 + 5 + 2$ Add from left to right.

$69 + 2 = 71$

(b) $35 \div 5 \cdot 6$ Divide first (start at left).

$7 \cdot 6 = 42$ Multiply.

(c) $9 + (20 - 4) \cdot 3$ Work inside parentheses first.

$9 + 16 \cdot 3$ Multiply.

$9 + 48 = 57$ Add last.

(d) $12 \cdot \sqrt{16} - 8(4)$ Find the square root first.

$12 \cdot 4 - 8(4)$ Multiply from left to right.

$48 - 32 = 16$ Subtract last.

Work Problem **3** at the Side. ▶

EXAMPLE 4 Using the Order of Operations

Use the order of operations to simplify each expression.

(a) $15 - 4 + 2$ Subtract first (start at left).

$11 + 2 = 13$ Add.

(b) $8 + (7 - 3) \div 2$ Work inside parentheses first.

$8 + 4 \div 2$ Divide.

$8 + 2 = 10$ Add last.

Add or subtract last.

(c) $4^2 \cdot 2^2 + (7 + 3) \cdot 2$ Work inside parentheses first.

$4^2 \cdot 2^2 + 10 \cdot 2$ Evaluate exponents.

$16 \cdot 4 + 10 \cdot 2$ Multiply from left to right.

$64 + 20 = 84$ Add last.

(d) $4 \cdot \sqrt{25} - 7 \cdot 2 + \frac{0}{5}$ Find the square root first.

Zero divided by any nonzero number is zero.

$4 \cdot 5 - 7 \cdot 2 + \frac{0}{5}$ Multiply or divide from left to right.

$20 - 14 + 0 = 6$ Add or subtract last.

Work Problem **4** at the Side. ▶

VOCABULARY TIP

Order of operations The first letter of the words "**P**lease **E**xcuse **M**y **D**ear **A**unt **S**ally" can help you remember the order of operations; Parentheses, Exponents, Multiplication and Division, Addition and Subtraction.

3 Simplify each expression.

GS **(a)** $4 + 5 + 2^2$ Evaluate exponents.

$4 + 5 + 2 \cdot 2$ Add left to right.

$4 + 5 + \_\_\_\_$

$9 + \_\_\_\_ = \_\_\_\_$

(b) $3^2 + 2^3$

(c) $60 \div \sqrt{36} \div 2$

(d) $8 + 6(14 \div 2)$

4 Simplify each expression.

GS **(a)** $12 - 6 + 4^2$ Evaluate exponents.

$12 - 6 + \_\_\_\_ \cdot \_\_\_\_$

$12 - 6 + 16$ Add/subtract left to right.

$6 + 16 = \_\_\_\_$

(b) $2^3 + 3^2 - (5 \cdot 3)$

(c) $20 \div 2 + (7 - 5)$

(d) $15 \cdot \sqrt{9} - 8 \cdot \sqrt{4}$

Answers

3. **(a)** 4; 4; 13 **(b)** 17 **(c)** 5 **(d)** 50
4. **(a)** 4; 4; 22 **(b)** 2 **(c)** 12 **(d)** 29

1.8 Exercises

FOR EXTRA HELP

Download the MyDashBoard App

MyMathLab®

CONCEPT CHECK *Identify the exponent and the base.*

1. 3^2
exponent ____
base ____

2. 2^3
exponent ____
base ____

3. 5^2
exponent ____
base ____

4. 4^2
exponent ____
base ____

Identify the exponent and the base and then simplify each expression. ***See Example 1.***

5. 8^2

6. 10^3

7. 15^2

8. 11^3

Use the Perfect Squares Table on the first page of this section to find each square root. ***See Example 2.***

9. $\sqrt{16}$

10 $\sqrt{25}$

11. $\sqrt{64}$

12. $\sqrt{36}$

13. $\sqrt{100}$

14. $\sqrt{49}$

15. $\sqrt{144}$

16. $\sqrt{225}$

CONCEPT CHECK *Decide whether each statement is* true *or* false. *If it is* false, *explain why.*

17. The expression 5^2 means that 2 is used as a factor 5 times.

18. $4^2 = 8$

19. $1^3 = 3$

20. $6^1 = 1$

Fill in each blank. ***See Example 2.***

21. $6^2 =$ ____ so $\sqrt{\quad} = 6$
GS $6 \cdot 6 =$ ____ so $\sqrt{\text{\_\_\_\_}} = 6$

22. $9^2 =$ ____ so $\sqrt{\quad} = 9$
GS $9 \cdot 9 =$ ____ so $\sqrt{\text{\_\_\_\_}} = 9$

23. $25^2 =$ ____ so $\sqrt{\quad} = 25$

24. $50^2 =$ ____ so $\sqrt{\quad} = 50$

25. $100^2 =$ ____ so $\sqrt{\quad} = 100$

26. $60^2 =$ ____ so $\sqrt{\quad} = 60$

27. Describe in your own words a perfect square. Of the two numbers 25 and 50, identify which is a perfect square and explain why.

28. Use the following list of words and phrases to write the four steps in the order of operations.

add	square root
exponents	subtract
multiply	divide

parentheses or other grouping symbols

CONCEPT CHECK *Decide whether each statement is* true *or* false. *If it is* false, *explain why.*

29. $3^2 + 8 - 5 = 12$

30. $5^2 + 5 - 6 = 24$

31. $6 + 8 \div 2 = 7$

32. $4 + 5(6 - 4) = 18$

Simplify each expression by using the order of operations. ***See Examples 3 and 4.***

33. $3^2 + 8 - 5$

$\underbrace{3^2} + 8 - 5$

$\underbrace{3 \cdot 3} + \underbrace{8 - 5}$

$9 + 3 =$

34. $5^2 + 5 - 6$

$\underbrace{5^2} + 5 - 6$

$\underbrace{5 \cdot 5} + 5 - 6$

$25 + 5 - 6 =$

35. $25 \div 5(8 - 4)$

36. $36 \div 18(7 - 3)$

37. $5 \cdot 3^2 + \frac{0}{8}$

38. $8 \cdot 3^2 - \frac{10}{2}$

39. $4 \cdot 1 + 8(9 - 2) + 3$

40. $3 \cdot 2 + 7(3 + 1) + 5$

41. $2^2 \cdot 3^3 + (20 - 15) \cdot 2$

42. $4^2 \cdot 5^2 + (20 - 9) \cdot 3$

43. $5\sqrt{36} - 2(4)$

44. $2 \cdot \sqrt{100} - 3(4)$

45. $8(2) + 3 \cdot 7 - 7 =$

46. $10(3) + 6 \cdot 5 - 20$

47. $2^3 \cdot 3^2 + 3(14 - 4)$

48. $3^2 \cdot 4^2 + 2(15 - 6)$

49. $7 + 8 \div 4 + \frac{0}{7}$

50. $6 + 8 \div 2 + \frac{0}{8}$

51. $3^2 + 6^2 + (30 - 21) \cdot 2$

52. $4^2 + 5^2 + (25 - 9) \cdot 3$

53. $7 \cdot \sqrt{81} - 5 \cdot 6$

GS

$$7 \cdot \underbrace{\sqrt{81}} - \underbrace{5 \cdot 6}$$
$$\underbrace{7 \cdot 9} - 30$$
$$63 - 30 =$$

54. $6 \cdot \sqrt{64} - 6 \cdot 5$

GS

$$6 \cdot \underbrace{\sqrt{64}} - \underbrace{6 \cdot 5}$$
$$\underbrace{6 \cdot 8} - 30$$
$$48 - 30 =$$

55. $8 \cdot 2 + 5(3 \cdot 4) - 6$

56. $5 \cdot 2 + 3(5 + 3) - 6$

57. $4 \cdot \sqrt{49} - 7(5 - 2)$

58. $3 \cdot \sqrt{25} - 6(3 - 1)$

59. $7(4 - 2) + \sqrt{9}$

60. $5(4 - 3) + \sqrt{9}$

61. $7^2 + 3^2 - 8 + 5$

62. $3^2 - 2^2 + 3 - 2$

63. $5^2 \cdot 2^2 + (8 - 4) \cdot 2$

64. $5^2 \cdot 3^2 + (30 - 20) \cdot 2$

65. $5 + 9 \div 3 + 6 \cdot 3$

66. $8 + 3 \div 3 + 6 \cdot 3$

67. $8 \cdot \sqrt{49} - 6(9 - 4)$

68. $8 \cdot \sqrt{49} - 6(5 + 3)$

69. $5^2 - 4^2 + 3 \cdot 6$

70. $3^2 + 6^2 - 5 \cdot 8$

71. $8 + 8 \div 8 + 6 + \frac{5}{5}$

GS

$8 + \underbrace{8 \div 8} + 6 + \frac{5}{5}$

$8 + \quad 1 \quad + 6 + 1 =$

72. $3 + 14 \div 2 + 7 + \frac{8}{8}$

GS

$3 + \underbrace{14 \div 2} + 7 + \frac{8}{8}$

$3 + \quad 7 \quad + 7 + 1 =$

73. $6 \cdot \sqrt{25} - 7(2)$

74. $8 \cdot \sqrt{36} - 4(6)$

75. $9 \cdot \sqrt{16} - 3 \cdot \sqrt{25}$

76. $6 \cdot \sqrt{81} - 3 \cdot \sqrt{49}$

77. $7 \div 1 \cdot 8 \cdot 2 \div (21 - 5)$

78. $12 \div 4 \cdot 5 \cdot 4 \div (15 - 13)$

79. $15 \div 3 \cdot 2 \cdot 6 \div (14 - 11)$

80. $9 \div 1 \cdot 4 \cdot 2 \div (11 - 5)$

81. $6 \cdot \sqrt{25} - 4 \cdot \sqrt{16}$

82. $10 \cdot \sqrt{49} - 4 \cdot \sqrt{64}$

83. $5 \div 1 \cdot 10 \cdot 4 \div (17 - 9)$

84. $15 \div 3 \cdot 8 \cdot 9 \div (12 - 8)$

85. $8 \cdot 9 \div \sqrt{36} - 4 \div 2 + (14 - 8)$

86. $3 - 2 + 5 \cdot 4 \cdot \sqrt{144} \div \sqrt{36}$

87. $2 + 1 - 2 \cdot \sqrt{1} + 4 \cdot \sqrt{81} - 7 \cdot 2$

88. $6 - 4 + 2 \cdot 9 - 3 \cdot \sqrt{225} \div \sqrt{25}$

89. $5 \cdot \sqrt{36} \cdot \sqrt{100} \div 4 \cdot \sqrt{9} + 8$

90. $9 \cdot \sqrt{36} \cdot \sqrt{81} \div 2 + 6 - 3 - 5$

Study Skills

TAKING LECTURE NOTES

OBJECTIVES

1. **Apply note taking strategies, such as writing problems as well as explanations.**
2. **Use appropriate abbreviations in notes.**

Study the set of sample math notes in this section, and read the comments about them. Then try to incorporate the techniques into your own math note taking in class.

January 2

Exponents

Exponents used to show repeated multiplication.

$3 \cdot 3 \cdot 3 \cdot 3$ can be written 3^4 — *exponent (how many times it's multiplied)*; *base (the number being multiplied)*

Read *3^2 as 3 to the 2nd power or 3 squared*

3^3 as 3 to the 3rd power or 3 cubed

3^4 as 3 to the 4th power

etc.

Simplifying an expression with exponents → *actually do the repeated multiplication*

2^3 means $2 \cdot 2 \cdot 2$ and $2 \cdot 2 \cdot 2 = 8$

☆ *Careful!* [*5^2 means $5 \cdot 5$ NOT $5 \cdot 2$*; *so $5^2 = 5 \cdot 5 = 25$ BUT $5^2 \neq 10$*]

Example	*Explanation*
Simplify $2^4 \cdot 3^2$	*Exponents mean multiplication.*
$2 \cdot 2 \cdot 2 \cdot 2 \cdot 3 \cdot 3$	*Use 2 as a factor 4 times.* *Use 3 as a factor 2 times.*
$16 \cdot 9$	*$2 \cdot 2 \cdot 2 \cdot 2$ is 16* *$3 \cdot 3$ is 9* → *$16 \cdot 9$ is 144*
144	*Simplified result is 144 (no exponents left)*

- The **date and title** of the day's lecture topic are always at the top of every page. **Always begin a new day with a new page.**
- Note the **definitions** of base and exponent are written in parentheses—don't trust your memory!
- **Skipping lines** makes the notes easier to read.
- See how the **direction word** (*simplify*) is emphasized and explained.
- A **star marks an important concept.** This is a warning to avoid future mistakes. **Note the underlining,** too, which highlights the importance.
- Notice the two columns, which allow for the example and its explanation to be close together. **Whenever you know you'll be given a series of steps to follow, try the two-column method.**
- Note the **brackets and arrows,** which clearly show how the problem is set up to be simplified.

Now Try This

Find one or two people in your math class to work with. Compare each other's lecture notes over a period of a week or so. Ask yourself the following questions as you examine the notes.

1. What are you doing in your notes to show the **main points** or larger concepts? (Such as underlining, boxing, using stars, capital letters, etc.)
2. In what ways do you **set off the explanations** for worked problems, examples, or smaller ideas (subpoints)? (Such as indenting, using arrows, circling or boxing)
3. What does **your instructor do** to show that he or she is moving from one idea to the next? (Such as saying "Next" or "Any questions," "Now," or erasing the board, etc.)
4. **How do you mark** that in your notes? (Such as skipping lines, using dashes or numbers, etc.)
5. What **explanations (in words) do you give yourself** in your notes, so when those new dendrites that you grew during lecture are fading, you can read your notes and still remember the new concepts later when you try to do your homework?
6. What **did you learn** by examining your classmates' notes?
   - ______________________________
   - ______________________________
   - ______________________________
7. What **will you try** in your own note taking? List **four** techniques that you will use next time you take notes in math class.
   - ______________________________
   - ______________________________
   - ______________________________
   - ______________________________

Why Are These Notes Brain Friendly?

The notes are **easy to look at,** and you know that the brain responds to things that are visually pleasing. Other techniques that are visually memorable are the use of spacing (the two columns), stars, underlining, and circling. All of these methods **allow your brain to take note of important concepts and steps.**

The notes are also **systematic,** which means that they use certain techniques regularly. This way, your brain easily recognizes the topic of the day, the signals that show an important point, and the steps to follow for procedures. When you develop a system that you always use in your notes, your notes are easy to understand later when you are reviewing for a test.

1.9 Reading Pictographs, Bar Graphs, and Line Graphs

OBJECTIVES

1. Read and understand a pictograph.
2. Read and understand a bar graph.
3. Read and understand a line graph.

We have all heard the saying "A picture is worth a thousand words," and there may be some truth in this. Today, so much information and data are being presented in the form of pictographs, circle graphs, bar graphs, and line graphs that it is important to be able to read and understand these tools.

OBJECTIVE 1 Read and understand a pictograph. A **pictograph** is a graph that uses pictures or symbols. It displays information that can be compared easily. However, since a symbol is used to represent a certain quantity, it can be difficult to determine the amount represented by a fraction of a symbol.

The pictograph below compares the number of U.S. postage stamps sold in the five top releases. In this pictograph, it is difficult to determine what fractional amount is represented by the partial stamps.

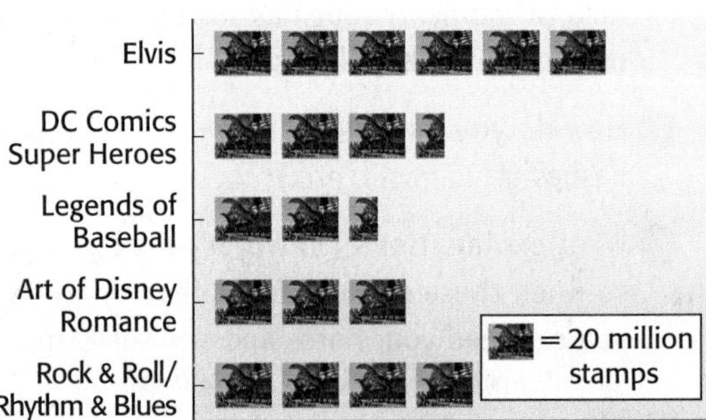

Source: United States Postal Service.

EXAMPLE 1 Using a Pictograph

Use the pictograph to answer each question. (Each stamp symbol represents 20 million stamps.)

(a) Which of the U.S. postage stamps shown has the lowest number sold?

The row representing Legends of Baseball has the fewest symbols. This means that the lowest number of U.S. postage stamps sold was Legends of Baseball.

(b) Approximately how many more Elvis stamps were sold than the Art of Disney Romance stamps?

The row representing the Elvis stamps has three more symbols than that for the Art of Disney stamps. This means that 3 • 20 million or 60,000,000 more Elvis stamps were sold than the Art of Disney Romance stamps.

◀ Work Problem 1 at the Side.

OBJECTIVE 2 Read and understand a bar graph. **Bar graphs** are useful for showing comparisons. For example, the following bar graph shows how many adults out of every 100 surveyed chose each money secret that they kept from their partner.

VOCABULARY TIP

Pictograph Every pictograph has a key. The key in the graph tells the number that each picture or symbol represents.

1 Use the pictograph to answer each question.

GS **(a)** Which of the U.S. stamps had the second greatest number of sales?

Each stamp symbol represents 20 million stamps. The greatest number of symbols (6) is for the Elvis stamp. The second greatest number of symbols is for the ____________ stamp.

(b) Approximately how many more Rock and Roll/Rhythm and Blues stamps were sold than Art of Disney Romance stamps?

VOCABULARY TIP

Bar graphs are typically used to display data that fit into different categories.

Answers

1. (a) 4; Rock and Roll/Rhythm & Blues
(b) about 20 million (or 20,000,000) more stamps

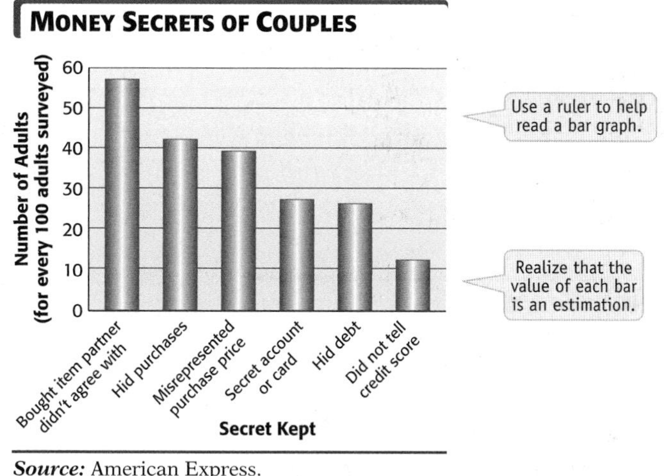

Source: American Express.

EXAMPLE 2 Using a Bar Graph

Use the bar graph to find the number of adults who picked "Bought item partner didn't agree with" as the secret they kept from their partner.

Use a ruler or straightedge to line up the top of the bar labeled "Bought item partner didn't agree with" with the numbers on the left edge of the graph, labeled "Number of Adults." We see that 57 out of 100 adults picked "Bought item partner didn't agree with."

Work Problem **2** at the Side. ▶

OBJECTIVE 3 **Read and understand a line graph.** A **line graph** is often used for showing a trend. The following line graph shows the U.S. Census Bureau predictions for U.S. population growth to the year 2100.

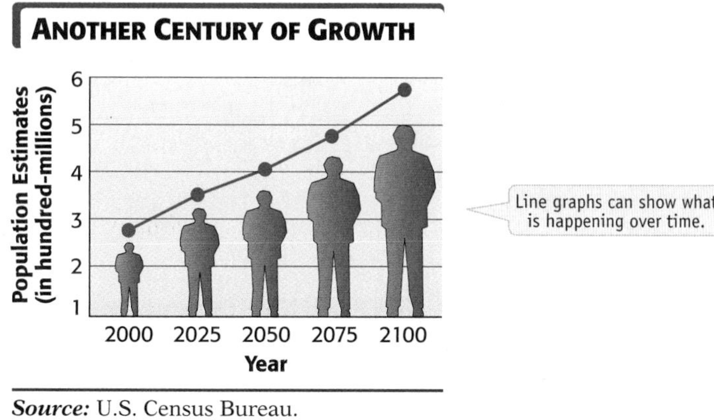

Source: U.S. Census Bureau.

EXAMPLE 3 Using a Line Graph

Use the line graph to answer each question.

(a) What trend or pattern is shown in the graph?
The population will continue to increase.

(b) What is the estimated population for 2025?

Use a ruler or straightedge to line up the dot above the year labeled 2025 on the horizontal line with the numbers along the left edge of the graph. Notice that the label on the left side says "in hundred-millions." Since the 2025 dot is halfway between 3 and 4, the population in 2025 is halfway between 3 • 100,000,000 and 4 • 100,000,000 or 300,000,000 and 400,000,000. That means that the predicted population in 2025 is about 350,000,000 people.

Work Problem **3** at the Side. ▶

2 Use the bar graph to find the approximate number of adults who picked each money secret they kept from their partner.

(a) Secret account or card

(b) Misrepresented purchase price

(c) Hid purchases

(d) Hid debt

(e) Did not tell credit score

VOCABULARY TIP

Line graphs can be used to show how something changes over time. These graphs often have peaks and valleys to indicate upward and downward trends.

3 Use the line graph to find the predicted population of the United States for each year.

GS **(a)** 2050

The dot above 2050 lines up with ____. This means that the predicted population will be ___ ___ ___, 000,000.

(b) 2075

(c) 2100

Answers

2. **(a)** 27 out of 100 **(b)** 39 out of 100 **(c)** 42 out of 100 **(d)** 26 out of 100 **(e)** 12 out of 100

3. **(a)** 4; 4; 0; 0; 400,000,000 **(b)** 475,000,000 **(c)** 575,000,000

1.9 Exercises

FOR EXTRA HELP

Download the MyDashBoard App

MyMathLab®

The following pictograph shows the number of retail stores for the seven companies with the greatest number of outlets. Use the pictograph to answer Exercises 1–8. ***See Example 1.***

SOMETHING IN STORE

While Walmart has the greatest amount of sales, it trails other chains in number of stores.

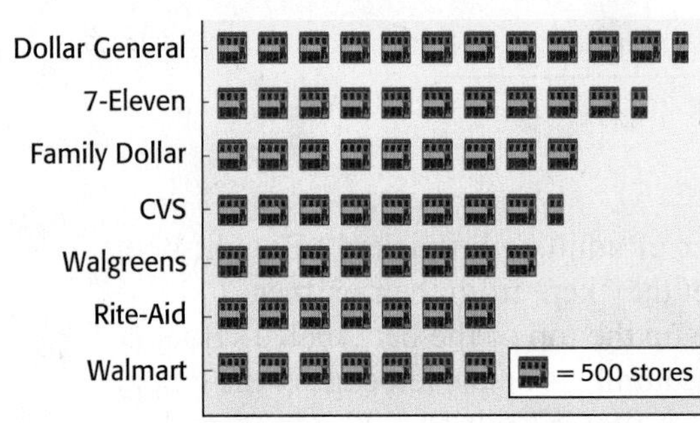

Source: T. D. Linx.

1. CONCEPT CHECK *Fill in the blanks.*
   The number of pictures or symbols for the Walgreens retail stores is ______. Since each symbol represents ______ stores, Walgreens has
   ______ • ______ = ______ stores.

2. CONCEPT CHECK *Fill in the blanks.*
   The number of pictures or symbols for the CVS retail stores is ______. Since each symbol represents ______ stores, CVS has
   ______ • ______ = ______ stores.

3. Find the number of Family Dollar retail stores.
   Family Dollar has 9 symbols.
   Each symbol is 500 stores
   $9 \cdot 500 =$ __________

4. Approximately how many retail stores does 7-Eleven have?
   7-Eleven has $10\frac{1}{2}$ symbols.
   Each symbol is 500 stores.
   $(10 \cdot 500) + \frac{1}{2} \cdot 500 =$ __________

5. Which company has the greatest number of retail stores? How many is that?

6. Which companies have the least number of retail stores? How many does each one have?

7. How many fewer stores does Family Dollar have than Dollar General?

8. How many more retail stores does Walgreens have than Walmart?

The following bar graph shows the results of a survey that was taken of 100 working adults to determine how they chose their careers. Use the bar graph to answer Exercises 11–16.
See Example 2.

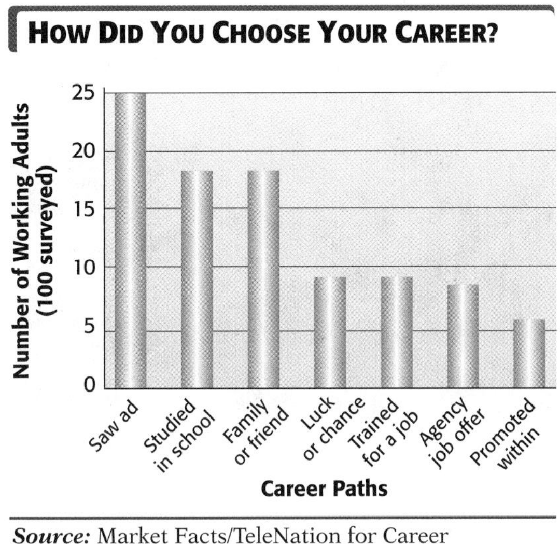

Source: Market Facts/TeleNation for Career Education Corporation.

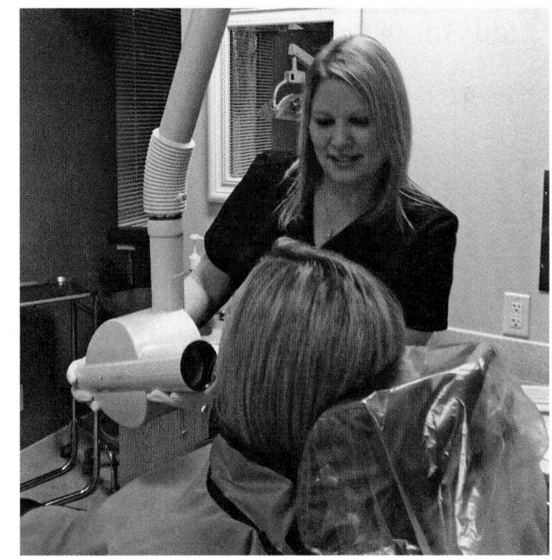

9. CONCEPT CHECK What is the number of working adults who were surveyed?

10. CONCEPT CHECK How many career paths were included in the survey?

11. How many people found their careers as a result of training for a job?

12. How many people found their careers because they studied for the career in school?

13. **(a)** Which career path was taken by the greatest number of people?

The bar graph shows that the greatest number of people ____________.

(b) How many people used this path?

From the bar graph ____________ people used this path.

14. **(a)** Which career path was taken by the least number of people?

The bar graph shows that the least number of people were ____________.

(b) How many people used this path?

From the bar graph, ____________ people used this path.

15. How many more people found their careers as a result of "Studied in school" than "Luck or chance"?

16. Find the total number of people who found their careers as a result of either "Studied in school" or "Trained for a job."

Sun Solar Products collected installation data and prepared the following line graph. Remembering that the solar installation data are shown in thousands of installations, use the line graph to answer Exercises 17–22. **See Example 3.**

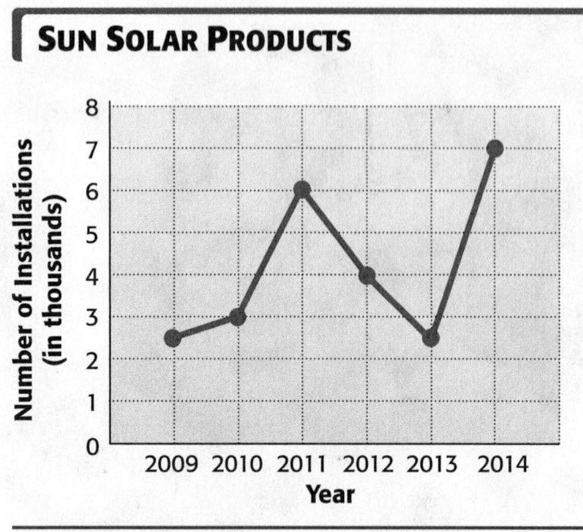

17. Which year had the greatest number of installations? How many installations were there?

The line graph shows that __________ had the greatest number.

There were 7 • 1000 = __________.

18. Which two years had the least number of installations? How many installations were there in each of those years?

The line graph shows that __________ and __________ had the least number.

The dots for both years are halfway between 2000 and 3000, or __________ installations.

19. Find the increase in the number of installations from 2013 to 2014.

20. Find the decrease in the number of installations from 2012 to 2013.

21. Give three possible explanations for the decrease in solar installations from 2011 to 2013.

22. Give three possible explanations for the increase in solar installations from 2013 to 2014.

Relating Concepts (Exercises 23–27) For Individual or Group Work

Getting a correct answer in mathematics always depends on following the order of operations. Insert grouping symbols (parentheses) so that each expression will result in the given number when simplified. ***Work Exercises 23–27 in order.***

23. $7 - 2 \cdot 3 - 6$; simplifies to 9

24. $4 + 2 \cdot 5 + 1$; simplifies to 36

25. $36 \div 3 \cdot 3 \cdot 4$; simplifies to 16

26. $56 \div 2 \cdot 2 \cdot 2 + \frac{0}{6}$; simplifies to 7

27. The Good Shepherd Ranch owns one section of land (one mile by one mile) shown below as Parcel 1. Parcels 2 and 3 are leased from neighbors.

(a) Use the order of operations to write an expression for the distance around the combined parcels.

(b) How many feet of barbed wire are needed for a three-strand barbed wire fence around the parcels?

(c) How many miles of barbed wire is this? (*Hint:* 1 mile = 5280 feet.)

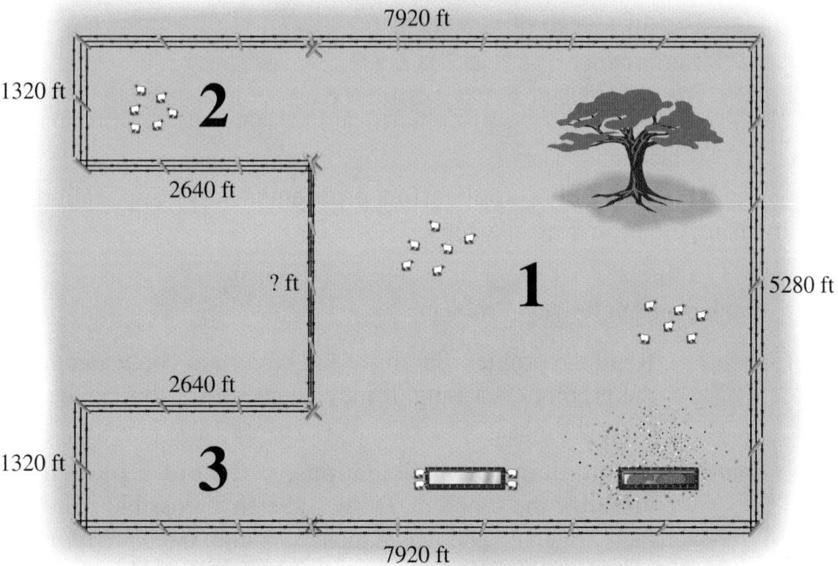

1.10 Solving Application Problems

OBJECTIVES

1. Find indicator words in application problems.
2. Solve application problems.
3. Estimate an answer.

VOCABULARY TIP

Indicator words These key words are useful in translating **word** problems from English into **math,** a necessary skill in becoming a good problem solver.

Most problems involving applications of mathematics are written in sentence form. You need to read the problem carefully to decide how to solve it.

OBJECTIVE 1 **Find indicator words in application problems.** As you read an application problem, look for **indicator words** that help you determine whether to use addition, subtraction, multiplication, or division. Some of these indicator words are shown here.

Addition	Subtraction	Multiplication	Division	Equals
plus	less	product	divided by	is
more	subtract	double	divided into	the same as
more than	subtracted from	triple	quotient	equals
added to	difference	times	goes into	equal to
increased by	less than	of	divide	yields
sum	fewer	twice	divided equally	results in
total	decreased by	twice as much	per	are
sum of	loss of			
increase of	minus			
gain of	take away			

CAUTION

The word *and* does not always indicate addition, so it does not appear as an indicator word in the table above. Notice how the "and" shows the location of different operation signs in the examples below.

The sum of 6 *and* 2 is $6 + 2$.
The difference of 6 *and* 2 is $6 - 2$.
The product of 6 *and* 2 is $6 \cdot 2$.
The quotient of 6 *and* 2 is $6 \div 2$.

OBJECTIVE 2 **Solve application problems.** Solve application problems by using the following six steps.

Solving an Application Problem

Step 1 **Read** the problem carefully and be certain you *understand* what the problem is asking. It may be necessary to read the problem several times.

Step 2 Before doing any calculations, **work out a plan** and try to visualize the problem. Draw a sketch if possible. Know which facts are given and which must be found. Use *indicator words* to help decide on the *plan* (whether you will need to add, subtract, multiply, or divide).

Always **estimate** the final answer.

Step 3 **Estimate** a *reasonable answer* by using rounding.

Step 4 **Solve** the problem by using the facts given and your plan.

Step 5 **State the answer.**

Step 6 **Check** your work. If the answer does not seem reasonable, begin again by reading the problem.

Check your answer to see if it is *reasonable.*

CAUTION

Do **NOT** make the mistake of trying the solve the problem before you know what is being asked.

OBJECTIVE 3 **Estimate an answer.** The six problem-solving steps give a systematic approach for solving word problems. Each of the steps is important, but special emphasis should be placed on Step 3, estimating a *reasonable answer*. Many times an "answer" just does not fit the problem.

What is a reasonable answer? Read the problem and try to determine the approximate size of the answer. Should the answer be part of a dollar, a few dollars, hundreds, thousands, or even millions of dollars? For example, if a problem asks for the cost of a man's shirt, would an answer of \$20 be reasonable? \$2000? \$2? \$200?

CAUTION

Always estimate the answer, then look at your final result to be sure it fits your estimate and is reasonable. This step will give greater success in problem solving.

Work Problem 1 at the Side. ▶

EXAMPLE 1 **Applying Division**

A community group has raised \$8260 for charity. Equal amounts are given to the Food Bank, Children's Center, Boy Scouts of America, and the Women's Shelter. How much did each group receive?

Step 1 **Read.** A reading of the problem shows that the four charities divided \$8260 equally.

Read and **understand** a problem before you begin.

Step 2 **Work out a plan.** The indicator words, ***divided equally,*** show that the amount each received can be found by dividing \$8260 by 4.

Step 3 **Estimate.** Round \$8260 to \$8000. Then $\$8000 \div 4 = \$2000$, so a reasonable answer would be a little greater than \$2000 each.

Step 4 **Solve.** Find the actual answer by dividing \$8260 by 4.

$$4\overline{)8260}\quad 2065$$

Step 5 **State the answer.** Each charity received \$2065.

Step 6 **Check.** The exact answer of \$2065 is reasonable, as \$2065 is close to the estimated answer of \$2000. Is the exact answer of \$2065 correct? Check by multiplying.

\$2065 ← Amount received by each charity
× 4 ← Number of charities
\$8260 ← Total raised; matches number given in problem

Remember: Check your work.

Work Problem 2 at the Side. ▶

1 Pick the most reasonable answer for each problem.

GS **(a)** A grocery clerk's hourly wage: \$1.40; \$14; \$140

____ is eliminated (low)

____ is eliminated (high)

____ is most reasonable

(b) The total length of five sport-utility vehicles: 8 ft; 18 ft; 80 ft; 800 ft

(c) The cost of heart bypass surgery: \$1000; \$100,000; \$10,000,000

2 Solve each problem.

(a) On a recent geology field trip, 84 fossils were collected. If the fossils are divided equally among John, Sean, Jenn, and Kara, how many fossils will each receive?

(b) This week there are 408 children attending a winter sports camp. If 12 children are assigned to each camp counselor, how many counselors are needed?

Answers

1. (a) \$1.40; \$140; \$14 **(b)** 80 ft **(c)** \$100,000

2. (a) 21 fossils **(b)** 34 counselors

3 Solve each problem.

(a) During the semester, Cindy Fong received the following points on examinations and quizzes: 92, 81, 83, 98, 15, 14, 15, and 12. Find her total points for the semester.

(b) Stephanie Dixon works at the telephone order desk of a catalog sales company. One week she had the following number of customer contacts: Monday, 78; Tuesday, 64; Wednesday, 118; Thursday, 102; and Friday, 196. How many customer contacts did she have that week?

Answers

3. **(a)** 410 points **(b)** 558 customer contacts

EXAMPLE 2 Applying Addition

One week, Andrea Abriani, operations manager, decided to total the stroller production at Safe T First Strollers. The daily figures were 7642 strollers on Monday, 8150 strollers on Tuesday, 7916 strollers on Wednesday, 8419 strollers on Thursday, and 7704 strollers on Friday. Find the total number of strollers for the week.

Step 1 **Read.** In this problem, the number of strollers for each day is given and the total strollers for the week must be found.

Step 2 **Work out a plan.** Add the daily stroller figures to arrive at the weekly total.

Step 3 **Estimate.** Because there were about 8000 strollers per day for a week of five days, a reasonable estimate would be $5 \cdot 8000 = 40{,}000$ strollers.

Step 4 **Solve.** Find the exact answer by adding the stroller numbers for the 5 days.

```
  39,831  ← Check by adding up.
   7 642
   8 150
   7 916
   8 419
 + 7 704
  39,831  ← Number of strollers for the week
```

Add from bottom to top to check addition.

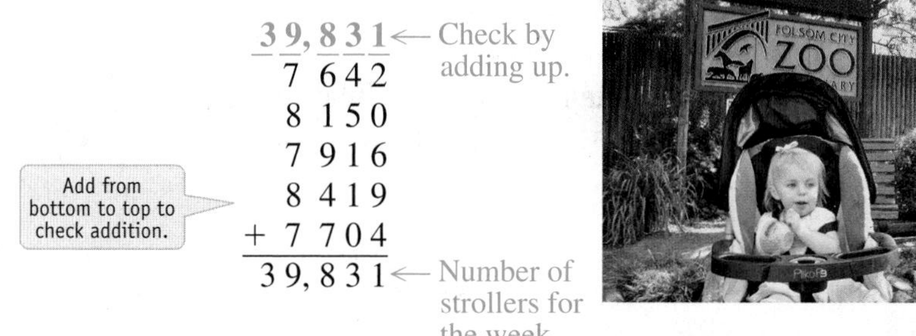

Step 5 **State the answer.** Abriani's total figure for the week was 39,831 strollers.

Step 6 **Check.** The exact answer of 39,831 strollers is close to the estimate of 40,000 strollers, so it is reasonable. Add up the columns to check the exact answer.

Calculator Tip

The calculator solution to **Example 2** uses chain calculations.

7642 (+) 8150 (+) 7916 (+) 8419 (+) 7704 (=) **39,831**

◀ **Work Problem 3 at the Side.**

EXAMPLE 3 Determining Whether Subtraction Is Necessary

The number of miles driven this year is 3028 fewer than the number driven last year. The miles driven last year was 16,735. Find the number of miles driven this year.

Step 1 **Read.** In this problem, the miles driven decreased from last year to this year. The miles driven last year and the decrease in miles driven are given. This year's miles driven must be found.

Step 2 **Work out a plan.** The indicator word, ***fewer,*** shows that subtraction must be used to find the number of miles driven this year.

Step 3 **Estimate.** Because the driving last year was about 17,000 miles, and the decrease in driving is about 3000 miles, a reasonable estimate would be $17{,}000 - 3000 = 14{,}000$ miles.

Continued on Next Page

Step 4 **Solve.** Find the exact answer by subtracting 3028 from 16,735.

$$\begin{array}{r} 16{,}735 \\ -\ \ 3\,028 \\ \hline 13{,}707 \end{array}$$

Step 5 **State the answer.** The driving this year is 13,707 miles.

Step 6 **Check.** The exact answer of 13,707 is reasonable, as it is close to the estimate of 14,000. Check by adding.

13,707 ← miles driven this year
\+ 3 028 ← decrease in miles driven
16,735 ← miles driven last year; matches number given in problem

Remember: Check your work.

Work Problem 4 at the Side. ▶

EXAMPLE 4 Solving a Two-Step Problem

In May, a landlord received $720 from each of eight tenants. After paying $2180 in expenses, how much rent money did the landlord have left?

Step 1 **Read.** The problem asks for the amount of rent remaining after expenses have been paid.

Step 2 **Work out a plan.** The wording *from each of eight tenants* indicates that the eight rents must be totaled. Since the rents are all the same, use multiplication to find the total rent received. Then, subtract expenses.

Step 3 **Estimate.** The amount of rent is about $700, making the total rent received about $700 • 8 = $5600. The expenses are about $2000. A reasonable estimate of the amount remaining is $5600 − $2000 = $3600.

Step 4 **Solve.** Find the exact amount by first multiplying $720 by 8 (the number of tenants).

$$\begin{array}{r} \$\ 720 \\ \times \quad\ 8 \\ \hline \$5760 \end{array}$$ ← Total rent

Then subtract the $2180 in expenses from $5760.

$$\begin{array}{r} \$5760 \\ -\ \$2180 \\ \hline \$3580 \end{array}$$

The exact answer is close to the estimate and reasonable.

Step 5 **State the answer.** The amount remaining is $3580.

Step 6 **Check.** The exact answer of $3580 is reasonable, since it is close to the estimated answer of $3600. Check the answer by adding the expenses to the amount remaining and then dividing by 8.

$3580 + $2180 = $5760

$$8\overline{)5760}\ \text{with quotient } \$720$$

← Matches the rent amount given in the problem

Always check your work.

Work Problem 5 at the Side. ▶

---

4 Solve each problem.

(a) Alaska is our largest state, with an area of 663,267 square miles. The second largest state is Texas, with an area of 268,580 square miles. How much larger is Alaska than Texas? (*Source:* U.S. Census Bureau.)

(b) The Antique Military Vehicle Collectors (AMVC) had $14,863 in their club treasury bank account. After writing a check for $1180 to rent a display hall, find the amount remaining in the club account.

5 Solve each problem.

GS **(a)** An automobile insurance company purchased 318 Samsung digital cameras at a cost of $129 each. After a rebate of $2470, find the final cost of the cameras.

318 • $____ each = $41,022

$______ − $2470 = $______

(b) An Internet book company had sales of 12,628 books with a profit of $6 for each book sold. If 863 books were returned, how much profit remains?

Answers

4. **(a)** 394,687 square miles **(b)** $13,683
5. **(a)** $129; $41,022; $38,552 **(b)** $70,590

1.10 Exercises

FOR EXTRA HELP

Download the MyDashBoard App

MyMathLab®

1. **CONCEPT CHECK** A problem asks for the number of hours a person worked each day. Which would *not* be a reasonable answer?

   (a) 4 hours **(b)** 8 hours **(c)** 25 hours

2. **CONCEPT CHECK** A problem asks for the hourly earnings of a part-time student employee. Which would *not* be a reasonable answer?

   (a) \$9/hour **(b)** \$100/hour **(c)** \$12/hour

3. **CONCEPT CHECK** A problem asks for the cost of lunch at a fast food restaurant. Which answer is reasonable?

   (a) \$5 **(b)** \$50 **(c)** \$500

4. **CONCEPT CHECK** You calculate the gas mileage for your car. Which answer is reasonable? (mpg = miles per gallon)

   (a) 5 mpg **(b)** 25 mpg **(c)** 125 mpg

Solve each application problem. First use front end rounding to estimate the answer. Then find the exact answer. ***See Examples 1–4.***

5. Last week, SUBWAY sold 602 Veggie Delite sandwiches, 935 ham sandwiches, 1328 turkey breast sandwiches, 757 roast beef sandwiches, and 1586 SUBWAY club sandwiches. Find the total number of sandwiches sold.

   Estimate: 600 + 900 + _____ + 800 + _____ = __________

   Exact:

6. During a recent week, Radio Flyer, Inc. manufactured 32,815 Model #18 wagons, 4875 steel miniwagons, 1975 wood 40-inch wagons, 15,308 scooters, and 9815 new-design plastic wagons. Find the total number of items manufactured.

   Estimate: _____ + 5000 + 2000 + 20,000 + _____ = __________

   Exact:

7. Paying ahead for a rental car can save 35% at Budget.com. They pay-at-the-counter base rate for a seven-day, full-size car rental is \$296. The rate for the same car when paying ahead using the Pay Now rate at Budget.com is \$192. How much is saved? (*Source:* Budget Car Rental.)

   Estimate: \$300 − \$200 = __________

   Exact:

8. The U.S. population will rise from 311 million today to 478 million by 2100. Find the expected increase in population. (*Source:* United Nations population division.)

   Estimate:

   Exact:

9. A packing machine can package 236 first-aid kits each hour. At this rate, find the number of first-aid kits packaged in 24 hours.

   Estimate:

   Exact:

10. If 450 admission tickets to a classic car show are sold each day, how many tickets are sold in a 12-day period?

    Estimate: 500 × 10

    Exact:

11. Clarence Hanks, coordinator of Toys for Tots, has collected 2628 toys. If his group can give the same number of toys to each of 657 children, how many toys will each child receive?

Estimate:

Exact:

12. If profits of \$680,000 are divided evenly among a firm's 1000 employees, how much money will each employee receive?

Estimate:

Exact:

13. Turn down the thermostat in the winter and you can save money and energy. In the upper Midwest, setting back the thermostat from 68° to 55° at night can save \$34 per month on fuel. Find the amount of money saved in five months.

Estimate:

Exact:

14. The cost of tuition and fees at a community college is \$785 per quarter. If Gale Klein has five quarters remaining, find the total amount that she will need for tuition and fees.

Estimate:

Exact:

The sesquicentennial anniversary (150 years) of the beginning of the Civil War occurred in 2011. The loss of life at the conclusion of the war still stands as the highest war casualty count in American history. The table below shows the number of deaths and cause of death for both the Southern and the Northern American States. Use these data to answer Exercises 15–20.

Union (Northern) Deaths	Confederate (Southern) Deaths
Battle 110,070	Battle 94,120
Disease 250,152	Disease 164,300

Source: *The Civil War, Strange and Fascinating Facts* by Burke Davis.

15. Find the total number of Union deaths in the Civil War.

Estimate: 100,000 + 300,000 = ____________

Exact:

16. Find the total number of Confederate deaths in the Civil War.

Estimate:

Exact:

17. How many more Union deaths resulted from disease than battle?

Estimate:

Exact:

18. How many more Confederate deaths were caused by disease than battle?

Estimate:

Exact:

19. Find the total loss of life (North and South) in the Civil War.

Estimate:

Exact:

20. How many more Union deaths than Confederate deaths were there in the Civil War?

Estimate:

Exact:

21. Ronda Biondi decides to establish a monthly budget. She will spend \$695 for rent, \$340 for food, \$435 for child care, \$240 for transportation, \$180 for other expenses, and she will put the remainder in savings. If her monthly take-home pay is \$2240, find her monthly savings.

Estimate:

Exact:

22. Robert Heisner had \$2874 in his bank account. He paid \$308 for auto repairs, \$580 for a dishwasher, and \$778 for an insurance payment. Find the amount remaining in his account.

Estimate:

Exact:

23. There are 43,560 square feet in one acre. How many square feet are there in 138 acres?

Estimate:

Exact:

24. The number of gallons of water polluted each day in an industrial area is 209,670. How many gallons of water are polluted each year? (Use a 365-day year.)

Estimate:

Exact:

The Internet was used to find the following minivan optional features and the price of each feature. Use this information to answer Exercises 25–28.

Safety and Exterior Options		Interior Options	
Option	**Cost**	**Option**	**Cost**
VIP Plus Security System	$299	Carpet floor mats	$321
Roof rack crossbars	$185	Cargo nets	$51
Mudguards	$99	Interface kit for iPod	$299
Alloy wheel locks	$67	Dual screen entertainment system	$1799
Paint protection	$395	Wireless headphones	$82
Lower body moulding	$209	XM Satellite Radio	$449

Source: www.edmunds.com

25. Find the total cost of all Safety and Exterior Options listed.

Estimate:

Exact:

26. Find the total cost of all Interior Options listed.

Estimate:

Exact:

27. A new-car dealer offers an option value package that includes VIP Plus Security System, alloy wheel locks, paint protection, and lower body moulding for $785. If Jill buys the value package instead of paying for each option separately, how much will she save?

Estimate:

Exact:

28. A new-car dealer offers an option package that includes VIP Plus Security System, roof rack crossbars, paint protection, carpet floor mats, cargo nets, and XM Satellite Radio for a total of $1495. How much will Samuel save if he buys the option package instead of paying for each option separately?

Estimate:

Exact:

29. The Enabling Supply House purchased 6 wheelchairs at $1256 each and 15 speech compression recorder-players at $895 each. Find the total cost.

Estimate:

Exact:

30. A college bookstore buys 17 laptop computers at $506 each and 13 printers at $482 each. Find the total cost.

Estimate:

Exact:

31. Being able to identify indicator words is helpful in determining how to solve an application problem. Write three indicator words for each of these operations: add, subtract, multiply, and divide. Write two indicator words that mean equals.

32. Identify and explain the six steps used to solve an application problem. You may refer to the text if you need help, but use your own words.

Solve each application problem. **See Examples 1–4.**

33. Steve Edwards, manager, decided to total his sales at SUBWAY. The daily sales figures were $2358 on Monday, $3056 on Tuesday, $2515 on Wednesday, $1875 on Thursday, $3978 on Friday, $3219 on Saturday, and $3008 on Sunday. Find his total sales for the week.

34. The numbers of visitors at a war veterans' memorial during one week are 5318; 2865; 4786; 1998; 3899; 2343; and 7221. Find the total attendance for the week.

35. A car weighs 2425 pounds. If its 582-pound engine is removed and replaced with a 634-pound engine, what will the car weigh?

36. Estelle Alan has $2324 in her preschool operating account. She spends $734 from this account, and then the class parents raise $568 in a rummage sale. Find the balance in the account after she deposits the money from the rummage sale.

37. In a recent survey of Reno/Lake Tahoe hotels, the cost per night at Harrah's in Reno was $45, while the cost at Harrah's in Lake Tahoe was $99 per night. Find the amount saved on a 7-night stay at Harrah's in Reno instead of staying at Harrah's in Lake Tahoe. (*Source:* Harrah's Casinos and Hotels.)

38. The most expensive hotel room in a recent study was the Ritz-Carlton at $645 per night, while the least expensive was Motel 6 at $74 per night. Find the amount saved in a 4-night stay at Motel 6 instead of staying at the Ritz-Carlton. (*Source:* Ritz-Carlton/Motel 6.)

39. A youth soccer association raised $7588 through fund-raising projects. After expenses of $838 were paid, the balance of the money was divided evenly among the 18 teams. How much did each team receive?

40. Feather Farms Egg Ranch collected 3545 eggs in the morning and 2575 eggs in the afternoon. If the eggs are packed in flats containing 30 eggs each, find the number of flats needed for packing.

41. A theater owner wants to provide enough seating for 1250 people. The main floor has 30 rows of 25 seats in each row. If the balcony has 25 rows, how many seats must be in each balcony row to satisfy the owner's seating requirements?

42. Jennie makes 24 grapevine wreaths per week to sell to gift shops. She works 40 weeks a year and packages six wreaths per box. If she ships equal quantities to each of five shops, find the number of boxes each store will receive.

Chapter 1 Summary

Key Terms

1.1

whole numbers The whole numbers are 0, 1, 2, 3, 4, 5, 6, 7, 8, and so on.

place value The place value of each digit in a whole number is determined by its position in the whole number.

table A table is a display of facts in rows and columns.

1.2

addition The process of finding the total is addition.

addends The numbers being added in an addition problem are addends.

sum (total) The answer in an addition problem is called the sum.

commutative property of addition The commutative property of addition states that the order of numbers in an addition problem can be changed without changing the sum.

associative property of addition The associative property of addition states that grouping the addition of numbers differently does not change the sum.

regrouping The process of regrouping is used in an addition problem when the sum of the digits in a column is greater than 9.

perimeter The perimeter is the distance around the outside edges of a flat figure.

1.3

minuend The number from which another number is being subtracted is the minuend.

subtrahend The subtrahend is the number being subtracted in a subtraction problem.

difference The answer in a subtraction problem is called the difference.

regrouping The process of regrouping is used in subtraction if a digit is less than the one directly below it.

1.4

factors The numbers being multiplied are called factors. For example, in $3 \times 4 = 12$, both 3 and 4 are factors.

product The answer in a multiplication problem is called the product.

commutative property of multiplication The commutative property of multiplication states that changing the order of the factors in a multiplication problem does not change the product.

associative property of multiplication The associative property of multiplication states that grouping the factors differently does not change the product.

chain multiplication problem A multiplication problem having more than two factors is a chain multiplication problem.

multiple The product of two whole number factors is a multiple of those numbers.

multiple of 10 A whole number that ends in 0, such as 10, 20, or 30; 100, 200, or 300.

partial products The products found when multiplying numbers having two or more digits.

1.5

dividend The number being divided by another number in a division problem is the dividend.

divisor The divisor is the number by which you are dividing in a division problem.

quotient The answer in a division problem is called the quotient.

undefined Dividing any number by 0 is *undefined*. Dividing by zero cannot be done.

short division A method of dividing a number by a one-digit divisor is short division.

remainder The remainder is the number left over when two numbers do not divide evenly.

1.6

long division The process of long division is used to divide by a number with more than one digit.

trial divisors A method of determining how many times the true divisor goes into the dividend by estimating.

trial quotient The quotient that results from dividing the dividend by the trial divisor.

1.7

rounding Rounding is used to find a number that is close to the original number, but easier to work with. Use the $\approx$ sign, which means "approximately equal to."

estimate An estimated answer is one that is close to the exact answer.

front end rounding Rounding to the highest possible place so that all the digits become zeros except the first one is front end rounding.

exponent The exponent is the small raised number (2) in the expression 3^2.

base The base is the number 3 in the expression 3^2.

1.8

square root The square root of a whole number is the number that can be multiplied by itself to produce the given number.

perfect square A number that is the square of a whole number is a perfect square.

order of operations For problems or expressions with more than one operation, the order of operations tells what to do first, second, and so on to get the correct answer.

1.9

pictograph A graph that uses pictures or symbols to show data is a pictograph.

bar graph A graph that uses bars of various heights to show quantity is a bar graph.

line graph A graph that uses dots connected by lines to show trends is a line graph.

1.10

indicator words Words in a problem that indicate the necessary operations—addition, subtraction, multiplication, or division—are indicator words.

New Symbols

$\approx$ This sign is used to show that an answer has been estimated. It means "is approximately equal to."

$\sqrt{}$ The symbol for square root.

5^2 The small raised 2 is an exponent; it tells how many times to use 5 (the base) as a factor in multiplication.

Test Your Word Power

See how well you have learned the vocabulary in this chapter.

1. When using **addends** you are performing
   A. division
   B. subtraction
   C. addition.

2. The subtrahend is the
   A. number being multiplied
   B. number being subtracted
   C. number being added.

3. A **factor** is
   A. one of two or more numbers being added
   B. one of two or more numbers being multiplied
   C. one of two or more numbers being divided.

4. The **divisor** is
   A. the number being multiplied
   B. always the largest number
   C. the number doing the dividing.

5. We use **rounding** to
   A. avoid solving a problem
   B. help estimate a reasonable answer
   C. find the remainder.

6. A **perfect square** is
   A. the square of a whole number
   B. the same as a square root
   C. similar to a perfect triangle.

Answers to Test Your Word Power

1. C; *Example:* In $2 + 3 = 5$, the 2 and the 3 are addends.

2. B; *Example:* In $5 - 4 = 1$, the 4 is the subtrahend.

3. B; *Example:* In $3 \times 5 = 15$, the numbers 3 and 5 are factors.

4. C; *Example:* In $8 \div 4 = 2$, $\frac{8}{4} = 2$, and $4\overline{)8}$ with quotient 2, the 4 is the divisor.

5. B; *Example:* We can use rounding to estimate our answer and then determine whether the exact answer is reasonable.

6. A; *Example:* 25 is a perfect square because $5^2 = 25$ and 5 is a whole number.

Quick Review

Concepts	*Examples*
1.1 Reading and Writing Whole Numbers Do not use the word *and* when writing a whole number. Commas help divide the periods or groups for ones, thousands, millions, and billions. A comma is not needed when a number has four digits or fewer.	795 is written *seven hundred ninety-five.* 9,768,002 is written *nine million, seven hundred sixty-eight thousand, two*

Concepts	Examples
1.2 Adding Whole Numbers Add from top to bottom, starting with the ones column and working left. To check, add from bottom to top.	(Add up to check.) 1140 687 26 9 + 418 } Addends 1140 Sum
1.2 Commutative Property of Addition Changing the order of the addends in an addition problem does not change the sum.	$2 + 4 = 6$ $4 + 2 = 6$ By the commutative property, the sum is the same.
1.2 Associative Property of Addition Grouping the addends differently when adding does not change the sum.	$(2 + 3) + 4 = 5 + 4 = 9$ $2 + (3 + 4) = 2 + 7 = 9$ By the associative property, the sum is the same.
1.3 Subtracting Whole Numbers Subtract the subtrahend from the minuend to get the difference, using regrouping when necessary. To check, add the difference to the subtrahend to get the minuend.	*Problem* 6 12 18 4 7 3 8 ← Minuend − 6 4 9 Subtrahend 4 0 8 9 Difference *Check* 4089 + 649 4738
1.4 Multiplying Whole Numbers Use ×, • (a raised dot), or parentheses to indicate multiplication. The numbers being multiplied are called *factors*. The multiplicand is being multiplied by the multiplier, giving the product. When the multiplier has more than one digit, partial products must be used and added to find the product.	3×4 or $3 \cdot 4$ or $(3)(4)$ or $3(4)$ 78 Multiplicand } Factors × 24 Multiplier 312 Partial product 156 Partial product (move one position left) 1872 Product
1.4 Commutative Property of Multiplication The product in a multiplication problem remains the same when the order of the factors is changed.	$3 \times 4 = 12$ $4 \times 3 = 12$ By the commutative property, the product is the same.
1.4 Associative Property of Multiplication Grouping the factors differently when multiplying does not change the product.	$(2 \times 3) \times 4 = 6 \times 4 = 24$ $2 \times (3 \times 4) = 2 \times 12 = 24$ By the associative property, the product is the same.
1.5 Dividing Whole Numbers ÷ and $\overline{)}$ mean divide. Also a —, as in $\frac{25}{5}$, means to divide the top number (dividend) by the bottom number (divisor).	22 ← Quotient Divisor → $4\overline{)88}$ ← Dividend 88 0 Dividend $\frac{88}{4} = 22$ ← Quotient $88 \div 4 = 22$ Dividend, Divisor, Quotient

Concepts	Examples
1.7 Rounding Whole Numbers **Rules for Rounding** *Step 1* Locate the place to be rounded, and draw a line under it. *Step 2* If the next digit to the right is 5 or more, increase the underlined digit by 1. If the next digit is 4 or less, do not change the underlined digit. *Step 3* Change all digits to the right of the underlined place to zeros.	Round 726 to the nearest ten. Next digit is 5 or more. 726 Tens place increases from 2 to 3. 726 rounds to 730. Round 1,498,586 to the nearest million. Next digit is 4 or less. 1,498,586 Millions place does not change. 1,498,586 rounds to 1,000,000.
1.7 Front End Rounding Front end rounding is rounding to the highest possible place so that all the digits become 0 except the first digit.	Round each number using front end rounding. 76 rounds to 80. 348 rounds to 300. 6512 rounds to 7000. 23,751 rounds to 20,000. 652,179 rounds to 700,000.
1.8 Order of Operations Problems may have several operations. Work these problems using the order of operations. **1.** Do all operations inside parentheses or other grouping symbols. **2.** Simplify any expressions with exponents and find any square roots $(\sqrt{\ })$. **3.** Multiply or divide proceeding from left to right. **4.** Add or subtract proceeding from left to right.	Simplify, using the order of operations. $7 \cdot \sqrt{9} - 4 \cdot 5$ Find the square root. $7 \cdot 3 - 4 \cdot 5$ Multiply from left to right. $21 - 20 = 1$ Subtract.
1.9 Reading Pictographs, Bar Graphs, and Line Graphs A *pictograph* uses pictures or symbols to show data.	When reading a pictograph, be certain that you determine the quantity represented by each picture or symbol.
A *bar graph* uses bars of various heights to show quantity.	When reading a bar graph, use a straightedge to line up the top of the bar with the numbers along the left edge of the graph.
A *line graph* uses dots connected by lines to show trends.	When reading a line graph, use a straightedge to line up the dot with the numbers along the left edge of the graph.

Concepts	Examples
1.10 Application Problems	
Steps for Solving an Application Problem	Manuel earns \$118 on Sunday, \$87 on Monday, and \$63 on Tuesday. Find his total earnings for the 3 days.
Step 1 **Read** the problem carefully, perhaps several times.	*Step 1* The earnings for each day are given, and the total for the 3 days must be found.
Step 2 **Work out a plan** before starting. Draw a sketch if possible.	*Step 2* Add the daily earnings to find the total.
Step 3 **Estimate** a reasonable answer.	*Step 3* Since the earnings were about \$100 + \$90 + \$60 = \$250, a reasonable estimate would be approximately \$250.
Step 4 **Solve** the problem.	*Step 4* $\begin{array}{r} \underline{\$268} \\ \$118 \\ 87 \\ \underline{+\ 63} \\ \$268 \end{array}$ Check by adding up; Total earnings
Step 5 **State the answer.**	*Step 5* Manuel's total earnings are \$268.
Step 6 **Check** your work. If the answer is not reasonable, start over.	*Step 6* The exact answer is reasonable, because it is close to the estimate of \$250.

Chapter 1 Review Exercises

If you need help with any of these Review Exercises, look in the section indicated inside the dark blue rectangles.

1.1 *Write the digits for the given period or group in each number.*

1. 6573
thousands
ones

2. 36,215
thousands
ones

3. 105,724
thousands
ones

4. 1,768,710,618
billions
millions
thousands
ones

Rewrite each number in words.

5. 728

6. 15,310

7. 319,215

8. 62,500,005

Rewrite each number in digits.

9. ten thousand, eight

10. two hundred million, four hundred fifty-five

1.2 *Add.*

11. $\begin{array}{r} 72 \\ +\ 38 \\ \hline \end{array}$

12. $\begin{array}{r} 54 \\ +\ 67 \\ \hline \end{array}$

13. $\begin{array}{r} 807 \\ 4606 \\ +\ \ 51 \\ \hline \end{array}$

14. $\begin{array}{r} 8215 \\ 9 \\ +\ 7433 \\ \hline \end{array}$

15. $\begin{array}{r} 2130 \\ 453 \\ 8107 \\ +\ \ 296 \\ \hline \end{array}$

16. $\begin{array}{r} 5684 \\ 218 \\ 2960 \\ +\ \ 983 \\ \hline \end{array}$

17. $\begin{array}{r} 5\,732 \\ 11{,}069 \\ 37 \\ 1\,595 \\ +\ 22{,}169 \\ \hline \end{array}$

18. $\begin{array}{r} 3\,451 \\ 12{,}286 \\ 43 \\ 1\,291 \\ +\ 32{,}784 \\ \hline \end{array}$

1.3 *Subtract.*

19. $\begin{array}{r} 64 \\ -\ 28 \\ \hline \end{array}$

20. $\begin{array}{r} 46 \\ -\ 19 \\ \hline \end{array}$

21. $\begin{array}{r} 375 \\ -\ 186 \\ \hline \end{array}$

22. $\begin{array}{r} 573 \\ -\ 389 \\ \hline \end{array}$

23. $\begin{array}{r} 7416 \\ -\ 567 \\ \hline \end{array}$

24. $\begin{array}{r} 5210 \\ -\ 883 \\ \hline \end{array}$

25. $\begin{array}{r} 2210 \\ -\ 1986 \\ \hline \end{array}$

26. $\begin{array}{r} 99{,}704 \\ -\ 73{,}838 \\ \hline \end{array}$

1.4 *Multiply.*

27. $\begin{array}{r} 7 \\ \times 7 \\ \hline \end{array}$

28. $\begin{array}{r} 8 \\ \times 0 \\ \hline \end{array}$

29. 8(4)

30. 8(8)

31. (5)(9)

32. (6)(7)

33. 7 • 8

34. 9 • 9

Work each chain multiplication.

35. $5 \times 4 \times 2$

36. $9 \times 1 \times 5$

37. $4 \times 4 \times 3$

38. $2 \times 2 \times 2$

39. (6)(0)(8)

40. (7)(1)(6)

41. 6 • 1 • 8

42. 7 • 7 • 0

Multiply.

43. $\begin{array}{r} 28 \\ \times\ 3 \\ \hline \end{array}$

44. $\begin{array}{r} 46 \\ \times\ 8 \\ \hline \end{array}$

45. $\begin{array}{r} 58 \\ \times\ 9 \\ \hline \end{array}$

46. $\begin{array}{r} 98 \\ \times\ 1 \\ \hline \end{array}$

47. $\begin{array}{r} 625 \\ \times\ \ 8 \\ \hline \end{array}$

48. $\begin{array}{r} 374 \\ \times\ \ 8 \\ \hline \end{array}$

49. $\begin{array}{r} 1349 \\ \times\ \ \ 4 \\ \hline \end{array}$

50. $\begin{array}{r} 9163 \\ \times\ \ \ 5 \\ \hline \end{array}$

51. $\begin{array}{r} 7456 \\ \times\ \ \ 2 \\ \hline \end{array}$

52. $\begin{array}{r} 2880 \\ \times\ \ \ 7 \\ \hline \end{array}$

53. $\begin{array}{r} 93{,}105 \\ \times\ \ \ \ 5 \\ \hline \end{array}$

54. $\begin{array}{r} 21{,}873 \\ \times\ \ \ \ 8 \\ \hline \end{array}$

55. $\begin{array}{r} 35 \\ \times\ 25 \\ \hline \end{array}$

56. $\begin{array}{r} 74 \\ \times\ 32 \\ \hline \end{array}$

57. $\begin{array}{r} 98 \\ \times\ 12 \\ \hline \end{array}$

58. $\begin{array}{r} 68 \\ \times\ 75 \\ \hline \end{array}$

59. $\begin{array}{r} 472 \\ \times\ 33 \\ \hline \end{array}$

60. $\begin{array}{r} 392 \\ \times\ 77 \\ \hline \end{array}$

61. $\begin{array}{r} 4051 \\ \times\ 219 \\ \hline \end{array}$

62. $\begin{array}{r} 1527 \\ \times\ 328 \\ \hline \end{array}$

Find each total cost.

63. 30 scientific calculators at $12 per calculator

64. 76 subscribers at $14 per subscription

65. 318 drill bit sets at $64 per set

66. 114 earplugs at $6 per earplug

Multiply by using the shortcut for multiples of 10.

67. $\begin{array}{r} 280 \\ \times\ 50 \\ \hline \end{array}$

68. $\begin{array}{r} 340 \\ \times\ 70 \\ \hline \end{array}$

69. $\begin{array}{r} 517 \\ \times\ 400 \\ \hline \end{array}$

70. $\begin{array}{r} 637 \\ \times\ 500 \\ \hline \end{array}$

71. $\begin{array}{r} 16{,}000 \\ \times\ 8\ 000 \\ \hline \end{array}$

72. $\begin{array}{r} 43{,}000 \\ \times\ 2\ 100 \\ \hline \end{array}$

1.5 *Divide. If the division is not possible, write "undefined."*

73. $20 \div 4$

74. $35 \div 5$

75. $42 \div 7$

76. $18 \div 9$

77. $\frac{54}{9}$

78. $\frac{36}{9}$

79. $\frac{49}{7}$

80. $\frac{0}{6}$

81. $\frac{148}{0}$

82. $\frac{0}{23}$

83. $\frac{64}{8}$

84. $\frac{81}{9}$

1.5-1.6 *Divide.*

85. $4\overline{)328}$

86. $3\overline{)294}$

87. $6\overline{)26{,}532}$

88. $76\overline{)26{,}752}$

89. $2704 \div 18$

90. $15{,}525 \div 125$

1.7 *Round as indicated.*

91. 817 to the nearest ten

92. 15,208 to the nearest hundred

93. 20,643 to the nearest thousand

94. 67,485 to the nearest ten-thousand

Round each number to the nearest ten, nearest hundred, and nearest thousand. Remember to round from the original number.

	Ten	Hundred	Thousand
95. 3487	______	______	______
96. 20,065	______	______	______
97. 98,201	______	______	______
98. 352,118	______	______	______

1.8 *Find each square root by using the Perfect Squares Table from this chapter.*

99. $\sqrt{16}$

100. $\sqrt{49}$

101. $\sqrt{144}$

102. $\sqrt{196}$

Identify the exponent and the base, and then simplify each expression.

103. 7^3

104. 3^6

105. 5^3

106. 4^5

Simplify each expression by using the order of operations.

107. $7^2 - 15$

108. $6^2 - 10$

109. $2 \cdot 3^2 \div 2$

110. $9 \div 1 \cdot 2 \cdot 2 \div (11 - 2)$

111. $\sqrt{9} + 2(3)$

112. $6 \cdot \sqrt{16} - 6 \cdot \sqrt{9}$

1.9 *The bar graph shows the number of parents out of 100 surveyed who nag their children about performing certain household chores.*

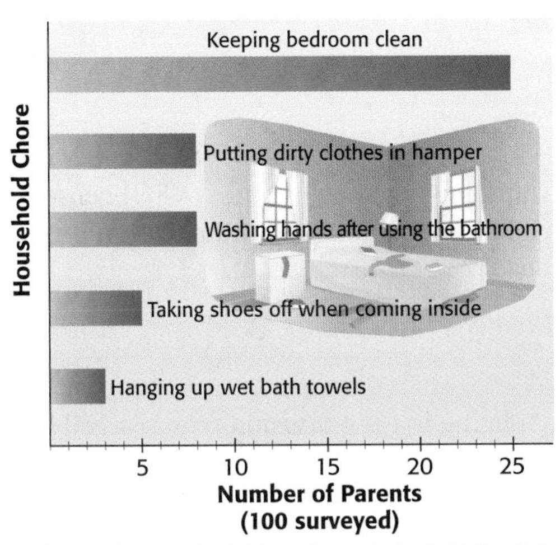

Source: Opinion Research Corporation for the Soap and Detergent Association.

113. How many parents nagged their children about washing hands after using the bathroom?

114. Find the number of parents who nagged their children about taking shoes off when coming inside.

115. Which household chore was nagged about by the greatest number of parents? How many were there?

116. Which household chore was nagged about by the least number of parents? How many did this?

1.10 *Solve each application problem. First use front end rounding to estimate the answer. Then find the exact answer.*

117. Bank of America processes 40 million checks each day. Find the number of checks processed by the bank in a year. Use a 365-day year. (*Source:* Bank of America.)

Estimate:

Exact:

118. A pulley on an evaporative cooler turns 1400 revolutions per minute. How many revolutions will the pulley turn in 60 minutes?

Estimate:

Exact:

119. The two most populated states are California, with 37,341,989 people, and Texas, with 25,268,418. Find the difference in population. (*Source:* U.S. Census Bureau.)

Estimate:

Exact:

120. The two least populated states are Alaska, with a population of 721,523, and Wyoming, with a population of 568,300. How many more people does Alaska have than Wyoming? (*Source:* U.S. Census Bureau.)

Estimate:

Exact:

121. The mechanic tells Kara Jantzi that the transmission on her Ford Escape has blown up. The cost of a new transaxle is \$2633, labor is 8 hours at \$90 per hour, and the sales tax is \$230. Find the total cost to replace her transmission. (*Source:* Undisclosed auto repair shop.)

Estimate:

Exact:

122. Moving to new home, Scott Samon rented a U-haul truck for \$55 plus \$2 per mile. After completing the move, the odometer shows that he has driven the truck 89 miles. Find the cost of the truck rental.

Estimate:

Exact:

123. Find the total cost if SUBWAY buys 32 baking ovens at \$1538 each and 28 warming ovens at \$887 each.

Estimate:

Exact:

124. A newspaper carrier has 62 customers who take the paper daily and 21 customers who take the paper on weekends only. A daily customer pays \$16 per month and a weekend-only customer pays \$7 per month. Find the total monthly collections.

Estimate:

Exact:

125. This holiday season, the average amount consumers plan to spend on holiday shopping for others is \$620. If they plan to spend \$107 on holiday shopping for themselves, how much more do they plan to spend on others than themselves? (*Source:* National Retail Federation.)

Estimate:

Exact:

126. Rachel Leach pays \$520 for rent and \$385 for her car payment. If she started with \$1924 in her bank account, how much remains in her account?

Estimate:

Exact:

127. A food canner uses 1 pound of pork for every 175 cans of pork and beans. How many pounds of pork are needed for 8750 cans?

Estimate:

Exact:

128. A stamping machine produces 986 license plates each hour. How long will it take to produce 32,538 license plates?

Estimate:

Exact:

129. Nitrogen sulfate is used in farming to enrich nitrogen-poor soil. If 625 pounds of nitrogen sulfate are spread per acre, how many acres can be spread with 32,500 pounds of nitrogen sulfate?

Estimate:

Exact:

130. Each home in a subdivision requires 180 feet of fencing. Find the number of homes that can be fenced with 5760 feet of fencing material.

Estimate:

Exact:

Mixed Review Exercises*

Perform the indicated operations.

131. 4(83)

132. 7(64)

133. $\begin{array}{r} 309 \\ -\ 56 \\ \hline \end{array}$

134. $\begin{array}{r} 835 \\ -\ 247 \\ \hline \end{array}$

135. $\begin{array}{r} 662 \\ +\ 379 \\ \hline \end{array}$

136. $\begin{array}{r} 789 \\ +\ 872 \\ \hline \end{array}$

137. $\begin{array}{r} 38{,}140 \\ -\ 6\,078 \\ \hline \end{array}$

138. $\begin{array}{r} 29{,}156 \\ -\ 4\,209 \\ \hline \end{array}$

139. $21 \div 7$

140. $\frac{42}{6}$

141. $\begin{array}{r} 7\,218 \\ 3 \\ 18 \\ 1\,791 \\ 82{,}623 \\ +\ 1\,982 \\ \hline \end{array}$

142. $\begin{array}{r} 3\,812 \\ 5 \\ 22 \\ 1\,836 \\ 75{,}134 \\ +\ 2\,369 \\ \hline \end{array}$

143. $\frac{9}{0}$

144. $\frac{7}{1}$

145. $27{,}600 \div 4$

*The order of exercises in this final group does not correspond to the order in which topics occur in the chapter. This random ordering should help you prepare for the chapter test in yet another way.

146. $18{,}480 \div 8$

147. $\begin{array}{r} 8430 \\ \times\ 128 \\ \hline \end{array}$

148. $\begin{array}{r} 21{,}702 \\ \times\quad 6 \\ \hline \end{array}$

149. $34\overline{)3672}$

150. $68\overline{)14{,}076}$

151. Rewrite 376,853 in words.

152. Rewrite 408,610 in words.

153. Round 8749 to the nearest hundred.

154. Round 400,503 to the nearest thousand.

Find each square root.

155. $\sqrt{64}$

156. $\sqrt{81}$

Find each total cost.

157. 308 pairs of knee guards at $18 per pair

158. 84 dishwashers at $370 per dishwasher

159. 208 baseball hats at $11 per hat

160. 607 boxes of avocados at $26 per box

Solve each application problem.

161. There are 52 playing cards in a deck. How many cards are there in nine decks?

162. A group of neighbors in Salem, Oregon founded an organization known as Salem Harvest. Picking fruits and vegetables to be distributed to those in need, 1700 volunteers picked 31 pounds of produce each. How many pounds did the volunteers pick? (*Source: AARP Magazine.*)

163. Push-type gasoline-powered lawn mowers cost $100 less than self-propelled mowers that you walk behind. If a self-propelled mower costs $380, find the cost of a push-type mower.

164. The Country Day School wants to raise $218,450 to construct and equip a computer lab. If $103,815 has already been raised, how much more is needed?

American River Raft Rentals lists the following daily raft rental fees. Notice that there is an additional $2 launch fee payable to the park system for each raft rented. Use this information to solve Exercises 165 and 166.

AMERICAN RIVER RAFT RENTALS

Size	Rental Fee	Launch Fee
4-person	$28	$2
6-person	$38	$2
10-person	$70	$2
12-person	$75	$2
16-person	$85	$2

Source: American River Raft Rentals.

165. On a recent Tuesday the following rafts were rented: 6 4-person; 15 6-person; 10 10-person; 3 12-person; and 2 16-person. Find the total receipts, including the $2 per-raft launch fee.

166. On the 4th of July the following rafts were rented: 38 4-person; 73 6-person; 58 10-person; 34 12-person; and 18 16-person. Find the total receipts, including the $2 per-raft launch fee.

1 World Trade Center will be the tallest building in the United States and one of the world's giants. The world's tallest building is the Burj Dubai in the United Arab Emirates. The pictogram below shows the height, in feet, of some of the tallest buildings in the world. Use the information to solve Exercises 167–170.

167. How much taller is the Burj Dubai than the Empire State Building?

168. How much taller will 1 WTC in New York be than the Petronas Tower in Malaysia?

169. **(a)** What is the combined height of all the buildings shown?

(b) Is the combined height of these six buildings greater or less than two miles? How much greater or less than two miles? (*Hint:* 1 mile = 5280 feet.)

170. The height of the Willis Tower in Chicago is equivalent to the length of how many football fields? (*Hint:* A football field is 100 yards long and 1 yard = 3 feet.)

The World of Giants
(Heights given in feet)

Source: Council on Tall Buildings and Urban Habitat; Silverstein Properties; The Port Authority of New York and New Jersey Associated Press.

Chapter 1 Test

The Chapter Test Prep Videos with test solutions are available on DVD, in MyMathLab, and on YouTube—search "LialBasicCollegeMath" and click on "Channels."

Write each number in words.

1. 9205

2. 25,065

3. Use digits to write four hundred twenty-six thousand, five.

Add.

4.
$$\begin{array}{r} 853 \\ 66 \\ 4022 \\ +\ 3589 \\ \hline \end{array}$$

5.
$$\begin{array}{r} 17{,}063 \\ 7 \\ 12 \\ 1\,505 \\ 93{,}710 \\ +\quad 333 \\ \hline \end{array}$$

Subtract.

6.
$$\begin{array}{r} 9009 \\ -\ 7964 \\ \hline \end{array}$$

7.
$$\begin{array}{r} 9075 \\ -\ 2869 \\ \hline \end{array}$$

Multiply

8. $7 \times 6 \times 4$

9. $57 \cdot 3000$

10. 85(19)

11.
$$\begin{array}{r} 7381 \\ \times\ 603 \\ \hline \end{array}$$

Divide. If the division is not possible, write "undefined."

12. $16\overline{)112{,}752}$

13. $\dfrac{835}{0}$

14. $19{,}241 \div 42$

15. $280\overline{)44{,}800}$

Round as indicated.

16. 6347 to the nearest ten

17. 76,489 to the nearest thousand

Simplify each expression.

18. $5^2 + 8(2)$

19. $7 \cdot \sqrt{64} - 14 \cdot 2$

Solve each application problem. First use front end rounding to estimate the answer. Then find the exact answer.

20. Judy Martinez collects the following monthly rents from the tenants in her fourplex: \$485, \$500, \$515, and \$425. After she pays expenses of \$785, how much does she have left?

Estimate:

Exact:

21. The major producer of ethanol made from corn is the United States. If 374 gallons of ethanol can be produced from the corn grown on one acre of land, how many acres are needed to produce 86,394 gallons. (*Source:* Earth Policy Institute.)

Estimate:

Exact:

22. Sadie Simms paid \$528 for tires, \$195 for brakes, and \$235 for a timing belt. If this money was withdrawn from her bank account, which had a beginning balance of \$1906, find her new balance.

Estimate:

Exact:

23. The technicians at a chicken ranch identify the sex of 48 baby chicks each minute for 4 hours in the morning and 36 baby chicks each minute for 3 hours in the afternoon. Find the total number of baby chicks identified. (*Source:* Discovery Channel, *Dirty Jobs.*)

Estimate:

Exact:

24. Explain in your own words the rules for rounding numbers. Give an example of rounding a number to the nearest ten-thousand.

25. List the six steps for solving application problems.

2 Multiplying and Dividing Fractions

Jelly Belly
original gourmet jelly bean
NEST COOKIES
cups all purpose flour
tsp. baking powder
tsp. salt
cup shortening
½ cup sugar
1 egg white
1 tsp. vanilla
2 Tbs. milk
2 cups shredded coconut
5 oz. Jelly Belly jelly beans assorted flavors
Heat oven to 375 F. Sift together flour, baking powder, and salt and set aside. In large bowl beat shortening, sugar, egg white, and vanilla until well blended. Add flour mixture and milk until blended. Stir in coconut.
Roll dough into a ball, divide in half. Roll 15 one-inch balls from each half and place on ungreased baking sheet. Make thumb print depression in center of each ball to form nest. Bake 6 minutes. Remove from oven and place 4 Jelly Belly beans in center of each cookie. Return to oven and bake 5 more minutes. Transfer cookies to wire rack to cool. Makes 30 cookies.

The recipe shown below uses Jelly Belly jelly beans and will make $2\frac{1}{2}$ dozen (30) cookies. But suppose you wanted to make 3 dozen, 10 dozen, or even $3\frac{1}{2}$ dozen cookies? In this chapter we discuss multiplication and division of fractions, which you need to know when cooking or baking.

2.1 Basics of Fractions

OBJECTIVES

1. Use a fraction to show how many parts of a whole are shaded.
2. Identify the numerator and denominator.
3. Identify proper and improper fractions.

In **Chapter 1** we discussed whole numbers. Many times, however, we find that parts of whole numbers are considered. One way to write parts of a whole is with **fractions.** Another way is with decimals, which is discussed in **Chapter 4.**

OBJECTIVE 1 Use a fraction to show how many parts of a whole are shaded. The number $\frac{1}{8}$ is a fraction that represents 1 of 8 equal parts. Read $\frac{1}{8}$ as "one eighth."

A fraction may be used to represent *part* of a whole.

EXAMPLE 1 Identifying Fractions

Use fractions to represent the shaded portions and the unshaded portions of each figure.

(a) The figure on the left has 6 equal parts. The 1 shaded part is represented by the fraction $\frac{1}{6}$. The *un*shaded part is $\frac{5}{6}$.

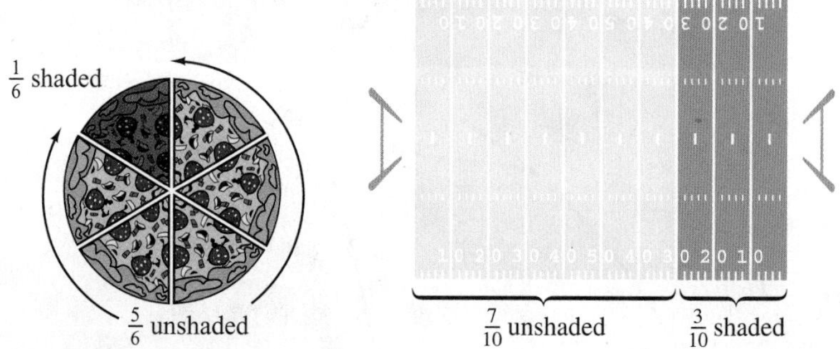

(b) The 3 shaded parts of the 10-part figure on the right are represented by the fraction $\frac{3}{10}$. The *un*shaded part is $\frac{7}{10}$.

◀ **Work Problem 1 at the Side.**

A fraction can also be used to represent more than one whole object.

EXAMPLE 2 Representing Fractions Greater Than 1

Use a fraction to represent the shaded part of each figure.

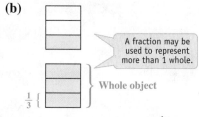

(a) $\frac{1}{4}$ Whole object

An area equal to 5 of the $\frac{1}{4}$ parts is shaded, so $\frac{5}{4}$ is shaded.

(b) Whole object $\frac{1}{3}$

A fraction may be used to represent more than 1 whole.

An area equal to 4 of the $\frac{1}{3}$ parts is shaded, so $\frac{4}{3}$ is shaded.

◀ **Work Problem 2 at the Side.**

OBJECTIVE 2 Identify the numerator and denominator. In the fraction $\frac{2}{3}$, the number 2 is the **numerator** and 3 is the **denominator**. The line (bar) between the numerator and the denominator is the *fraction bar.*

$$\text{Fraction bar} \rightarrow \frac{2 \leftarrow \text{Numerator}}{3 \leftarrow \text{Denominator}}$$

---

1 Write fractions for the shaded portions and the unshaded portions of each figure.

(a)

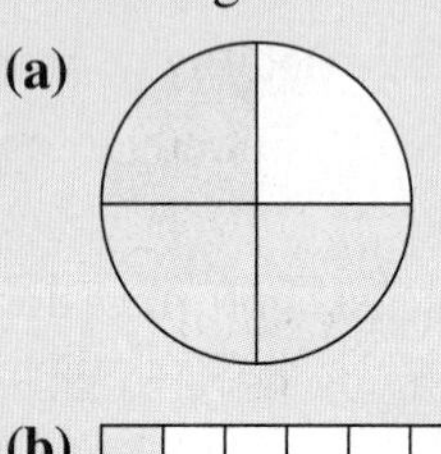

(b)

VOCABULARY TIP

Fraction The word *fraction* actually comes from the Latin *fractio*, which means "to break." Mathematically speaking, the bottom number of a fraction tells us how many parts a whole is broken into.

2 Write fractions for the shaded portions of each figure.

(a)

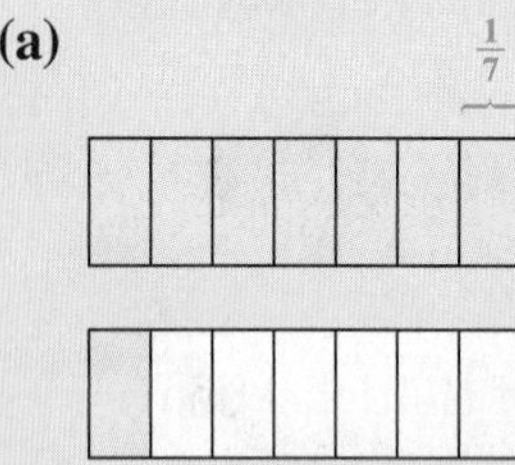

(b)

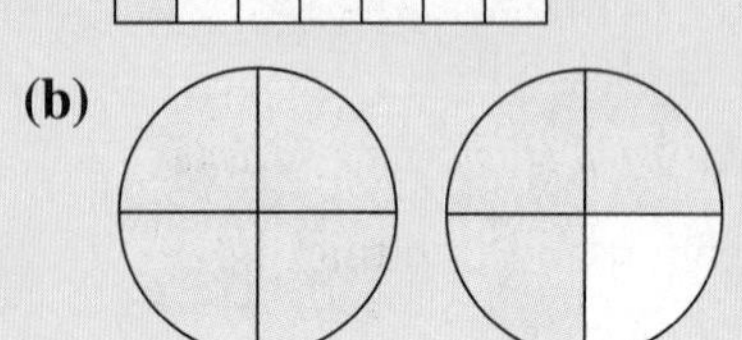

Answers

1. (a) $\frac{3}{4}$; $\frac{1}{4}$ (b) $\frac{1}{6}$; $\frac{5}{6}$
2. (a) $\frac{8}{7}$ (b) $\frac{7}{4}$

Numerators and Denominators

The **denominator** of a fraction shows the number of equivalent parts in the whole, and the **numerator** shows how many parts are being considered.

Note

A fraction bar, —, is one of the division symbols. Because division by 0 is undefined, a fraction with a denominator of 0 is also undefined.

EXAMPLE 3 Identifying Numerators and Denominators

Identify the numerator and denominator in each fraction.

(a) $\frac{3}{4}$ **(b)** $\frac{8}{5}$

$\frac{3}{4}$ ← Numerator / ← Denominator

$\frac{8}{5}$ ← Numerator / ← Denominator

Work Problem 3 at the Side. ▶

OBJECTIVE 3 Identify proper and improper fractions. Fractions can be identified as *proper* or *improper* fractions.

Proper and Improper Fractions

If the numerator of a fraction is *less* than the denominator, the fraction is a **proper fraction.** A proper fraction is less than 1 whole.

If the numerator is *greater than or equal to* the denominator, the fraction is an **improper fraction.** An improper fraction is greater than or equal to 1 whole.

Proper Fractions	Improper Fractions
$\frac{5}{8}$ $\frac{3}{5}$ $\frac{23}{24}$	$\frac{6}{5}$ $\frac{10}{10}$ $\frac{115}{112}$

EXAMPLE 4 Classifying Types of Fractions

(a) Identify all proper fractions in this list.

$$\frac{3}{4} \quad \frac{5}{9} \quad \frac{17}{5} \quad \frac{9}{7} \quad \frac{3}{3} \quad \frac{12}{25} \quad \frac{1}{9} \quad \frac{5}{3}$$

Proper fractions have a numerator that is less than the denominator. The proper fractions are shown below.

$\frac{3}{4}$ ← 3 is less than 4. $\frac{5}{9}$ $\frac{12}{25}$ $\frac{1}{9}$ (A proper fraction is less than 1.)

(b) Identify all improper fractions in the list in part (a).

Improper fractions have a numerator that is equal to or greater than the denominator. The improper fractions are shown below.

$\frac{17}{5}$ ← 17 is greater than 5. $\frac{9}{7}$ $\frac{3}{3}$ $\frac{5}{3}$ (An improper fraction is equal to or greater than 1.)

Work Problem 4 at the Side. ▶

VOCABULARY TIP

Denominator The first letter "D" of the word denominator can help you remember that it is the number "down under" the fraction bar.

VOCABULARY TIP

Numerator The numerator lies on top of the fraction bar and is the *number* of parts being considered.

3 Identify the numerator and the denominator. Draw a picture with shaded parts to show each fraction. Drawings may vary, but should have the correct number of parts.

(a) $\frac{2}{3}$ **(b)** $\frac{1}{4}$

(c) $\frac{8}{5}$ **(d)** $\frac{5}{2}$

VOCABULARY TIP

Proper fraction The everyday definition of *proper* means the correct way of doing something. If the numerator is less than the denominator, the fraction is considered in proper form, or proper for short.

4 From the following group of fractions:

$$\frac{2}{3} \quad \frac{4}{3} \quad \frac{3}{4} \quad \frac{8}{8} \quad \frac{3}{1} \quad \frac{1}{3}$$

GS **(a)** list all proper fractions;

$\frac{\_\_}{3}$ $\frac{\_\_}{4}$ $\frac{1}{\_\_}$

GS **(b)** list all improper fractions.

$\frac{4}{\_\_}$ $\frac{\_\_}{8}$ $\frac{3}{\_\_}$

Answers

3. (a) N: 2; D: 3 (b) N: 1; D: 4

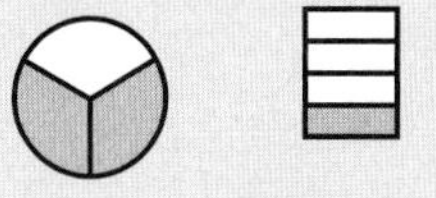

(c) N: 8; D: 5

(d) N: 5; D: 2

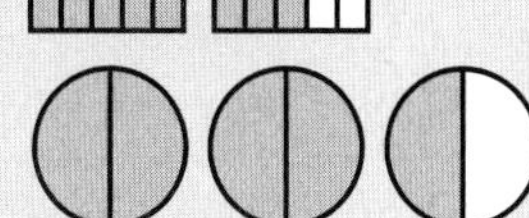

4. (a) $\frac{2}{3}$; $\frac{3}{4}$; $\frac{1}{3}$ (b) $\frac{4}{3}$; $\frac{8}{8}$; $\frac{3}{1}$

2.1 Exercises

FOR EXTRA HELP

Download the MyDashBoard App

MyMathLab®

CONCEPT CHECK *Identify the numerator and denominator.*

	Numerator	Denominator		Numerator	Denominator
1. $\frac{4}{5}$	______	______	**2.** $\frac{5}{6}$	______	______
3. $\frac{9}{8}$	______	______	**4.** $\frac{7}{5}$	______	______

CONCEPT CHECK *Fill in the blanks to complete each sentence.*

5. The fraction $\frac{3}{8}$ represents ______ of the ______ equal parts into which a whole is divided.

6. The fraction $\frac{7}{16}$ represents ______ of the ______ equal parts into which a whole is divided.

7. The fraction $\frac{5}{24}$ represents ______ of the ______ equal parts into which a whole is divided.

8. The fraction $\frac{24}{32}$ represents ______ of the ______ equal parts into which a whole is divided.

*Write fractions to represent the shaded and unshaded portions of each figure. See **Examples 1 and 2**.*

9.

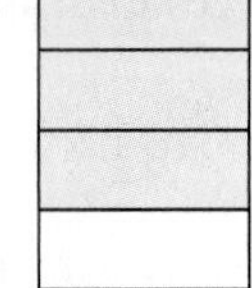

10.

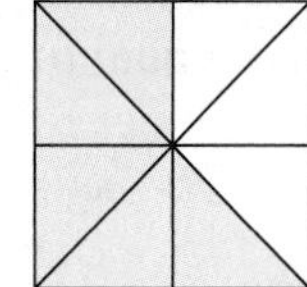

11.

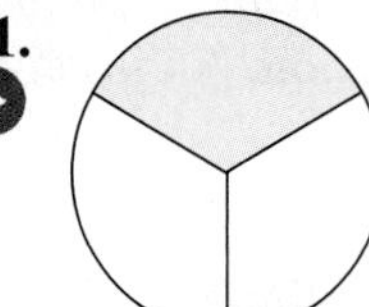

12.

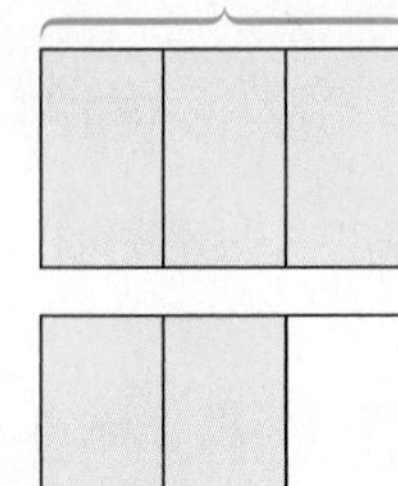

13.

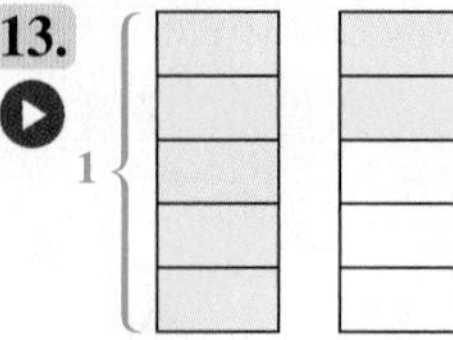

14.

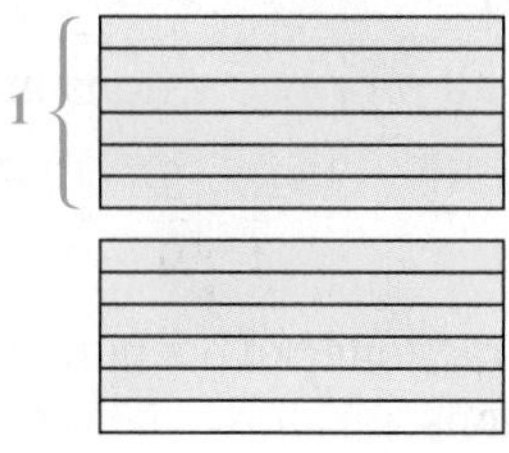

15. What fraction of these 6 bills has a life span of 2 years or greater? What fraction has a life of 4 years or less? What fraction has a life of 9 years?

A BILL'S LIFE

A $1 bill lasts about 18 months as compared with the average lifespan of other denominations:

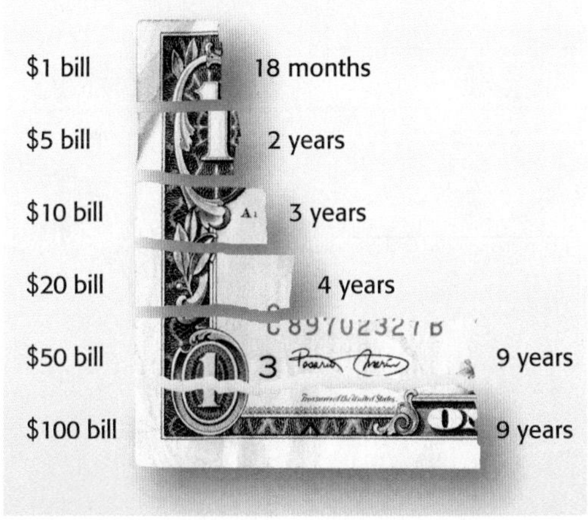

Source: Federal Reserve System; Bureau of Engraving and Printing.

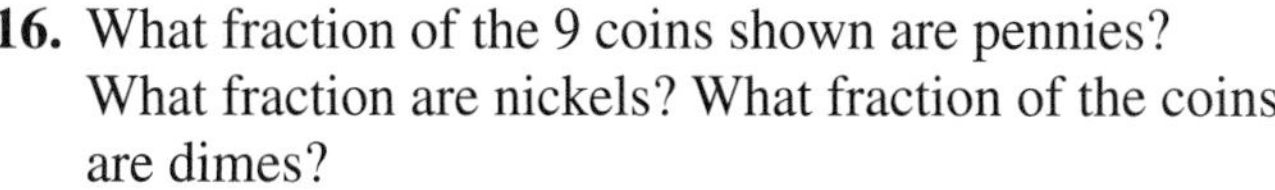

16. What fraction of the 9 coins shown are pennies? What fraction are nickels? What fraction of the coins are dimes?

17. In an American Sign Language (ASL) class of 25 students, 8 are hearing impaired. What fraction of the students are hearing impaired?

8 ← hearing impaired students (numerator)

____ ← total students (denominator)

18. A supermarket has 215 shopping carts. If 76 of the shopping carts are in the parking lot and the rest are in the store, what fraction are in the store?

215 total shopping carts
− 76 in parking lot
139 shopping carts in store

Fraction of carts in store: $\dfrac{139}{\_\_\_\_}$

19. There are 520 rooms in a hotel. If 217 of the rooms are reserved for nonsmokers, what fraction of the rooms are for smokers?

20. Of the 46 employees at the college bookstore, 15 are full-time while the rest are part-time student help. What fraction of the bookstore employees work part-time?

List the proper and improper fractions in each group. **See Example 4.**

	Proper	Improper
21. $\frac{8}{5}$ $\frac{1}{3}$ $\frac{5}{8}$ $\frac{6}{6}$ $\frac{12}{2}$ $\frac{7}{16}$	________	________
22. $\frac{1}{3}$ $\frac{3}{8}$ $\frac{16}{12}$ $\frac{10}{8}$ $\frac{6}{6}$ $\frac{3}{4}$	________	________
23. $\frac{3}{4}$ $\frac{3}{2}$ $\frac{5}{5}$ $\frac{9}{11}$ $\frac{7}{15}$ $\frac{19}{18}$	________	________
24. $\frac{12}{12}$ $\frac{15}{11}$ $\frac{13}{12}$ $\frac{11}{8}$ $\frac{17}{17}$ $\frac{19}{12}$	________	________

25. Write a fraction of your own choice. Label the parts of the fraction and write a sentence describing what each part represents. Draw a picture with shaded parts showing your fraction.

26. Give one example of a proper fraction and one example of an improper fraction. What determines whether a fraction is proper or improper? Draw pictures with shaded parts showing these fractions.

Study Skills

HOMEWORK: HOW, WHY, AND WHEN

OBJECTIVES

1. **Select an appropriate strategy for homework.**
2. **Use textbook features effectively.**

It is best for your brain if you keep up with the reading and homework in your math class. Remember that the more times you work with the information, the more dendrites you grow! So, give yourself every opportunity to read, work problems, and review your mathematics.

You have two choices for reading your math textbook. Read the short descriptions below and decide which will be best for you.

Preview before Class; Read Carefully after Class

Maddy learns best by listening to her instructor explain things. She "gets it" when she sees the instructor work problems on the board. She likes to ask questions in class and put the information in her notes. She has learned that it helps if she has *previewed* the section before the lecture, so she knows generally what to expect in class. *But after the class instruction*, when Maddy gets home, she finds that she can understand the math textbook easily. She remembers what her instructor said, and she can double-check her notes if she gets confused. So, Maddy does her **careful** reading of the section in her text **after** hearing the classroom lecture on the topic.

Read Carefully before Class

De'Lore, on the other hand, feels he learns well by reading on his own. He prefers to read the section and try working the example problems before coming to class. That way, he already knows what the instructor is going to talk about. Then, he can follow the instructor's examples more easily. It is also easier for him to take notes in class. De'Lore likes to have his questions answered right away, which he can do if he has already read the chapter section. So, De'Lore **carefully** reads the section in his text **before** he hears the classroom lecture on the topic.

Notice that there is **no one right way** to work with your textbook. You always must figure out what works best for you. Note also that both Maddy and De'Lore work with one section at a time. **The key is that you read the textbook regularly!** The rest of this activity will give you some ideas of how to make the most of your reading.

Try the following steps as you **read** your math textbook.

- Read slowly. Read only one section—or even part of a section—at a time.
- Do the sample problems in the margins **as you go.** Check them right away. The answers are at the bottom of the page.
- If your mind wanders, work problems on separate paper and write explanations in your own words.
- Make study cards as you read each section. Pay special attention to the yellow and blue boxes in the book. Make cards for new vocabulary, rules, procedures, formulas, and sample problems.
- **NOW,** you are ready to do your homework assignment!

Why Are These Reading Techniques Brain Friendly?

The steps at the left encourage you to be **actively working with the material** in your text. Your brain grows dendrites when it is doing something.

These methods require you to **try several different techniques,** not just the same thing over and over. Your brain loves variety!

Also, the techniques allow you to **take small breaks** in your learning. Those rest periods are crucial for good dendrite growth.

Study Skills

Continued from page 119

Why Are These Homework Suggestions Brain Friendly?

Your brain will grow dendrites as you study the worked examples in the text and **try doing them yourself** on separate paper. So, when you see similar problems in the homework, you will already have dendrites to work from.

Giving yourself a practice test by trying to remember the steps (without looking at your card) is an excellent way to reinforce what you are learning.

Correcting errors right away is how you learn and reinforce the correct procedures. It is hard to unlearn a mistake, so always check to see that you are on the right track.

Now Try This

Which steps for reading this book will be most helpful for you?

1. ______________________________
2. ______________________________
3. ______________________________

Homework

Instructors assign homework so you can grow your own dendrites (learn the material) and then coat the dendrites with myelin through practice (remember the material). Really! In learning, you get good at what you practice. So, completing homework every day will strengthen your neural network and prepare you for exams.

If you have read each section in your textbook according to the steps above, you will probably encounter few difficulties with the exercises in the homework. Here are some additional suggestions that will help you succeed with the homework.

- If you **have trouble with a problem,** find a similar worked example in the section. Pay attention to *every line* of the worked example to see how to get from step to step. Work it yourself too, on separate paper; don't just look at it.
- If it is **hard to remember the steps** to follow for certain procedures, write the steps on a separate card. Then write a short explanation of each step. Keep the card nearby while you do the exercises, but try *not* to look at it.
- If you **aren't sure you are working the assigned exercises correctly,** choose two or three odd-numbered problems that are a similar type and work them. Then check the answers in the Answers section of your book and see if you are doing them correctly. If you aren't, go back to the section in the text and review the examples and find out how to correct your errors. Finally, when you are sure you understand, try the assigned problems again.
- **Make sure you do some homework every day,** even if the math class does not meet each day!

Now Try This

What are your biggest homework concerns?

List your two main concerns and a **brain friendly solution** for each one.

1. **Concern:** ______________________________

   Solution: ______________________________

2. **Concern:** ______________________________

   Solution: ______________________________

2.2 Mixed Numbers

OBJECTIVES

1. Identify mixed numbers.
2. Write mixed numbers as improper fractions.
3. Write improper fractions as mixed numbers.

Suppose you had three whole trays of muffins and half of another tray. You would state this as a whole number and a fraction.

OBJECTIVE 1 **Identify mixed numbers.** When a whole number and a fraction are written together, the result is a **mixed number.** For example, the mixed number

$$3\frac{1}{2} \quad \text{represents} \quad 3+\frac{1}{2}$$

or 3 wholes and $\frac{1}{2}$ of a whole. Read $3\frac{1}{2}$ as "three and one-half." As this figure shows, the mixed number $3\frac{1}{2}$ is equal to the improper fraction $\frac{7}{2}$.

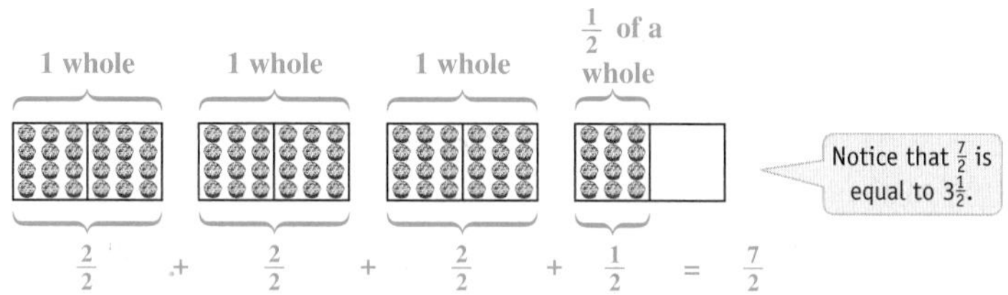

Work Problem 1 at the Side. ▶

OBJECTIVE 2 **Write mixed numbers as improper fractions.** Use the following steps to write $3\frac{1}{2}$ as an improper fraction without drawing a figure.

Step 1 Multiply 3 and 2.

$$3\frac{1}{2} \quad 3 \cdot 2 = 6$$

Step 2 Add 1 to the product.

$$3\frac{1}{2} \quad 6 + 1 = 7$$

Step 3 Use 7, from Step 2, as the numerator of the improper fraction and 2 as the denominator.

$$3\frac{1}{2} = \frac{7}{2}$$

Same denominator

In summary, use the following steps to *write a mixed number as an improper fraction.*

Writing a Mixed Number as an Improper Fraction

Step 1 *Multiply* the denominator of the fraction and the whole number.

Step 2 *Add* to this product the numerator of the fraction.

Step 3 Write the result of Step 2 as the *numerator* of the improper fraction and the original denominator as the *denominator*.

1 **(a)** Use these diagrams to write $1\frac{2}{3}$ as an improper fraction.

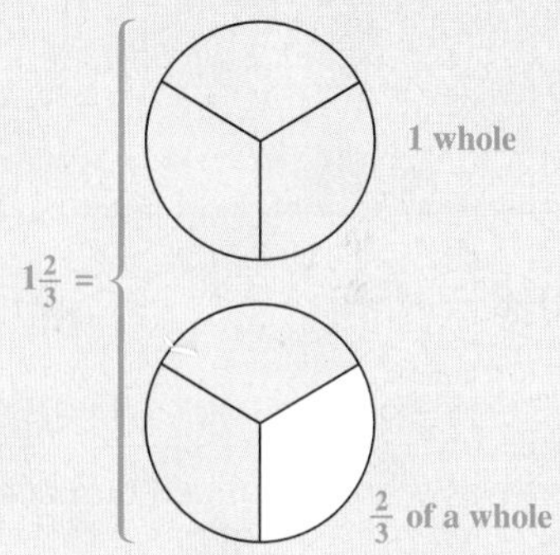

GS **(b)** Use these diagrams to write $2\frac{1}{4}$ as an improper fraction.

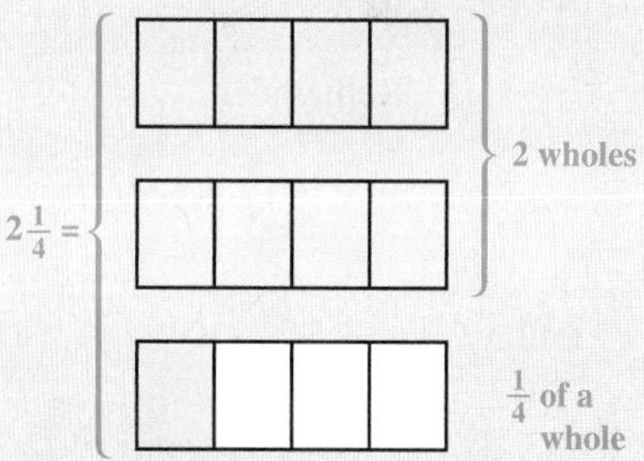

Since each of these diagrams is divided into ____ pieces, the denominator will be ____. The number of pieces shaded is ____.

VOCABULARY TIP

Mixed number A memory tip is to think "whole number times the bottom plus the top."

Answers

1. **(a)** $\frac{5}{3}$ **(b)** 4; 4; 9; $\frac{9}{4}$

2 Write as improper fractions.

GS (a) $6\frac{1}{2}$

$6 \cdot$ ____ = ____

$12 + 1 = 13$

$6\frac{1}{2} = \frac{\_\_\_\_}{\_\_\_\_}$

(b) $7\frac{3}{4}$

(c) $4\frac{7}{8}$

(d) $8\frac{5}{6}$

3 Write as whole or mixed numbers.

GS (a) $\frac{6}{5}$

Divide ____ by ____

$$\begin{array}{r} 1 \leftarrow \text{Whole number part} \\ 5\overline{)6} \\ \underline{5} \\ 1 \leftarrow \text{Remainder} \end{array}$$

So $\frac{6}{5} = 1\frac{1}{\_\_\_\_}$

(b) $\frac{9}{4}$

(c) $\frac{35}{5}$

(d) $\frac{78}{7}$

Answers

2. (a) 2; 12; $\frac{13}{2}$ (b) $\frac{31}{4}$ (c) $\frac{39}{8}$ (d) $\frac{53}{6}$

3. (a) 6; 5; $1\frac{1}{5}$ (b) $2\frac{1}{4}$ (c) 7 (d) $11\frac{1}{7}$

EXAMPLE 1 **Writing a Mixed Number as an Improper Fraction**

Write $7\frac{2}{3}$ as an improper fraction (numerator greater than denominator).

Step 1 $7\frac{2}{3}$ $7 \cdot 3 = 21$ Multiply 7 and 3.

Step 2 $7\frac{2}{3}$ $21 + 2 = 23$ Add 2. The numerator of the improper fractions is 23.

Step 3 $7\frac{2}{3} = \frac{23}{3}$ Use the same denominator.

Always use the same denominator.

◀ **Work Problem 2 at the Side.**

OBJECTIVE 3 **Write improper fractions as mixed numbers.** Write an improper fraction as a mixed number as follows.

Writing an Improper Fraction as a Mixed Number

Write an **improper fraction** as a mixed number by dividing the numerator by the denominator. The quotient is the whole number (of the mixed number), the remainder is the numerator of the fraction part, and the denominator stays the same.

EXAMPLE 2 **Writing Improper Fractions as Mixed Numbers**

Write each improper fraction as a mixed number.

Divide numerator by denominator.

(a) $\frac{17}{5}$ Divide 17 by 5. ⟶

$$\begin{array}{r} 3 \leftarrow \textbf{Whole number part} \\ 5\overline{)17} \\ \underline{15} \\ 2 \leftarrow \text{Remainder} \end{array}$$

The quotient **3** is the whole number part of the mixed number. The remainder 2 is the numerator of the fraction, and the denominator stays as **5**.

$\frac{17}{5} = 3\frac{2}{5}$ ← Remainder

The denominator stays the same.

We can check this by using a diagram in which $\frac{17}{5}$ is shaded.

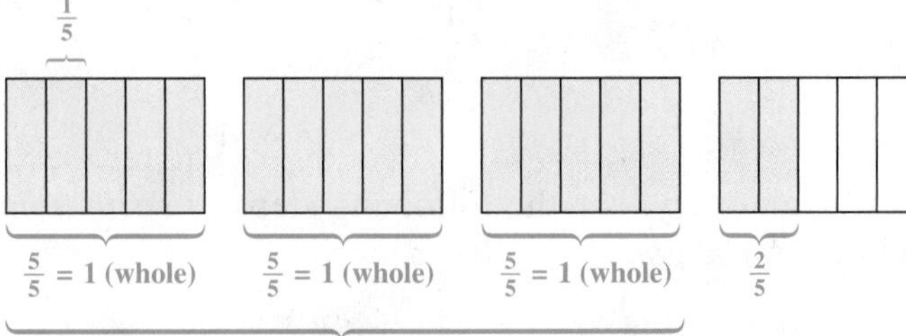

(b) $\frac{24}{4}$ Divide 24 by 4. ⟶

$$\begin{array}{r} 6 \\ 4\overline{)24} \\ \underline{24} \\ 0 \leftarrow \text{No remainder} \end{array}$$

so $\frac{24}{4} = 6$

No remainder, so no fraction part; just a whole number.

◀ **Work Problem 3 at the Side.**

2.2 Exercises

FOR EXTRA HELP

MyMathLab®

CONCEPT CHECK *Decide whether each statement is* true *or* false. *If it is* false, *explain why.*

1. The fraction $\frac{12}{12}$ is an improper fraction.

2. The fraction $\frac{2}{3}$ is a proper fraction.

3. The mixed number $7\frac{2}{5}$ can be changed to the improper fraction $\frac{14}{5}$.

4. Some mixed numbers cannot be changed to an improper fraction.

5. The mixed number $6\frac{1}{2}$ written as an improper fraction is $\frac{12}{2}$.

6. The mixed number $5\frac{5}{6}$ written as an improper fraction is $\frac{35}{6}$.

Write each mixed number as an improper fraction. **See Example 1.**

7. $1\frac{1}{4}$

8. $2\frac{1}{2}$

9. $4\frac{3}{5}$

10. $8\frac{1}{4}$

11. $8\frac{1}{2}$

12. $1\frac{7}{11}$

13. $10\frac{1}{8}$

GS

Find the new numerator:

$10 \cdot 8 = 80$ Multiply.

$80 + 1 = 81$ Add.

$10\frac{1}{8} =$

14. $12\frac{2}{3}$

GS

Find the new numerator:

$12 \cdot 3 = 36$ Multiply.

$36 + 2 = 38$ Add.

$12\frac{2}{3} =$

15. $10\frac{3}{4}$

16. $3\frac{3}{8}$

17. $5\frac{4}{5}$

18. $2\frac{8}{9}$

19. $8\frac{3}{5}$

20. $3\frac{4}{7}$

21. $4\frac{10}{11}$

22. $11\frac{5}{8}$

23. $32\frac{3}{4}$

24. $15\frac{3}{10}$

25. $18\frac{5}{12}$

26. $19\frac{8}{11}$

27. $17\frac{14}{15}$

28. $9\frac{5}{16}$

29. $7\frac{19}{24}$

30. $9\frac{7}{12}$

CONCEPT CHECK *Decide whether each statement is* true *or* false. *If it is* false, *explain why.*

31. The improper fraction $\frac{4}{3}$ written as a mixed number is $1\frac{1}{4}$.

32. An improper fraction cannot always be written as a whole number or mixed number.

33. Some improper fractions can be written as a whole number with no fraction part.

34. The improper fraction $\frac{48}{6}$ can be written as the whole number 8.

Write each improper fraction as a whole or mixed number. **See Example 2.**

35. $\frac{4}{3}$

$$\begin{array}{r} 1 \leftarrow \text{Whole number part} \\ 3\overline{)4} \\ \underline{3} \\ 1 \leftarrow \text{Remainder} \end{array}$$

36. $\frac{11}{9}$

$$\begin{array}{r} 1 \leftarrow \text{Whole number part} \\ 9\overline{)11} \\ \underline{9} \\ 2 \leftarrow \text{Remainder} \end{array}$$

37. $\frac{9}{4}$

38. $\frac{7}{2}$

39. $\frac{54}{6}$

40. $\frac{63}{9}$

41. $\frac{38}{5}$

42. $\frac{33}{7}$

43. $\frac{63}{4}$

44. $\frac{19}{5}$

45. $\frac{47}{9}$

46. $\frac{65}{9}$

47. $\frac{65}{8}$

48. $\frac{37}{6}$

49. $\frac{84}{5}$

50. $\frac{92}{3}$

51. $\frac{112}{4}$

52. $\frac{117}{9}$

53. $\frac{183}{7}$

54. $\frac{212}{11}$

55. Your classmate asks you how to change a mixed number to an improper fraction. Write a couple of sentences and give an example to show her how this is done.

56. Explain in a sentence or two how to change an improper fraction to a mixed number. Give an example to show how this is done.

Write each mixed number as an improper fraction.

57. $250\frac{1}{2}$

58. $185\frac{3}{4}$

59. $333\frac{1}{3}$

60. $138\frac{4}{5}$

61. $522\frac{3}{8}$

62. $622\frac{1}{4}$

Write each improper fraction as a whole or mixed number.

63. $\dfrac{617}{4}$

64. $\dfrac{760}{8}$

65. $\dfrac{2565}{15}$

66. $\dfrac{2915}{16}$

67. $\dfrac{3917}{32}$

68. $\dfrac{5632}{64}$

Relating Concepts (Exercises 69–74) For Individual or Group Work

Knowing the basics of fractions is necessary in problem solving.
Work Exercises 69–74 in order.

69. Which of these fractions are proper fractions?

$\dfrac{2}{3}\quad \dfrac{4}{5}\quad \dfrac{8}{5}\quad \dfrac{3}{4}\quad \dfrac{6}{6}\quad \dfrac{7}{10}$

70. (a) The proper fractions in **Exercise 69** are the ones where the __________ is less than the __________.

(b) Draw a picture with shaded parts to show each proper fraction in **Exercise 69.**

(c) The proper fractions in **Exercise 69** are all (*less/greater*) than 1.

71. Which of these fractions are improper fractions?

$\dfrac{5}{5}\quad \dfrac{3}{4}\quad \dfrac{10}{3}\quad \dfrac{2}{3}\quad \dfrac{5}{6}\quad \dfrac{6}{5}$

72. (a) The improper fractions in **Exercise 71** are the ones where the __________ is equal to or greater than the __________.

(b) Draw a picture with shaded parts to show each improper fraction in **Exercise 71.**

(c) The improper fractions in **Exercise 71** are all equal to or (*less/greater*) than 1.

73. Identify which of these fractions can be written as whole or mixed numbers, and then write them as whole or mixed numbers.

$\dfrac{5}{3}\quad \dfrac{7}{8}\quad \dfrac{7}{7}\quad \dfrac{11}{6}\quad \dfrac{4}{5}\quad \dfrac{15}{16}$

74. (a) The fractions that can be written as whole or mixed numbers in **Exercise 73** are (*proper/improper*) fractions, and their value is always (*less than/greater than or equal to*) 1.

(b) Draw a picture with shaded parts to show each whole or mixed number in **Exercise 73.**

2.3 Factors

OBJECTIVES

1. Find factors of a number.
2. Identify prime numbers and composite numbers.
3. Find prime factorizations.

VOCABULARY TIP

Factors of a number A *factor* is a whole number that divides *exactly* into a whole number, leaving *no* remainder.

1 Find all the whole number factors of each number.

GS **(a)** 18

1, 2, ____, 6, ____, 18

(b) 16

(c) 36

(d) 80

VOCABULARY TIP

Prime numbers Because a prime number can be divided evenly only by itself and 1, it is the basic building block of all numbers.

2 Which of the following are prime?

4, 7, 9, 13, 17, 19, 29, 33

Answers

1. **(a)** 3; 9; 1, 2, 3, 6, 9, 18 **(b)** 1, 2, 4, 8, 16 **(c)** 1, 2, 3, 4, 6, 9, 12, 18, 36 **(d)** 1, 2, 4, 5, 8, 10, 16, 20, 40, 80
2. 7, 13, 17, 19, 29

OBJECTIVE 1 **Find factors of a number.** You will recall that numbers multiplied to give a product are called **factors.** Because $2 \cdot 5 = 10$, both 2 and 5 are factors of 10. The numbers 1 and 10 are also factors of 10, because $1 \cdot 10 = 10$. The various tests for divisibility show that 1, 2, 5, and 10 are the only whole number factors of 10. The products $2 \cdot 5$ and $1 \cdot 10$ are called **factorizations** of 10.

Note

The tests to decide whether one number is divisible by another number were shown in **Chapter 1.** You might want to review these. The tests that you will use most often are those for 2, 3, 5, and 10.

EXAMPLE 1 Using Factors

Find all possible two-number factorizations of each number.

(a) 12

$1 \cdot 12 = 12 \qquad 2 \cdot 6 = 12 \qquad 3 \cdot 4 = 12$

The factors of 12 are 1, 2, 3, 4, 6, and 12.

(b) 60

$1 \cdot 60 = 60 \qquad 2 \cdot 30 = 60$
$3 \cdot 20 = 60 \qquad 4 \cdot 15 = 60$
$5 \cdot 12 = 60 \qquad 6 \cdot 10 = 60$

The factors of a number all divide evenly into that number.

The factors of 60 are 1, 2, 3, 4, 5, 6, 10, 12, 15, 20, 30, and 60.

◀ Work Problem 1 at the Side.

OBJECTIVE 2 **Identify prime numbers and composite numbers.** Whole numbers that have only two factors are called **prime numbers.** They can only be divided evenly by themselves and 1.

Prime Numbers

A **prime number** is a whole number that has exactly ***two different*** factors, ***itself*** and ***1.***

The number 3 is a prime number, since it can be divided evenly only by itself and 1. The number 8 is **not** a prime number (it is composite), since 8 can be divided evenly by 2 and 4, as well as by itself and 1.

CAUTION

A prime number has ***only two*** different factors, itself and 1. The number 1 is not a prime number because it does not have *two different* factors; the only factor of 1 is 1.

EXAMPLE 2 Finding Prime Numbers

Which of the following numbers are prime?

2 5 11 15 27

A prime number can be divided evenly only by the number itself and by 1.

The number 15 can be divided by 3 and 5, so it is **not** prime. Also, because 27 can be divided by 3 and 9, then 27 is **not** prime. The other numbers in the list, 2, 5, and 11, are divisible only by themselves and 1, so they are prime.

◀ Work Problem 2 at the Side.

Composite Numbers

A number that is not prime is called a **composite number.** The numbers 0 and 1 are neither prime nor composite.

EXAMPLE 3 Identifying Composite Numbers

Which of the following numbers are composite?

(a) 6

Because 6 has factors of 2 and 3, as well as 6 and 1, the number 6 is composite.

(b) 11

The number 11 has only two factors, 11 and 1. It is *not* composite. (It is a prime number.)

(c) 25

Because 25 has a factor of 5, as well as 25 and 1, 25 is composite.

Work Problem 3 at the Side. ▶

OBJECTIVE 3 Find prime factorizations. For reference, here are the prime numbers less than 50.

2	3	5	7	11
13	17	19	23	29
31	37	41	43	47

These are the prime numbers less than 50.

The **prime factorization** of a number can be especially useful when we are adding or subtracting fractions and need to find a common denominator or write a fraction in lowest terms.

Prime Factorization

A **prime factorization** of a number is a factorization in which every factor is a *prime number.*

EXAMPLE 4 Determining the Prime Factorization

Find the prime factorization of 12.

Try to divide 12 by the first prime, 2.

$$12 \div 2 = 6,$$

↑ First prime

so

$$12 = 2 \cdot 6.$$

Try to divide 6 by the prime, 2.

$$6 \div 2 = 3,$$

so

$$12 = 2 \cdot \underbrace{2 \cdot 3}.$$

↑ Factorization of 6

Because all factors are prime, the prime factorization of 12 is

$$2 \cdot 2 \cdot 3.$$

All these factors are prime numbers.

Work Problem 4 at the Side. ▶

VOCABULARY TIP

Composite number A number that is not a prime number is a composite number.

3 Which of these numbers are composite?

2, 4, 5, 6, 8, 10, 11, 13, 19, 21, 27, 28, 33, 36, 42

VOCABULARY TIP

Prime factorization Every whole number has its own unique prime factorization. The order of the factors may differ, but the factors themselves will be unique to that number.

4 Find the prime factorization of each number.

GS **(a)** 8

2 • ___ • ___

GS **(b)** 28

2 • ___ • ___

(c) 18

(d) 40

Answers

3. 4, 6, 8, 10, 21, 27, 28, 33, 36, 42

4. **(a)** 2; 2; 2 • 2 • 2 **(b)** 2; 7; 2 • 2 • 7
(c) 2 • 3 • 3 **(d)** 2 • 2 • 2 • 5

CAUTION

All prime numbers are odd numbers except the number 2. Be careful, though, because *all odd numbers are not prime numbers.* For example, 9, 15, and 21 are odd numbers, but are *not* prime numbers.

EXAMPLE 5 Factoring by Using the Division Method

Find the prime factorization of 48.

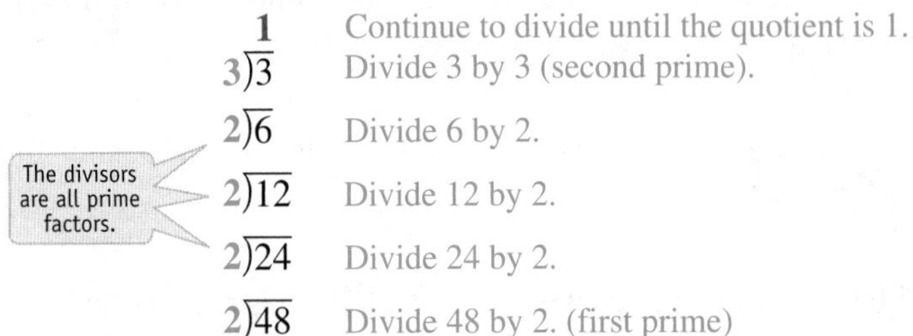

1 Continue to divide until the quotient is 1.
3)3 Divide 3 by 3 (second prime).
2)6 Divide 6 by 2.
2)12 Divide 12 by 2.
2)24 Divide 24 by 2.
2)48 Divide 48 by 2. (first prime)
The divisors are all prime factors.

Because all factors (divisors) are prime, the prime factorization of 48 is

$$2 \cdot 2 \cdot 2 \cdot 2 \cdot 3.$$

In **Chapter 1,** we wrote $2 \cdot 2 \cdot 2 \cdot 2$ as 2^4, so the prime factorization of 48 can be written, using exponents, as

$$48 = 2 \cdot 2 \cdot 2 \cdot 2 \cdot 3 = 2^4 \cdot 3.$$

◀ **Work Problem 5 at the Side.**

Note

When using the division method of factoring, the last quotient found is 1. The "1" is never used as a prime factor because 1 is neither prime nor composite. Besides, 1 times any number is the number itself.

EXAMPLE 6 Using Exponents with Prime Factorization

Find the prime factorization of 225.

1	Continue to divide until the quotient is 1.
5)5	Divide 5 by 5.
5)25	25 is not divible by 3; use 5.
3)75	Divide 75 by 3.
3)225	225 is not divisible by 2; use 3.

All the divisors are prime factors.

1 is not a prime factor.

Write the prime factorization.

$$225 = 3 \cdot 3 \cdot 5 \cdot 5$$

Or, using exponents,

$$225 = 3^2 \cdot 5^2$$

◀ **Work Problem 6 at the Side.**

5 Find the prime factorization of each number. Write the factorization with exponents.

GS **(a)** 36

$2 \cdot 2 \cdot 3 \cdot$ ____

$2^2 \cdot$ ____

(b) 54

(c) 60

(d) 81

6 Write the prime factorization of each number using exponents.

(a) 48

(b) 44

(c) 90

(d) 120

(e) 180

Answers

5. (a) 3; 3^2; $2^2 \cdot 3^2$ **(b)** $2 \cdot 3^3$
(c) $2^2 \cdot 3 \cdot 5$ **(d)** 3^4

6. (a) $2^4 \cdot 3$ **(b)** $2^2 \cdot 11$
(c) $2 \cdot 3^2 \cdot 5$ **(d)** $2^3 \cdot 3 \cdot 5$
(e) $2^2 \cdot 3^2 \cdot 5$

Another method of factoring is a *factor tree.*

EXAMPLE 7 **Factoring by Using a Factor Tree**

Find the prime factorization of each number using a factor tree.

(a) 30

Try to divide by the first prime, 2. Write the factors under the 30. Circle the 2, since it is a prime.

Since 15 cannot be divided evenly by 2, try the next prime, 3.

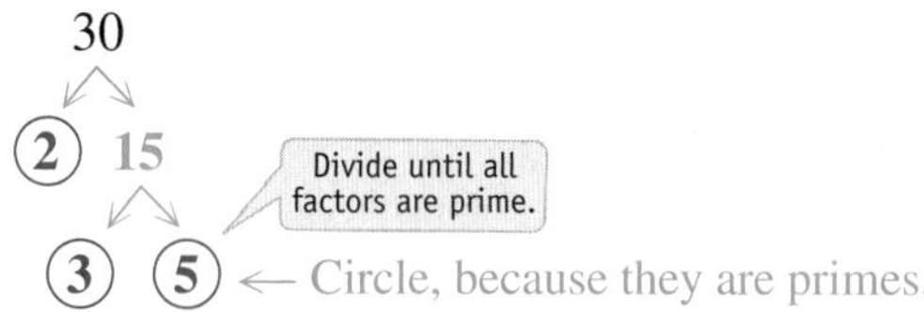

No uncircled factors remain, so the prime factorization (the circled factors) has been found.

$$30 = \mathbf{2 \cdot 3 \cdot 5}$$

(b) 24

Divide by 2.

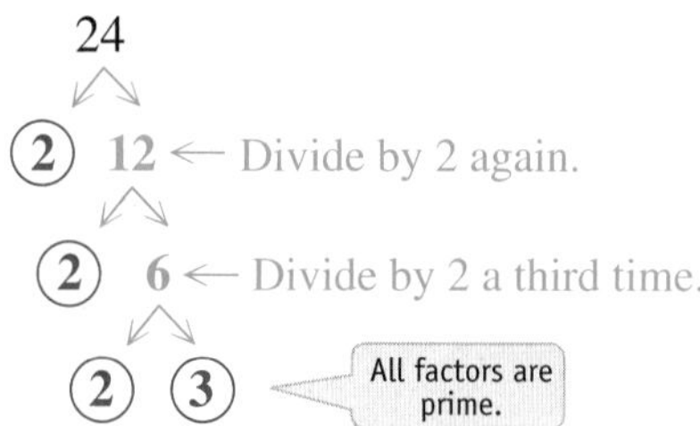

$24 = \mathbf{2 \cdot 2 \cdot 2 \cdot 3}$ or, using exponents, $24 = \mathbf{2^3 \cdot 3}$

(c) 45

Because 45 cannot be divided by 2, try 3.

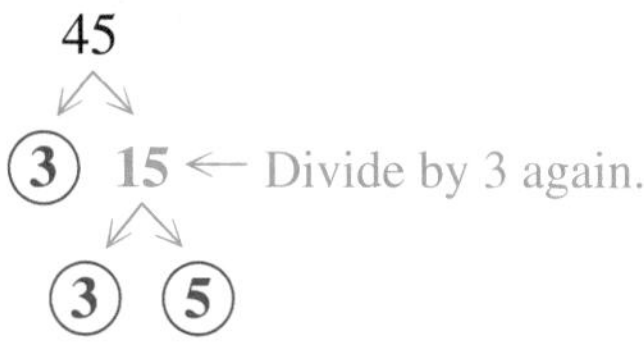

$45 = \mathbf{3 \cdot 3 \cdot 5}$ or, using exponents, $45 = \mathbf{3^2 \cdot 5}$

Note

The diagrams used in **Example 7** look like tree branches, and that is why this method is referred to as using a factor tree.

Work Problem 7 at the Side. ▶

7 Complete each factor tree and give the prime factorization.

(a) 28

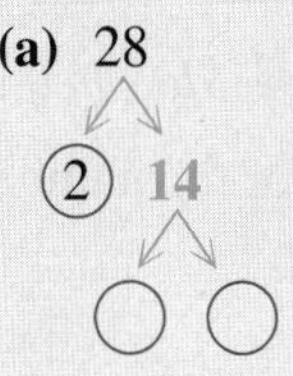

(b) 35

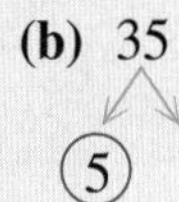

(c) 78

Answers

7. (a) 28

2 14

2 7

$28 = 2 \cdot 2 \cdot 7 = 2^2 \cdot 7$

(b) 35

5 7

$35 = 5 \cdot 7$

(c) 78

2 39

3 13

$78 = 2 \cdot 3 \cdot 13$

2.3 Exercises

FOR EXTRA HELP

Download the MyDashBoard App

MyMathLab®

CONCEPT CHECK *Decide whether the the following are* true *or* false.

1. All the factors for the number 8 are 2, 4, 6, and 8.

2. All the factors for the number 12 are 1, 2, 3, 4, 6, and 12.

3. All the factors for the number 15 are 1, 3, 5, and 15.

4. All the factors for the number 28 are 2, 4, 7, and 14.

Find all the factors of each number. **See Example 1.**

5. 48

6. 30

7. 56 Factors are: 1, 2, 4, 7, 8, 14, 28, ____

8. 72 Factors are: 1, 2, 3, 4, 6, 8, 9, 12, 18, 24, 36, ____

9. 36

10. 20

11. 40

12. 60

13. 64

14. 84

15. 82

16. 39

CONCEPT CHECK *Decide whether each number is* prime *or* composite.

17. 6

18. 9

19. 5

20. 16

21. 10

22. 13

23. 19

24. 17

25. 25

26. 48

27. 47

28. 45

CONCEPT CHECK *Find the prime factorization of each number. Circle the letter of the correct answer.*

29. 40
(a) $2 \cdot 4 \cdot 6$
(b) $2^3 \cdot 5$
(c) $2 \cdot 4 \cdot 5$

30. 36
(a) $3 \cdot 12$
(b) $2^2 \cdot 6$
(c) $2^2 \cdot 3^2$

31. 100
(a) $2^2 \cdot 5^2$
(b) $5^2 \cdot 4$
(c) $2 \cdot 5 \cdot 10$

Find the prime factorization of each number. Write answers with exponents when repeated factors appear. **See Examples 4–7.**

32. 8

33. 6

34. 20

35. 25
$5 \cdot \_ = 25$

36. 56
$\_ \cdot \_ \cdot 2 \cdot 7 = 56$

37. 68

38. 70

39. 72

40. 64

41. 44

42. 104

43. 100

44. 112

45. 125

46. 135

47. 180

48. 300

49. 320

50. 480

51. 360

52. 400

53. Give a definition in your own words of both a prime number and a composite number. Give three examples of each. Which whole numbers are neither prime nor composite?

54. With the exception of the number 2, all prime numbers are odd numbers. Nevertheless, all odd numbers are not prime numbers. Explain why these statements are true.

55. Explain the difference between finding all possible factors of 24 and finding the prime factorization of 24.

56. Use the division method to find the prime factorization of 36. Can you divide by 3s before you divide by 2s? Does the order of division change the answers?

Find the prime factorization of each number. Write answers using exponents.

57. 350

58. 640

59. 960

60. 1000

61. 1560

62. 2000

63. 1260

64. 2200

Relating Concepts (Exercises 65–70) For Individual or Group Work

An understanding of factors and factorization will be needed to solve fraction problems.
Work Exercises 65–70 in order.

65. A prime number is a whole number that has exactly two different factors, itself and 1. List all prime numbers less than 50.

66. Explain what it is about the numbers in **Exercise 65** that makes them prime.

67. The number 2 is an even number and a prime number. Can any other even numbers be prime numbers? Explain.

68. Can a multiple of a prime number be prime (for example, 6, 9, 12, and 15 are multiples of 3)? Explain.

69. Find the prime factorization of 2100. Do not use exponents in your answer.

70. Write the answer to **Exercise 69** using exponents for repeated factors.

2.4 Writing a Fraction in Lowest Terms

OBJECTIVES

1. Tell whether a fraction is written in lowest terms.
2. Write a fraction in lowest terms using common factors.
3. Write a fraction in lowest terms using prime factors.
4. Determine whether two fractions are equivalent.

When working problems involving fractions, we must often compare two fractions to determine whether they represent the same portion of a whole. Look at the two cases of soda.

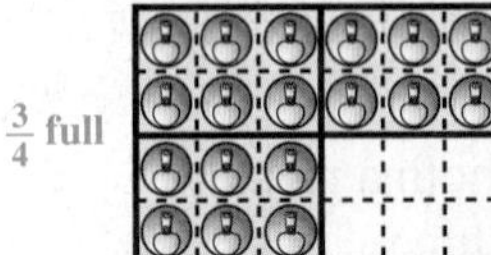

$\frac{3}{4}$ full

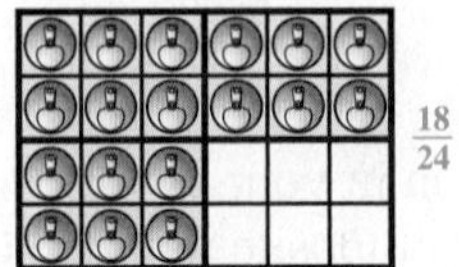

$\frac{18}{24}$ full

The cases of soda show areas that are $\frac{3}{4}$ full and $\frac{18}{24}$ full. Because the full areas are equivalent (the same portion), the fractions $\frac{3}{4}$ and $\frac{18}{24}$ are **equivalent fractions.** They each represent the same portion of the whole.

$$\frac{3}{4} = \frac{18}{24}$$

Because the numbers 18 and 24 both have 6 as a factor, 6 is called a **common factor** of the numbers. Other common factors of 18 and 24 are 1, 2, and 3.

◀ Work Problem 1 at the Side.

VOCABULARY TIP

Equivalent fractions The word *equivalent* means being of equal value. So, equivalent fractions have the same value, even though they may look different.

VOCABULARY TIP

Common factor A number also called the common divisor.

1 Decide whether the number in blue is a common factor of the other two numbers.

(a) 6, 12; 2 **(b)** 32, 64; 8

(c) 32, 56; 16 **(d)** 75, 81; 1

OBJECTIVE 1 Tell whether a fraction is written in lowest terms. The fraction $\frac{3}{4}$ is written in *lowest terms* because the numerator and denominator have no common factor other than 1. However, the fraction $\frac{18}{24}$ is ***not*** in lowest terms because its numerator and denominator have common factors of 6, 3, 2, and 1.

Writing a Fraction in Lowest Terms

A fraction is written in **lowest terms** when the numerator and denominator have no common factor other than 1.

VOCABULARY TIP

Lowest terms The process of reducing a fraction to lowest terms is also called *simplifying* the fraction.

EXAMPLE 1 Understanding Lowest Terms

Are the following fractions in lowest terms?

(a) $\frac{3}{8}$ — No common factors other than 1.

The numerator and denominator have no common factor other than 1, so the fraction is in lowest terms.

(b) $\frac{21}{36}$ — 3 is a common factor of 21 and 36.

The numerator and denominator have a common factor of 3, so the fraction is **not in lowest terms.**

◀ Work Problem 2 at the Side.

2 Are the following fractions in lowest terms?

(a) $\frac{4}{5}$ **(b)** $\frac{6}{18}$

(c) $\frac{9}{15}$ **(d)** $\frac{17}{46}$

OBJECTIVE 2 Write a fraction in lowest terms using common factors. There are two common methods for writing a fraction in lowest terms. These methods are shown in the next examples. The first method works best when the numerator and denominator are small numbers.

Answers

1. **(a)** yes **(b)** yes **(c)** no **(d)** yes
2. **(a)** yes **(b)** no **(c)** no **(d)** yes

EXAMPLE 2 **Writing Fractions in Lowest Terms**

Write each fraction in lowest terms.

(a) $\frac{18}{24}$

The greatest common factor of 18 and 24 is 6. Divide both numerator and denominator by 6.

$$\frac{18}{24} = \frac{18 \div 6}{24 \div 6} = \frac{3}{4}$$

Divide numerator and denominator by the greatest common factor, 6.

(b) $\frac{30}{50} = \frac{30 \div 10}{50 \div 10} = \frac{3}{5}$ Divide both numerator and denominator by 10.

(c) $\frac{24}{42} = \frac{24 \div 6}{42 \div 6} = \frac{4}{7}$ Divide both numerator and denominator by 6.

(d) $\frac{60}{72}$

Suppose we thought that 4 was the greatest common factor of 60 and 72. Dividing by 4 would give

$$\frac{60}{72} = \frac{60 \div 4}{72 \div 4} = \frac{15}{18}. \leftarrow \text{Not in lowest terms}$$

But $\frac{15}{18}$ is **not** in lowest terms, because 15 and 18 have a common factor of 3. So we divide by 3.

$$\frac{15}{18} = \frac{15 \div 3}{18 \div 3} = \frac{5}{6}$$

Continue dividing until there is no common factor other than 1.

The fraction $\frac{60}{72}$ could have been written in lowest terms in one step by dividing by 12, the greatest common factor of 60 and 72.

$$\frac{60}{72} = \frac{60 \div 12}{72 \div 12} = \frac{5}{6} \leftarrow \text{Same answer as above}$$

Continue dividing until the fraction is in lowest terms.

Note

Dividing the numerator and denominator by the same number results in an equivalent fraction.

In **Example 2,** we wrote fractions in lowest terms by dividing by a common factor. This method is summarized in the following steps.

The Method of Dividing by a Common Factor

Step 1 Find a number that will divide evenly into both the numerator and denominator. This number is a ***common factor.***

Step 2 ***Divide*** both numerator and denominator by the common factor.

Step 3 ***Check*** to see whether the new fraction has any common factors (besides 1). If it does, repeat *Steps 2 and 3*. If the only common factor is 1, the fraction is in lowest terms.

Work Problem 3 at the Side. ▶

3 Write in lowest terms.

GS **(a)** $\frac{8}{16}$

$$\frac{8}{16} = \frac{8 \div 8}{16 \div \_\_\_} = \frac{1}{\_\_\_}$$

GS **(b)** $\frac{9}{12}$

$$\frac{9}{12} = \frac{9 \div \_\_\_}{12 \div 3} = \frac{\_\_\_}{4}$$

(c) $\frac{28}{42}$

(d) $\frac{30}{80}$

(e) $\frac{16}{40}$

Answers

3. (a) 8; 2; $\frac{1}{2}$ **(b)** 3; 3; $\frac{3}{4}$ **(c)** $\frac{2}{3}$ **(d)** $\frac{3}{8}$ **(e)** $\frac{2}{5}$

OBJECTIVE 3 Write a fraction in lowest terms using prime factors. The method of writing a fraction in lowest terms by division works well for fractions with small numerators and denominators. For larger numbers, when common factors are not obvious, use the method of *prime factors,* which is shown in the next example.

EXAMPLE 3 Using Prime Factors

Write each fraction in lowest terms.

(a) $\frac{24}{42}$

Write the prime factorization of both numerator and denominator. See the previous section for help.

$$\frac{24}{42} = \frac{2 \cdot 2 \cdot 2 \cdot 3}{2 \cdot 3 \cdot 7}$$

Just as with the method used in **Example 2,** divide both numerator and denominator by any common factors. Write a 1 by each factor that has been divided.

2 ÷ 2 is 1 3 ÷ 3 is 1

$$\frac{24}{42} = \frac{\overset{1}{\cancel{2}} \cdot 2 \cdot 2 \cdot \overset{1}{\cancel{3}}}{\underset{1}{\cancel{2}} \cdot \underset{1}{\cancel{3}} \cdot 7}$$

Multiply the remaining factors in both numerator and denominator.

$$\frac{24}{42} = \frac{1 \cdot 2 \cdot 2 \cdot 1}{1 \cdot 1 \cdot 7} = \frac{4}{7}$$

$\frac{4}{7}$ is equivalent to $\frac{24}{42}$ but is written in lowest terms.

Finally, $\frac{24}{42}$ written in lowest terms is $\frac{4}{7}$.

(b) $\frac{162}{54}$

Write the prime factorization of both numerator and denominator.

$$\frac{162}{54} = \frac{2 \cdot 3 \cdot 3 \cdot 3 \cdot 3}{2 \cdot 3 \cdot 3 \cdot 3}$$

Now divide by the common factors. ***Do not forget to write the 1s.***

$$\frac{162}{54} = \frac{\overset{1}{\cancel{2}} \cdot \overset{1}{\cancel{3}} \cdot \overset{1}{\cancel{3}} \cdot \overset{1}{\cancel{3}} \cdot 3}{\underset{1}{\cancel{2}} \cdot \underset{1}{\cancel{3}} \cdot \underset{1}{\cancel{3}} \cdot \underset{1}{\cancel{3}}}$$

Remember to write in the 1s when dividing by a common factor.

$$= \frac{1 \cdot 1 \cdot 1 \cdot 1 \cdot 3}{1 \cdot 1 \cdot 1 \cdot 1} = \frac{3}{1} = 3$$

(c) $\frac{18}{90}$

$$\frac{18}{90} = \frac{\overset{1}{\cancel{2}} \cdot \overset{1}{\cancel{3}} \cdot \overset{1}{\cancel{3}}}{\underset{1}{\cancel{2}} \cdot \underset{1}{\cancel{3}} \cdot \underset{1}{\cancel{3}} \cdot 5} = \frac{1 \cdot 1 \cdot 1}{1 \cdot 1 \cdot 1 \cdot 5} = \frac{1}{5}$$

All factors of the numerator were divided, and $1 \cdot 1 \cdot 1 = 1$.

Continued on Next Page

In **Example 3,** we wrote fractions in lowest terms using prime factors. This method is summarized as follows.

The Method of Prime Factors

Step 1 Write the ***prime factorization*** of both numerator and denominator.

Step 2 Use slashes to show you are ***dividing*** both numerator and denominator by common factors.

Step 3 ***Multiply*** the remaining factors in the numerator and denominator.

Work Problem 4 at the Side. ▶

OBJECTIVE 4 Determine whether two fractions are equivalent. The next example shows how to decide whether two fractions are equivalent.

EXAMPLE 4 Determining Whether Two Fractions Are Equivalent

Determine whether each pair of fractions is equivalent. In other words, do both fractions represent the same part of a whole?

(a) $\frac{16}{48}$ and $\frac{24}{72}$

Use the method of prime factors to write each fraction in lowest terms.

$$\frac{16}{48} = \frac{\overset{1}{\cancel{2}} \cdot \overset{1}{\cancel{2}} \cdot \overset{1}{\cancel{2}} \cdot \overset{1}{\cancel{2}}}{\underset{1}{\cancel{2}} \cdot \underset{1}{\cancel{2}} \cdot \underset{1}{\cancel{2}} \cdot \underset{1}{\cancel{2}} \cdot 3} = \frac{1 \cdot 1 \cdot 1 \cdot 1}{1 \cdot 1 \cdot 1 \cdot 1 \cdot 3} = \frac{1}{3}$$

Equivalent $\left(\frac{1}{3} = \frac{1}{3}\right)$

$$\frac{24}{72} = \frac{\overset{1}{\cancel{2}} \cdot \overset{1}{\cancel{2}} \cdot \overset{1}{\cancel{2}} \cdot \overset{1}{\cancel{3}}}{\underset{1}{\cancel{2}} \cdot \underset{1}{\cancel{2}} \cdot \underset{1}{\cancel{2}} \cdot \underset{1}{\cancel{3}} \cdot 3} = \frac{1 \cdot 1 \cdot 1 \cdot 1}{1 \cdot 1 \cdot 1 \cdot 1 \cdot 3} = \frac{1}{3}$$

(b) $\frac{32}{52}$ and $\frac{64}{112}$

$$\frac{32}{52} = \frac{\overset{1}{\cancel{2}} \cdot \overset{1}{\cancel{2}} \cdot 2 \cdot 2 \cdot 2}{\underset{1}{\cancel{2}} \cdot \underset{1}{\cancel{2}} \cdot 13} = \frac{2 \cdot 2 \cdot 2}{1 \cdot 1 \cdot 13} = \frac{8}{13}$$

Not equivalent $\left(\frac{8}{13} \neq \frac{4}{7}\right)$

$$\frac{64}{112} = \frac{\overset{1}{\cancel{2}} \cdot \overset{1}{\cancel{2}} \cdot \overset{1}{\cancel{2}} \cdot \overset{1}{\cancel{2}} \cdot 2 \cdot 2}{\underset{1}{\cancel{2}} \cdot \underset{1}{\cancel{2}} \cdot \underset{1}{\cancel{2}} \cdot \underset{1}{\cancel{2}} \cdot 7} = \frac{1 \cdot 1 \cdot 1 \cdot 1 \cdot 2 \cdot 2}{1 \cdot 1 \cdot 1 \cdot 1 \cdot 7} = \frac{4}{7}$$

(c) $\frac{75}{15}$ and $\frac{60}{12}$

$$\frac{75}{15} = \frac{\overset{1}{\cancel{3}} \cdot \overset{1}{\cancel{5}} \cdot 5}{\underset{1}{\cancel{3}} \cdot \underset{1}{\cancel{5}}} = \frac{1 \cdot 1 \cdot 5}{1 \cdot 1} = 5$$

Equivalent (5 = 5)

$$\frac{60}{12} = \frac{\overset{1}{\cancel{2}} \cdot \overset{1}{\cancel{2}} \cdot \overset{1}{\cancel{3}} \cdot 5}{\underset{1}{\cancel{2}} \cdot \underset{1}{\cancel{2}} \cdot \underset{1}{\cancel{3}}} = \frac{1 \cdot 1 \cdot 1 \cdot 5}{1 \cdot 1 \cdot 1} = 5$$

Work Problem 5 at the Side. ▶

4 Use the method of prime factors to write each fraction in lowest terms.

GS **(a)** $\frac{12}{36}$

$$\frac{12}{36} = \frac{2 \cdot 2 \cdot 3}{2 \cdot \_\_\_ \cdot \_\_\_ \cdot 3}$$

$$\frac{\overset{1}{\cancel{2}} \cdot \overset{1}{\cancel{2}} \cdot \overset{1}{\cancel{3}}}{\underset{1}{\cancel{2}} \cdot \underset{1}{\cancel{2}} \cdot \underset{1}{\cancel{3}} \cdot 3} = \frac{\_\_\_}{3}$$

GS **(b)** $\frac{32}{56} = \frac{2 \cdot 2 \cdot 2 \cdot 2 \cdot \_\_\_}{2 \cdot 2 \cdot \_\_\_ \cdot 7}$

$$\frac{\overset{1}{\cancel{2}} \cdot \overset{1}{\cancel{2}} \cdot \overset{1}{\cancel{2}} \cdot 2 \cdot 2}{\underset{1}{\cancel{2}} \cdot \underset{1}{\cancel{2}} \cdot \underset{1}{\cancel{2}} \cdot 7} = \frac{4}{\_\_\_}$$

(c) $\frac{74}{111}$ **(d)** $\frac{124}{340}$

5 Is each pair of fractions equivalent?

(a) $\frac{24}{48}$ and $\frac{36}{72}$

(b) $\frac{45}{60}$ and $\frac{50}{75}$

(c) $\frac{20}{4}$ and $\frac{110}{22}$

(d) $\frac{120}{220}$ and $\frac{180}{320}$

Answers

4. (a) 2; 3; $\frac{1}{3}$ **(b)** 2; 2; 7; $\frac{4}{7}$
(c) $\frac{2}{3}$ **(d)** $\frac{31}{85}$

5. (a) equivalent **(b)** not equivalent
(c) equivalent **(d)** not equivalent

2.4 Exercises

FOR EXTRA HELP

Download the MyDashBoard App

MyMathLab®

CONCEPT CHECK *Underline the correct answer or fill in the blank in each of the following.*

1. A number can be divided by 2 if the number is an (*odd*/*even*) number.

2. A number can be divided by 5 if the number ends in ______ or ______.

3. Any number can be divided by 10 if the number ends in ______.

4. If the sum of a number's digits is divisible by ______, the number is divisible by 3.

Put a ✓ mark in the blank if the number at the left is divisible by the number at the top. Put an ✗ in the blank if the number is not divisible by the number at the top.

	2	3	5	10		2	3	5	10
5. 60	______	______	______	______	**6.** 90	______	______	______	______
7. 48	______	______	______	______	**8.** 36	______	______	______	______
9. 160	______	______	______	______	**10.** 175	______	______	______	______
11. 138	______	______	______	______	**12.** 150	______	______	______	______

CONCEPT CHECK *Decide* true *or* false *whether each pair of fractions is equivalent.*

13. $\frac{6}{8} = \frac{3}{4}$ **14.** $\frac{7}{12} = \frac{1}{2}$ **15.** $\frac{3}{8} = \frac{5}{16}$ **16.** $\frac{4}{12} = \frac{1}{3}$

Write each fraction in lowest terms. **See Example 2.**

17. $\frac{15}{25}$ GS

$\frac{15}{25} = \frac{15 \div 5}{25 \div 5} =$

18. $\frac{32}{48}$ GS

$\frac{32}{48} = \frac{32 \div 16}{48 \div 16} =$

19. $\frac{36}{42}$ **20.** $\frac{22}{33}$

21. $\frac{56}{64}$ **22.** $\frac{21}{35}$ **23.** $\frac{180}{210}$ **24.** $\frac{72}{80}$

25. $\frac{72}{126}$ **26.** $\frac{73}{146}$ **27.** $\frac{12}{600}$ **28.** $\frac{8}{400}$

29. $\frac{96}{132}$ **30.** $\frac{165}{180}$ **31.** $\frac{60}{108}$ **32.** $\frac{112}{128}$

Write the numerator and denominator of each fraction as a product of prime factors and divide by the common factors. Then write the fraction in lowest terms. ***See Example 3.***

33. $\frac{18}{24}$ **34.** $\frac{16}{64}$ **35.** $\frac{35}{40}$

36. $\frac{20}{32}$ **37.** $\frac{90}{180}$ **38.** $\frac{36}{48}$

39. $\frac{36}{12}$ **40.** $\frac{192}{48}$

41. $\frac{72}{225}$ **42.** $\frac{65}{234}$

Write each fraction in lowest terms. Then state whether the fractions are equivalent or not equivalent. ***See Example 4.***

43. $\frac{3}{6}$ and $\frac{18}{36}$ $\frac{3 \div 3}{6 \div 3} = \frac{\_\_\_}{\_\_\_}$ and $\frac{18 \div \_\_\_}{36 \div \_\_\_} = \frac{\_\_\_}{\_\_\_}$

44. $\frac{3}{8}$ and $\frac{27}{72}$ $\frac{3 \div \_\_\_}{8 \div \_\_\_} = \frac{\_\_\_}{\_\_\_}$ and $\frac{27 \div 9}{72 \div 9} = \frac{\_\_\_}{\_\_\_}$

45. $\frac{10}{24}$ and $\frac{12}{30}$

46. $\frac{15}{35}$ and $\frac{18}{40}$ **47.** $\frac{15}{24}$ and $\frac{35}{52}$ **48.** $\frac{21}{33}$ and $\frac{9}{12}$

49. $\frac{14}{16}$ and $\frac{35}{40}$ **50.** $\frac{27}{90}$ and $\frac{24}{80}$ **51.** $\frac{48}{6}$ and $\frac{72}{8}$

52. $\frac{33}{11}$ and $\frac{72}{24}$ **53.** $\frac{25}{30}$ and $\frac{65}{78}$ **54.** $\frac{24}{72}$ and $\frac{30}{90}$

55. What does it mean when a fraction is written in lowest terms? Give three examples.

56. Explain what equivalent fractions are, and give an example of a pair of equivalent fractions. Show that they are equivalent.

Write each fraction in lowest terms.

57. $\frac{160}{256}$ **58.** $\frac{363}{528}$ **59.** $\frac{238}{119}$ **60.** $\frac{570}{95}$

Study Skills

USING STUDY CARDS

OBJECTIVES

1. **Create study cards for all new terms.**
2. **Create study cards for new procedures.**

You may have used "flash cards" in other classes before. In math, study cards can be helpful, too. However, they are different because the main things to remember in math are *not* necessarily terms and definitions; they are *sets of steps to follow* to solve problems (and how to know which set of steps to follow) and *concepts about how math works* (principles). So, the cards will look different but will be just as useful.

In this two-part activity, you will find four types of study cards to use in math. Look carefully at what kinds of information to put on them and where to put it. Then use them the way you would any flash card:

- to quickly review when you have a few minutes,
- to do daily reviews,
- to review before a test.

Remember, the most helpful thing about study cards is making them. While you are making them, you have to do the kind of thinking that is most brain friendly which improves your neural network of dendrites. After each card description you will find an assignment to try. It is marked **NOW TRY THIS.**

New Vocabulary Cards

For **new vocabulary cards,** put the word (spelled correctly) and the page number where it is found on the front of the card. On the back, write:

- the definition (in your own words if possible),
- an example, an exception (if there are any),
- any related words, and
- a sample problem (if appropriate).

Prime Numbers *p. 126*

Front of Card

Definition: Whole numbers that can only be divided by themselves and 1.

★Must divide evenly, NO remainders!

Ex: 2, 3, 5, 7, 11, 13, 17 are the first few primes.

→ NOT 0 or 1

→ Used in factoring

→ Related word: composite number

Back of Card

Why Are Study Cards Brain Friendly?

- Making cards is **active.**
- Cards are **visually** appealing.
- **Repetition** is good for your brain.

For details see Using Study Cards Revisited following **Section 2.7.**

Now Try This

List four new vocabulary words/concepts you need to learn right now. Make a card for each one.

______________ ______________ ______________ ______________

Procedure ("Steps") Cards

For **procedure cards,** write the name of the procedure at the top on the front of the card. Then write each step *in words*. If you need to know abbreviations for some words, include them along with the whole words written out. On the back, put an example of the procedure, showing each step you need to take. You can review by looking at the front and practicing a new worked example, or by looking at the back and remembering what the procedure is called and what the steps are.

Front of Card

Writing a fraction in lowest terms using prime factors
Use this method with larger denominators.

Step 1: *Write prime factorization of numerator and denominator.*

Step 2: *Divide out all common factors.*

Step 3: *Multiply remaining factors.*

Back of Card

Example: Write this fraction in lowest terms.

$$\frac{64}{112} = \frac{2 \cdot 2 \cdot 2 \cdot 2 \cdot 2 \cdot 2}{2 \cdot 2 \cdot 2 \cdot 2 \cdot 7}$$

← *Prime factors of 64.*
← *Prime factors of 112.*

Divide out common factors of 2 →

$$= \frac{\overset{1}{\cancel{2}} \cdot \overset{1}{\cancel{2}} \cdot \overset{1}{\cancel{2}} \cdot \overset{1}{\cancel{2}} \cdot 2 \cdot 2}{\underset{1}{\cancel{2}} \cdot \underset{1}{\cancel{2}} \cdot \underset{1}{\cancel{2}} \cdot \underset{1}{\cancel{2}} \cdot 7} = \frac{1 \cdot 1 \cdot 1 \cdot 1 \cdot 2 \cdot 2}{1 \cdot 1 \cdot 1 \cdot 1 \cdot 7} = \frac{4}{7}$$

multiply remaining factors · *lowest terms*

Now Try This

What procedure are you learning right now? Make a "steps" card for it.

Procedure: ____

Summary Exercises *Fraction Basics*

CONCEPT CHECK *Write fractions to represent the shaded and unshaded portions of each figure.*

1. 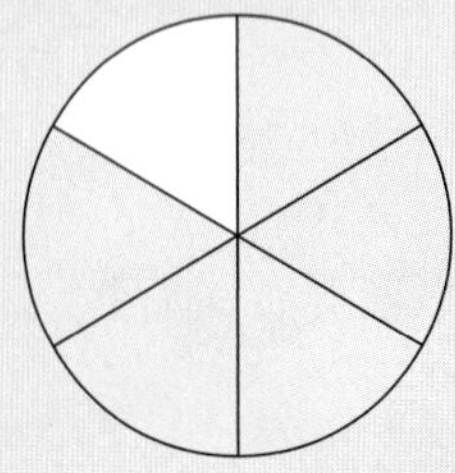

2. 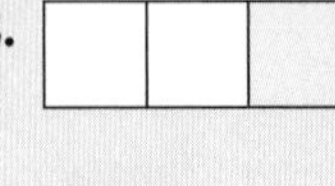

3. 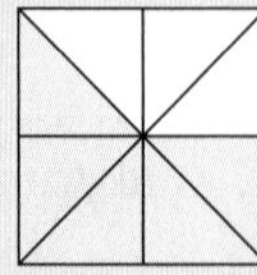

CONCEPT CHECK *Identify the numerator and denominator.*

	Numerator	Denominator		Numerator	Denominator
4. $\frac{3}{4}$	________	________	**5.** $\frac{8}{5}$	________	________

CONCEPT CHECK *List the proper and improper fractions.*

	Proper	Improper
6. $\frac{8}{2}$ $\frac{3}{5}$ $\frac{16}{7}$ $\frac{8}{8}$ $\frac{4}{25}$ $\frac{1}{32}$	________	________

The bar graph below shows the countries that have won the Boston Marathon in the Men's Wheelchair Division. Use the graph to answer Exercises 7–10. Write fractions in lowest terms.

Boston Marathon–Men's Wheelchair Division

In the past, Ernst Van Dyk of South Africa has won the Boston Marathon in the Men's Wheelchair Division 9 times, while Franz Nietispach of Switzerland and Jim Knaub of the United States have each won 5 times. The most winning countries are:

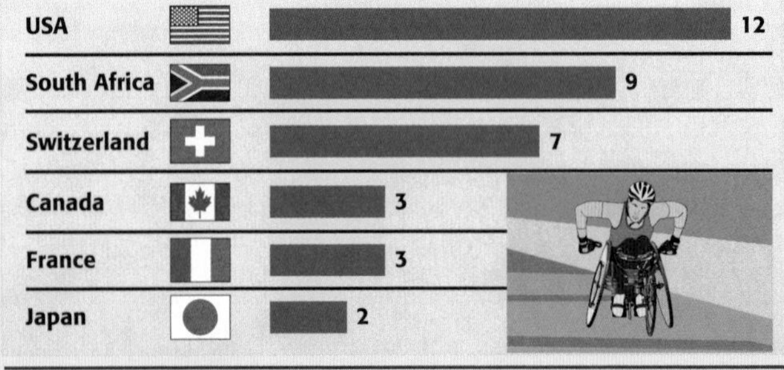

Source: Wikipedia.

7. What fraction of the winners were from Switzerland?

8. What fraction of the winners were not from either France or South Africa?

9. What fraction of the winners were from either Japan or the United States?

10. What fraction of the winners were not from Canada?

Write each improper fraction as a mixed number or whole number.

11. $\frac{5}{2}$ **12.** $\frac{11}{8}$ **13.** $\frac{9}{7}$ **14.** $\frac{8}{3}$

15. $\frac{64}{8}$ **16.** $\frac{45}{9}$ **17.** $\frac{36}{5}$ **18.** $\frac{47}{10}$

Write each mixed number as an improper fraction.

19. $3\frac{1}{3}$

20. $5\frac{3}{8}$

21. $6\frac{4}{5}$

22. $10\frac{3}{5}$

23. $12\frac{3}{4}$

24. $4\frac{10}{13}$

25. $11\frac{5}{6}$

26. $23\frac{5}{8}$

Write the prime factorization of each number using exponents.

27. 10

28. 55

29. 36

30. 81

31. 280

32. 360

Write each fraction in lowest terms.

33. $\frac{4}{12}$

34. $\frac{7}{14}$

35. $\frac{4}{16}$

36. $\frac{15}{25}$

37. $\frac{18}{24}$

38. $\frac{30}{36}$

39. $\frac{56}{64}$

40. $\frac{6}{300}$

41. $\frac{25}{200}$

42. $\frac{125}{225}$

43. $\frac{88}{154}$

44. $\frac{70}{126}$

Write the numerator and denominator of each fraction as a product of prime factors and divide by the common factors. Then write the fraction in lowest terms.

45. $\frac{24}{36}$

46. $\frac{80}{160}$

47. $\frac{126}{42}$

48. $\frac{96}{112}$

2.5 Multiplying Fractions

OBJECTIVES

1. Multiply fractions.
2. Use a multiplication shortcut.
3. Multiply a fraction and a whole number.
4. Find the area of a rectangle.

1 Use these figures to find $\frac{1}{4}$ of $\frac{1}{2}$.

GS

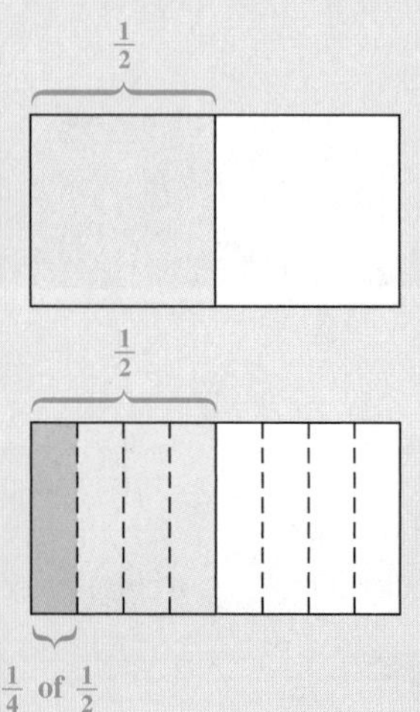

$\frac{1}{4}$ of $\frac{1}{2}$ is $\frac{\_\_\_}{\_\_\_}$.

Answer

1. $\frac{1}{8}$

OBJECTIVE 1 **Multiply fractions.** Suppose that you give $\frac{1}{2}$ of your Energy Bar to your kickboxing partner Ally. Then Ally gives $\frac{1}{2}$ of her share to Jake. How much of the Energy Bar does Jake get to eat?

Start with a sketch showing the Energy Bar cut in half (2 equal pieces).

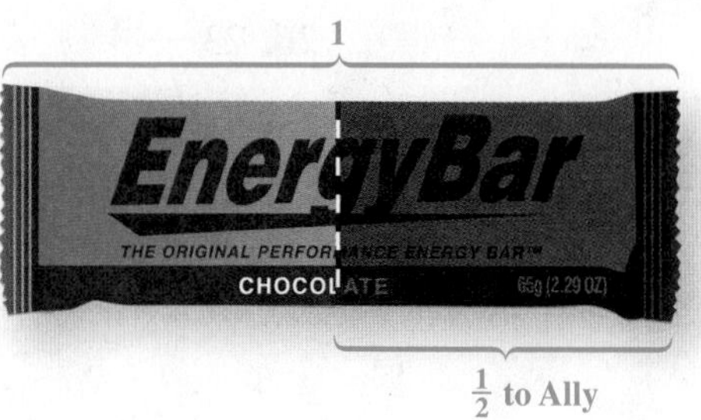

Next, take $\frac{1}{2}$ of the shaded area. (Here we are dividing $\frac{1}{2}$ into 2 equal parts and shading one darker than the other.)

The sketch shows that Jake gets $\frac{1}{4}$ of the Energy Bar.

Jake gets $\frac{1}{2}$ of $\frac{1}{2}$ of the Energy Bar. When used between two fractions, the word **of** tells us to multiply.

$$\frac{1}{2} \text{ of } \frac{1}{2} \quad \text{means} \quad \frac{1}{2} \cdot \frac{1}{2}$$

Jake's share of the Energy Bar is

$$\frac{1}{2} \cdot \frac{1}{2} = \frac{1}{4}.$$

◀ **Work Problem 1 at the Side.**

The rule for multiplying fractions follows.

Multiplying Fractions

Multiply two fractions by multiplying the numerators and multiplying the denominators.

Use this rule to find the product of $\frac{2}{3}$ and $\frac{1}{3}$, that is, to multiply $\frac{2}{3}$ by $\frac{1}{3}$.

$$\frac{2}{3} \cdot \frac{1}{3} = \frac{2 \cdot 1}{3 \cdot 3} \begin{matrix} \leftarrow \text{Multiply numerators.} \\ \leftarrow \text{Multiply denominators.} \end{matrix}$$

Finish multiplying.

$$\frac{2}{3} \cdot \frac{1}{3} = \frac{2 \cdot 1}{3 \cdot 3} = \frac{2}{9} \begin{matrix} \leftarrow 2 \cdot 1 = 2 \\ \leftarrow 3 \cdot 3 = 9 \end{matrix}$$

Multiply numerators.

Multiply denominators.

Check that the final result is in lowest terms. $\frac{2}{9}$ is in lowest terms because 2 and 9 have no common factor other than 1.

EXAMPLE 1 Multiplying Fractions

Multiply. Write answers in lowest terms.

(a) $\frac{5}{8} \cdot \frac{3}{4}$

Multiply the numerators and multiply the denominators.

$$\frac{5}{8} \cdot \frac{3}{4} = \frac{5 \cdot 3}{8 \cdot 4} = \frac{15}{32}$$

Already in lowest terms.

Notice that 15 and 32 have no common factors other than 1, so the answer is in lowest terms.

(b) $\frac{4}{7} \cdot \frac{2}{5}$

$$\frac{4}{7} \cdot \frac{2}{5} = \frac{4 \cdot 2}{7 \cdot 5} = \frac{8}{35} \leftarrow \text{Lowest terms}$$

(c) $\frac{5}{8} \cdot \frac{3}{4} \cdot \frac{1}{2}$

$$\frac{5}{8} \cdot \frac{3}{4} \cdot \frac{1}{2} = \frac{5 \cdot 3 \cdot 1}{8 \cdot 4 \cdot 2} = \frac{15}{64} \leftarrow \text{Lowest terms}$$

Work Problem 2 at the Side. ▶

OBJECTIVE 2 Use a multiplication shortcut. A **multiplication shortcut** that can be used with fractions is shown in **Example 2.**

EXAMPLE 2 Using the Multiplication Shortcut

Multiply $\frac{5}{6}$ and $\frac{9}{10}$. Write the answer in lowest terms.

$$\frac{5}{6} \cdot \frac{9}{10} = \frac{5 \cdot 9}{6 \cdot 10} = \frac{45}{60} \leftarrow \text{Not in lowest terms}$$

The numerator and denominator have a common factor other than 1, so write the prime factorization of each number.

$$\frac{5}{6} \cdot \frac{9}{10} = \frac{5 \cdot 9}{6 \cdot 10} = \frac{5 \cdot 3 \cdot 3}{2 \cdot 3 \cdot 2 \cdot 5}$$

Write the prime factorization of each number.

Continued on Next Page

2 Multiply. Write answers in lowest terms.

GS **(a)** $\frac{1}{2} \cdot \frac{3}{4}$

$$\frac{1}{2} \cdot \frac{3}{4} = \frac{3}{\_\_\_} \begin{matrix} \leftarrow 1 \cdot 3 \\ \leftarrow 2 \cdot 4 \end{matrix}$$

(b) $\frac{3}{5} \cdot \frac{1}{3}$

(c) $\frac{5}{6} \cdot \frac{1}{2} \cdot \frac{1}{8}$

(d) $\frac{1}{2} \cdot \frac{3}{4} \cdot \frac{3}{8}$

Answers

2. (a) 8; $\frac{3}{8}$ **(b)** $\frac{1}{5}$ **(c)** $\frac{5}{96}$ **(d)** $\frac{9}{64}$

Next, divide by the common factors of 5 and 3.

$$\frac{5}{6} \cdot \frac{9}{10} = \frac{5 \cdot 9}{6 \cdot 10} = \frac{\overset{1}{\cancel{5}} \cdot \overset{1}{\cancel{3}} \cdot 3}{2 \cdot \underset{1}{\cancel{3}} \cdot 2 \cdot \underset{1}{\cancel{5}}}$$

Finally, multiply the remaining factors in the numerator and in the denominator.

$$\frac{5}{6} \cdot \frac{9}{10} = \frac{1 \cdot 1 \cdot 3}{2 \cdot 1 \cdot 2 \cdot 1} = \frac{3}{4} \leftarrow \text{Lowest terms}$$

As a shortcut, instead of writing the prime factorization of each number, find the product of $\frac{5}{6}$ and $\frac{9}{10}$ as follows.

First, divide by 5, a common factor of both 5 and 10.

$$\frac{\overset{1}{\cancel{5}}}{6} \cdot \frac{9}{\underset{2}{\cancel{10}}}$$

Divide 5 by 5 to get 1.

Divide 10 by 5 to get 2.

Next, divide by 3, a common factor of both 6 and 9.

$$\frac{\overset{1}{\cancel{5}}}{\underset{2}{\cancel{6}}} \cdot \frac{\overset{3}{\cancel{9}}}{\underset{2}{\cancel{10}}}$$

Divide 9 by 3 to get 3.

Divide 6 by 3 to get 2.

Finally, multiply numerators and multiply denominators.

$$\frac{1 \cdot 3}{2 \cdot 2} = \frac{3}{4}$$

CAUTION

When using the multiplication shortcut, you are dividing a numerator and a denominator by a common factor. Be certain that you divide a numerator and a denominator ***by the same number.*** If you do all possible divisions, your answer will be in lowest terms.

EXAMPLE 3 Using the Multiplication Shortcut

Use the multiplication shortcut to find each product. Write the answers in lowest terms and as mixed numbers where possible.

(a) $\frac{6}{11} \cdot \frac{7}{8}$

Divide both 6 and 8 by their common factor of 2. Notice that 7 and 11 have no common factor. Then multiply.

$$\frac{\overset{3}{\cancel{6}}}{11} \cdot \frac{7}{\underset{4}{\cancel{8}}} = \frac{3 \cdot 7}{11 \cdot 4} = \frac{21}{44} \leftarrow \text{Lowest terms}$$

(b) $\frac{7}{10} \cdot \frac{20}{21}$

Divide a numerator and a denominator by the same number.

Divide 7 and 21 by 7, then divide 10 and 20 by 10.

$$\frac{\overset{1}{\cancel{7}}}{\underset{1}{\cancel{10}}} \cdot \frac{\overset{2}{\cancel{20}}}{\underset{3}{\cancel{21}}} = \frac{1 \cdot 2}{1 \cdot 3} = \frac{2}{3} \leftarrow \text{Lowest terms}$$

Continued on Next Page

(c) $\frac{35}{12} \cdot \frac{32}{25}$

$$\frac{\overset{7}{\cancel{35}}}{\underset{3}{\cancel{12}}} \cdot \frac{\overset{8}{\cancel{32}}}{\underset{5}{\cancel{25}}} = \frac{7 \cdot 8}{3 \cdot 5} = \frac{56}{15} \text{ or } 3\frac{11}{15} \leftarrow \text{Mixed number}$$

(d) $\frac{2}{3} \cdot \frac{8}{15} \cdot \frac{3}{4}$

$$\frac{\overset{1}{\cancel{2}}}{\underset{1}{\cancel{3}}} \cdot \frac{\overset{4}{\cancel{8}}}{15} \cdot \frac{\overset{1}{\cancel{3}}}{\underset{\underset{1}{\cancel{2}}}{\cancel{4}}} = \frac{1 \cdot 4 \cdot 1}{1 \cdot 15 \cdot 1} = \frac{4}{15} \leftarrow \text{Lowest terms}$$

This shortcut is especially helpful when the fractions involve large numbers.

Note

There is no specific order that must be used when dividing numerators and denominators, as long as both a numerator and a denominator are divided by the *same* number each time.

Work Problem 3 at the Side. ▶

OBJECTIVE 3 Multiply a fraction and a whole number. The rule for multiplying a fraction and a whole number follows.

Multiplying a Whole Number and a Fraction

Multiply a whole number and a fraction by writing the whole number as a fraction with a denominator of 1.

For example, write the whole numbers 8, 10, and 25 as follows.

$$8 = \frac{8}{1} \qquad 10 = \frac{10}{1} \qquad 25 = \frac{25}{1}$$ Write the whole number over 1.

EXAMPLE 4 Multiplying by Whole Numbers

Multiply. Write answers in lowest terms and as whole numbers where possible.

(a) $8 \cdot \frac{3}{4}$

Write 8 as $\frac{8}{1}$ and multiply.

$$8 \cdot \frac{3}{4} = \frac{\overset{2}{\cancel{8}}}{1} \cdot \frac{3}{\underset{1}{\cancel{4}}} = \frac{2 \cdot 3}{1 \cdot 1} = \frac{6}{1} = 6$$ $\frac{6}{1}$ is the same as $6 \div 1$, which equals 6.

Continued on Next Page

3 Use the multiplication shortcut to find each product.

GS (a) $\frac{3}{4} \cdot \frac{2}{5}$

$$\frac{3}{4} \cdot \frac{2}{5} =$$

(b) $\frac{6}{11} \cdot \frac{33}{21}$

(c) $\frac{20}{4} \cdot \frac{3}{40} \cdot \frac{1}{3}$

(d) $\frac{18}{17} \cdot \frac{1}{36} \cdot \frac{2}{3}$

Answers

3. (a) 2; 1; $\frac{3}{\underset{2}{\cancel{4}}} \cdot \frac{\overset{1}{\cancel{2}}}{5} = \frac{3}{10}$

(b) $\frac{\overset{2}{\cancel{6}}}{\underset{1}{\cancel{11}}} \cdot \frac{\overset{3}{\cancel{33}}}{\underset{7}{\cancel{21}}} = \frac{6}{7}$

(c) $\frac{\overset{1}{\cancel{20}}}{4} \cdot \frac{\overset{1}{\cancel{3}}}{\underset{2}{\cancel{40}}} \cdot \frac{1}{\underset{1}{\cancel{3}}} = \frac{1}{8}$

(d) $\frac{\overset{1}{\cancel{18}}}{17} \cdot \frac{1}{\underset{\underset{1}{\cancel{2}}}{\cancel{36}}} \cdot \frac{\overset{1}{\cancel{2}}}{3} = \frac{1}{51}$

4 Multiply. Write answers in lowest terms and as whole numbers or mixed numbers where possible.

GS (a) $8 \cdot \frac{1}{8}$

$$\frac{\cancel{8}}{\_\_\_} \times \frac{1}{\cancel{8}_1} = \frac{1}{1} =$$

VOCABULARY TIP

Rectangle The formula for finding the area of a rectangle works for any figure with four straight sides and four right angles.

(b) $\frac{3}{4} \cdot 5 \cdot \frac{5}{3}$

(c) $\frac{3}{5} \cdot 40$

(d) $\frac{3}{25} \cdot \frac{5}{11} \cdot 99$

Answers

4. (a) 1; 1; 1 (b) $6\frac{1}{4}$ (c) 24 (d) $5\frac{2}{5}$

(b) $15 \cdot \frac{5}{6}$

$$15 \cdot \frac{5}{6} = \frac{\overset{5}{\cancel{15}}}{1} \cdot \frac{5}{\underset{2}{\cancel{6}}} = \frac{5 \cdot 5}{1 \cdot 2} = \frac{25}{2} = 12\frac{1}{2}$$

◀ **Work Problem 4 at the Side.**

OBJECTIVE 4 Find the area of a rectangle. To find the area of a rectangle (the amount of surface inside the rectangle), use this formula.

Area of a Rectangle

The area of a rectangle is equal to the length multiplied by the width.

Area = length • width

For example, the rectangle shown here has an area of 12 square feet (12 ft^2).

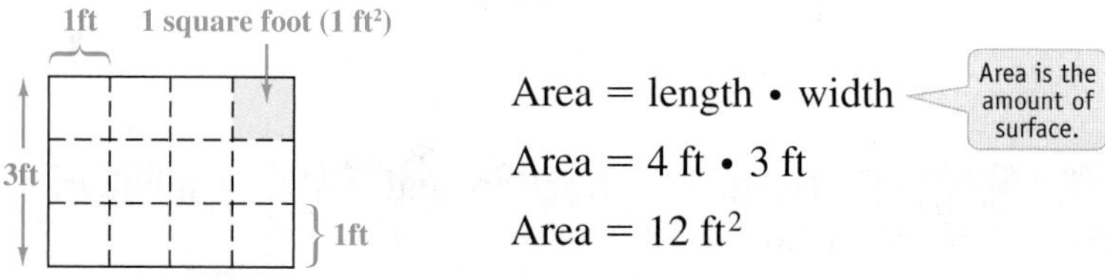

Other units for measuring area are square inches (in.^{2}), square yards (yd^2), and square miles (mi^2). (See **Chapter 8** for more information on area.)

EXAMPLE 5 Applying Fraction Skills

To find the area of each rectangle, multiply its length by its width.

(a) Find the area of each rectangular shower tile.

$\frac{3}{4}$ ft

$\frac{11}{12}$ ft

Area = length • width

$$\text{Area} = \frac{11}{12} \cdot \frac{3}{4}$$

$$= \frac{11}{\underset{4}{\cancel{12}}} \cdot \frac{\overset{1}{\cancel{3}}}{4}$$ Divide numerator and denominator by 3, so 3 ÷ 3 is 1, and 12 ÷ 3 is 4.

$$= \frac{11}{16} \text{ square foot (ft}^2\text{)}$$

Continued on Next Page

(b) Find the area of this rectangular SUV running board.

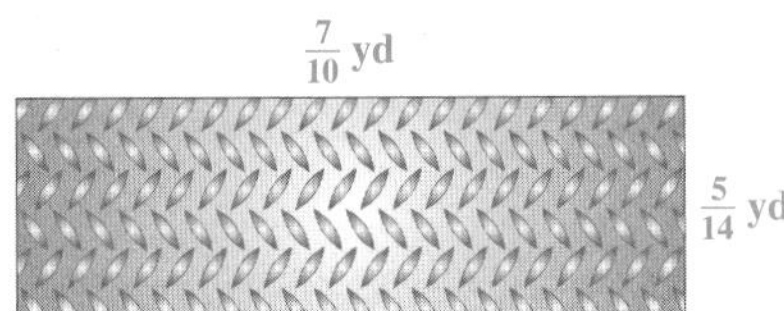

Multiply the length by the width.

$$\text{Area} = \frac{7}{10} \cdot \frac{5}{14}$$

$$= \frac{\overset{1}{\cancel{7}}}{\underset{2}{\cancel{10}}} \cdot \frac{\overset{1}{\cancel{5}}}{\underset{2}{\cancel{14}}} \quad \text{Divide 7 and 14 by 7. Divide 10 and 5 by 5.}$$

$$= \frac{1}{4} \text{ square yard (yd}^2\text{)}$$

Work Problem 5 at the Side. ▶

5 Find the area of each rectangle.

(a)

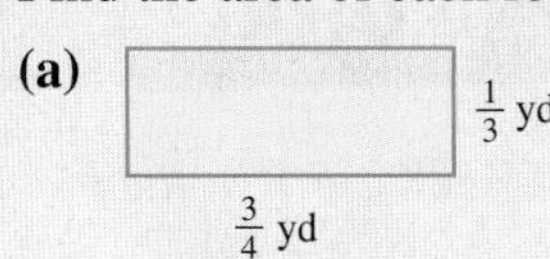

(b) a community college campus that is $\frac{3}{8}$ mile by $\frac{1}{3}$ mile

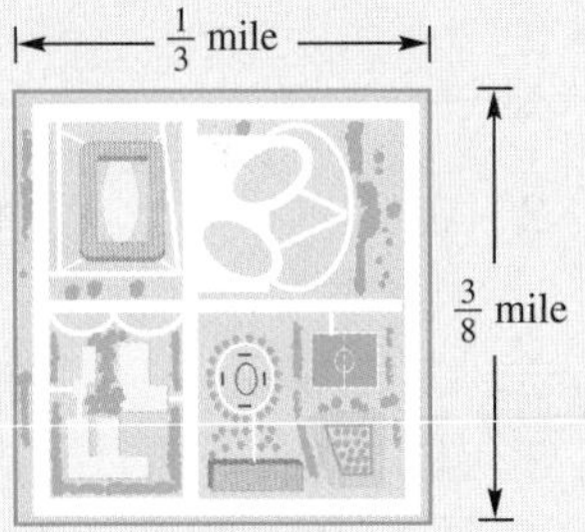

(c) a parcel of land that is $\frac{9}{7}$ mile by $\frac{7}{12}$ mile

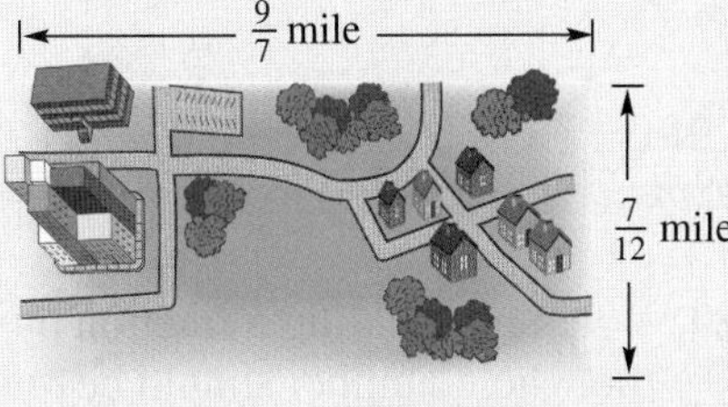

Answers

5. **(a)** $\frac{1}{4}$ yd^2 **(b)** $\frac{1}{8}$ mi^2 **(c)** $\frac{3}{4}$ mi^2

2.5 Exercises

FOR EXTRA HELP

MyMathLab®

CONCEPT CHECK *Fill in the blanks with the correct response.*

1. To multiply two or more fractions, you __________ the numerators and you multiply the __________.

2. To write a fraction answer in lowest terms, you must divide both the __________ and __________ by a common factor.

3. A shortcut when multiplying fractions is to __________ both a numerator and a __________ by the same number.

4. Using the shortcut when multiplying fractions should result in an answer that is in __________.

Multiply. Write answers in lowest terms. ***See Examples 1–3.***

5. $\frac{1}{3} \cdot \frac{3}{4}$

6. $\frac{2}{5} \cdot \frac{3}{4}$

7. $\frac{2}{7} \cdot \frac{1}{5}$

8. $\frac{2}{3} \cdot \frac{1}{2}$

9. $\frac{8}{5} \cdot \frac{15}{32}$

10. $\frac{5}{9} \cdot \frac{4}{3}$

11. $\frac{2}{3} \cdot \frac{7}{12} \cdot \frac{9}{14}$

$$\frac{\overset{1}{\cancel{2}}}{\underset{1}{\cancel{3}}} \cdot \frac{\overset{1}{\cancel{7}}}{\underset{4}{\cancel{12}}} \cdot \frac{\overset{\overset{1}{\cancel{3}}}{\cancel{9}}}{\underset{\underset{1}{\cancel{2}}}{\cancel{14}}} =$$

12. $\frac{7}{8} \cdot \frac{16}{21} \cdot \frac{1}{2}$

$$\frac{\overset{1}{\cancel{7}}}{\underset{1}{\cancel{8}}} \cdot \frac{\overset{\overset{1}{\cancel{2}}}{\cancel{16}}}{\underset{3}{\cancel{21}}} \cdot \frac{1}{\underset{1}{\cancel{2}}} =$$

13. $\frac{3}{4} \cdot \frac{5}{6} \cdot \frac{2}{3}$

14. $\frac{5}{8} \cdot \frac{16}{25}$

15. $\frac{6}{11} \cdot \frac{22}{15}$

16. $\frac{14}{25} \cdot \frac{65}{48} \cdot \frac{15}{28}$

17. $\frac{35}{64} \cdot \frac{32}{15} \cdot \frac{27}{72}$

18. $\frac{16}{25} \cdot \frac{35}{32} \cdot \frac{15}{64}$

19. $\frac{39}{42} \cdot \frac{7}{13} \cdot \frac{7}{24}$

CONCEPT CHECK *Decide whether each of the following is* true *or* false. *If* false, *explain why.*

20. When multiplying a fraction by a whole number, the whole number should be rewritten as the number over 1.

21. $\frac{4}{5} \cdot 8 = \frac{\overset{1}{\cancel{4}}}{5} \cdot \frac{1}{\underset{2}{\cancel{8}}} = \frac{1}{10}$

Multiply. Write answers in lowest terms and as whole or mixed numbers where possible. ***See Example 4.***

22. $5 \cdot \frac{4}{5}$

23. $20 \cdot \frac{3}{4}$

$$\frac{\overset{5}{\cancel{20}}}{1} \cdot \frac{3}{\underset{1}{\cancel{4}}} = \frac{15}{1} =$$

24. $36 \cdot \frac{2}{3}$

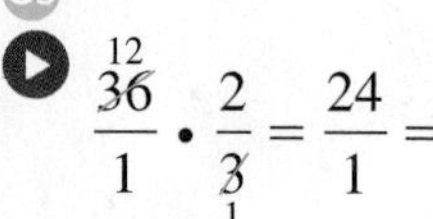

$$\frac{\overset{12}{\cancel{36}}}{1} \cdot \frac{2}{\underset{1}{\cancel{3}}} = \frac{24}{1} =$$

25. $36 \cdot \frac{5}{8} \cdot \frac{9}{15}$

26. $30 \cdot \frac{3}{10}$

27. $100 \cdot \frac{21}{50} \cdot \frac{3}{4}$

28. $400 \cdot \frac{7}{8}$

29. $\frac{2}{5} \cdot 200$

30. $\frac{6}{7} \cdot 245$

31. $142 \cdot \frac{2}{3}$

32. $\frac{12}{25} \cdot 430$

33. $\frac{28}{21} \cdot 640 \cdot \frac{15}{32}$

34. $\frac{21}{13} \cdot 520 \cdot \frac{7}{20}$

35. $\frac{54}{38} \cdot 684 \cdot \frac{5}{6}$

36. $\frac{76}{43} \cdot 473 \cdot \frac{5}{19}$

Find the area of each rectangle. ***See Example 5.***

37.

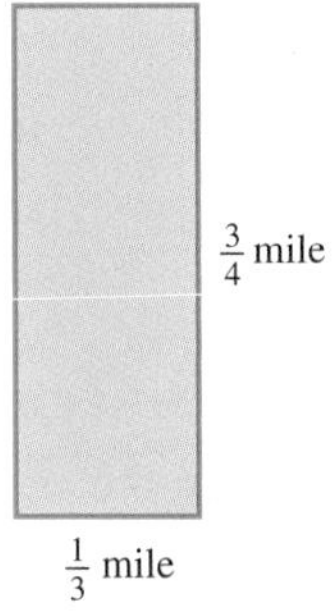

38.
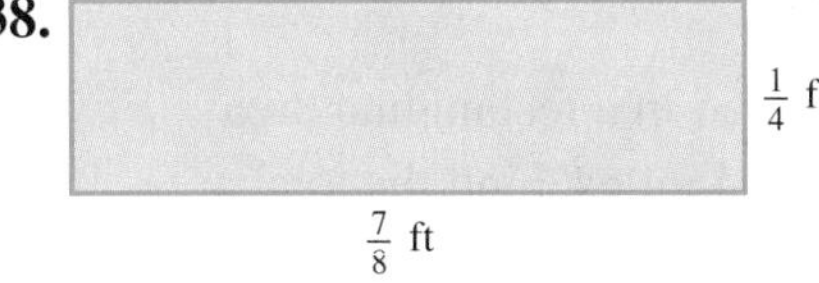

39.
$\frac{3}{4}$ meter
12 meters

40.
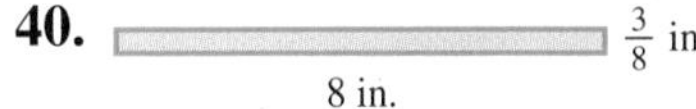

41.
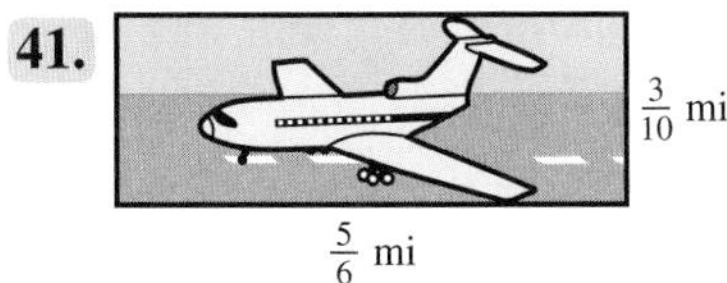

42.
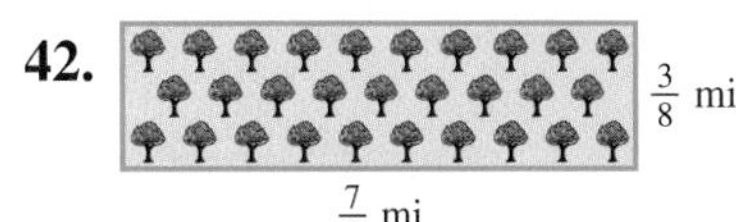

43. Write in your own words the rule for multiplying fractions. Make up an example problem to show how this works.

44. A useful shortcut when multiplying fractions is to divide a numerator and a denominator by the same number. Describe how this works and give an example.

Find the area of each rectangle in these application problems. Write answers in lowest terms and as whole or mixed numbers where possible. ***See Example 5.***

45. Find the area of a heating-duct grill having a length of 2 yd and a width of $\frac{3}{4}$ yd.
Hint: Area = length • width.

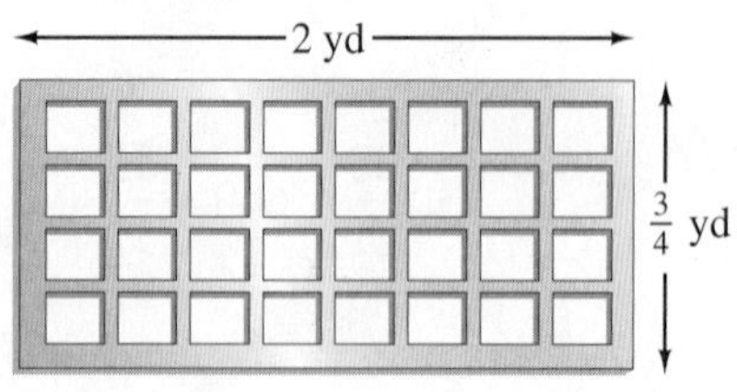

46. Find the area of a HD television having a width of 2 yd and a height of $\frac{15}{16}$ yd.

47. A wildfire is contained in a rectangular area measuring $\frac{7}{8}$ mile by 4 miles. Find the total area of the containment.

48. A rectangular flood plain is $\frac{3}{4}$ mile wide by 7 miles long. Find the area of the flood plain.

49. The Sunny Side Soccer Park is $\frac{1}{4}$ mile long and $\frac{3}{16}$ mile wide, while the Creek Side Soccer Park is $\frac{3}{8}$ mile long and $\frac{1}{8}$ mile wide. Which park has the larger area?

50. The Rocking Horse Ranch is $\frac{3}{4}$ mile long and $\frac{2}{3}$ mile wide. The Silver Spur Ranch is $\frac{5}{8}$ mile long and $\frac{4}{5}$ mile wide. Which ranch has the larger area?

Relating Concepts (Exercises 51–56) For Individual or Group Work

Front end rounding can be used to estimate an answer when multiplying fractions. ***Work Exercises 51–56 in order.*** *Round exact answers to the nearest whole number.*

The bar graph shows the number of supermarkets in several states. The greatest number are in California, while the least number are in Wyoming.

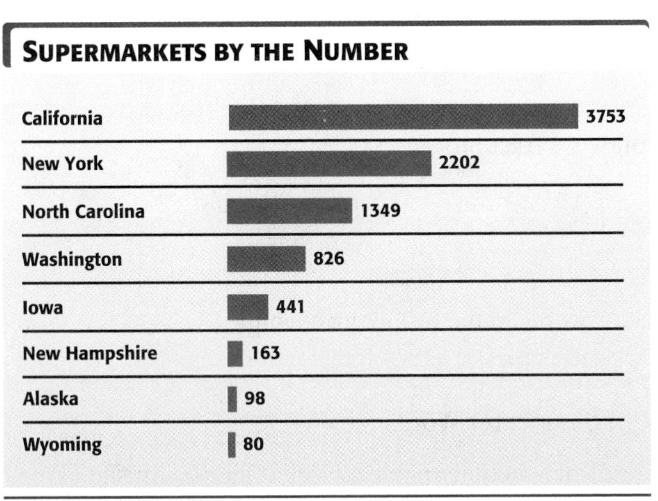

Source: The Nielsen Company.

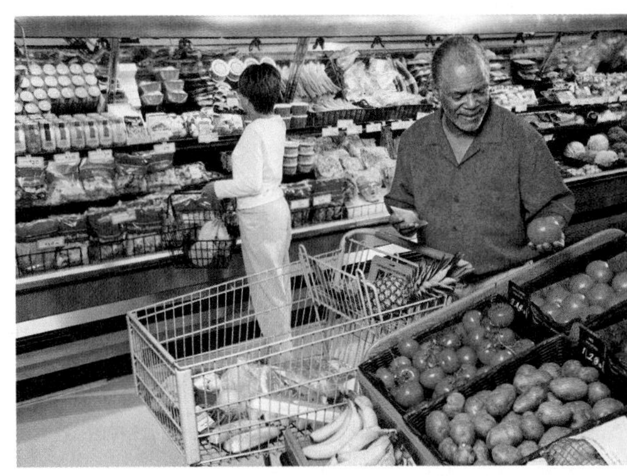

51. Use front end rounding to estimate the total number of supermarkets in these states.

4000 + 2000 + 1000 + 800 + 400 + 200 + 100 + 80 =

______________.

52. Find the exact total number of supermarkets in these states.

3753 + 2202 + 1349 + 826 + 441 + 163 + 98 + 80 =

______________.

53. If $\frac{4}{5}$ of the supermarkets in New York are in medium to large population areas, use front end rounding to estimate, and then find the exact number of stores in these population areas.

Estimate:

Exact:

54. If $\frac{3}{8}$ of the supermarkets in New Hampshire are in shopping centers, use front end rounding to estimate, and then find the exact number of stores in these locations.

Estimate:

Exact:

Compare the estimated answers and the exact answers in ***Exercises 53 and 54.*** *How can you get an estimated answer that is closer to the exact answer? Try rounding the number of stores to some multiple of the denominator in the fraction that has two nonzero digits.*

55. Refer to **Exercise 53.** Round the number of supermarkets in New York to some multiple of the denominator in $\frac{4}{5}$ that has *two* nonzero digits. Now estimate the answer, showing your work.

56. Refer to **Exercise 54.** Round the number of supermarkets in New Hampshire to some multiple of the denominator in $\frac{3}{8}$ that has *two* nonzero digits. Now estimate the answer, showing your work.

2.6 Applications of Multiplication

OBJECTIVE

1 Solve fraction application problems using multiplication.

1 Solve each problem. Use the six problem solving steps.

GS **(a)** Erich and Sabrina Means are saving $\frac{3}{8}$ of their income for the down payment on their first home. If they have a combined annual income of \$81,576, how much can they save in a year?

Estimate:

Round 81,576 to 82,000.

$\frac{1}{2} \cdot \frac{82{,}000}{1} =$ ______

Solve: to find the exact answer.

$\frac{3}{\cancel{8}_{1}} \cdot \frac{\overset{10{,}197}{\cancel{81{,}576}}}{1} =$ ______

State the answer: \$30,591 can be saved in a year.

Check: Exact answer is close to estimate.

(b) A retiring firefighter will receive $\frac{5}{8}$ of her highest annual salary as retirement income. If her highest annual salary is \$62,504, how much will she receive as retirement income?

Answers

1. (a) *Estimate:* \$41,000; *Exact:* \$30,591
   (b) \$39,065

OBJECTIVE 1 Solve fraction application problems using multiplication. Many application problems are solved by multiplying fractions. Use the following indicator words for multiplication.

product
double
triple
times
of (when "of" follows a fraction)
twice
twice as much

Always look for indicator words.

Look for these indicator words in the following examples.

EXAMPLE 1 Applying Indicator Words

Lois Stevens gives $\frac{1}{10}$ of her income to her church. One month she earned \$2980. How much did she give to the church that month?

Step 1 **Read** the problem. The problem asks us to find the amount of money given to the church.

Step 2 **Work out a plan.** Stevens gave $\frac{1}{10}$ ***of*** her income. The indicator word is ***of***. When it follows a fraction, the word ***of*** indicates multiplication, so find the amount given to the church by multiplying $\frac{1}{10}$ and \$2980.

Step 3 **Estimate** a reasonable answer. Round the income of \$2980 to \$3000. Then divide \$3000 by 10 to find $\frac{1}{10}$ of the income (one of 10 equal parts). Our estimate is $\$3000 \div 10 = \$300$. (Recall the shortcut for dividing by 10; drop one 0 from the dividend.)

Step 4 **Solve** the problem.

$$\text{amount} = \frac{1}{\cancel{10}_{1}} \cdot \frac{\overset{298}{\cancel{2980}}}{1} = \frac{298}{1} = 298$$

Divide by the common factor of 10.

Step 5 **State the answer.** Stevens gave \$298 to her church that month.

Step 6 **Check.** The exact answer, \$298, is close to our estimate of \$300.

◀ Work Problem **1** at the Side.

EXAMPLE 2 Solving a Fraction Application Problem

Of the 39 students in Sharon Martin's high school economics class, $\frac{2}{3}$ plan to go to college. How many plan to go to college?

Step 1 **Read** the problem. The problem asks us to find the number of students who plan to go to college.

Continued on Next Page

Step 2 **Work out a plan.** Reword the problem to read

$\frac{2}{3}$ of the students plan to go to college.
↑
Indicator word for multiplication when it follows a fraction

Step 3 **Estimate** a reasonable answer. Round the number of students in the class from 39 to 40. Then, $\frac{1}{2}$ of 40 is 20. Since $\frac{2}{3}$ is more than $\frac{1}{2}$, our estimate is that "more than 20 students" plan to go to college.

Step 4 **Solve** the problem. Find the number who plan to go to college by multiplying $\frac{2}{3}$ and 39.

$$\text{number who plan to go} = \frac{2}{3} \cdot 39$$

$$= \frac{2}{\cancel{3}_1} \cdot \frac{\cancel{39}^{13}}{1} = \frac{26}{1} = 26$$

Step 5 **State the answer.** 26 students plan to go to college.

Check that the answer is reasonable.

Step 6 **Check.** The exact answer, 26, fits our estimate of "more than 20."

Work Problem **2** at the Side. ▶

2 At Sid's Pharmacy, $\frac{5}{16}$ of the prescriptions are paid by a third party (insurance company). If 3696 prescriptions are filled, find the number paid by a third party. Use the six problem-solving steps.

EXAMPLE 3 Finding a Fractional Part of a Fraction

In her will, a woman divides her estate into 6 equal parts. Five of the 6 parts are given to relatives. Of the sixth part, $\frac{1}{3}$ goes to the Salvation Army. What fraction of her total estate goes to the Salvation Army?

Step 1 **Read** the problem. The problem asks for the fraction of an estate that goes to the Salvation Army.

Step 2 **Work out a plan.** Reword the problem to read

the Salvation Army gets $\frac{1}{3}$ of $\frac{1}{6}$.
↑
Indicator word for multiplication when it follows a fraction

Step 3 Estimate a reasonable answer. If the estate is divided into 6 equal parts and each of these parts was divided into 3 equal parts, we would have 6 • 3 = 18 equal parts. Our estimate is $\frac{1}{18}$.

Step 4 **Solve** the problem. The Salvation Army gets $\frac{1}{3}$ of $\frac{1}{6}$. (Indicator word)

To find the fraction that the Salvation Army is to receive, multiply $\frac{1}{3}$ and $\frac{1}{6}$.

$$\text{fraction to Salvation Army} = \frac{1}{3} \cdot \frac{1}{6}$$

$$= \frac{1}{18}$$

Step 5 **State the answer.** The Salvation Army gets $\frac{1}{18}$ of the total estate.

Step 6 **Check.** The exact answer, $\frac{1}{18}$, matches our estimate.

Remember to check your work.

Work Problem **3** at the Side. ▶

3 At our college, $\frac{1}{3}$ of the students speak a foreign language. Of those speaking a foreign language, $\frac{3}{4}$ speak Spanish. What fraction of the students speak Spanish? Use the six problem-solving steps.

GS

$\frac{1}{3} \cdot \frac{3}{4}$

$\frac{1}{\cancel{3}} \cdot \frac{\cancel{3}^{1}}{\_\_\_\_} = \_\_\_\_\_\_$

Answers

2. 1155 prescriptions

3. 1; 4; $\frac{1}{4}$ speak Spanish

4 Solve each problem using the six problem-solving steps. Use the circle graph in **Example 4.**

(a) What fraction of the children buy food from vending machines?

GS **(b)** What number of children buy food from vending machines?

Estimate: $\frac{1}{5}$ of 2500 is 500

Solve: $\frac{1}{\cancel{5}_1} \cdot \frac{\cancel{2500}^{500}}{1} = $ ____

State the answer: 500 children

Check: Answer is the same as the estimate.

(c) What fraction of the children buy food from a convenience store or street vendor?

(d) What number of children buy food from a convenience store or street vendor?

Answers

4. **(a)** $\frac{1}{5}$ **(b)** 5; 500 children
(c) $\frac{1}{10}$ **(d)** 250 children

EXAMPLE 4 Using Fractions with a Circle Graph

The circle graph, or pie chart, shows where children 8 to 17 years of age are most likely to make food purchases when away from home. If 2500 children were in the survey, find the number of children who buy food in the school cafeteria.

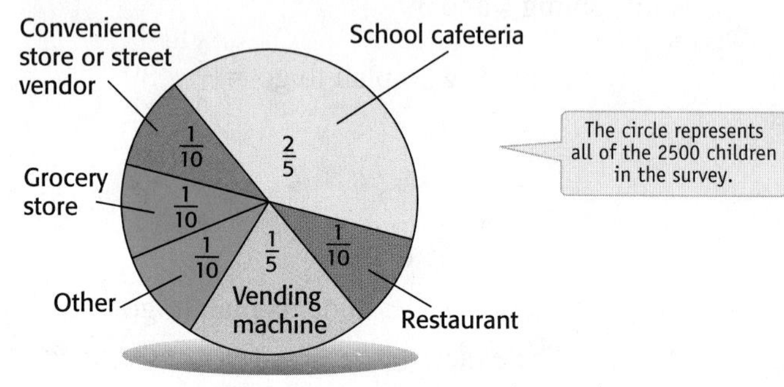

Source: Pursuant Inc. for American Dietetic Association Foundation.

Step 1 **Read** the problem. The problem asks for the number of children who buy food in the school cafeteria.

Step 2 **Work out a plan.** Reword the problem to read

$\frac{2}{5}$ of 2500 children buy food in the school cafeteria.

Indicator word for multiplication when it follows a fraction

Step 3 **Estimate** a reasonable answer. $\frac{1}{2}$ of 2500 people is 1250 people. $\frac{2}{5}$ is less than $\frac{1}{2}$, so our estimate is "less than 1250 people."

Step 4 **Solve** the problem. Find the number who buy food in the school cafeteria by multiplying $\frac{2}{5}$ and 2500.

$$\text{number in school cafeteria} = \frac{2}{5} \cdot 2500$$

$$= \frac{2}{5} \cdot \frac{2500}{1}$$

$$= \frac{2}{\cancel{5}_1} \cdot \frac{\cancel{2500}^{500}}{1}$$

Divide both numerator and denominator by 5.

$$= \frac{1000}{1} = 1000$$

Step 5 **State the answer.** 1000 children buy food in the school cafeteria.

Step 6 **Check.** The exact answer, 1000 children, fits our estimate of "less than 1250 children."

◀ Work Problem **4** at the Side.

2.6 Exercises

FOR EXTRA HELP

MyMathLab®

1. CONCEPT CHECK Circle the words that are indicator words for multiplication.

more than	of	sum
times	twice	difference
triple	greater than	product
less than	all together	quotient
divided evenly	after a decrease	twice as much

2. CONCEPT CHECK Underline the correct answer. The final step when solving an application problem is to (*solve the problem/check your work*),

3. CONCEPT CHECK Fill in the blank. When you multiply length by width you are finding the ______ of a rectangular surface.

4. CONCEPT CHECK Fill in the blank. When calculating area, the length and the width must be in the same units of measurement. If the measurements are both in miles, the answer will be in ______ miles and shown as mi^2.

Solve each application problem. Look for indicator words. ***See Examples 1–4.***

5. A digital photo frame measures $\frac{3}{4}$ ft by $\frac{2}{3}$ ft. Find its area.

Hint: Area = length • width

6. The rectangular floor of Darby's dog house measures $\frac{14}{15}$ yd by $\frac{3}{4}$ yd. Find its area.

7. A cookie sheet is $\frac{4}{3}$ ft by $\frac{2}{3}$ ft. Find its area.

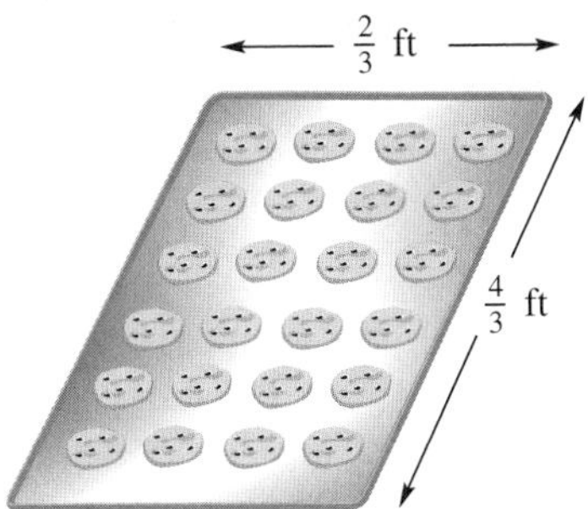

8. Each day there are 16 million people who shop at flea markets. If $\frac{2}{5}$ of these people purchase produce at the flea market, how many purchase produce? (*Source:* National Flea Market Association.)

9. Pete is helping Collin make a rectangular mahogany lamp table for Carolyn's birthday. Find the area of the top of the table if it is $\frac{4}{5}$ yd long by $\frac{3}{8}$ yd wide.

10. The average person consumes 160 bowls of cereal each year. If $\frac{3}{10}$ of the cereal is eaten in the summer months, how many bowls of cereal are eaten in the summer months? (*Source:* Target Stores.)

11. Dan Crump had expenses of $6848 during one semester of college. His part-time job provided $\frac{3}{8}$ of the amount he needed. How much did he earn on his job?

12. The average household does 400 loads of wash each year. If $\frac{3}{8}$ of the wash loads are done in the winter months, how many wash loads are done in the winter months? (*Source:* Target Stores.)

13. The city with the most expensive daily parking fee is New York City (Midtown) at $40. The daily parking fee in Boston (third most expensive) is $\frac{7}{8}$ as much as New York City. Find the daily parking fee in Boston. (*Source:* Colliers International.)

14. The daily parking fee in Boston (third most expensive) is $35. In San Francisco (fourth most expensive), the daily parking fee is $\frac{4}{5}$ the cost of Boston. How much is the daily parking fee in San Francisco? (*Source:* Colliers International.)

15. At the Garlic Festival Fun Run, $\frac{7}{12}$ of the runners are women. If there are 1560 runners, how many of the runners are

(a) women?

(b) men?

16. A hotel has 408 rooms. Of these rooms, $\frac{9}{17}$ are nonsmoking rooms. How many rooms are

(a) nonsmoking?

(b) smoking?

Most electric cars can go 100 miles before they need to be recharged, which could take up to eight hours. The circle graph below shows how long people would be willing to wait to recharge an electric car. Use this information to work Exercises 17–20.

TIME TO RECHARGE

How long would you wait to recharge your electric car?

1020 people surveyed

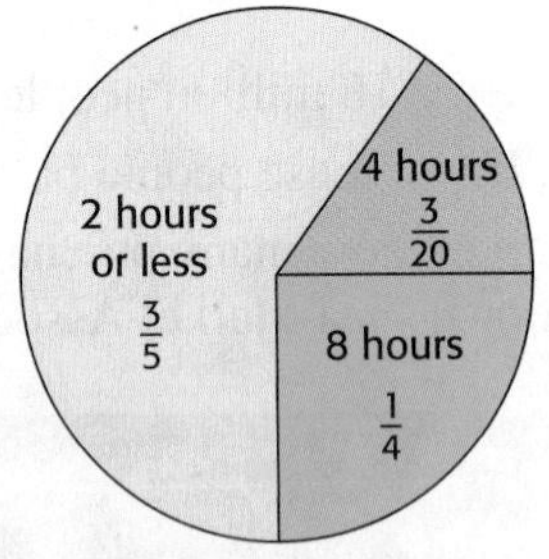

Source: Deloitte Touch Tohmatsu Limited [DTTL].

17. Which response was given by the least number of people? How many people gave this response?

18. Which response was given by the greatest number of people? How many gave this response?

19. Find the fraction and the total number of people who would be willing to wait 4 hours or less to recharge the car.

20. Find the fraction and the total number of people who would be willing to wait 4 hours or more to recharge the car.

21. Without actually adding the fractions given for all the groups in the "Time to Recharge" circle graph, explain why their sum has to be 1.

22. Refer to **Exercise 21.** Suppose you added all the fractions for the groups and did not get 1 as an answer. List some possible explanations.

The table shows the earnings for the Owens family last year, and the circle graph shows how they spent their earnings. Use this information to answer Exercises 23–28.

Month	Earnings	Month	Earnings
January	\$6075	July	\$7040
February	\$5812	August	\$5232
March	\$6488	September	\$5670
April	\$6030	October	\$7012
May	\$5820	November	\$6465
June	\$6398	December	\$7958

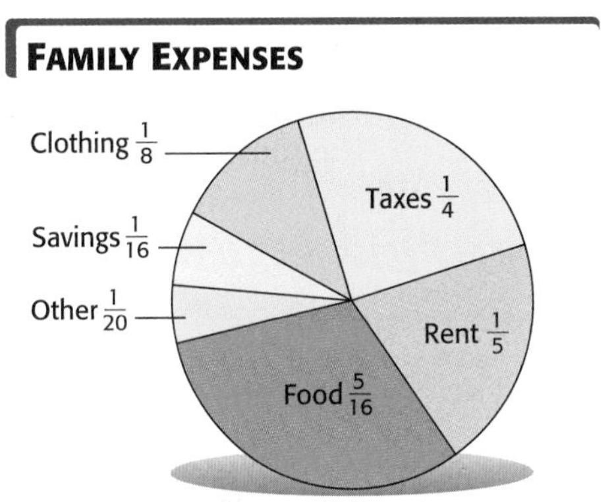

23. Find the Owens family's total income for the year.

24. How much of their annual earnings went to taxes?

25. Find the amount of their rent for the year.

26. How much did they spend for food during the year?

27. Find their annual savings.

28. How much of their annual income was spent on clothing?

29. Here is how one student solved a multiplication problem. Find the error and solve the problem correctly.

$$\frac{9}{10} \times \frac{20}{21} = \frac{\overset{3}{\cancel{9}}}{\underset{1}{\cancel{10}}} \times \frac{\overset{2}{\cancel{20}}}{\underset{3}{\cancel{21}}} = \frac{6}{3} = 2$$

30. When two whole numbers are multiplied, the product is always larger than the numbers being multiplied. When two proper fractions are multiplied, the product is always smaller than the numbers being multiplied. Are these statements true? Why or why not?

Solve each application problem.

31. The cost of laser eye surgery in the United States is \$2000 for each eye. The same surgery in Thailand is $\frac{3}{8}$ of this amount. What is the cost of this procedure in Thailand? (*Source: Reader's Digest.*)

32. A knee replacement in the United States costs \$36,300, while in Mexico the same procedure costs $\frac{3}{50}$ of this amount. Find the cost of a knee replacement in Mexico. (*Source: Reader's Digest.*)

33. A collector of scale model World War II ships wants to know the length of a $\frac{1}{128}$ scale model of a ship that was 256 feet in length. Find the length of the scale model. (*Source:* Lilliput Motor Company, LTD.)

34. United Parcel Service (UPS) has trucks built to their specifications and they weigh 10,000 pounds. The company is testing a new truck that weighs $\frac{1}{10}$ less than the currently used trucks in order to save on fuel. How much does the test truck weigh?

35. LaDonna Washington is running for city council. She needs to get $\frac{2}{3}$ of her votes from senior citizens and 27,000 votes in all to win. How many votes does she need from voters other than the senior citizens?

36. The average debt of a graduate with a bachelor's degree from a four-year college is \$24,400. If $\frac{5}{16}$ of the amount was borrowed in the last year of college, find the amount borrowed in the first years. (*Source:* U.S. Department of Education.)

37. A will states that $\frac{7}{8}$ of an estate is to be divided among relatives. Of the remaining estate, $\frac{1}{4}$ goes to the American Cancer Society. What fraction of the estate goes to the American Cancer Society?

38. A couple has $\frac{2}{5}$ of their total investments in real estate. Of the remaining investments, $\frac{1}{3}$ is invested in bonds. What fraction of the total investments is in bonds?

2.7 Dividing Fractions

OBJECTIVES

1. Find the reciprocal of a fraction.
2. Divide fractions.
3. Solve application problems in which fractions are divided.

OBJECTIVE 1 **Find the reciprocal of a fraction.** To divide fractions, we need to know how to find the **reciprocal** of a fraction.

Reciprocal of a Fraction

Two numbers are reciprocals of each other if their product is 1. To find the reciprocal of a fraction, switch the numerator and denominator.

For example, the reciprocal of $\frac{3}{4}$ is $\frac{4}{3}$.

Fraction $\frac{3}{4} \times \frac{4}{3}$ Reciprocal

Note

Notice that you invert, or "flip," a fraction to find its reciprocal.

EXAMPLE 1 Finding Reciprocals

Find the reciprocal of each fraction.

Flip a fraction to find the reciprocal.

(a) The reciprocal of $\frac{1}{4}$ is $\mathbf{\frac{4}{1}}$ because $\frac{1}{4} \cdot \frac{4}{1} = \frac{4}{4} = 1$

(b) The reciprocal of $\frac{2}{3}$ is $\mathbf{\frac{3}{2}}$ because $\frac{2}{3} \cdot \frac{3}{2} = \frac{6}{6} = 1$

(c) The reciprocal of $\frac{3}{5}$ is $\mathbf{\frac{5}{3}}$ because $\frac{3}{5} \cdot \frac{5}{3} = \frac{15}{15} = 1$

(d) The reciprocal of 8 is $\mathbf{\frac{1}{8}}$ because $\frac{8}{1} \cdot \frac{1}{8} = \frac{8}{8} = 1$ Think of 8 as $\frac{8}{1}$.

Work Problem 1 at the Side. ▶

Note

Every number has a reciprocal except 0. The number 0 has no reciprocal because there is no number that can be multiplied by 0 to get 1.

$$0 \cdot (\text{reciprocal}) = 1$$

There is no number to use here that will give an answer of 1. When you multiply by 0, you always get 0.

VOCABULARY TIP

Reciprocal Taking the reciprocal of a number involves alternating, or switching the numerator and denominator.

1 Find the reciprocal of each fraction.

GS **(a)** $\frac{4}{5} \times \frac{5}{\_\_}$

GS **(b)** $\frac{3}{8} \times \frac{\_\_}{3}$

(c) $\frac{9}{4}$

(d) 16

Answers

1. **(a)** 4; $\frac{5}{4}$ **(b)** 8; $\frac{8}{3}$ **(c)** $\frac{4}{9}$ **(d)** $\frac{1}{16}$

In **Chapter 1,** we saw that the division problem $12 \div 3$ asks how many 3s are in 12. In the same way, the division problem $\frac{2}{3} \div \frac{1}{6}$ asks how many $\frac{1}{6}$s are in $\frac{2}{3}$. The figure illustrates $\frac{2}{3} \div \frac{1}{6}$.

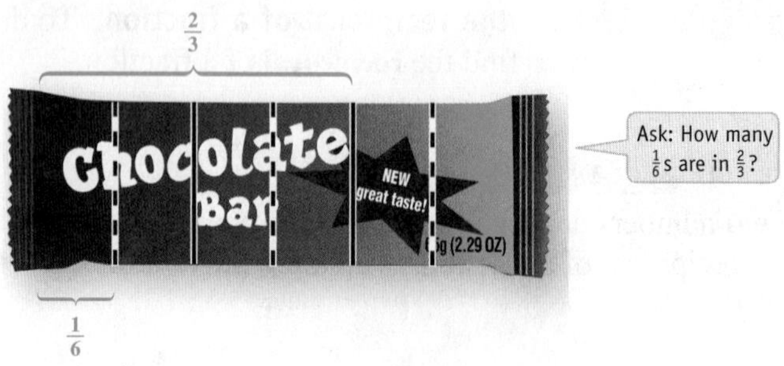

The figure shows that there are 4 of the $\frac{1}{6}$ pieces in $\frac{2}{3}$, or

$$\frac{2}{3} \div \frac{1}{6} = 4.$$

OBJECTIVE 2 Divide fractions. We will use reciprocals to divide fractions.

Dividing Fractions

To divide two fractions, multiply the first fraction by the reciprocal of the divisor (the second fraction).

EXAMPLE 2 Dividing One Fraction by Another

Divide. Write answers in lowest terms and as mixed numbers where possible.

(a) $\frac{7}{8} \div \frac{15}{16}$

The reciprocal of $\frac{15}{16}$ is $\frac{16}{15}$. (Be sure to find the reciprocal of the divisor (the second fraction).)

Reciprocals

$$\frac{7}{8} \div \frac{15}{16} = \frac{7}{8} \cdot \frac{16}{15}$$

Change division to multiplication. (Use $\frac{16}{15}$, the reciprocal of $\frac{15}{16}$, and multiply.)

$$= \frac{7}{\underset{1}{\cancel{8}}} \cdot \frac{\overset{2}{\cancel{16}}}{15}$$ Divide the numerator and denominator by the common factor of 8.

$$= \frac{7 \cdot 2}{1 \cdot 15}$$ Multiply.

$$= \frac{14}{15}$$ ← Lowest terms

Continued on Next Page

(b) $\dfrac{\frac{4}{5}}{\frac{3}{10}}$

$$\frac{\frac{4}{5}}{\frac{3}{10}} = \frac{4}{5} \div \frac{3}{10}$$ Rewrite by using the ÷ symbol for division.

$$= \frac{4}{\cancel{5}_1} \cdot \frac{\cancel{10}^2}{3}$$ The reciprocal of $\frac{3}{10}$ is $\frac{10}{3}$. Change ÷ to "•". Divide the numerator and denominator by the common factor of 5.

$$= \frac{4 \cdot 2}{1 \cdot 3}$$ Multiply.

$$= \frac{8}{3} = 2\frac{2}{3}$$ Rewrite the answer as a mixed number.

CAUTION

Be certain that the divisor fraction is changed to its reciprocal *before* you divide numerators and denominators by common factors.

Work Problem 2 at the Side. ▶

EXAMPLE 3 **Dividing with a Whole Number**

Divide. Write all answers in lowest terms and as whole or mixed numbers where possible.

(a) $5 \div \frac{1}{4}$ (*Hint:* How many quarters are in \$5?)

Write 5 as $\frac{5}{1}$. Next, use the reciprocal of $\frac{1}{4}$, which is $\frac{4}{1}$.

$$5 \div \frac{1}{4} = \frac{5}{1} \cdot \frac{4}{1}$$ Reciprocal of $\frac{1}{4}$ is $\frac{4}{1}$. (Reciprocals)

$$= \frac{5 \cdot 4}{1 \cdot 1}$$ Multiply.

$$= \frac{20}{1} = 20$$ Rewrite the answer as a whole number.

Continued on Next Page

2 Divide. Write answers in lowest terms and as mixed numbers where possible.

GS **(a)** $\frac{1}{4} \div \frac{2}{3}$

$$\frac{1}{4} \cdot \frac{3}{\_\_\_} = \frac{\_\_\_}{8}$$

(b) $\frac{3}{8} \div \frac{5}{8}$

GS **(c)** $\dfrac{\frac{2}{3}}{\frac{4}{5}} = \frac{2}{3} \div \frac{4}{5}$

$$= \frac{\cancel{2}^1}{3} \cdot \frac{\_\_\_}{\cancel{4}_2}$$

$$= \frac{1 \cdot \_\_\_}{3 \cdot 2} =$$

(d) $\dfrac{\frac{5}{6}}{\frac{7}{12}}$

Answers

2. **(a)** 2; 3; $\frac{3}{8}$ **(b)** $\frac{3}{5}$ **(c)** 5; 5; $\frac{5}{6}$ **(d)** $1\frac{3}{7}$

3 Divide. Write answers in lowest terms and as whole or mixed numbers where possible.

GS (a) $10 \div \frac{1}{2}$

$= \frac{\_\_\_}{1} \cdot \frac{2}{1}$

$= \frac{10 \cdot 2}{1 \cdot 1} =$

(b) $6 \div \frac{6}{7}$

(c) $\frac{4}{5} \div 6$

(d) $\frac{3}{8} \div 4$

4 Solve each problem using the six problem-solving steps.

(a) How many $\frac{5}{6}$-ounce dispensers can be filled with 40 ounces of eye drops?

(b) Find the number of $\frac{4}{5}$-quart bottles that can be filled from a 120-quart cask.

Answers

3. (a) 10; 20 (b) 7 (c) $\frac{2}{15}$ (d) $\frac{3}{32}$

4. (a) 48 dispensers (b) 150 bottles

(b) $\frac{2}{3} \div 6$

Write 6 as $\frac{6}{1}$. The reciprocal of $\frac{6}{1}$ is $\frac{1}{6}$.

$$\frac{2}{3} \div \frac{6}{1} = \frac{2}{3} \cdot \frac{1}{6}$$

Reciprocals

Careful! Divide out common factors **after** changing to the reciprocal and to multiplication.

$$= \frac{\overset{1}{\cancel{2}}}{3} \cdot \frac{1}{\underset{3}{\cancel{6}}}$$ Divide the numerator and denominator by 2, then multiply.

$$= \frac{1 \cdot 1}{3 \cdot 3} = \frac{1}{9}$$ Lowest terms

◀ Work Problem **3** at the Side.

OBJECTIVE 3 Solve application problems in which fractions are divided. Many application problems require division of fractions. Recall from **Chapter 1** that typical indicator words for division are ***goes into, per, divide, divided by, divided equally,*** and ***divided into.***

EXAMPLE 4 Applying Fraction Skills

Goldie, the manager of the Burnside Deli, must fill a 10-gallon kosher dill pickle crock with salt brine. She has only a $\frac{2}{3}$-gallon container to use. How many times must she fill the $\frac{2}{3}$-gallon container and empty it into the 10-gallon crock?

Step 1 **Read** the problem. We need to find the number of times Goldie needs to use a $\frac{2}{3}$-gallon container in order to fill a 10-gallon crock.

Step 2 **Work out a plan.** We can solve the problem by finding the number of times $\frac{2}{3}$ goes into 10.

Step 3 **Estimate** a reasonable answer. Round $\frac{2}{3}$ gallon to 1 gallon. In order to fill the 10-gallon container, she would have to use the 1-gallon container 10 times, so our estimate is 10.

Step 4 **Solve** the problem.

Reciprocals

$$10 \div \frac{2}{3} = \frac{10}{1} \cdot \frac{3}{2}$$ The reciprocal of $\frac{2}{3}$ is $\frac{3}{2}$. Change "÷" to "•."

$$= \frac{\overset{5}{\cancel{10}}}{1} \cdot \frac{3}{\underset{1}{\cancel{2}}}$$ Divide the numerator and denominator by 2, and then multiply.

$$= \frac{15}{1} = 15$$

Step 5 **State the answer.** Goldie must fill the container 15 times.

Step 6 **Check.** The exact answer, 15 times, is reasonably close to our estimate of 10 times.

◀ Work Problem **4** at the Side.

EXAMPLE 5 Applying Fraction Skills

At the Happi-Time Day Care Center, $\frac{6}{7}$ of the total budget goes to classroom operation. If there are 18 classrooms and each one receives the same amount, what fraction of the operating amount does each classroom receive?

Step 1 **Read** the problem. Since $\frac{6}{7}$ of the total budget must be split into 18 parts, we must find the fraction of the classroom operating amount received by each classroom.

Step 2 **Work out a plan.** We must divide the fraction of the total budget going to classroom operation $\left(\frac{6}{7}\right)$ by the number of classrooms (18).

Step 3 **Estimate** a reasonable answer. Round $\frac{6}{7}$ to 1. If all of the operating expenses (1 whole) were divided between 18 classrooms, each classroom would receive $\frac{1}{18}$ of the operating expenses, our estimate.

Step 4 **Solve** the problem. We solve by dividing $\frac{6}{7}$ by 18.

$$\frac{6}{7} \div 18 = \frac{6}{7} \div \frac{18}{1}$$ Write 18 as $\frac{18}{1}$.

$$= \frac{\overset{1}{\cancel{6}}}{7} \cdot \frac{1}{\underset{3}{\cancel{18}}}$$ The reciprocal of $\frac{18}{1}$ is $\frac{1}{18}$. Change "÷" to "•". Divide the numerator and denominator by 6.

$$= \frac{1}{21}$$ Multiply.

Step 5 **State the answer.** Each classroom receives $\frac{1}{21}$ of the total budget.

Step 6 **Check.** The exact answer, $\frac{1}{21}$, is close to our estimate of $\frac{1}{18}$.

Work Problem 5 at the Side. ▶

---

5 Solve each problem using the six problem-solving steps.

GS **(a)** The top 12 employees at Mayfield Manufacturing will divide $\frac{3}{4}$ of the annual bonus money. What fraction of the bonus money will each employee receive?

Estimate: $1 \div 12$ is $\frac{1}{12}$

Solve: $\frac{3}{4} \div 12 = \frac{3}{4} \div \frac{12}{1}$

$= \frac{\overline{}3}{4} \cdot \frac{1}{\cancel{12}}$

$= \underline{}$

State the answer: = ____ of the bonus money

Check: Answer is close to estimate.

(b) A winning lottery ticket was purchased by 8 employees of United States Marketing and Promotions (USMP). They will donate $\frac{1}{5}$ of the total winnings to pay the medical expenses of a fellow employee, and then divide the remaining winnings evenly. What fraction of the prize money will each receive?

Answers

5. (a) 1; 4; $\frac{1}{16}$ of the bonus money

(b) $\frac{1}{10}$ of the lottery prize money

2.7 Exercises

FOR EXTRA HELP

Download the MyDashBoard App

MyMathLab®

CONCEPT CHECK *Fill in the blank.*

1. When you invert or flip a fraction, you have the __________ of the fraction.

2. To find the reciprocal of a whole number, you must first write the whole number over __________, and then invert it.

3. To divide by a fraction, you must first __________ the divisor and then change division to __________.

4. After completing a fraction division problem, it is best to write the answer in __________ terms.

CONCEPT CHECK *Find the reciprocal of each number.*

5. $\frac{3}{8}$

6. $\frac{2}{5}$

7. $\frac{5}{6}$

8. $\frac{12}{7}$

9. $\frac{8}{5}$

10. $\frac{13}{20}$

11. 4

12. 10

Divide. Write answers in lowest terms and as whole or mixed numbers where possible. See ***Examples 2 and 3.***

13. $\frac{1}{2} \div \frac{3}{4}$

14. $\frac{5}{8} \div \frac{7}{8}$

15. $\frac{7}{8} \div \frac{1}{3}$

16. $\frac{7}{8} \div \frac{3}{4}$

17. $\frac{3}{4} \div \frac{5}{3}$

18. $\frac{4}{5} \div \frac{9}{4}$

19. $\frac{7}{9} \div \frac{7}{36}$

20. $\frac{5}{8} \div \frac{5}{16}$

21. $\frac{15}{32} \div \frac{5}{64}$

22. $\frac{7}{12} \div \frac{14}{15}$

23. $\dfrac{\frac{13}{20}}{\frac{4}{5}}$

24. $\dfrac{\frac{9}{10}}{\frac{3}{5}}$

25. $\dfrac{\frac{5}{6}}{\frac{25}{24}}$

$\frac{5}{6} \div \frac{25}{24}$

$\frac{\overset{1}{\cancel{5}}}{\underset{1}{\cancel{6}}} \cdot \frac{\overset{4}{\cancel{24}}}{\underset{5}{\cancel{25}}} =$

26. $\dfrac{\frac{28}{15}}{\frac{21}{5}}$

$\frac{28}{15} \div \frac{21}{5}$

$\frac{\overset{4}{\cancel{28}}}{\underset{3}{\cancel{15}}} \cdot \frac{\overset{1}{\cancel{5}}}{\underset{3}{\cancel{21}}} =$

27. $12 \div \frac{2}{3}$

28. $7 \div \frac{1}{4}$

29. $\dfrac{18}{\frac{3}{4}}$

30. $\dfrac{12}{\frac{3}{4}}$

31. $\dfrac{\frac{4}{7}}{8}$

32. $\dfrac{\frac{7}{10}}{3}$

Solve each application problem by using division. ***See Examples 4 and 5.***

33. Veterinarian Jasmine Cato has $\frac{8}{9}$ quart of medication. She prescribes this medication for 4 horses in her care. If she divides the medication evenly, how much will each horse receive?

$$\frac{8}{9} \div 4 = \frac{\overset{2}{\cancel{8}}}{9} \cdot \frac{1}{\underset{1}{\cancel{4}}} =$$

34. Harold Pishke, barber, has 15 quarts of conditioning shampoo. If he wants to put this shampoo into $\frac{3}{8}$-quart containers, how many containers can be filled?

$$\frac{15}{1} \div \frac{3}{8} = \frac{\overset{5}{\cancel{15}}}{1} \cdot \frac{8}{\underset{1}{\cancel{3}}} =$$

35. Some college roommates want to make pancakes for their neighbors. They need 5 cups of flour, but have only a $\frac{1}{3}$-cup measuring cup. How many times will they need to fill their measuring cup?

36. How many $\frac{3}{8}$ pound bags of Jelly Belly jelly beans can be filled with 408 pounds of jelly beans? (*Source:* Jelly Belly Candy Company.)

37. How many $\frac{1}{8}$-ounce eye drop dispensers can be filled with 11 ounces of eye drops?

38. It is estimated that each guest at a party will eat $\frac{5}{16}$ pound of peanuts. How many guests can be served with 10 pounds of peanuts?

39. Metal fasteners used in furniture assembly weigh $\frac{5}{32}$ pound and are packaged in 25-pound cartons. Find the number of fasteners in each carton.

40. The new residential subdivision features large $\frac{3}{4}$-acre lots. How many lots are there in the 210- acre- subdivision? (Do not consider streets, sidewalks, or other improvements.)

41. Your classmate is confused on how to divide by a fraction. Write a short note telling him how this should be done.

42. If you multiply positive proper fractions, the product is less than the fractions multiplied. When you divide by a proper fraction, is the quotient less than the fractions in the problem? Prove your answer with examples.

Solve each application problem using multiplication or division.

43. The recipe for a Jelly Belly Express loafcake calls for $\frac{3}{4}$ pound of Jelly Belly jelly beans in assorted colors. If you want to make 16 cakes, how many pounds of Jelly Belly jelly beans will you need?

$$\frac{3}{4} \cdot 16 = \frac{3}{\cancel{4}_{1}} \cdot \frac{\cancel{16}^{4}}{1} =$$

44. In a recent study, it was found that one month after leaving the hospital only $\frac{7}{8}$ of the 1520 heart attack patients were still taking the life-saving drugs prescribed for them. How many of these patients were still taking their drugs? (*Source:* Dr. Michael Ho, Denver Veterans Medical Center.)

45. Broadly Plumbing finds that $\frac{3}{4}$ can of pipe joint compound is needed for each new home. How many homes can be plumbed with 156 cans of compound?

46. The Auto Technology Center at American River Community College purchases differential fluid in drums. If each car serviced needs $\frac{2}{3}$ gallon of lubricant, how many cars can be serviced with a 50-gallon drum?

47. In recordings of 186 patient visits, doctors failed to mention a new drug's side effects or how long to take the drug in $\frac{2}{3}$ of the visits.

(a) How many visits did doctors fail to discuss these issues with patients?

(b) How many times did they discuss them? (*Source:* Dr. Deijieng Tam, UCLA.)

48. Hans Bueff has completed $\frac{4}{5}$ of his kayak excursion down the Colorado River. If the trip is a total of 285 miles,

(a) how many miles has he gone?

(b) how many miles remain?

49. A dish towel manufacturer requires $\frac{3}{8}$ yard of cotton fabric for each towel. Find the number of dish towels that can be made from 912 yards of fabric.

50. A local McDonald's expects 300 applicants but only has jobs for $\frac{1}{60}$ of those who apply. How many job openings are these?

Relating Concepts (Exercises 51–56) For Individual or Group Work

Many application problems are solved using multiplication and division of fractions.
Work Exercises 51–54 in order.

51. Perhaps the most common indicator word for multiplication is the word *of* (when it follows a fraction). Circle the words in the list below that are also indicator words for multiplication.

more than	per
double	twice
times	product
less than	difference
equals	twice as much

52. Circle the words in the list below that are indicator words for division.

fewer	sum of
goes into	divide
per	quotient
equals	double
loss of	divided by

53. To divide two fractions, multiply the first fraction by the _________ of the second fraction.

54. Find the reciprocals for each number.

$\frac{3}{4}$ $\frac{7}{8}$ 5 $\frac{12}{19}$

The size of an antique U.S.A. postage stamp is shown here. Use this to answer Exercises 55 and 56.

55. **(a)** Explain how to find the perimeter (distance around the edges) of any flat, equal-sided, 3-, 4-, 5-, or 6-sided figure using multiplication.

(b) Find the perimeter of the stamp using multiplication.

56. Find the area of the postage stamp. Explain how to find the area of any rectangle.

Study Skills

USING STUDY CARDS REVISITED

OBJECTIVES

1. **Create study cards for difficult problems.**
2. **Create study cards of quiz problems.**

This is the second part of the Study Cards activity. As you get further into a chapter, you can choose particular problems that will serve as a good test review. Here are two more types of study cards that will help you.

Tough Problems Card

When you are doing your homework and find yourself saying, "This is really hard," or "I'm having trouble with this," make a **tough problem** study card! On the front, write out the procedure to work the type of problem *in words*. If there are special notes (like what *not* to do), include them. On the back, work at least one example; make sure you label what you are doing.

Front of Card

Warning: Division is NOT commutative. The order in which you write the numbers DOES matter.

Example: $1\frac{1}{3} \div 4 = \frac{\cancel{4}^{1}}{3} \cdot \frac{1}{\cancel{4}_{1}} = \frac{1}{3}$

But $4 \div 1\frac{1}{3} = \frac{\cancel{4}^{1}}{1} \cdot \frac{3}{\cancel{4}_{1}} = \frac{3}{1} = 3$

Very different answers!

In an application problem, do NOT assume the numbers are given in the correct order. Use estimation to check that the answer is reasonable!

Back of Card

Maite painted 4 windows using $1\frac{1}{3}$ cans of paint. How much paint did she use on each window?

Try $4 \div 1\frac{1}{3}$ (round)

$4 \div 1 = 4$ cans on each window ← Estimate

Not reasonable — she only used $1\frac{1}{3}$ cans in all!

Need to find: paint on each window

$1\frac{1}{3}$ cans $\div$ $4 = \frac{\cancel{4}^{1}}{3} \cdot \frac{1}{\cancel{4}_{1}} = \frac{1}{3}$ can

Reasonable!

Now Try This

Choose three types of difficult problems, and work them out on *study cards*. Be sure to put the words for solving the problem on one side and the worked problem on the other side.

Practice Quiz Cards

Make up a few **quiz cards** for each type of problem you learn, and use them to prepare for a test. Choose two or three problems from the different sections of the chapter. Be sure you don't just choose the easiest problems! Put the problem **with the direction words** (like *solve, simplify, estimate*) on the front, and work the problem on the back. If you like, put the page number from the text there, too. When you review, you work the problem on a separate paper, and check it by looking at the back.

Front of Card

Solve this application problem.

Tiffany's monthly income is $1275. She spends $\frac{2}{5}$ of her income on rent and utilities. How much does she pay for rent and utilities?

Back of Card

Proper fraction followed by "of" indicates multiplication.

She spends $\frac{2}{5}$ of her income

$$\frac{2}{5} \cdot 1275 = \frac{2}{\cancel{5}_1} \cdot \frac{\cancel{1275}^{255}}{1} = \frac{510}{1} = 510$$

Tiffany spends $510 on rent and utilities.

Why Are Study Cards Brain Friendly?

First, making the study cards is an **active technique** that really gets your dendrites growing. You have to make decisions about what is most important and how to put it on the card. This kind of thinking is more in depth than just memorizing, and as a result, you will understand the concepts better and remember them longer.

Second, the cards are **visually appealing** (if you write neatly and try some color). Your brain responds to pleasant visual images, and again, you will remember longer and may even be able to "picture in your mind" how your cards look. This will help you during tests.

Third, because study cards are small and portable, you can review them easily whenever you have a few minutes. Even while you're waiting for a bus or have a few minutes between classes you can take out your cards and read them to yourself. Your **brain really benefits from repetition**; each time you review your cards your dendrites are growing thicker and stronger. After a while, the information will become automatic and you will remember it for a long time.

2.8 Multiplying and Dividing Mixed Numbers

OBJECTIVES

1. Estimate the answer and multiply mixed numbers.
2. Estimate the answer and divide mixed numbers.
3. Solve application problems with mixed numbers.

Earlier in this chapter we worked with mixed numbers—a whole number and a fraction written together. Many of the fraction problems you encounter in everyday life involve mixed numbers.

OBJECTIVE 1 Estimate the answer and multiply mixed numbers. When multiplying mixed numbers, it is a good idea to estimate the answer first.

To estimate the answer, round each mixed number to the nearest whole number. If the numerator is *half* of the denominator or *more,* round up the whole number part. If the numerator is *less* than half the denominator, leave the whole number as it is.

$1\frac{5}{8}$ ← 5 is more than 4. ← Half of 8 is 4. $1\frac{5}{8}$ rounds up to 2.

$3\frac{2}{5}$ ← 2 is less than $2\frac{1}{2}$. ← Half of 5 is $2\frac{1}{2}$. $3\frac{2}{5}$ rounds to 3.

Round mixed numbers to the nearest whole number when estimating.

◀ Work Problem 1 at the Side.

1 Round each mixed number to the nearest whole number.

GS **(a)** $4\frac{2}{3}$ ← Half of 3 is $1\frac{1}{2}$

2 is (*less than/more than*) half of 3.

$4\frac{2}{3}$ rounds to ____.

GS **(b)** $3\frac{2}{5}$ ← Half of 5 is $2\frac{1}{2}$

2 is (*less than/more than*) half of 5.

$3\frac{2}{5}$ rounds to ____.

(c) $5\frac{3}{4}$ **(d)** $4\frac{7}{12}$

(e) $1\frac{1}{2}$ **(f)** $8\frac{4}{9}$

After estimating the answer, multiply the mixed numbers by using the following steps.

Multiplying Mixed Numbers

Step 1 *Change* each mixed number to an improper fraction.

Step 2 *Multiply* as fractions.

Step 3 *Simplify* the answer, which means to write it in *lowest terms,* and change it to a mixed number or whole number where possible.

EXAMPLE 1 Multiplying Mixed Numbers

First estimate the answer. Then multiply to get an exact answer. Simplify your answers.

(a) $2\frac{1}{2} \cdot 3\frac{1}{5}$

Estimate the answer by rounding the mixed numbers.

$2\frac{1}{2}$ rounds to 3 and $3\frac{1}{5}$ rounds to 3

$3 \cdot 3 = 9$ Estimated answer

To find the exact answer, change each mixed number to an improper fraction.

Step 1

$2\frac{1}{2} = \frac{5}{2}$ and $3\frac{1}{5} = \frac{16}{5}$

Continued on Next Page

Answers

1. (a) more than; 5 (b) less than; 3 (c) 6 (d) 5 (e) 2 (f) 8

Next, multiply.

Step 1 / Step 2 / Step 3

$$2\frac{1}{2} \cdot 3\frac{1}{5} = \frac{5}{2} \cdot \frac{16}{5} = \frac{\overset{1}{\cancel{5}}}{\underset{1}{\cancel{2}}} \cdot \frac{\overset{8}{\cancel{16}}}{\underset{1}{\cancel{5}}} = \frac{1 \cdot 8}{1 \cdot 1} = \frac{8}{1} = 8$$

Remember: Change to improper fractions, then divide out common factors, and finally multiply.

The estimated answer is 9 and the exact answer is 8. The exact answer is reasonable.

(b) $3\frac{5}{8} \cdot 4\frac{4}{5}$

$3\frac{5}{8}$ rounds to 4 and $4\frac{4}{5}$ rounds to 5

$4 \cdot 5 = 20$ Estimated answer

Now find the exact answer.

Step 1 / Step 2

$$3\frac{5}{8} \cdot 4\frac{4}{5} = \frac{29}{8} \cdot \frac{24}{5} = \frac{29}{\underset{1}{\cancel{8}}} \cdot \frac{\overset{3}{\cancel{24}}}{5} = \frac{29 \cdot 3}{1 \cdot 5} = \frac{87}{5}$$

Step 3

$$\frac{87}{5} = 17\frac{2}{5}$$

Simplify this answer by writing it as a mixed number.

The estimate was 20, so the exact answer of $17\frac{2}{5}$ is reasonable.

(c) $1\frac{3}{5} \cdot 3\frac{1}{3}$

$1\frac{3}{5}$ rounds to 2 and $3\frac{1}{3}$ rounds to 3

$2 \cdot 3 = 6$ Estimated answer

The exact answer is shown below.

$$1\frac{3}{5} \cdot 3\frac{1}{3} = \frac{8}{\underset{1}{\cancel{5}}} \cdot \frac{\overset{2}{\cancel{10}}}{3} = \frac{8 \cdot 2}{1 \cdot 3} = \frac{16}{3} = 5\frac{1}{3}$$

The estimate was 6, so the exact answer of $5\frac{1}{3}$ is reasonable.

Work Problem 2 at the Side. ▶

2 First estimate the answer. Then multiply to find the exact answer. Simplify your answers.

GS **(a)** $3\frac{1}{4} \cdot 6\frac{2}{3}$

↓ ↓

___ • ___

= ___ *estimate*

Exact: $3\frac{1}{4} \cdot 6\frac{2}{3}$

$$\frac{13}{\underset{1}{\cancel{4}}} \cdot \frac{\overset{5}{\cancel{20}}}{3} = \frac{65}{3} = \_\_\_\_$$

(b) $4\frac{2}{3} \cdot 2\frac{3}{4}$

↓ ↓

___ • ___

= ___ *estimate*

(c) $3\frac{3}{5} \cdot 4\frac{4}{9}$

↓ ↓

___ • ___

= ___ *estimate*

(d) $5\frac{1}{4} \cdot 3\frac{3}{5}$

↓ ↓

___ • ___

= ___ *estimate*

Answers

2. **(a)** *Estimate:* 3; 7; 21; *Exact:* $21\frac{2}{3}$

(b) *Estimate:* $5 \cdot 3 = 15$; *Exact:* $12\frac{5}{6}$

(c) *Estimate:* $4 \cdot 4 = 16$; *Exact:* 16

(d) *Estimate:* $5 \cdot 4 = 20$; *Exact:* $18\frac{9}{10}$

OBJECTIVE 2 Estimate the answer and divide mixed numbers. Just as you did when multiplying mixed numbers, it is also a good idea to estimate the answer when dividing mixed numbers. To divide mixed numbers, use the following steps.

Dividing Mixed Numbers

Step 1 *Change* each mixed number to an improper fraction.

Step 2 Use the *reciprocal* of the second fraction (divisor).

Step 3 *Change* division to multiplication.

Step 4 *Simplify* the answer, which means to write it in *lowest terms,* and change it to a mixed number or whole number where possible.

Note

Recall that the reciprocal of a fraction is found by interchanging the numerator and the denominator.

EXAMPLE 2 Dividing Mixed Numbers

First estimate the answer. Then divide to find the exact answer. Simplify your exact answers.

(a) $2\frac{2}{5} \div 1\frac{1}{2}$

First estimate the answer by rounding each mixed number to the nearest whole number.

$$2\frac{2}{5} \div 1\frac{1}{2}$$

Rounded

$$2 \div 2 = 1 \quad \text{Estimated answer}$$

To find the exact answer, first change each mixed number to an improper fraction.

Step 1

$$2\frac{2}{5} \div 1\frac{1}{2} = \frac{12}{5} \div \frac{3}{2}$$

Change mixed numbers to improper fractions.

Next, use the reciprocal of the second fraction and change division to multiplication.

Step 2 *Step 3* *Step 4*

$$\frac{12}{5} \div \frac{3}{2} = \frac{\overset{4}{\cancel{12}}}{5} \cdot \frac{2}{\underset{1}{\cancel{3}}} = \frac{4 \cdot 2}{5 \cdot 1} = \frac{8}{5} = 1\frac{3}{5} \quad \text{Exact answer simplified}$$

Reciprocals

Remember: Use the reciprocal of the *divisor* (the *second* fraction).

The estimate was 1, so the exact answer of $1\frac{3}{5}$ is reasonable.

Continued on Next Page

(b) $8 \div 3\frac{3}{5}$

$$8 \div 3\frac{3}{5} \quad \text{Rounded} \quad 8 \div 4 = 2 \quad \text{Estimate}$$

Now find the exact answer.

$$8 \div 3\frac{3}{5} = \frac{8}{1} \div \frac{18}{5} = \frac{\overset{4}{\cancel{8}}}{1} \cdot \frac{5}{\underset{9}{\cancel{18}}} = \frac{20}{9} = 2\frac{2}{9}$$

Reciprocals

Write 8 as $\frac{8}{1}$.

Divide out common factors only *after* you have changed to the reciprocal and are multiplying.

The estimate was 2, so the exact answer of $2\frac{2}{9}$ is reasonable.

(c) $4\frac{3}{8} \div 5$

$$4\frac{3}{8} \div 5 \quad \text{Rounded} \quad 4 \div 5 = \frac{4}{1} \div \frac{5}{1} = \frac{4}{1} \cdot \frac{1}{5} = \frac{4}{5} \quad \text{Estimate}$$

Reciprocals

The exact answer is shown below.

$$4\frac{3}{8} \div 5 = \frac{35}{8} \div \frac{5}{1} = \frac{\overset{7}{\cancel{35}}}{8} \cdot \frac{1}{\underset{1}{\cancel{5}}} = \frac{7}{8}$$

Reciprocals

Write 5 as $\frac{5}{1}$.

The estimate was $\frac{4}{5}$, so the exact answer of $\frac{7}{8}$ is reasonable.

Work Problem 3 at the Side. ▶

OBJECTIVE 3 Solve application problems with mixed numbers. The next two examples show how to solve application problems involving mixed numbers.

EXAMPLE 3 Applying Multiplication Skills

The local Habitat for Humanity chapter is looking for 11 contractors who will each donate $3\frac{1}{4}$ days of labor to a community building project. How many days of labor will be donated in all?

Step 1 **Read** the problem. The problem asks for the total days of labor donated by the 11 contractors.

Step 2 **Work out a plan.** Multiply the number of contractors (11) and the amount of labor that each donates ($3\frac{1}{4}$ days).

Continued on Next Page

3 First estimate the answer. Then divide to find the exact answer. Simplify all answers.

GS **(a)** $3\frac{1}{8} \div 6\frac{1}{4}$

$\underline{3} \div \underline{6}$

$=$ ____ *estimate*

Exact: $3\frac{1}{8} \div 6\frac{1}{4} =$

$$\frac{25}{8} \div \frac{25}{4} = \frac{\overset{1}{\cancel{25}}}{\underset{2}{\cancel{8}}} \cdot \frac{\overset{1}{\cancel{4}}}{\underset{1}{\cancel{25}}} =$$

(b) $10\frac{1}{3} \div 2\frac{1}{2}$

____ $\div$ ____

$=$ ____ *estimate*

(c) $8 \div 5\frac{1}{3}$

____ $\div$ ____

$=$ ____ *estimate*

(d) $13\frac{1}{2} \div 18$

____ $\div$ ____

$=$ ____ *estimate*

Answers

3. **(a)** *Estimate:* $3 \div 6 = \frac{1}{2}$; *Exact:* $\frac{1}{2}$

(b) *Estimate:* $10 \div 3 = 3\frac{1}{3}$; *Exact:* $4\frac{2}{15}$

(c) *Estimate:* $8 \div 5 = 1\frac{3}{5}$; *Exact:* $1\frac{1}{2}$

(d) *Estimate:* $14 \div 18 = \frac{7}{9}$; *Exact:* $\frac{3}{4}$

4 Use the six problem-solving steps. Simplify the answer. If one car requires $2\frac{5}{8}$ quarts of paint, find the number of quarts needed to paint 16 cars.

Estimate: ____ • 16 = 48

Solve: $2\frac{5}{8} \cdot 16 = \frac{21}{8} \cdot \frac{16}{1}$

$= \frac{21}{\cancel{8}_1} \cdot \frac{\cancel{16}^2}{1}$

State the answer: ____ quarts of paint are needed.

Check: Answer is close to estimate.

5 Use the six problem-solving steps. Simplify all answers.

(a) The manufacture of one outboard engine propeller requires $4\frac{3}{4}$ pounds of brass. How many propellers can be manufactured from 57 pounds of brass?

(b) Jack Armstrong Trucking uses $21\frac{3}{4}$ quarts of motor oil for each oil change on his diesel engine truck. Find the number of oil changes that can be made with 609 quarts of oil.

Answers

4. *Estimate:* 3 • 16 = 48; *Exact:* 42 quarts
5. (a) *Estimate:* 57 ÷ 5 ≈ 11; *Exact:* 12 propellers
(b) *Estimate:* 600 ÷ 22 ≈ 27; *Exact:* 28 oil changes

Step 3 **Estimate** a reasonable answer. Round $3\frac{1}{4}$ days to 3 days. Multiply 3 days by 11 contractors (3 • 11) to get an estimate of 33 days.

Step 4 **Solve** the problem. Find the exact answer.

$$11 \cdot 3\frac{1}{4} = 11 \cdot \frac{13}{4}$$

$$= \frac{11}{1} \cdot \frac{13}{4} = \frac{143}{4} = 35\frac{3}{4}$$

Step 5 **State the answer.** The community building project will receive $35\frac{3}{4}$ days of donated labor.

Always check to see if the answer is close to the estimate.

Step 6 **Check.** The exact answer, $35\frac{3}{4}$ days, is close to our estimate of 33 days.

◀ **Work Problem 4 at the Side.**

EXAMPLE 4 Applying Division Skills

A dome tent for backpacking requires $7\frac{1}{4}$ yards of nylon material. How many tents can be made from $21\frac{3}{4}$ yards of material?

Step 1 **Read** the problem. The problem asks how many tents can be made from $21\frac{3}{4}$ yards of material.

Step 2 **Work out a plan.** Divide the number of yards of cloth ($21\frac{3}{4}$ yd) by the number of yards needed for one tent ($7\frac{1}{4}$ yd).

Step 3 **Estimate** a reasonable answer.

$$21\frac{3}{4} \div 7\frac{1}{4}$$

↓ Rounded ↓

$$22 \div 7 \approx 3 \text{ tents} \quad \text{Estimate}$$

Recall ≈ means "approximately equal to."

Step 4 **Solve** the problem.

$$21\frac{3}{4} \div 7\frac{1}{4} = \frac{87}{4} \div \frac{29}{4}$$

$$= \frac{\cancel{87}^3}{\cancel{4}_1} \cdot \frac{\cancel{4}^1}{\cancel{29}_1} = \frac{3}{1} = 3 \quad \text{Matches estimate}$$

Step 5 **State the answer.** 3 tents can be made from $21\frac{3}{4}$ yards of cloth.

Step 6 **Check.** The exact answer, 3, matches our estimate.

◀ **Work Problem 5 at the Side.**

Note

When rounding mixed numbers to estimate the answer to a problem, the estimated answer usually varies somewhat from the exact answer. However, the importance of the estimated answer is that it will show you whether your exact answer is reasonable or not.

2.8 EXERCISES

FOR EXTRA HELP

MyMathLab®

CONCEPT CHECK *Decide whether each statement is* true *or* false. *If it is* false, *say why.*

1. When multiplying two mixed numbers, the reciprocal of the second mixed number must be used.

2. If you were dividing a mixed number by the whole number 10, the reciprocal of 10 would be $\frac{10}{1}$.

3. To round mixed numbers before estimating the answer, decide whether the numerator of the fraction part is less than or more than half of the denominator.

4. When rounding mixed numbers to estimate the answer to a problem, the estimated answer can vary quite a bit from the exact answer. However, it can still show whether the exact answer is reasonable.

First estimate the answer. Then multiply to find the exact answer. Simplify all answers.
See Example 1.

5. $4\frac{1}{2} \cdot 1\frac{3}{4}$

GS

Estimate:

___ • ___ = ___

Exact:

6. $2\frac{1}{2} \cdot 2\frac{1}{4}$

GS

Estimate:

___ • ___ = ___

Exact:

7. $1\frac{2}{3} \cdot 2\frac{7}{10}$

Estimate:

___ • ___ = ___

Exact:

8. $4\frac{1}{2} \cdot 2\frac{1}{4}$

Estimate:

___ • ___ = ___

Exact:

9. $3\frac{1}{9} \cdot 1\frac{2}{7}$

Estimate:

___ • ___ = ___

Exact:

10. $6\frac{1}{4} \cdot 3\frac{1}{5}$

Estimate:

___ • ___ = ___

Exact:

11. $8 \cdot 6\frac{1}{4}$

Estimate:

___ • ___ = ___

Exact:

12. $6 \cdot 2\frac{1}{3}$

Estimate:

___ • ___ = ___

Exact:

13. $4\frac{1}{2} \cdot 2\frac{1}{5} \cdot 5$

Estimate:

___ • ___ • ___ = ___

Exact:

14. $5\frac{1}{2} \cdot 1\frac{1}{3} \cdot 2\frac{1}{4}$

Estimate:

___ • ___ • ___ = ___

Exact:

15. $3 \cdot 1\frac{1}{2} \cdot 2\frac{2}{3}$

Estimate:

___ • ___ • ___ = ___

Exact:

16. $\frac{2}{3} \cdot 3\frac{2}{3} \cdot \frac{6}{11}$

Estimate:

___ • ___ • ___ = ___

Exact:

CONCEPT CHECK *Circle the letter of the correct answer.*

17. $3\frac{1}{4} \cdot 7\frac{5}{8}$

(a) 21 **(b)** 32 **(c)** 28 **(d)** 24

18. $5\frac{3}{4} \cdot 2\frac{2}{5}$

(a) 12 **(b)** 18 **(c)** 10 **(d)** 15

19. $2\frac{1}{8} \div 1\frac{3}{4}$

(a) 2 **(b)** 1 **(c)** $1\frac{1}{2}$ **(d)** 3

20. $8\frac{3}{5} \div 2\frac{3}{8}$

(a) 4 **(b)** $3\frac{1}{2}$ **(c)** $4\frac{1}{2}$ **(d)** 3

First estimate the answer. Then divide to find the exact answer. Simplify all answers.
See Example 2.

21. $1\frac{1}{4} \div 3\frac{3}{4}$

Estimate:

___ ÷ ___ = ___

Exact:

22. $1\frac{1}{8} \div 2\frac{1}{4}$

Estimate:

___ ÷ ___ = ___

Exact:

23. $2\frac{1}{2} \div 3$

GS

Estimate:

___ ÷ ___ = ___

Exact: $\frac{5}{2} \div \frac{3}{1}$

$\frac{5}{2} \cdot \frac{1}{3} =$

24. $2\frac{3}{4} \div 2$

GS

Estimate:

___ ÷ ___ = ___

Exact: $2\frac{3}{4} \div \frac{2}{1}$

$\frac{11}{4} \cdot \frac{1}{2} = \frac{11}{8} =$

25. $9 \div 2\frac{1}{2}$

Estimate:

___ ÷ ___ = ___

Exact:

26. $5 \div 1\frac{7}{8}$

Estimate:

___ ÷ ___ = ___

Exact:

27. $\frac{5}{8} \div 1\frac{1}{2}$

Estimate:

___ ÷ ___ = ___

Exact:

28. $\frac{3}{4} \div 2\frac{1}{2}$

Estimate:

___ ÷ ___ = ___

Exact:

29. $1\frac{7}{8} \div 6\frac{1}{4}$

Estimate:

___ ÷ ___ = ___

Exact:

30. $8\frac{2}{5} \div 3\frac{1}{2}$

Estimate:

___ ÷ ___ = ___

Exact:

31. $5\frac{2}{3} \div 6$

Estimate:

___ ÷ ___ = ___

Exact:

32. $5\frac{3}{4} \div 2$

Estimate:

___ ÷ ___ = ___

Exact:

For Exercises 33–50, first estimate the answer. Then find each exact answer and simplify it if possible. ***See Examples 3 and 4.*** *Use the recipe for Carrot Cake Cupcakes to work Exercises 33–36.*

CARROT CAKE CUPCAKES

12 paper bake cups
$1\frac{3}{4}$ cups flour
1 cup packed brown sugar
1 tsp. baking powder
1 tsp. baking soda
1 tsp. ground cinnamon
$\frac{1}{2}$ tsp. salt
1 cup shredded carrots
$\frac{3}{4}$ cup applesauce
$\frac{1}{3}$ cup vegetable oil
1 large egg
$\frac{1}{2}$ tsp. vanilla extract
1 container (16 oz.) ready-to-spread cream cheese frosting
Shredded coconut, tinted green
3 oz. Jelly Belly jelly beans, Orange Sherbet flavor

Preheat oven to 350° F. Place 12 bake cups in a muffin pan; set aside. In a large bowl, using a wire whisk, stir together flour, brown sugar, baking powder, baking soda, cinnamon, and salt. In a medium bowl, combine carrots, applesauce, oil, egg, and vanilla until blended. Add carrot mixture to flour mixture, stir well. Spoon batter into bake cups, filling $\frac{2}{3}$ full. Bake until toothpick inserted in center comes out clean, 20–25 minutes. Cool cupcakes in pan 10 minutes; remove to wire rack and cool completely. Frost with cream cheese frosting. To make carrot design on cupcakes, place gourmet Jelly Belly jelly beans in a carrot shape on each cupcake, top with green coconut for carrot top. Makes 12 cupcakes.

33. If 30 cupcakes are baked ($2\frac{1}{2}$ times the recipe), find the amount of each ingredient.

(a) Applesauce

Estimate:

Exact:

(b) Salt

Estimate:

Exact:

(c) Flour

Estimate:

Exact:

34. If 18 cupcakes are baked ($1\frac{1}{2}$ times the recipe), find the amount of each ingredient.

(a) Flour

Estimate:

Exact:

(b) Applesauce

Estimate:

Exact:

(c) Vegetable oil

Estimate:

Exact:

35. How much of each ingredient is needed if you bake one-half of the recipe?

(a) Vanilla extract

Estimate:

Exact:

(b) Applesauce

Estimate:

Exact:

(c) Flour

Estimate:

Exact:

36. How much of each ingredient is needed if you bake one-third of the recipe?

(a) Flour

Estimate:

Exact:

(b) Salt

Estimate:

Exact:

(c) Applesauce

Estimate:

Exact:

37. A new condominium conversion project requires $11\frac{3}{4}$ gallons of paint for each unit. How many units can be painted with 329 gallons of paint?

Estimate:

Exact:

38. According to an old English system of time units, a moment is $1\frac{1}{2}$ minutes. How many moments are there in an 8-hour work day? (8 hours = 480 minutes). (*Source:* hightechscience.org/funfacts.htm)

Estimate:

Exact:

39. A manufacturer of floor jacks is ordering steel tubing to make the handles for the jack shown below. How much steel tubing is needed to make 45 of these jacks? The symbol for inches is ". For example, 5" means 5 inches. (*Source:* Harbor Freight Tools.)

Estimate:

Exact:

40. A wheelbarrow manufacturer uses handles made of hardwood. Find the amount of wood that is necessary to make 182 handles. The longest dimension shown in the advertisement below is the handle length. (*Source:* Harbor Freight Tools.)

Estimate:

Exact:

2-TON COMPACT FLOOR JACK LOT NO. 36119

4000 LB. CAPACITY

- $19\frac{1}{2}$" handle
- Lifts 5" to $15\frac{1}{4}$"
- 21" x $9\frac{1}{2}$" x 6"
- Compact size & lightweight for portability—perfect for the trunk

6.0 CUBIC FT.
WHEEL BARROW
LOT NO. 46852
• Steel construction with hardwood handles
• 14" tubeless pneumatic tire
• Fully rolled edge for added tray strength
• Overall dimensions: 61 1/2" L x 27" W x 24.9" H

41. Write the three steps for multiplying mixed numbers. Use your own words.

42. Refer to **Exercise 41.** In your own words, write the additional step that must be added to the rule for multiplying mixed numbers to make it the rule for dividing mixed numbers.

43. The average cell phone contains $\frac{1}{1000}$ ounce of gold worth about $\$1\frac{2}{5}$. If you could extract the gold from all of the 130 million cell phones junked each year, how much money would you have from the sale of the gold? (*Source:* Discovery Channel, *Gold Rush Alaska.*)

Estimate:

Exact:

44. The manager of the flooring department at The Home Depot determines that each apartment unit requires $62\frac{1}{2}$ square yards of carpet. Find the number of apartment units that can be carpeted with 6750 square yards of carpet.

Estimate:

Exact:

45. Foxworthy Forest Products owns a truck that holds $1\frac{1}{4}$ cords of firewood. How many trips will he need to make to deliver 140 cords of firewood?

Estimate:

Exact:

46. BAE Roofing has completed an agreement to re-roof homes in a 20-year old residential development. Each home needs $31\frac{1}{2}$ squares of roofing material and the company has 1827 squares of material in inventory. How many homes will they be able to re-roof with this material?

Estimate:

Exact:

Use the information on bottle jacks in the advertisement to answer Exercises 47–48. The " symbol is for inches.

47. A mechanic needs a hydraulic lift that will raise a car 4 times as high as the standard jack shown.

(a) How high must it lift?

(b) Will a mechanic 6 feet tall be able to fit under the vehicle without bending down? *Hint:* 1 ft = 12 in.

(a) *Estimate:*

Exact:

(b)

12 TON
HEAVY DUTY
INDUSTRIAL
BOTTLE
JACKS
These all-purpose hydraulic jacks easily handle lifting jobs up to 12 tons, even at 45 degree and 90 degree angles. Rugged steel construction. Jacks feature twist tops for even more height. Certified to meet ASME standards.
Base dimensions:
5-1/4" x 5-3/8"
CENTRAL HYDRAULICS
17-3/4" MAXIMUM HEIGHT
Item 93378
STANDARD JACK
15-1/16" MAXIMUM HEIGHT
Item 93376
LOW PROFILE
$19.99
SAVE 33%
REGULAR PRICE $29.99

48. A race car driver needs a jack that will raise a vehicle only $\frac{1}{3}$ as high as the low-profile jack pictured.

(a) How high must it lift?

(b) Will a 6-inch part fit under the car?

(a) *Estimate:*

Exact:

(b)

49. A shark can swim $5\frac{1}{2}$ times faster than a person. If a person can swim $6\frac{1}{8}$ miles per hour, how fast can a shark swim? (*Source:* Discovery Channel, *Animal Planet.*)

Estimate:

Exact:

50. A flooring contractor needs $24\frac{2}{7}$ boxes of tile to cover a kitchen floor. If there are 24 homes in a subdivision, how many boxes of tile are needed to cover all of the floors?

Estimate:

Exact:

Study Skills

REVIEWING A CHAPTER

OBJECTIVES

1. **Use the Chapter Quick Review to see examples of every type of problem.**
2. **Create study cards for vocabulary.**
3. **Practice by doing review and mixed review exercises.**
4. **Take the Chapter Test as a practice test.**

This activity is really about **preparing for tests.** Some of the suggestions are ideas that you will learn to use a little later in the term, but get started trying them out now. Often, the first chapters in your math textbook will be review, so it is good to practice some of the study techniques on material that is not too challenging.

Chapter Reviewing Techniques

Use these **chapter reviewing techniques.**

▶ **Make a study card for each vocabulary word and concept.** Include a definition, an example, a sketch, and a page reference. Include the symbol or formula if there is one. See the *Using Study Cards* activity for a quick look at some sample study cards.

▶ **Go back to the section** to find more explanations or information about any new vocabulary, formulas, or symbols.

▶ **Use the Chapter Summary** to see examples of each type of problem. Do not expect it to substitute for reading and working through the whole chapter! First, take the Test Your Word Power quiz to check your understanding of new vocabulary. The answers follow the quiz. Then read the Quick Review. **Pay special attention to the dark blue headings.** Check the explanations for the solutions to problems given. Try to think about how all the topics in the **whole chapter** are related.

▶ **Study your lecture notes** to see what your instructor has emphasized in class. Then review that material in your text.

▶ **Do the Review Exercises** to practice every type of problem.

- ✓ Check your answers **after** you're done with each **section of exercises.**
- ✓ If you get stuck on a problem, **first** check the Chapter Quick Review. If that doesn't clear up your confusion, then check the section and your lecture notes.
- ✓ Pay attention to **direction words** for the problems, such as *simplify, round, solve*, and *estimate*.
- ✓ Make **study cards for especially difficult problems.**

▶ **Do the Mixed Review exercises.** This is a good check to see if you can still do the problems when they are in mixed-up order. **Check your answers carefully** in the Answers section in the back of your book. Are your answers **exact** and **complete?** Make sure you are **labeling** answers correctly, using the right **units.** For example, does your answer need to include *$*, cm^2, *ft*, and so on?

- **Take the Chapter Test as if it is a real test.** If your instructor has skipped sections in the chapter, ask which problems to skip on the test before you start.
  - ✓ **Time yourself** just as you would for a real test.
  - ✓ **Use a calculator or notes** just as you would be permitted to (or not) on a real test.
  - ✓ **Take the test in one sitting,** just like a real test is given in one sitting.
  - ✓ **Show all your work.** Practice showing your work just the way your instructor has asked you to show it.
  - ✓ **Practice neatness.** Can someone else follow your steps?
  - ✓ **Check your answers** in the back of the book.

Notice that reviewing a chapter will take some time. Remember that it takes time for dendrites to grow! You cannot grow a good network of dendrites by rushing through a review in one night. But if you use the suggestions over a few days or evenings, you will notice that you understand the material more thoroughly and remember it longer.

Now Try This

Follow the reviewing techniques listed above for your next test. For each technique, write a comment about how it worked for you in the spaces below.

1. **Make a study card for each vocabulary word and concept.**

2. **Go back to the section** to find more explanations or information.

3. **Take the Test Your Word Power quiz and use the Quick Review** to review each concept in the chapter.

4. **Study your lecture notes** to see what your instructor has emphasized in class.

5. **Do the Review Exercises,** following the specific suggestions on the previous page.

6. **Do the Mixed Review exercises.**

7. **Take the Chapter Test** as if it is a real test.

Why Are These Review Activities Brain Friendly?

You have already become familiar with the features of your textbook. This activity requires you to make good use of them. Your **brain needs repetition** to strengthen dendrites and the connections between them. By following the steps outlined here, you will be reinforcing the concepts, procedures, and skills you need to use for tests (and for the next chapters).

The combination of techniques provides repetition in different ways. That **promotes good branching of dendrites** instead of just relying on one branch, or route, to connect to the other dendrites. A thorough review of each chapter will **solidify your dendrite connections.** It will help you be sure that you understand the concepts **completely and accurately.** Also, taking the Chapter Test will **simulate the testing situation,** which gives you practice in test taking conditions.

Study Skills

TIPS FOR TAKING MATH TESTS

OBJECTIVES

1. Apply suggestions to tests and quizzes.
2. Develop a set of "best practices" to apply while testing.

Improving Your Test Score

To Improve Your Test Score	Comments
Come prepared with a pencil, eraser, and calculator, if allowed. If you are easily distracted, sit in the corner farthest from the door.	***Working in pencil lets you erase,*** keeping your work neat and readable.
Scan the entire test, note the point value of different problems, and plan your time accordingly. Allow at least five minutes to check your work at the end of the testing time.	If you have 50 minutes to do 20 problems, $50 \div 20 = 2.5$ minutes per problem. ***Spend less time on easy ones,*** more time on problems with higher point values.
Read directions carefully, and circle any significant words. When you finish a problem, read the directions again to make sure you did what was asked.	***Pay attention to announcements*** written on the board or made by your instructor. Ask if you don't understand. You don't want to get problems wrong because you misread the directions!
Show your work. Most math teachers give partial credit if some of the steps in your work are correct, even if the final answer is wrong. ***Write neatly.*** If you like to scribble when first working or checking a problem, do it on scratch paper.	***If your instructor can't read your writing, you won't get credit for it.*** If you need more space to work, ask if you can use extra pieces of paper that you hand in with your test paper.
Check that the ***answer to an application problem is reasonable*** and makes sense. Read the problem again to make sure you've answered the question.	***Use common sense.*** Can the father really be seven years old? Would a month's rent be $32,140? Label your answer: $, years, inches, etc.
To check for careless errors, you need to ***rework the problem again, without looking at your previous work.*** Cover up your work with a piece of scratch paper, and pretend you are doing the problem for the first time. Then compare the two answers.	If you just "look over" your work, your mind can make the same mistake again without noticing it. Reworking the problem from the beginning ***forces you to rethink it.*** If possible, use a different method to solve the problem the second time.
Do not try to review up until the last minute before the test. Instead, go for a walk, do some deep breathing, and arrive just in time for the test. Ignore other students.	Listening to anxious classmates before the test ***may cause you to panic.*** Moderate exercise and deep breathing will calm your mind.

Reducing Anxiety

To Reduce Anxiety	Comments
Do a "knowledge dump" as soon as you get the test. Write important notes to yourself in a corner of the test paper: formulas, or common errors you want to watch out for.	Writing down tips and things that you've memorized ***lets you relax;*** you won't have to worry about forgetting those things and can refer to them as needed.
Do the easy problems first in order to build confidence. If you feel your anxiety starting to build, *immediately* stop for a minute, close your eyes, and take several slow, deep breaths.	Greater confidence helps you ***get the easier problems correct.*** Anxiety causes shallow breathing, which leads to confusion and reduced concentration. Deep breathing calms you.
As you work on more difficult problems, ***notice your "inner voice."*** You may have negative thoughts such as, "I can't do it," or "who cares about this test anyway." In your mind, yell, "STOP" and take several deep, slow breaths. Or, replace the negative thought with a positive one.	Here are ***examples of positive statements.*** Try writing one of them on the top of your test paper. • I know I can do it. • I can do this one step at a time. • I've studied hard, and I'll do the best I can.
If you still can't solve a difficult problem when you come back to it the second time, ***make a guess and do not change it.*** In this situation, your first guess is your best bet. Do not change an answer just because you're a little unsure. ***Change it only if you find an obvious mistake.***	If you are thinking about changing an answer, be sure you have a good reason for changing it. If you cannot find a specific error, leave your first answer alone. ***When the tests are returned, check to see if changing answers helped or hurt you.***
Read the harder problems twice. Write down *anything* that might help solve the problem: a formula, a picture, etc. If you still can't get it, circle the problem and ***come back to it later***. Do *not* erase any of the things you wrote down.	If you know even a *little* bit about the problem, write it down. The ***answer may come to you*** as you work on it, or you may get partial credit. Don't spend too long on any one problem. Your subconscious mind will work on the tough problem while you go on with the test.
Ignore students who finish early. Use the entire test time. *You do not get extra credit for finishing early.* Use the extra time to rework problems and correct careless errors.	Students who leave early are often the ones who didn't study or who are too anxious to continue working. If they bother you, ***sit as far from the door as possible.***

Why Are These Suggestions Brain Friendly?

Several suggestions address anxiety. **Reducing anxiety allows your brain to make the connections between dendrites;** in other words, you can think clearly.

Remember that **your brain continues to work on a difficult problem** even if you skip it and go on to the next one. Your subconscious mind will come through for you if you are open to the idea!

Some of the suggestions ask you to **use your common sense.** Follow the directions, show your work, write neatly, and pay attention to whether your answers really make sense.

Chapter 2 Summary

Key Terms

2.1

numerator The number above the fraction bar in a fraction is called the numerator. It shows how many of the equivalent parts are being considered.

denominator The number below the fraction bar in a fraction is called the denominator. It shows the number of equal parts in a whole.

proper fraction In a proper fraction, the numerator is smaller than the denominator. The fraction is less than 1.

improper fraction In an improper fraction, the numerator is greater than or equal to the denominator. The fraction is equal to or greater than 1.

2.2

mixed number A mixed number includes a fraction and a whole number written together.

2.3

factors Numbers that are multiplied to give a product are factors.

factorizations The numbers that can be multiplied to give a specific number (product) are factorizations of that number.

prime number A prime number is a whole number other than 0 and 1 that has exactly two factors, itself and 1.

composite number A composite number has at least one factor other than itself and 1.

prime factorization In a prime factorization, every factor is a prime number.

2.4

equivalent fractions Two fractions are equivalent when they represent the same portion of a whole.

common factor A common factor is a number that can be divided evenly into two or more whole numbers.

lowest terms A fraction is written in lowest terms when its numerator and denominator have no common factor other than 1.

2.5

multiplication shortcut When multiplying or dividing fractions, the process of dividing a numerator and denominator by a common factor can be used as a shortcut.

2.7

reciprocal Two numbers are reciprocals of each other if their product is 1. To find the reciprocal of a fraction, interchange the numerator and the denominator. Zero does not have a reciprocal.

New Formula

Area of a rectangle: Area = length • width

Test Your Word Power

See how well you have learned the vocabulary in this chapter.

1. A **numerator** is
   A. a number greater than 5
   B. the number above the fraction bar in a fraction
   C. the number below the fraction bar in a fraction.

2. A **proper fraction**
   A. has a value less than 1
   B. has a whole number and a fraction
   C. has a value greater than 1.

3. A **mixed number** is
   A. less than 1
   B. a whole number and a fraction written together
   C. a number multiplied by another number.

4. A **factor** is
   A. one of two or more numbers that are added to get another number
   B. the answer in division
   C. one of two or more numbers that are multiplied to get another number.

5. A whole number greater than 1 is **prime** if
   A. it cannot be factored
   B. it has just one factor
   C. it has only itself and 1 as factors.

6. A **common factor** can
   A. only be divided by itself and 1
   B. be divided evenly into two or more whole numbers
   C. never be divided by 2.

7. A fraction is in **lowest terms** when
   A. it cannot be divided
   B. its numerator and denominator have no common factor other than 1
   C. it has a value less than 1.

8. To find the **reciprocal** of a fraction,
   A. multiply it by itself
   B. interchange the numerator and the denominator
   C. change it to an improper fraction.

Answers to Test Your Word Power

1. B; *Example:* In $\frac{3}{8}$, the numerator is 3.

2. A; *Example:* $\frac{1}{2}$, $\frac{3}{4}$, and $\frac{7}{8}$ are all proper fractions with a value less than 1.

3. B; *Example:* $2\frac{3}{8}$ and $5\frac{3}{4}$ are mixed numbers.

4. C; *Example:* Since $3 \cdot 5 = 15$, the numbers 3 and 5 are factors of 15.

5. C; *Example:* 3, 5, and 11 are prime numbers; 4, 8, and 12 are composite numbers.

6. B; *Example:* 3 is a common factor of both 6 and 9 because it can be evenly divided into each of them.

7. B; *Example:* $\frac{3}{8}$, $\frac{4}{5}$, and $\frac{5}{6}$ are in lowest terms but $\frac{6}{8}$, $\frac{3}{6}$, and $\frac{2}{4}$ are not.

8. B; *Example:* The reciprocal of $\frac{3}{8}$ is $\frac{8}{3}$, the reciprocal of $\frac{25}{4}$ is $\frac{4}{25}$, and the reciprocal of 6 or $\frac{6}{1}$ is $\frac{1}{6}$.

Quick Review

Concepts	*Examples*
2.1 Types of Fractions	
Proper Numerator smaller than denominator; a value less than 1	$\frac{2}{3}$ $\frac{3}{4}$ $\frac{15}{16}$ $\frac{1}{8}$ Proper fractions
Improper Numerator equal to or greater than denominator; a value equal to or greater than 1	$\frac{17}{8}$ $\frac{19}{12}$ $\frac{11}{2}$ $\frac{5}{3}$ $\frac{7}{7}$ Improper fractions

Concepts	Examples

2.2 Converting Fractions

Mixed to Improper Multiply denominator by whole number, add numerator, and place over denominator.

$$7\frac{2}{3} = \frac{23}{3} \leftarrow (3 \cdot 7) + 2$$

Same denominator

Improper to Mixed Divide numerator by denominator and place remainder over denominator.

$$\frac{17}{5} = 3\frac{2}{5} \leftarrow \text{Remainder}$$

Same denominator

$$\begin{array}{r} 3 \\ 5\overline{)17} \\ \underline{15} \\ 2 \end{array}$$

17 ← Divide numerator by denominator.

2 ← Remainder

2.3 Prime Numbers

Determine whether a whole number is evenly divisible only by itself and 1. (By definition, 0 and 1 are not prime.)

The prime numbers less than 50 are 2, 3, 5, 7, 11, 13, 17, 19, 23, 29, 31, 37, 41, 43, 47.

2.3 Finding the Prime Factorization of a Number

Divide each factor by a prime number using a diagram that forms the shape of tree branches.

Find the prime factorization of 30. Use a factor tree.

Prime factors are circled.

$30 = \mathbf{2 \cdot 3 \cdot 5}$

2.4 Writing Fractions in Lowest Terms

Divide the numerator and denominator by the greatest common factor.

Write $\frac{30}{42}$ in lowest terms.

$$\frac{30}{42} = \frac{30 \div 6}{42 \div 6} = \frac{5}{7}$$

2.5 Multiplying Fractions

1. Multiply the numerators and multiply the denominators.
2. Write answers in lowest terms if the multiplication shortcut was not used.

Multiply.

$$\frac{6}{11} \cdot \frac{7}{8} = \frac{\overset{3}{\cancel{6}}}{11} \cdot \frac{7}{\underset{4}{\cancel{8}}} = \frac{3 \cdot 7}{11 \cdot 4} = \frac{21}{44}$$

2.7 Finding the Reciprocal

To find the reciprocal of a fraction, interchange the numerator and denominator.

Find the reciprocal of each fraction.

$\frac{3}{4}$ The reciprocal of $\frac{3}{4}$ is $\mathbf{\frac{4}{3}}$.

$\frac{8}{5}$ The reciprocal of $\frac{8}{5}$ is $\mathbf{\frac{5}{8}}$.

9 The reciprocal of 9 is $\mathbf{\frac{1}{9}}$.

Concepts	Examples
2.7 Dividing Fractions Use the reciprocal of the divisor (second fraction) and change division to multiplication.	Divide. $\frac{25}{36} \div \frac{15}{18} = \frac{\overset{5}{\cancel{25}}}{\underset{2}{\cancel{36}}} \cdot \frac{\overset{1}{\cancel{18}}}{\underset{3}{\cancel{15}}} = \frac{5 \cdot 1}{2 \cdot 3} = \frac{5}{6}$ Reciprocals
2.8 Multiplying Mixed Numbers First estimate the answer. Then follow these steps. ***Step 1*** *Change* each mixed number to an improper fraction. ***Step 2*** *Multiply.* ***Step 3*** *Simplify* the answer, which means to write it in *lowest terms,* and change it to a mixed number or whole number where possible.	First estimate the answer. Then multiply to get the exact answer. *Estimate:* $1\frac{3}{5} \cdot 3\frac{1}{3}$ Rounded $2 \cdot 3 = \mathbf{6}$ *Exact:* $1\frac{3}{5} \cdot 3\frac{1}{3} = \frac{8}{\underset{1}{\cancel{5}}} \cdot \frac{\overset{2}{\cancel{10}}}{3}$ $= \frac{8 \cdot 2}{1 \cdot 3}$ $= \frac{16}{3} = \mathbf{5\frac{1}{3}}$ Close to estimate
2.8 Dividing Mixed Numbers First estimate the answer. Then follow these steps. ***Step 1*** *Change* each mixed number to an improper fraction. ***Step 2*** Use the *reciprocal* of the second fraction (divisor). ***Step 3*** *Change* division to multiplication. ***Step 4*** *Simplify* the answer, which means to write it in *lowest terms,* and change it to a mixed number or whole number where possible.	First estimate the answer. Then divide to get the exact answer. *Estimate:* $3\frac{5}{9} \div 2\frac{2}{5}$ Rounded $4 \div 2 = \mathbf{2}$ *Exact:* $3\frac{5}{9} \div 2\frac{2}{5} = \frac{32}{9} \div \frac{12}{5}$ $= \frac{\overset{8}{\cancel{32}}}{9} \cdot \frac{5}{\underset{3}{\cancel{12}}} = \frac{40}{27}$ $= \mathbf{1\frac{13}{27}}$ Close to estimate

Chapter 2 Review Exercises

2.1 *Write the fraction that represents each shaded portion.*

1. 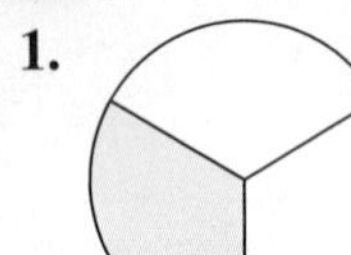

2. 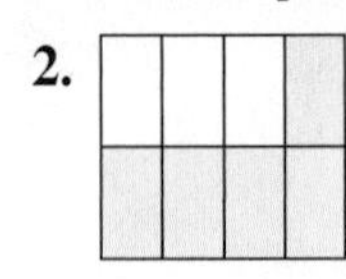

3. 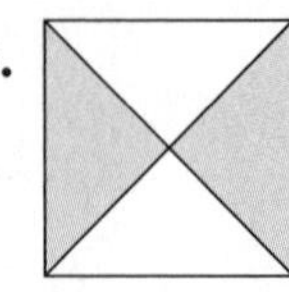

List the proper and improper fractions in each group.

	Proper	Improper
4. $\frac{1}{8}$ $\frac{4}{3}$ $\frac{5}{5}$ $\frac{3}{4}$ $\frac{2}{3}$	______	______
5. $\frac{6}{5}$ $\frac{15}{16}$ $\frac{16}{13}$ $\frac{1}{8}$ $\frac{5}{3}$	______	______

2.2 *Write each mixed number as an improper fraction. Write each improper fraction as a mixed number.*

6. $4\frac{3}{4}$ **7.** $9\frac{5}{6}$ **8.** $\frac{27}{8}$ **9.** $\frac{63}{5}$

2.3 *Find all factors of each number.*

10. 6 **11.** 24 **12.** 55 **13.** 90

Write the prime factorization of each number by using exponents.

14. 27 **15.** 150 **16.** 420

Simplify each expression.

17. 5^2 **18.** $6^2 \cdot 2^3$ **19.** $8^2 \cdot 3^3$ **20.** $4^3 \cdot 2^5$

2.4 *The purity of gold is regulated by law and is the same in all parts of the world. Exercises 21–24 show the purity of 24-kt (karat), 18-kt, 14-kt, and 10-kt gold by comparing the parts of gold to the parts of alloy (metals other than gold). Write a fraction in lowest terms to show the portion that is gold. (Source:* Costco Wholesale.)

21. 24 kt. = (24 parts gold, 0 parts alloy)

22. 18 kt. = (18 parts gold, 6 parts alloy)

23. 14 kt. = (14 parts gold, 10 parts alloy)

24. 10 kt. = (10 parts gold, 14 parts alloy)

Write the numerator and denominator of each fraction as a product of prime factors. Then, write the fraction in lowest terms.

25. $\dfrac{25}{60}$

26. $\dfrac{384}{96}$

Decide whether each pair of fractions is equivalent or not equivalent, using the method of prime factors.

27. $\dfrac{3}{4}$ and $\dfrac{48}{64}$

28. $\dfrac{5}{8}$ and $\dfrac{70}{120}$

29. $\dfrac{2}{3}$ and $\dfrac{360}{540}$

2.5–2.8 *Multiply. Write answers in lowest terms, and as mixed numbers or whole numbers where possible.*

30. $\dfrac{4}{5} \cdot \dfrac{3}{4}$

31. $\dfrac{3}{10} \cdot \dfrac{5}{8}$

32. $\dfrac{70}{175} \cdot \dfrac{5}{14}$

33. $\dfrac{44}{63} \cdot \dfrac{3}{11}$

34. $\dfrac{5}{16} \cdot 48$

35. $\dfrac{5}{8} \cdot 1000$

Divide. Write answers in lowest terms, and as mixed numbers or whole numbers where possible.

36. $\dfrac{2}{3} \div \dfrac{1}{2}$

37. $\dfrac{5}{6} \div \dfrac{1}{2}$

38. $\dfrac{\frac{15}{18}}{\frac{10}{30}}$

39. $\dfrac{\frac{3}{4}}{\frac{3}{8}}$

40. $7 \div \dfrac{7}{8}$

41. $18 \div \dfrac{3}{4}$

42. $\dfrac{5}{8} \div 3$

43. $\dfrac{2}{3} \div 5$

44. $\dfrac{\frac{12}{13}}{3}$

Find the area of each rectangle.

45.

$2\frac{3}{4}$ ft

$\frac{1}{2}$ ft

46.

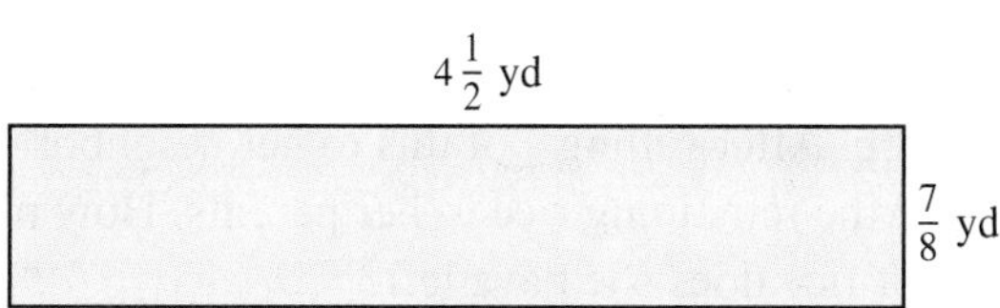

47. Ceramic tile is being installed on the floor of a meeting hall that is 108 ft long and $72\frac{3}{4}$ ft wide. Find the area.

48. Find the area of a display shelf that is a rectangle measuring 6 ft long and $\frac{11}{12}$ ft wide.

First estimate the answer. Then multiply or divide to find the exact answer. Simplify all answers.

49. $5\frac{1}{2} \bullet 1\frac{1}{4}$

Estimate:

___ • ___ = ___

Exact:

50. $2\frac{1}{4} \bullet 7\frac{1}{8} \bullet 1\frac{1}{3}$

Estimate:

___ • ___ • ___ = ___

Exact:

51. $15\frac{1}{2} \div 3$

Estimate:

___ ÷ ___ = ___

Exact:

52. $4\frac{3}{4} \div 6\frac{1}{3}$

Estimate:

___ ÷ ___ = ___

Exact:

Solve each application problem by using the six problem-solving steps.

53. Blue Diamond Almonds has 320 tons of almonds. How many $\frac{5}{8}$ ton bins will be needed to store the almonds?

54. The founding partner of Professional Networking owns $\frac{2}{5}$ of the business, while 4 other equal partners own the rest. What fraction of the total business is owned by each of the other partners?

55. How many window-blind pull cords can be made from $157\frac{1}{2}$ yards of cord if $4\frac{3}{8}$ yards of cord are needed for each blind? First estimate, and then find the exact answer.

Estimate:

Exact:

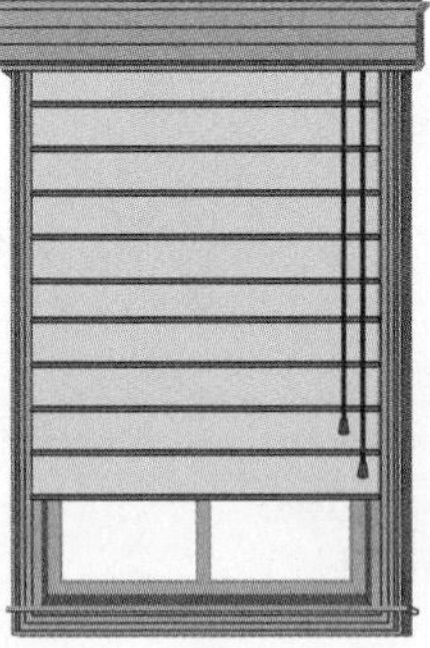

56. A gallon of water weighs $8\frac{1}{3}$ pounds. Find the weight of the water in two 50-gallon aquariums. First estimate, and then find the exact answer.

Estimate:

Exact:

57. Ebony Wilson purchased 100 pounds of rice at the food co-op. After selling $\frac{1}{4}$ of this to her neighbor, she gave $\frac{2}{3}$ of the remaining rice to her parents. How many pounds of rice does she have left?

58. Sheila Spinney, a recent college graduate, receives a salary of \$2976 each month. She pays $\frac{3}{8}$ of this amount in taxes, social security, and a retirement plan. Of the remainder, $\frac{9}{10}$ goes for basic living expenses. How much money remains?

59. The results of a back-to-school fundraiser to purchase school supplies showed that 6 schools will divide $\frac{7}{8}$ of the total amount raised. What fraction of the total amount raised will each school receive?

60. In a morning of deep-sea fishing, 5 fishermen catch $\frac{4}{5}$ ton of salmon. If they divide the fish evenly, how much will each receive?

Mixed Review Exercises

Multiply or divide as indicated. Simplify all answers.

61. $\frac{1}{2} \cdot \frac{3}{4}$

62. $\frac{2}{3} \cdot \frac{3}{5}$

63. $12\frac{1}{2} \cdot 2\frac{1}{2}$

64. $8\frac{1}{3} \cdot 3\frac{2}{5}$

65. $\frac{\frac{4}{5}}{8}$

66. $\frac{\frac{5}{8}}{4}$

67. $\frac{15}{31} \cdot 62$

68. $3\frac{1}{4} \div 1\frac{1}{4}$

Write each mixed number as an improper fraction. Write each improper fraction as a mixed number.

69. $\frac{8}{5}$

70. $\frac{153}{4}$

71. $5\frac{2}{3}$

72. $38\frac{3}{8}$

Write the numerator and denominator of each fraction as a product of prime factors; then write the fraction in lowest terms.

73. $\frac{8}{12}$

74. $\frac{108}{210}$

Write each fraction in lowest terms.

75. $\frac{75}{90}$

76. $\frac{48}{72}$

77. $\frac{44}{110}$

78. $\frac{87}{261}$

Solve each application problem.

79. The label of the Roundup Weed and Grass Killer says to mix $2\frac{1}{2}$ ounces of the product with 1 gallon of water. How many ounces are needed for 50 gallons of water? First estimate, then find the exact answer. (*Source:* Roundup Weed and Grass Killer.)

Estimate:

Exact:

80. Valley Farms purchased some diesel fuel additive. The instructions say to use $7\frac{1}{4}$ quarts of additive for each tank of fuel. How many quarts are needed for $9\frac{1}{3}$ tanks? First estimate, then find the exact answer.

Estimate:

Exact:

81. The antique U.S.A. stamp is $1\frac{3}{4}$ in. by $\frac{7}{8}$ in. Find its area.

82. A patio table top is $\frac{7}{8}$ yard by $2\frac{1}{4}$ yards. What is the area?

Chapter 2 Test

The Chapter Test Prep Videos with test solutions are available on DVD, in MyMathLab, and on YouTube—search "LialBasicCollegeMath" and click on "Channels."

Write a fraction to represent each shaded portion.

1.

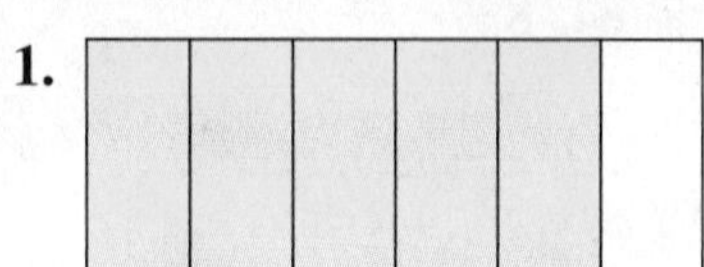

2.

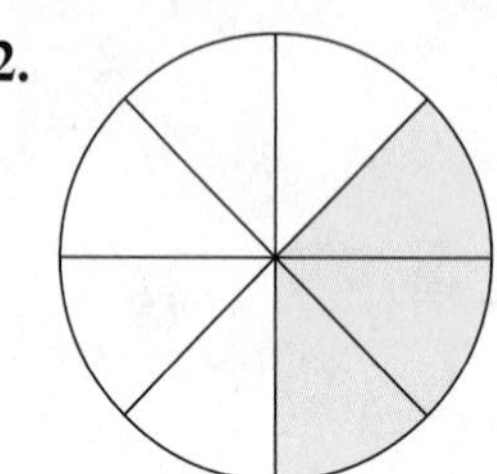

3. Identify all the proper fractions in this list:

$\frac{2}{3}$ $\frac{4}{4}$ $\frac{6}{7}$ $\frac{5}{2}$ $\frac{1}{4}$ $\frac{5}{8}$ $\frac{30}{18}$

4. Write $3\frac{3}{8}$ as an improper fraction.

5. Write $\frac{123}{4}$ as a mixed number.

6. Find all factors of 18.

Find the prime factorization of each number. Write the answers using exponents.

7. 45

8. 144

9. 500

Write each fraction in lowest terms.

10. $\frac{36}{48}$

11. $\frac{60}{72}$

12. The method of prime factors is used to write a fraction in lowest terms. Briefly explain how this is done. Use the fraction $\frac{56}{84}$ to show how this works.

13. Explain how to multiply fractions. What additional steps must be taken when dividing fractions?

Multiply or divide. Write answers in lowest terms, and as mixed numbers or whole numbers where possible.

14. $\frac{3}{4} \cdot \frac{4}{9}$

15. $54 \cdot \frac{2}{3}$

16. A rectangular barbecue grill is $\frac{15}{16}$ yard by $\frac{4}{9}$ yard. Find the area of the grill.

17. The Sierra College Conservation Club planted 8760 Douglas Fir seedlings. If $\frac{3}{8}$ of these seedlings are not expected to survive, find the number of seedlings that do survive.

18. $\frac{3}{4} \div \frac{5}{6}$

19. $\frac{7}{\frac{4}{9}}$

20. To complete a custom-designed cabinet, oak trim pieces must be cut exactly $2\frac{1}{4}$ inches long so that they can be used as dividers in a spice rack. Find the number of pieces that can be cut from a piece of oak that is 54 inches in length.

First estimate the answer. Then find the exact answer. Simplify all answers.

21. $4\frac{1}{8} \cdot 3\frac{1}{2}$

Estimate:

Exact:

22. $1\frac{5}{6} \cdot 4\frac{1}{3}$

Estimate:

Exact:

23. $9\frac{3}{5} \div 2\frac{1}{4}$

Estimate:

Exact:

24. $\frac{8\frac{1}{2}}{1\frac{3}{4}}$

Estimate:

Exact:

25. A new vaccine is synthesized at the rate of $2\frac{1}{2}$ grams per day. How many grams can be synthesized in $12\frac{1}{4}$ days?

Estimate:

Exact:

Chapters 1–2 Cumulative Review Exercises

Name the digit that has the given place value in each number.

1. 783
hundreds
tens

2. 8,621,785
millions
ten-thousands

Add, subtract, multiply, or divide as indicated.

3. $\begin{array}{r} 71 \\ 23 \\ 47 \\ +\ 36 \\ \hline \end{array}$

4. $\begin{array}{r} 82{,}121 \\ 5\,468 \\ 316 \\ +\ 61{,}294 \\ \hline \end{array}$

5. $\begin{array}{r} 6537 \\ -\ 2085 \\ \hline \end{array}$

6. $\begin{array}{r} 4{,}819{,}604 \\ -\ 1{,}597{,}783 \\ \hline \end{array}$

7. $\begin{array}{r} 83 \\ \times 9 \\ \hline \end{array}$

8. $9 \cdot 4 \cdot 2$

9. $\begin{array}{r} 3784 \\ \times \quad 573 \\ \hline \end{array}$

10. $\begin{array}{r} 563 \\ \times \quad 800 \\ \hline \end{array}$

11. $\frac{63}{7}$

12. $18\overline{)136{,}458}$

13. $33{,}886 \div 4$

14. $492\overline{)10{,}850}$

Round each number to the nearest ten, nearest hundred, and nearest thousand.

	Ten	Hundred	Thousand
15. 6583	________	________	________
16. 76,271	________	________	________

Simplify.

17. $2^5 - 6(4)$

18. $\sqrt{36} - 2 \cdot 3 + 5$

Solve each application problem using the six problem-solving steps.

19. A Gettysburg tour van uses 9 gallons of fuel on a half-day tour and 17 gallons of fuel on a full-day tour. Find the total number of gallons of fuel used in 26 half-day tours and 18 full-day tours.

20. As a result of unvaccinated travelers and children, there have been outbreaks of several diseases. Last year there were 4519 cases of mumps and 21,291 cases of pertussis (whooping cough) in the United States. How many more cases of pertussis than mumps were there? (*Source:* Centers for Disease Control and Prevention.)

21. A typical adult loses 100 hairs a day out of approximately 120,000 hairs. If the lost hairs were not replaced, find the number of hairs remaining after two years. (1 year = 365 days.)

22. A group of 22 health care workers will divide 4136 work hours evenly this month. Find the number of hours each will work.

23. The standard for low-flow toilets in the United States is $1\frac{3}{5}$ gallons of water per flush. Find the number of gallons of water used in 160 flushes. (*Source:* Wikipedia, The Free Encyclopedia.)

24. A welder needs angle iron pieces that are $3\frac{1}{3}$ inches long. Find the number of pieces that can be cut from a piece of angle iron that is 70 inches in length.

Write proper or improper for each fraction.

25. $\frac{2}{3}$

26. $\frac{6}{6}$

27. $\frac{9}{18}$

Write each mixed number as an improper fraction. Write each improper fraction as a whole or mixed number.

28. $3\frac{3}{8}$

29. $6\frac{2}{5}$

30. $\frac{14}{7}$

31. $\frac{103}{8}$

Find the prime factorization of each number. Write answers using exponents.

32. 72

33. 126

34. 350

Simplify each expression.

35. $4^2 \cdot 2^2$

36. $2^3 \cdot 6^2$

37. $2^3 \cdot 4^2 \cdot 5$

Write each fraction in lowest terms.

38. $\frac{42}{48}$

39. $\frac{24}{36}$

40. $\frac{30}{54}$

Multiply or divide as indicated. Simplify all answers.

41. $\frac{1}{2} \cdot \frac{3}{4}$

42. $30 \cdot \frac{2}{3} \cdot \frac{3}{5}$

43. $7\frac{1}{2} \cdot 3\frac{1}{3}$

44. $\frac{3}{5} \div \frac{5}{8}$

45. $\frac{7}{8} \div 1\frac{1}{2}$

46. $3 \div 1\frac{1}{4}$

Math in the Media

RECIPES

Rachael Ray is a Food Network television host, bestselling cookbook author, and the editor of her own lifestyle magazine. Rachael Ray's recipes can be found on her Web site, www.rachaelray.com, on her television show, or in one of her many cookbooks. The recipe at the side is from her book *Rachael Ray: 30-Minute Meals.*

Alaska Burgers

1 pound 93% lean ground beef
$\frac{1}{2}$ medium Spanish onion, minced or processed
4 shakes Worcestershire sauce
$\frac{1}{4}$ teaspoon allspice (1 good pinch)
$\frac{1}{2}$ teaspoon ground cumin (two good pinches)
Cracked black pepper
$\frac{1}{3}$ pound brick of smoked cheddar cheese, cut into $\frac{1}{2}$ inch slices

4 fresh, crusty onion rolls
Thick-sliced tomato and lettuce to top

Mix beef, onion, Worcestershire, allspice, cumin, and black pepper in a bowl. Separate a quarter of the mixture. Take a slice of the smoked cheese and place it in the middle of the mixture. Form the pattie shape around the cheese filling. Patties should be no more than $\frac{3}{4}$ inch thick. Repeat with rest of mixture to have a total of 4 patties.

Heat a nonstick griddle or frying pan to medium hot. Cook burgers 5 to 6 minutes on each side. Meat should be cooked through and cheese melted. Check each burger with an instant-read thermometer for an internal temp of 170°F for well done if undercooking concerns you. Or cut into one and check the color of the meat.

Salt burgers after preparation to your taste. (Salting beef before cooking draws out juices and flavor.) Top with tomato slices and lettuce. Serves 4.

1. Following the recipe, **(a)** what is the weight of one $\frac{1}{2}$-inch slice of cheese, and **(b)** what is the thickness of a $\frac{1}{3}$-pound brick of smoked cheddar cheese?

2. According to the recipe, **(a)** how many teaspoons of allspice are in 8 good pinches, and **(b)** how many teaspoons of ground cumin are in 7 good pinches?

3. Suppose you are preparing Alaska Burgers for 18 guests. By what factor will you change the ingredient amounts?

4. You know that of the 15 guests at your next party, 5 large eaters will eat $1\frac{1}{2}$ burgers each, 5 children will eat $\frac{1}{2}$ burger each, and the rest of the guests will each eat 1 burger. **(a)** How many burgers will you need, and **(b)** by what factor will you change the ingredient amounts?

5. Fill in the blanks with the ingredient amounts needed to make 9 servings of Alaska Burgers

Lean ground beef ____________

Spanish onions ____________

Worcestershire sauce ____________

Allspice ____________

Ground cumin ____________

Cheddar cheese ____________

6. If you have $5\frac{3}{4}$ pounds of beef, how many servings of Alaska Burgers can be prepared?

3 Adding and Subtracting Fractions

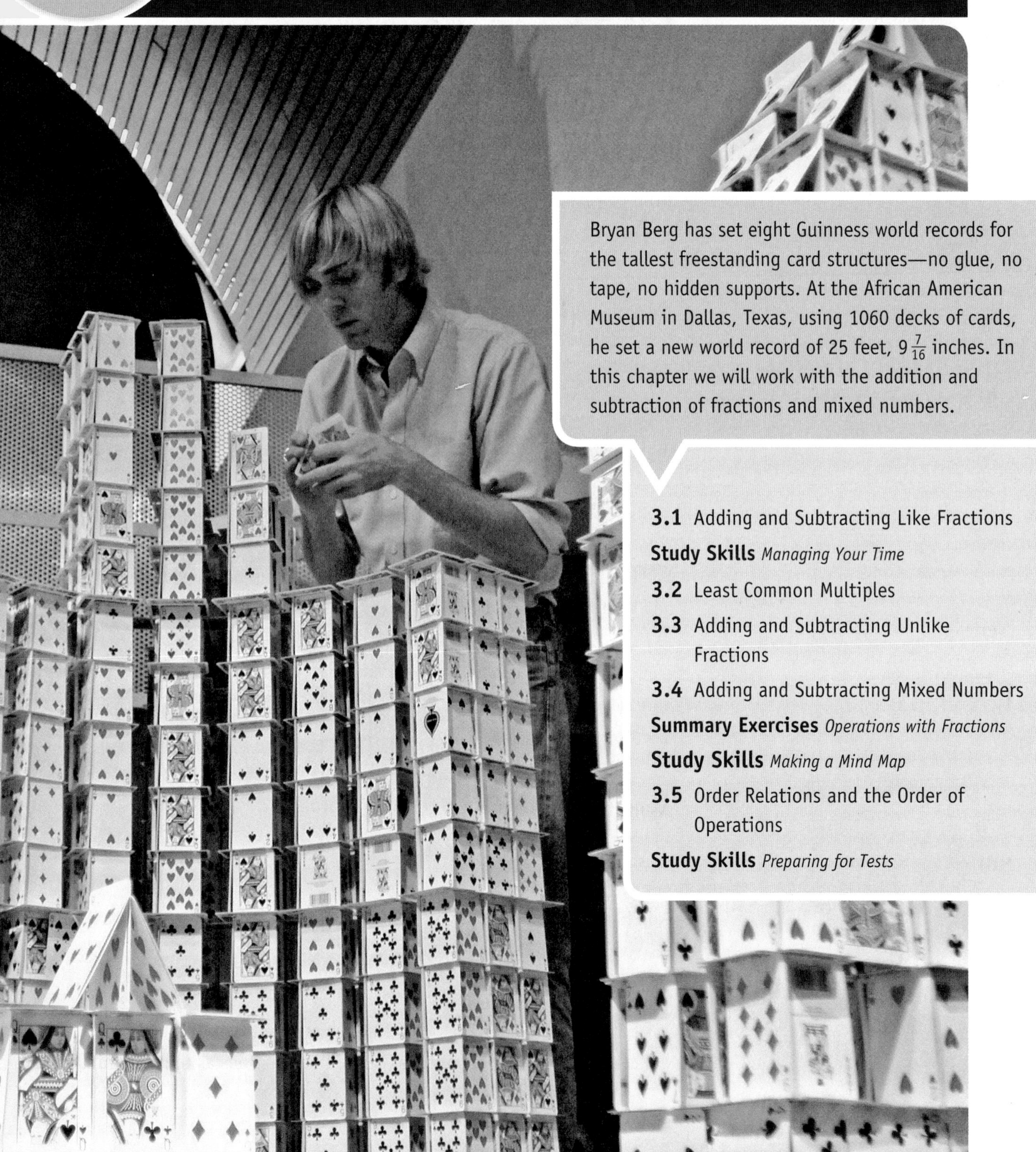

Bryan Berg has set eight Guinness world records for the tallest freestanding card structures—no glue, no tape, no hidden supports. At the African American Museum in Dallas, Texas, using 1060 decks of cards, he set a new world record of 25 feet, $9\frac{7}{16}$ inches. In this chapter we will work with the addition and subtraction of fractions and mixed numbers.

3.1 Adding and Subtracting Like Fractions

OBJECTIVES

1. **Define like and unlike fractions.**
2. **Add like fractions.**
3. **Subtract like fractions.**

In **Chapter 2** we looked at the basics of fractions and then practiced with multiplication and division of fractions and mixed numbers. In this chapter we will work with addition and subtraction of fractions and mixed numbers.

OBJECTIVE 1 **Define like and unlike fractions.** Fractions with the same denominators are **like fractions.** Fractions with different denominators are **unlike fractions.**

EXAMPLE 1 Identifying Like and Unlike Fractions

(a) $\frac{3}{4}, \frac{1}{4}, \frac{5}{4}, \frac{6}{4}$, and $\frac{4}{4}$ are **like** fractions.

All denominators are the same.

(b) $\frac{7}{12}$ and $\frac{12}{7}$ are **unlike** fractions.

Unlike fractions have different denominators.

Denominators are different.

Note

Like fractions have the ***same*** denominator.

◀ **Work Problem 1 at the Side.**

OBJECTIVE 2 **Add like fractions.** The figures below show you how to add the fractions $\frac{2}{7}$ and $\frac{4}{7}$.

$\frac{2}{7}$ + $\frac{4}{7}$ = $\frac{6}{7}$

As the figures show,

$$\frac{2}{7} + \frac{4}{7} = \frac{6}{7}.$$

Add like fractions as follows.

Adding Like Fractions

Step 1 Add the numerators to find the numerator of the sum.

Step 2 Write the denominator of the like fractions as the denominator of the sum.

Step 3 Write the sum in lowest terms.

1 Next to each pair of fractions write *like* or *unlike.*

(a) $\frac{2}{5}$ $\frac{3}{5}$ ______

(b) $\frac{2}{3}$ $\frac{3}{4}$ ______

(c) $\frac{7}{12}$ $\frac{11}{12}$ ______

(d) $\frac{3}{8}$ $\frac{3}{16}$ ______

Answers

1. **(a)** like **(b)** unlike **(c)** like **(d)** unlike

EXAMPLE 2 Adding Like Fractions

Add and write the sum in lowest terms.

(a) $\frac{1}{5}+\frac{2}{5}$

Add numerators.

$$\frac{1}{5}+\frac{2}{5}=\frac{1+2}{5}=\frac{3}{5}$$ ← Sum of numerators; ← Same denominator

$\frac{3}{5}$ is already in lowest terms.

(b) $\frac{1}{12}+\frac{7}{12}+\frac{1}{12}$

Fractions are ready to add if they are *like* fractions.

Add numerators.

Step 1 $\frac{1+7+1}{12}$

Step 2 $=\frac{9}{12}$ ← Sum of numerators; ← Same denominator

Step 3 $=\frac{9\div 3}{12\div 3}=\frac{3}{4}$ ← Lowest terms

CAUTION

Fractions may be added ***only*** if they have like denominators.

Work Problem 2 at the Side. ▶

2 Add and write the sums in lowest terms.

GS (a) $\frac{3}{8}+\frac{1}{8}$

$\frac{3+1}{8}=\frac{\quad}{8}=\frac{\quad}{2}$

(b) $\frac{2}{9}+\frac{5}{9}$

(c) $\frac{3}{16}+\frac{1}{16}$

(d) $\frac{3}{10}+\frac{1}{10}+\frac{4}{10}$

OBJECTIVE 3 Subtract like fractions. The figures below show $\frac{7}{8}$ broken into $\frac{4}{8}$ and $\frac{3}{8}$.

Subtracting $\frac{3}{8}$ from $\frac{7}{8}$ gives the answer $\frac{4}{8}$, or

$$\frac{7}{8}-\frac{3}{8}=\frac{4}{8}.$$

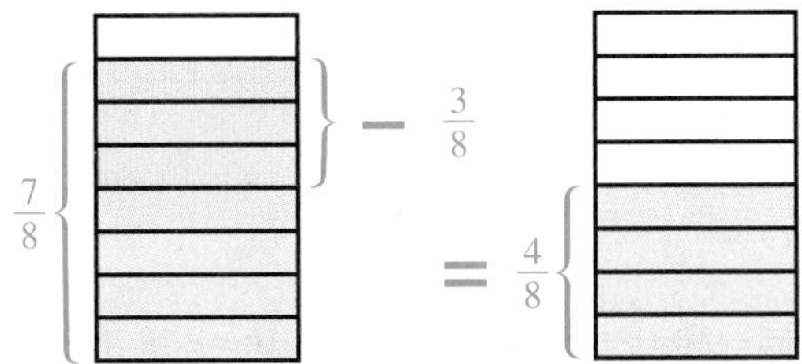

Answers

2. (a) 4; $\frac{1}{2}$ (b) $\frac{7}{9}$ (c) $\frac{1}{4}$ (d) $\frac{4}{5}$

3 Find the difference and simplify.

GS (a) $\frac{5}{6} - \frac{1}{6}$

$\frac{5-1}{6} = \frac{}{6} = \frac{}{3}$

(b) $\frac{16}{10}$
$- \frac{7}{10}$

(c) $\frac{15}{3} - \frac{5}{3}$

(d) $\frac{25}{32}$
$- \frac{6}{32}$

Answers

3. (a) $4; \frac{2}{3}$ (b) $\frac{9}{10}$ (c) $3\frac{1}{3}$ (d) $\frac{19}{32}$

Write $\frac{4}{8}$ in lowest terms.

$$\frac{7}{8} - \frac{3}{8} = \frac{4 \div 4}{8 \div 4} = \frac{1}{2}$$

The steps for subtracting like fractions are very similar to those for adding like fractions.

Subtracting Like Fractions

Step 1 Subtract the numerators to find the numerator of the difference.

Step 2 Write the denominator of the like fractions as the denominator of the difference.

Step 3 Write the answer in lowest terms.

EXAMPLE 3 Subtracting Like Fractions

Find the difference and simplify the answer.

(a) $\frac{15}{16} - \frac{3}{16}$

Subtract numerators.

Step 1 $\frac{15}{16} - \frac{3}{16} = \frac{15-3}{16}$

Step 2 $= \frac{12}{16}$ ← Difference of numerators ← Same denominator

Step 3 $= \frac{12 \div 4}{16 \div 4} = \frac{3}{4}$ ← Lowest terms

(b) $\frac{13}{4} - \frac{6}{4}$ Fractions are ready to subtract if they are *like* fractions.

Subtract numerators.

$\frac{13}{4} - \frac{6}{4} = \frac{13-6}{4}$ ← Difference of numerators ← Same denominator

$= \frac{7}{4}$

To simplify the answer, write $\frac{7}{4}$ as a mixed number.

$\frac{7}{4} = 1\frac{3}{4}$ Always simplify the answer.

CAUTION

Fractions may be subtracted ***only*** if they have like denominators.

◀ Work Problem **3** at the Side.

3.1 Exercises

FOR EXTRA HELP

Download the MyDashBoard App

MyMathLab®

CONCEPT CHECK *Next to each pair of fractions write* like *or* unlike.

1. $\frac{3}{8}\ \frac{5}{8}$

2. $\frac{5}{16}\ \frac{1}{4}$

3. $\frac{3}{5}\ \frac{3}{4}$

4. $\frac{5}{12}\ \frac{7}{12}$

CONCEPT CHECK *Fill in the blank.*

5. In order to add or subtract fractions, they must be ___________ fractions.

6. After adding or subtracting like fractions, the answer should be written in ___________.

Find the sum and simplify the answer. ***See Example 2.***

7. $\frac{3}{8}+\frac{2}{8}$

8. $\frac{1}{5}+\frac{3}{5}$

9. $\frac{1}{4}+\frac{1}{4}$

10. $\frac{9}{10} + \frac{3}{10}$

11. $\frac{13}{12} + \frac{5}{12}$

12. $\frac{2}{9} + \frac{1}{9}$

13. $\frac{7}{12}+\frac{3}{12}=\frac{7+3}{12}=\frac{10}{12}=$

14. $\frac{4}{15}+\frac{2}{15}+\frac{5}{15}=\frac{4+2+5}{15}=$

15. $\frac{3}{8}+\frac{7}{8}+\frac{2}{8}$

16. $\frac{4}{9}+\frac{1}{9}+\frac{7}{9}$

17. $\frac{2}{54}+\frac{8}{54}+\frac{12}{54}$

18. $\frac{7}{64}+\frac{15}{64}+\frac{20}{64}$

Find the difference and simplify the answer. ***See Example 3.***

19. $\frac{7}{8}-\frac{4}{8}$

20. $\frac{2}{3}-\frac{1}{3}$

21. $\frac{10}{11}-\frac{4}{11}$

22. $\frac{4}{5}-\frac{3}{5}$

23. $\frac{9}{10}-\frac{3}{10}=\frac{9-3}{10}=\frac{6}{10}=$

24. $\frac{7}{14}-\frac{3}{14}=\frac{7-3}{14}=\frac{4}{14}=$

25. $\frac{31}{21} - \frac{7}{21}$

26. $\frac{43}{24} - \frac{13}{24}$

27. $\frac{27}{40} - \frac{19}{40}$

28. $\frac{38}{55} - \frac{16}{55}$

29. $\frac{47}{36}-\frac{5}{36}$

30. $\frac{76}{45}-\frac{21}{45}$

31. $\frac{73}{60}-\frac{7}{60}$

32. $\frac{181}{100}-\frac{31}{100}$

33. In your own words, write an explanation of how to add like fractions. Use three steps in your explanation.

34. Describe in your own words the difference between *like* fractions and *unlike* fractions. Give three examples of each type.

Solve each application problem. Write answers in lowest terms.

35. The Fair Oaks Save the Bluffs Committee raised $\frac{2}{9}$ of their target goal last year and another $\frac{5}{9}$ of the goal this year. What fraction of their goal has been raised?

36. After an initial payment to a winner at an arcade fundraiser, the organization still owed the winner $\frac{7}{10}$ of her total winnings. If the organization pays the winner another $\frac{3}{10}$ of the winnings, what fraction is still owed?

37. GS Julie Circle, a landscaper, is working on a commercial job and completed only $\frac{1}{8}$ of the irrigation system in the first week. If she completed $\frac{5}{8}$ of the system in the second week, what fraction of the irrigation system has she completed?

$$\frac{1}{8}+\frac{5}{8}=\frac{6}{8}=$$

38. GS On a wild shopping spree, Angie Gragg spent $\frac{3}{24}$ of the day shopping in the morning and another $\frac{5}{24}$ of the day shopping after lunch. What fraction of the day did she spend shopping?

$$\frac{3}{24}+\frac{5}{24}=\frac{8}{24}=$$

39. An organic farmer purchased $\frac{9}{10}$ acre of land one year and $\frac{3}{10}$ acre the next year. She then planted carrots on $\frac{7}{10}$ acre of the land and squash on the remainder. How much land is planted with squash?

40. A forester planted $\frac{5}{12}$ acre in seedlings in the morning and $\frac{11}{12}$ acre in the afternoon. That night, $\frac{7}{12}$ acre of seedlings were destroyed by frost. How many acres of seedlings remain?

Study Skills
MANAGING YOUR TIME

OBJECTIVES

1. Create a semester schedule.
2. Create a "to do" list.

Many college students find themselves juggling a difficult schedule and multiple responsibilities. Perhaps you are going to school, working part time, and managing family demands. Here are some tips to help you develop good time management skills and habits.

- **Read the syllabus for each class.** Check on class policies, such as attendance, late homework, and make-up tests. Find out how you are graded. Keep the syllabus in your notebook.
- **Make a semester or quarter calendar.** Put test dates and major due dates for *all* your classes on the same calendar. That way you will see which weeks are the really busy ones. Try using a different color pen for each class. Your brain responds well to the use of color. A semester calendar is on the next page.
- **Make a weekly schedule.** After you fill in your classes and other regular responsibilities (such as work, picking up kids from school, etc.), block off some study periods during the day that you can guarantee you will use for studying. Aim for 2 hours of study for each 1 hour you are in class.
- **Make "To Do" lists.** Then use them by crossing off the tasks as you complete them. You might even number them in the order they need to be done (most important ones first).
- **Break big assignments into smaller chunks.** They won't seem so big that way. Make deadlines for each small part so you stay on schedule.
- **Give yourself small breaks in your studying.** Do not try to study for hours at a time! Your brain needs rest between periods of learning. Try to give yourself a 10 minute break each hour or so. You will learn more and remember it longer.
- **If you get off schedule, just try to get back on schedule tomorrow.** We all slip from time to time. All is not lost! Make a new "to do" list and start doing the most important things first.
- **Get help when you need it.** Talk with your instructor during office hours. Also, most colleges have a Learning Center, tutoring center, or counseling office. If you feel lost and overwhelmed, ask for help. Someone can help you decide what to do first and what to spend your time on right away.

Which two or three of the suggestions above will you try this week? How do you think they will help you?

1. ______________________________

2. ______________________________

3. ______________________________

Why Are These Techniques Brain Friendly?

Your brain appreciates some order. It enjoys a little routine, for example, choosing the same study time and place each day. You will find that you quickly settle in to your reading or homework.

Also, your brain **functions better when you are calm.** Too much rushing around at the last minute to get your homework and studying done sends hostile chemicals to your brain and makes it more difficult for you to learn and remember. So, a little planning can really pay off.

Building rest into your schedule is good for your brain. Remember, it takes time for dendrites to grow.

We've suggested using color on your calendars. This too, is brain friendly. Remember, your brain **likes pleasant colors and visual material** that are nice to look at. Messy and hard to read calendars will not be helpful, and you probably won't look at them often.

Study Skills

Continued from page 203

SEMESTER CALENDAR

WEEK	MON	TUES	WED	THUR	FRI	SAT	SUN
1							
2							
3							
4							
5							
6							
7							
8							
9							
10							
11							
12							
13							
14							
15							
16							

3.2 Least Common Multiples

OBJECTIVES

1. Find the least common multiple (LCM).
2. Find the least common multiple using multiples of the largest number.
3. Find the least common multiple using prime factorization.
4. Find the least common multiple using an alternative method.
5. Write a fraction with an indicated denominator.

Only *like* fractions can be added or subtracted. So, we must rewrite *unlike* fractions as *like* fractions before we can add or subtract them.

OBJECTIVE 1 Find the least common multiple (LCM). We can rewrite unlike fractions as like fractions by finding the *least common multiple* of the denominators.

> **Least Common Multiple (LCM)**
>
> The **least common multiple (LCM)** of two whole numbers is the smallest whole number divisible by both of those numbers.

EXAMPLE 1 Finding the Least Common Multiple (LCM)

Find the least common multiple of 6 and 9.

First, find the multiples of 6.

$6 \cdot 1$	$6 \cdot 2$	$6 \cdot 3$	$6 \cdot 4$	$6 \cdot 5$	$6 \cdot 6$	$6 \cdot 7$	$6 \cdot 8$
6,	12,	18,	24,	30,	36,	42,	48, . . .

(The three dots at the end of the list show that the list continues in the same pattern without stopping.) Now, find the multiples of 9.

$9 \cdot 1$	$9 \cdot 2$	$9 \cdot 3$	$9 \cdot 4$	$9 \cdot 5$	$9 \cdot 6$	$9 \cdot 7$	$9 \cdot 8$
9,	18,	27,	36,	45,	54,	63,	72, . . .

The smallest number found in *both* lists is 18, so 18 is the **least common multiple** of 6 and 9; the number 18 is the smallest whole number divisible by both 6 and 9.

Multiples of 6: 6, 12, 18, 24, 30, 36, 42, 48, . . .

Multiples of 9: 9, 18, 27, 36, 45, 54, 63, 72, . . .

The *smallest* number in *both* lists is the LCM.

18 is the smallest number found in both lists. 18 is the least common multiple (LCM) of 6 and 9.

Work Problem 1 at the Side. ▶

OBJECTIVE 2 Find the least common multiple using multiples of the largest number. There are several ways to find the least common multiple. If the numbers are small, the least common multiple can often be found by inspection. Can you think of a number that can be divided evenly by both 3 and 4? The number 12 will work; it is the least common multiple of 3 and 4. You can also find the least common multiple by writing multiples of the larger number.

In this case, 4 is larger than 3, so write the multiples of 4.

4, 8, 12, 16, 20, . . .

Now, check each multiple of 4 to see if it is divisible by 3.

4 is *not* divisible by 3
8 is *not* divisible by 3
12 *is* divisible by 3

The first multiple of 4 that is divisible by 3 is 12, so 12 is the least common multiple of 3 and 4.

VOCABULARY TIP

Least common multiple (LCM) A common multiple can be found by simply multiplying together the numbers under consideration. To make simplifying fractions easier, it's helpful to find the **least common multiple (LCM).** This is the *smallest* number into which the original numbers divide.

1 (a) List the multiples of 5.

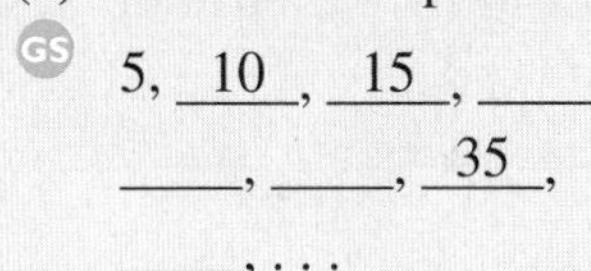

(b) List the multiples of 8.

8, ___, ___, ___,
___, ___, ___, . . .

(c) Find the least common multiple of 5 and 8.

Answers

1. (a) 5, 10, 15, 20, 25, 30, 35, 40, . . .
(b) 8, 16, 24, 32, 40, 48, 56, . . .
(c) 40

2 Use multiples of the larger number to find the least common multiple in each set of numbers.

GS **(a)** 2 and 5

Multiples of 5 are:

5, 10, 15, 20, . . .

What is the first multiple of 5 that is divisible by 2? ____

So the LCM of 2 and 5 is ____.

(b) 3 and 9

(c) 6 and 8

(d) 4 and 7

3 Use prime factorization to find the LCM for each pair of numbers.

GS **(a)** 15 and 18

$15 = 3 \cdot 5$

$18 =$ ____ • ____ • ____

LCM = ____ • ____ • ____ • ____ = ____

(b) 12 and 20

Answers

2. (a) 10, 10 (b) 9 (c) 24 (d) 28

3. (a) $15 = 3 \cdot ⑤$, $18 = ② \cdot ③ \cdot ③$, LCM = $2 \cdot 3 \cdot 3 \cdot 5 = 90$

(b) $12 = ② \cdot ② \cdot ③$, $20 = 2 \cdot 2 \cdot ⑤$, LCM = $2 \cdot 2 \cdot 3 \cdot 5 = 60$

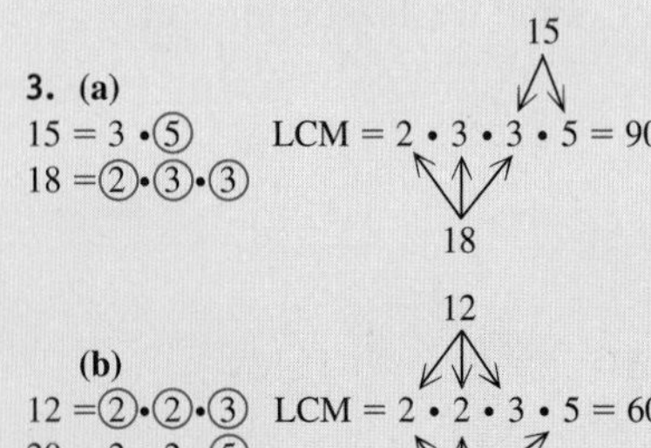

EXAMPLE 2 Finding the Least Common Multiple (LCM)

Use multiples of the larger number to find the least common multiple of 6 and 9.

Start by writing the first few multiples of 9.

Multiples of 9

9, 18, 27, 36, 45, 54, . . .

Now check each multiple of 9 to see if it is divisible by 6. The first multiple of 9 that is divisible by 6 is 18.

9, 18, 27, 36, 45, 54, . . .

First multiple divisible by 6, because $18 \div 6 = 3$

The least common multiple (LCM) of 6 and 9 is 18.

◀ Work Problem **2** at the Side.

OBJECTIVE 3 Find the least common multiple using prime factorization. **Example 2** shows how to find the least common multiple of two numbers by making a list of the multiples of the *larger* number. Although this method works well if both numbers are fairly small, it is usually easier to find the least common multiple for larger numbers by using *prime factorization,* as shown in the next example. (See **Chapter 2** for further review.)

EXAMPLE 3 Using Prime Factorization to Find the LCM

Use prime factorization to find the least common multiple of 9 and 12.

Start by finding the prime factorization of each number.

$9 = 3 \cdot 3$

$12 = 2 \cdot 2 \cdot 3$

Prime factorizations of 9 and 12.

Circle the factors where they appear the greatest number of times in either factorization.

$9 = ③ \cdot ③$ — 3 appears most often in this factorization.

$12 = ② \cdot ② \cdot 3$ — 2 appears most often in this factorization.

The LCM is the product of the circled factors.

Factors of 9

$\text{LCM} = 3 \cdot 3 \cdot 2 \cdot 2 = 36$ — The product of the circled factors is the LCM.

Factors of 12

The product of the prime factors, 36, is the least common multiple. Check to see that 36 is divisible by 9 (yes) and by 12 (yes). The smallest whole number divisible by both 9 and 12 is 36.

CAUTION

Notice that we did ***not*** repeat the factors that 9 and 12 have in common. In this case, the **3** in $2 \cdot 2 \cdot \mathbf{3} = 12$ was ***not*** used because 3 is already included in $3 \cdot 3 = 9$.

◀ Work Problem **3** at the Side.

EXAMPLE 4 Using Prime Factorization

Find the least common multiple of 12, 18, and 20.

Find the prime factorization of each number. Then use the prime factors to build the LCM.

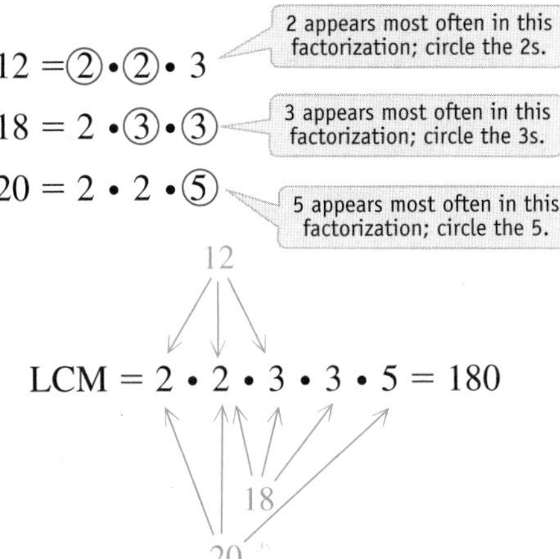

12 =②•②• 3
2 appears most often in this factorization; circle the 2s.
18 = 2 •③•③
3 appears most often in this factorization; circle the 3s.
20 = 2 • 2 •⑤
5 appears most often in this factorization; circle the 5.
12

LCM = 2 • 2 • 3 • 3 • 5 = 180

18

20

Check to see that 180 is divisible by 12 (yes) and by 18 (yes) and by 20 (yes). This smallest whole number divisible by 12, 18, and 20 is 180. The LCM is 180.

Work Problem 4 at the Side. ▶

EXAMPLE 5 Finding the Least Common Multiple

Find the least common multiple for each set of numbers.

(a) 5, 6, 35

Find the prime factorization for each number.

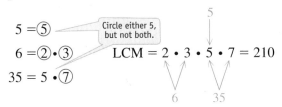

5 =⑤ (Circle either 5, but not both.)

6 =②•③

35 = 5 •⑦

LCM = 2 • 3 • 5 • 7 = 210

The least common multiple of 5, 6, and 35 is 210.

(b) 10, 20, 24

Find the prime factorization for each number.

10 = 2 • 5

20 = 2 • 2 •⑤

24 =②•②•②•③

LCM = 2 • 2 • 2 • 3 • 5 = 120

The least common multiple of 10, 20, and 24 is 120.

Work Problem 5 at the Side. ▶

OBJECTIVE 4 Find the least common multiple using an alternative method. Some people like the following ***alternative method*** for finding the least common multiple for larger numbers. Try both methods, and ***use the one you prefer.*** As a review, a list of the first few prime numbers follows.

First few prime numbers ⟶ 2, 3, 5, 7, 11, 13, 17

4 Find the least common multiple of the denominators in each set of fractions.

(a) $\frac{3}{8}$ and $\frac{6}{5}$

(b) $\frac{5}{6}$ and $\frac{1}{14}$

(c) $\frac{4}{9}$, $\frac{5}{18}$, and $\frac{7}{24}$

5 Find the least common multiple for each set of numbers.

GS **(a)** 4, 8, 9

4 = 2 • 2

8 = 2 • 2 • 2

9 = 3 • 3

LCM = ____ • ____ • ____ • ____ • ____ = ____

(b) 3, 6, 8

(c) 15, 20, 30, 40

Answers

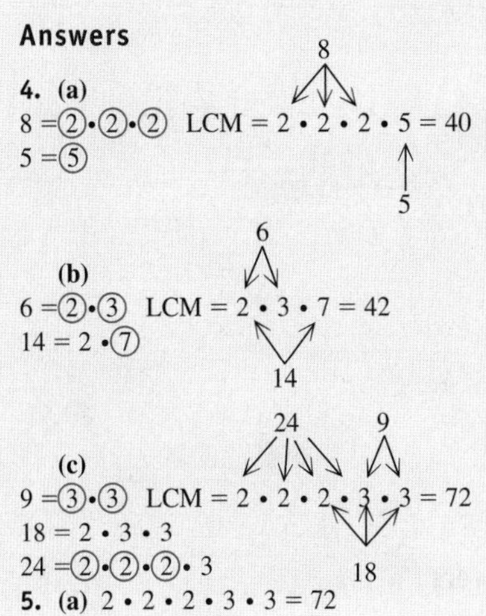

4. (a)
8 =②•②•② LCM = 2 • 2 • 2 • 5 = 40
5 =⑤
(b)
6 =②•③ LCM = 2 • 3 • 7 = 42
14 = 2 •⑦
(c)
9 =③•③ LCM = 2 • 2 • 2 • 3 • 3 = 72
18 = 2 • 3 • 3
24 =②•②•②• 3

5. (a) 2 • 2 • 2 • 3 • 3 = 72 **(b)** 24 **(c)** 120

6 In the following problems, the divisions have already been worked out. Multiply the prime numbers on the left to find the least common multiple.

GS **(a)**

$$\begin{array}{r|rr} 2 & 6 & \cancel{15} \\ 3 & 3 & 15 \\ 5 & \cancel{1} & 5 \\ & 1 & 1 \end{array}$$

2 • ____ • ____ = ____

(b)

$$\begin{array}{r|rr} 2 & 20 & 36 \\ 2 & 10 & 18 \\ 3 & \cancel{5} & 9 \\ 3 & \cancel{5} & 3 \\ 5 & 5 & \cancel{1} \\ & 1 & 1 \end{array}$$

Answers

6. **(a)** 3; 5; 30 **(b)** 180

EXAMPLE 6 **Alternative Method for Finding the Least Common Multiple**

Find the least common multiple for each set of numbers.

(a) 14 and 21

Start by trying to divide 14 and 21 by the first prime number, which is 2. Use the following shortcut.

$$\begin{array}{r|rr} 2 & 14 & \cancel{21} \\ & 7 & 21 \end{array}$$

14 divided by 2 is 7.

21 cannot be divided evenly by 2, so bring it down.

Because 21 cannot be divided evenly by 2, cross out 21 and bring it down. Divide by 3, the second prime.

$$\begin{array}{r|rr} 2 & 14 & \cancel{21} \\ 3 & \cancel{7} & 21 \\ & 7 & 7 \end{array}$$

7 cannot be divided evenly by 3, so bring it down.

21 divided by 3 is 7.

Since 7 cannot be divided evenly by the third prime, 5, skip 5. Divide by 7, the fourth prime.

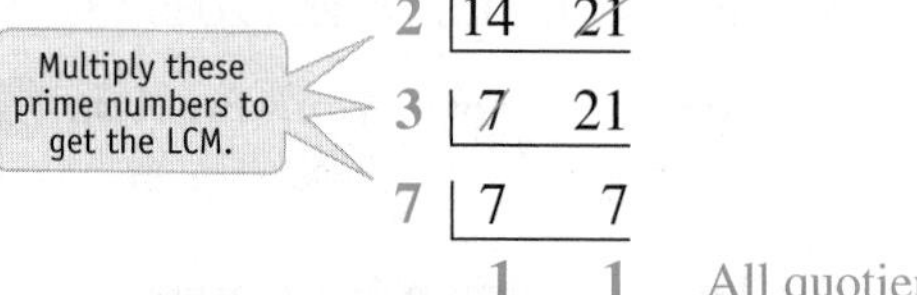

Multiply these prime numbers to get the LCM.
2 | 14 21
3 | 7 21
7 | 7 7
1 1

All quotients are 1.

When all quotients are 1, multiply the prime numbers on the left side.

least common multiple = $2 \cdot 3 \cdot 7 = 42$

The least common multiple of 14 and 21 is 42.

(b) 6, 15, 18

Divide by 2.

$$\begin{array}{r|rrr} 2 & 6 & \cancel{15} & 18 \\ & 3 & 15 & 9 \end{array}$$

Cross out 15 and bring it down.

Divide by 3.

$$\begin{array}{r|rrr} 2 & 6 & \cancel{15} & 18 \\ 3 & 3 & 15 & 9 \\ & 1 & 5 & 3 \end{array}$$

Divide by 3 again, since the remaining 3 can be divided.

$$\begin{array}{r|rrr} 2 & 6 & \cancel{15} & 18 \\ 3 & 3 & 15 & 9 \\ 3 & \cancel{1} & \cancel{5} & 3 \\ & 1 & 5 & 1 \end{array}$$

Finally, divide by 5.

$$\begin{array}{r|rrr} 2 & 6 & \cancel{15} & 18 \\ 3 & 3 & 15 & 9 \\ 3 & \cancel{1} & \cancel{5} & 3 \\ 5 & \cancel{1} & 5 & \cancel{1} \\ & 1 & 1 & 1 \end{array}$$

All quotients are 1.

Multiply the prime numbers on the left side.

$2 \cdot 3 \cdot 3 \cdot 5 = 90$ ← Least common multiple

◄ **Work Problem 6 at the Side.**

EXAMPLE 7 Find the Least Common Multiple Using Either Method

Find the least common multiple of 12, 21, and 24. For **(a)** use the prime factorization method. Then, for **(b)** use the alternative method.

(a) Find the prime factorization for each number.

$$12 = 2 \cdot 2 \cdot \textcircled{3}$$
$$21 = 3 \cdot \textcircled{7}$$
$$24 = \textcircled{2} \cdot \textcircled{2} \cdot \textcircled{2} \cdot 3$$

The product of the circled factors is the LCM.

$$\text{LCM} = 2 \cdot 2 \cdot 2 \cdot 3 \cdot 7 = 168$$

(b) Use the alternative method to find the LCM.

2	12	21	24
2	6	21	12
2	3	21	6
3	3	21	3
7	1	7	1
	1	1	1

All quotients are 1.

The product of the prime numbers on the left is the LCM.

$$\text{LCM} = 2 \cdot 2 \cdot 2 \cdot 3 \cdot 7 = 168$$

Work Problem 7 at the Side. ▶

7 Find the least common multiple of each set of numbers. Use whichever method you prefer.

GS **(a)** 3, 6, 10

2	3	6	10
3	3	3	5
5	___	1	___
	1	1	1

$2 \cdot$ ____ $\cdot 5 =$ ____

(b) 15, 40

(c) 9, 24

(d) 8, 21, 24

OBJECTIVE 5 Write a fraction with an indicated denominator. Before adding or subtracting *unlike* fractions, you must find the least common multiple, which is then used as the denominator of the fractions.

EXAMPLE 8 Writing a Fraction with an Indicated Denominator

Write the fraction $\frac{2}{3}$ with a denominator of 15.

Find a numerator, so that these fractions are equivalent.

$$\frac{2}{3} = \frac{?}{15}$$

To find the new numerator, first divide **15** by **3.**

$$\frac{2}{3} = \frac{?}{15} \qquad 15 \div 3 = 5$$

Multiply both numerator and denominator of the fraction $\frac{2}{3}$ by 5.

$$\frac{2}{3} = \frac{2 \cdot 5}{3 \cdot 5} = \frac{10}{15}$$

Multiplying by $\frac{5}{5}$ is the same as multiplying by 1, because $\frac{5}{5} = 1$.

This process is just the opposite of writing a fraction in lowest terms. Check the answer by writing $\frac{10}{15}$ in lowest terms; you should get $\frac{2}{3}$ again.

Answers

7. **(a)** 1; 5; 3; 30 **(b)** 120 **(c)** 72 **(d)** 168

8 Rewrite each fraction with the indicated denominator.

(a) $\frac{1}{4} = \frac{?}{16}$

(b) $\frac{5}{3} = \frac{?}{15}$

(c) $\frac{7}{16} = \frac{?}{32}$

(d) $\frac{6}{11} = \frac{?}{33}$

Answers

8. (a) $\frac{4}{16}$ **(b)** $\frac{25}{15}$ **(c)** $\frac{14}{32}$ **(d)** $\frac{18}{33}$

EXAMPLE 9 Writing Fractions with a New Denominator

Rewrite each fraction with the indicated denominator.

(a) $\frac{3}{8} = \frac{?}{48}$

Divide 48 by 8, to get 6. Now multiply both the numerator and the denominator of $\frac{3}{8}$ by 6.

$$\frac{3}{8} = \frac{3 \cdot 6}{8 \cdot 6} = \frac{18}{48}$$ Multiply numerator and denominator by 6.

Multiplying a number by 1 does *not* change the number, and $\frac{6}{6} = 1$.

That is, $\frac{3}{8} = \frac{18}{48}$. As a check, write $\frac{18}{48}$ in lowest terms; you should get $\frac{3}{8}$ again.

(b) $\frac{7}{6} = \frac{?}{42}$

Divide 42 by 6, to get 7. Next, multiply both the numerator and the denominator of $\frac{7}{6}$ by 7.

$$\frac{7}{6} = \frac{7 \cdot 7}{6 \cdot 7} = \frac{49}{42}$$ Multiply numerator and denominator by 7.

Multiplying by $\frac{7}{7}$ is the same as multiplying by 1.

This shows that $\frac{7}{6} = \frac{49}{42}$. As a check, write $\frac{49}{42}$ in lowest terms. Did you get $\frac{7}{6}$ again?

Note

In **Example 8,** on the previous page, the fraction $\frac{2}{3}$ was multiplied by $\frac{5}{5}$. In **Example 9,** the fraction $\frac{3}{8}$ was multiplied by $\frac{6}{6}$ and the fraction $\frac{7}{6}$ was multiplied by $\frac{7}{7}$. The fractions, $\frac{5}{5}$, $\frac{6}{6}$, and $\frac{7}{7}$ are all equal to 1.

$$\frac{5}{5} = 1 \qquad \frac{6}{6} = 1 \qquad \frac{7}{7} = 1$$

Recall that any number multiplied by 1 is the number itself.

◀ Work Problem **8** at the Side.

3.2 Exercises

FOR EXTRA HELP

Download the MyDashBoard App

MyMathLab®

CONCEPT CHECK *Write either* true *or* false *for each statement. If* false, *explain why.*

1. The least common multiple (LCM) of 4 and 8 is 8.

2. The least common multiple (LCM) of 6 and 5 is 25.

3. The least common multiple (LCM) of 9 and 4 is 36.

4. The least common multiple (LCM) of 3 and 7 is 28.

Use multiples of the larger number to find the least common multiple in each set of numbers. ***See Examples 1 and 2.***

5. 3 and 6

6. 2 and 4

7. 3 and 5

8. 3 and 7

9. 4 and 9

10. 6 and 8

GS

Multiples of 8

8 16 24 32 40

What is the first multiple divisible by 6? ____

11. 12 and 16

GS

Multiples of 16

16 32 48 64 80

What is the first multiple divisible by 16? ____

12. 25 and 75

13. 20 and 50

Find the least common multiple of each set of numbers. Use any method. ***See Examples 3–7.***

14. 4, 10

GS

Prime factorizations

4 = ②•②

10 = 2 •⑤

LCM = ____ • ____ • ____ = ____

15. 8, 10

GS

Prime factorizations

8 = ②•②•②

10 = 2 •⑤

LCM = ____ • ____ • ____ • ____ = ____

16. 12, 20

17. 9 and 15

18. 6, 9, 12

19. 20, 24, 30

20. 8, 9, 12, 18

21. 4, 6, 8, 10

22. 12, 15, 18, 20

23. 6, 8, 9, 27, 36

24. 8, 10, 12, 16, 36

25. 5, 6, 8, 25, 30

Rewrite each fraction with a denominator of 24. **See Examples 8 and 9.**

26. $\frac{2}{3} =$

27. $\frac{3}{8} =$

28. $\frac{3}{4} =$

29. $\frac{5}{12} =$

30. $\frac{5}{6} =$

31. $\frac{7}{8} =$

32. CONCEPT CHECK Circle the fraction that is equivalent to $\frac{2}{3}$.

$\frac{7}{8}$ $\frac{3}{4}$ $\frac{12}{16}$ $\frac{8}{12}$

33. CONCEPT CHECK Circle the fraction that is equivalent to $\frac{3}{4}$.

$\frac{4}{5}$ $\frac{18}{22}$ $\frac{21}{28}$ $\frac{9}{15}$

34. CONCEPT CHECK Circle the fraction that is equivalent to $\frac{9}{4}$.

$\frac{36}{16}$ $\frac{35}{20}$ $\frac{21}{8}$ $\frac{15}{6}$

35. CONCEPT CHECK Circle the fraction that is equivalent to $\frac{7}{8}$.

$\frac{14}{15}$ $\frac{21}{24}$ $\frac{8}{7}$ $\frac{35}{48}$

Rewrite each fraction with the indicated denominator.

36. $\frac{1}{2} = \frac{}{6}$

37. $\frac{2}{3} = \frac{}{9}$

38. $\frac{3}{4} = \frac{}{16}$

39. $\frac{7}{8} = \frac{}{32}$

40. $\frac{8}{5} = \frac{}{20}$

41. $\frac{3}{16} = \frac{}{64}$

42. $\frac{5}{8} = \frac{}{40}$

43. $\frac{9}{7} = \frac{}{56}$

44. $\frac{3}{2} = \frac{}{64}$

45. $\frac{7}{4} = \frac{}{48}$

GS

$\frac{7}{4} = \frac{?}{48}$

Divide 48 by 4 to get ______.

Now multiply 7 by ______ to get the new numerator, which is ______.

46. $\frac{5}{6} = \frac{}{120}$

GS

$\frac{5}{6} = \frac{?}{120}$

Divide 120 by 6 to get ______.

Now multiply 5 by ______ to get the new numerator, which is ______.

47. $\frac{8}{11} = \frac{}{132}$

48. $\frac{4}{15} = \frac{}{165}$

49. $\frac{3}{16} = \frac{}{144}$

50. $\frac{7}{16} = \frac{}{112}$

51. There are several methods for finding the least common multiple (LCM). Do you prefer the method using multiples of the largest number, the method using prime factorizations, or the alternative method for finding the least common multiple? Why? Would you ever use the other methods?

52. Explain in your own words how to write a fraction with an indicated denominator. As part of your explanation, show how to change $\frac{3}{4}$ to a fraction having 12 as a denominator. Also explain how you could check your answer.

Find the least common multiple of the denominators of each pair of fractions.

53. $\frac{25}{400}, \frac{38}{1800}$

54. $\frac{53}{600}, \frac{115}{4000}$

55. $\frac{109}{1512}, \frac{23}{392}$

56. $\frac{61}{810}, \frac{37}{1170}$

Relating Concepts (Exercises 57–64) For Individual or Group Work

Most people think that addition and subtraction of fractions are more difficult than multiplication and division of fractions. This is probably because a common denominator must be used. ***Work Exercises 57–64 in order.***

57. Fractions with the same denominators are __________ fractions and fractions with different denominators are __________ fractions.

58. To subtract like fractions, first subtract the __________ to find the numerator of the difference. Write the denominator of the like fractions as the __________ of the difference. Finally, write the answer in __________ terms.

59. The __________ common multiple (LCM) of two numbers is the (*smallest/largest*) whole number divisible by both those numbers.

60. The following shows the common multiples for both 8 and 10. What is the least common multiple for these two numbers?

Multiples of 8: 8, 16, 24, 32, 40, 48, 56, 64, 72, 80, 88, . . .

Multiples of 10: 10, 20, 30, 40, 50, 60, 70, 80, 90, . . .

Find the least common multiple for each set of numbers. Use whichever method you like best.

61. 5, 7, 14, 10

62. 25, 18, 30, 5

63. Explain why the least common multiple for 8, 3, 5, 4, and 10 is not 240. Find the least common multiple.

64. Explain why the least common multiple of 55 and 1760 is 1760.

3.3 Adding and Subtracting Unlike Fractions

OBJECTIVES

1. Add unlike fractions.
2. Add unlike fractions vertically.
3. Subtract unlike fractions.
4. Subtract unlike fractions vertically.

OBJECTIVE 1 **Add unlike fractions.** In this section, we add and subtract unlike fractions. To add unlike fractions, we must first change them to like fractions (fractions with the same denominator). For example, the figures below show $\frac{3}{8}$ and $\frac{1}{4}$.

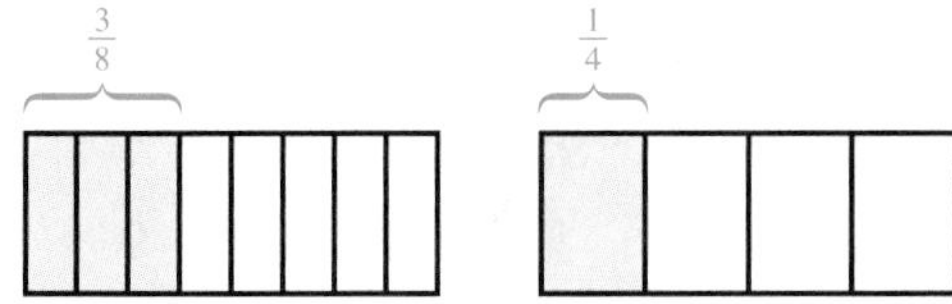

These fractions can be added by changing them to like fractions. Make like fractions by changing $\frac{1}{4}$ to the equivalent fraction $\frac{2}{8}$.

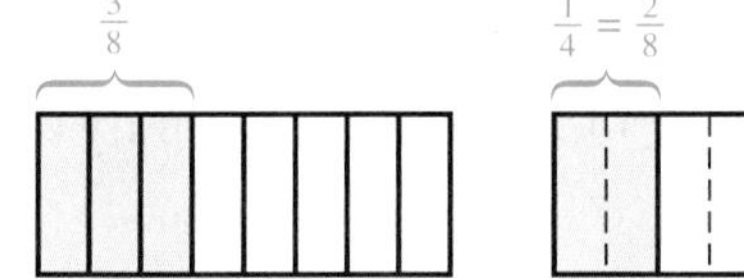

Now you can add the fractions.

Becomes

$$\frac{3}{8}+\frac{1}{4}=\frac{3}{8}+\frac{2}{8}=\frac{5}{8}$$

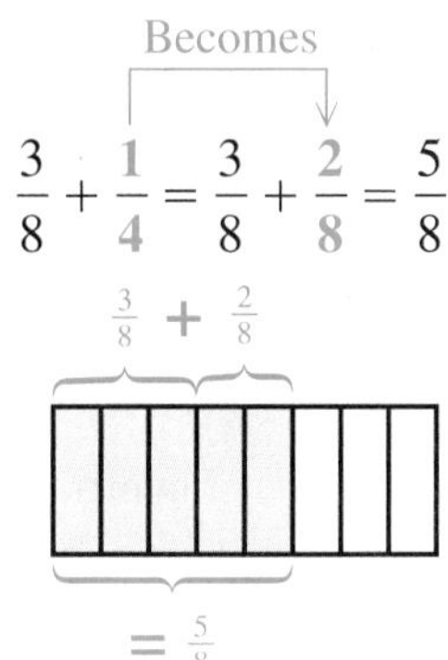

Use the following steps to add or subtract unlike fractions.

VOCABULARY TIP

LCM and LCD The least common multiple **(LCM)** and the least common denominator **(LCD)** are similar. When the LCM of two or more numbers is found and then used as a new denominator, it is called the least common denominator (LCD). You must find the LCD in order to add or subtract unlike fractions.

Adding or Subtracting Unlike Fractions

Step 1 Rewrite the *unlike fractions* as *like fractions* with the least common multiple as their new denominator. This new denominator is called the **least common denominator (LCD).**

Step 2 Add or subtract as with like fractions.

Step 3 Simplify the answer by writing it in lowest terms and as a whole or mixed number where possible.

EXAMPLE 1 Adding Unlike Fractions

Add $\frac{2}{3}$ and $\frac{1}{9}$.

The least common multiple of 3 and 9 is 9, so first rewrite the fractions as like fractions with a denominator of 9. This is the ***least common denominator (LCD)*** of 3 and 9.

Continued on Next Page

1 Add.

GS (a) $\frac{1}{2} + \frac{3}{8}$

$\frac{\_\_\_}{8} + \frac{3}{8} = \frac{\_\_\_}{8}$

(b) $\frac{3}{4} + \frac{1}{8}$

(c) $\frac{3}{5} + \frac{3}{10}$

(d) $\frac{1}{12} + \frac{5}{6}$

2 Add. Simplify all answers.

GS (a) $\frac{3}{10} + \frac{1}{5}$

$\frac{3}{10} + \frac{\_\_\_}{10} = \frac{\_\_\_}{\_\_\_} = \frac{\_\_\_}{\_\_\_}$

(b) $\frac{5}{8} + \frac{1}{3}$

(c) $\frac{1}{10} + \frac{1}{3} + \frac{1}{6}$

Answers

1. (a) 4; $\frac{7}{8}$ (b) $\frac{7}{8}$ (c) $\frac{9}{10}$ (d) $\frac{11}{12}$

2. (a) 2; $\frac{5}{10}$; $\frac{1}{2}$ (b) $\frac{23}{24}$ (c) $\frac{3}{5}$

Step 1 $\frac{2}{3} = \frac{?}{9}$

Divide 9 by 3, getting 3. Next, multiply numerator and denominator by 3.

$$\frac{2}{3} = \frac{2 \cdot 3}{3 \cdot 3} = \frac{6}{9}$$ ◄ $\frac{6}{9}$ is equivalent to $\frac{2}{3}$.

Now, add the like fractions $\frac{6}{9}$ and $\frac{1}{9}$.

Becomes

Both fractions must have the *same* denominator **before** you add them.

Step 2 $\frac{2}{3} + \frac{1}{9} = \frac{6}{9} + \frac{1}{9} = \frac{6+1}{9} = \frac{7}{9}$

Step 3 Step 3 is not needed because $\frac{7}{9}$ is already in lowest terms.

◄ **Work Problem 1 at the Side.**

EXAMPLE 2 Adding Fractions

Add each pair of fractions using the three steps. Simplify all answers.

(a) $\frac{1}{3} + \frac{1}{6}$

The least common multiple of 3 and 6 is 6. Rewrite both fractions as fractions with a least common denominator of 6.

Rewritten as like fractions

Step 1 $\frac{1}{3} + \frac{1}{6} = \frac{2}{6} + \frac{1}{6}$ ◄ 6 is the LCD (least common denominator).

Add numerators.

Step 2 $\frac{2}{6} + \frac{1}{6} = \frac{2+1}{6} = \frac{3}{6}$ ← Sum of numerators / ← Least common denominator

Step 3 $\frac{3}{6} = \frac{1}{2}$ ← Lowest terms

(b) $\frac{6}{15} + \frac{3}{10}$

The least common multiple of 15 and 10 is 30, so rewrite both fractions with a least common denominator of 30.

Rewritten as like fractions

Step 1 $\frac{6}{15} + \frac{3}{10} = \frac{12}{30} + \frac{9}{30}$ ◄ The least common multiple of the denominators, 30, is also the LCD.

Add numerators.

Step 2 $\frac{12}{30} + \frac{9}{30} = \frac{12+9}{30} = \frac{21}{30}$

Step 3 $\frac{21}{30} = \frac{7}{10}$ ← Lowest terms

◄ **Work Problem 2 at the Side.**

OBJECTIVE 2 Add unlike fractions vertically. Fractions can also be added vertically (one fraction written below the other).

EXAMPLE 3 Vertical Addition of Fractions

Add the following fractions vertically.

(a)

$$\begin{array}{rcccl} \frac{3}{8} & = & \frac{3 \cdot 3}{8 \cdot 3} & = & \frac{9}{24} \\ +\frac{7}{12} & = & \frac{7 \cdot 2}{12 \cdot 2} & = & \frac{14}{24} \\ \hline & & & & \frac{23}{24} \end{array}$$

24 is the LCD.

Rewritten as like fractions

23 ← Add the numerators.

24 ← Denominator is 24, the LCD.

$\frac{23}{24}$ is in simplest form.

(b)

$$\begin{array}{rcccl} \frac{2}{9} & = & \frac{2 \cdot 4}{9 \cdot 4} & = & \frac{8}{36} \\ +\frac{1}{4} & = & \frac{1 \cdot 9}{4 \cdot 9} & = & \frac{9}{36} \\ \hline & & & & \frac{17}{36} \end{array}$$

36 is the LCD.

Rewritten as like fractions

17 ← Add the numerators.

36 ← Denominator is 36, the LCD.

$\frac{17}{36}$ is in simplest form.

Work Problem 3 at the Side. ▶

OBJECTIVE 3 Subtract unlike fractions. The next example shows subtraction of unlike fractions.

EXAMPLE 4 Subtracting Unlike Fractions

Subtract. Simplify all answers.

As with addition, rewrite unlike fractions with a least common denominator.

(a) $\frac{3}{4} - \frac{3}{8}$

Rewritten as like fractions

Step 1 $\frac{3}{4} - \frac{3}{8} = \frac{6}{8} - \frac{3}{8}$ The LCD is 8.

Subtract numerators.

Step 2 $\frac{6}{8} - \frac{3}{8} = \frac{6-3}{8} = \frac{3}{8}$ ← Difference of numerators ← Least common denominator

Step 3 Not needed because $\frac{3}{8}$ is in lowest terms.

Continued on Next Page

3 Add the following fractions vertically.

GS **(a)**

$$\begin{array}{rcccl} \frac{5}{8} & = & \frac{5 \cdot 3}{8 \cdot \_\_} & = & \frac{15}{\_\_} \\ +\frac{1}{12} & = & \frac{1 \cdot 2}{12 \cdot 2} & = & \frac{2}{24} \\ \hline & & & & \frac{\_\_}{\_\_} \end{array}$$

(b)

$$\begin{array}{r} \frac{7}{16} \\ +\frac{1}{4} \\ \hline \end{array}$$

(c)

$$\begin{array}{r} \frac{1}{8} \\ \frac{5}{24} \\ +\frac{7}{16} \\ \hline \end{array}$$

Answers

3. (a) 3; 24; $\frac{17}{24}$ (b) $\frac{11}{16}$ (c) $\frac{37}{48}$

4 Subtract. Simplify all answers.

GS **(a)** $\frac{5}{8} - \frac{1}{4}$

$\frac{5}{8} - \frac{\_\_\_}{8} =$

(b) $\frac{4}{5} - \frac{3}{4}$

5 Subtract vertically. Simplify all answers.

GS **(a)** $\frac{7}{8} = \frac{7 \cdot \_\_}{8 \cdot 3} = \frac{21}{\_\_\_}$

$-\frac{2}{3} = \frac{2 \cdot 8}{3 \cdot 8} = \frac{16}{24}$

(b) $\frac{5}{6}$

$-\frac{1}{12}$

Answers

4. (a) 2; $\frac{3}{8}$ (b) $\frac{1}{20}$

5. (a) 3; 24; $\frac{5}{24}$ (b) $\frac{3}{4}$

(b) $\frac{3}{4} - \frac{7}{12}$

Rewritten as like fractions

Step 1 $\frac{3}{4} - \frac{7}{12} = \frac{9}{12} - \frac{7}{12}$

Subtract numerators.

Step 2 $\frac{9}{12} - \frac{7}{12} = \frac{9-7}{12} = \frac{2}{12}$ ← Subtract the numerators. ← Denominator is 12, the LCD.

Step 3 $\frac{2}{12} = \frac{1}{6}$ ← Lowest terms

Always simplify the final answer.

◀ **Work Problem 4 at the Side.**

OBJECTIVE 4 Subtract unlike fractions vertically.

EXAMPLE 5 Vertical Subtraction of Fractions

Subtract the following fractions vertically. Simplify all answers.

(a) $\frac{4}{5} = \frac{4 \cdot 8}{5 \cdot 8} = \frac{32}{40}$

$-\frac{3}{8} = \frac{3 \cdot 5}{8 \cdot 5} = \frac{15}{40}$ } Rewritten as like fractions

$\frac{17}{40}$ ← Subtract numerators. ← Denominator is 40, the LCD.

$\frac{17}{40}$ is in simplest form.

(b) $\frac{3}{7} = \frac{3 \cdot 12}{7 \cdot 12} = \frac{36}{84}$

$-\frac{5}{12} = \frac{5 \cdot 7}{12 \cdot 7} = \frac{35}{84}$ } Rewritten as like fractions

$\frac{1}{84}$ ← Subtract numerators. ← Denominator is 84, the LCD.

$\frac{1}{84}$ is in simplest form.

◀ **Work Problem 5 at the Side.**

3.3 Exercises

FOR EXTRA HELP

Download the MyDashBoard App

MyMathLab®

CONCEPT CHECK *Fill in the blank with the correct response.*

1. To add or subtract unlike fractions, the first step is to rewrite the fractions as ________________ fractions.

2. To rewrite unlike fractions as like fractions, you must find the ________________ (LCD).

Add the following fractions. Simplify all answers. ***See Examples 1–3.***

3. $\frac{3}{4}+\frac{1}{8}$

4. $\frac{1}{6}+\frac{2}{3}$

5. $\frac{2}{3}+\frac{2}{9}$

6. $\frac{3}{7}+\frac{1}{14}$ LCD is 14.

$\frac{6}{14}+\frac{1}{14}$

$=\frac{7}{14}=$ simplest form

7. $\frac{9}{20}+\frac{3}{10}$ LCD is 20.

$\frac{9}{20}+\frac{6}{20}$

$=\frac{15}{20}=$ simplest form

8. $\frac{5}{8}+\frac{1}{4}$

9. $\frac{3}{5}+\frac{3}{8}$

10. $\frac{5}{7}+\frac{3}{14}$

11. $\frac{2}{9}+\frac{5}{12}$

12. $\frac{1}{4}+\frac{2}{9}+\frac{1}{3}$ LCD is 36.

$\frac{9}{36}+\frac{8}{36}+\frac{12}{36}=\frac{\quad}{36}$ Already in simplest form

13. $\frac{3}{7}+\frac{2}{5}+\frac{1}{10}$ LCD is 70.

$\frac{30}{70}+\frac{28}{70}+\frac{7}{70}$

$=\frac{65}{70}=$ simplest form

14. $\frac{3}{10}+\frac{2}{5}+\frac{3}{20}$

15. $\frac{1}{3}+\frac{3}{8}+\frac{1}{4}$

16. $\frac{4}{15}+\frac{1}{6}+\frac{1}{3}$

17. $\frac{5}{12}+\frac{2}{9}+\frac{1}{6}$

18. $\frac{2}{3}+\frac{1}{6}$

19. $\frac{1}{4}+\frac{1}{8}$

20. $\frac{7}{12}+\frac{1}{8}$

21. $\frac{5}{12}+\frac{1}{16}$

22. $\frac{3}{7}+\frac{1}{3}$

Subtract the following fractions. Simplify all answers. ***See Example 4.***

23. $\frac{5}{6} - \frac{1}{3}$ LCD is 6.

GS $\frac{5}{6} - \frac{2}{6} = \frac{3}{6} =$ simplest form

24. $\frac{3}{4} - \frac{5}{8}$ LCD is 8.

GS $\frac{6}{8} - \frac{5}{8} =$ simplest form

25. $\frac{2}{3} - \frac{1}{6}$

26. $\frac{5}{8} - \frac{1}{4}$

27. $\frac{2}{3} - \frac{1}{5}$

28. $\frac{5}{6} - \frac{7}{9}$

29. $\frac{5}{12} - \frac{1}{4}$

30. $\frac{5}{7} - \frac{1}{3}$

31. $\frac{8}{9} - \frac{7}{15}$

32. $\frac{4}{5} - \frac{1}{3}$

33. $\frac{7}{8} - \frac{4}{5}$

34. $\frac{5}{8} - \frac{1}{3}$

35. $\frac{5}{12} - \frac{1}{16}$

36. $\frac{7}{12} - \frac{1}{3}$

Solve each application problem.

4-Piece Chisel Set
LOT NO. 42429
• Cutting-edge widths of 1/4", 1/2", 3/4", and 1"
• Heat-treated, high-carbon steel
• Straight bevel
SALE! $4.97
REGULAR PRICE $7.99

Use the newspaper advertisement for this 4-piece chisel set to answer Exercises 37–38. (Source: Harbor Freight Tools.)

37. Find the difference in the cutting-edge width of the two chisels with the widest blades. The symbol " is for inches.

38. Find the difference in the cutting-edge width of the two chisels with the narrowest blades. The " symbol is for inches.

39. A sports and entertainment center has $\frac{4}{5}$ of its total area devoted to seating of fans and guests. If $\frac{3}{8}$ of the seating area is used for general admission seating and the rest for reserved seating, find the fraction of the total area used for reserved seating.

40. A dairy farmer must vaccinate $\frac{5}{8}$ of her cows this week. If she vaccinates $\frac{3}{16}$ of the herd on Monday, and $\frac{1}{4}$ of the herd on Wednesday, what fraction of the herd remains to be vaccinated?

41. When installing cabinets for The Home Depot, Sarah Bryn must be certain that the proper type and size of mounting screw is used. Find the total length of the screw shown.

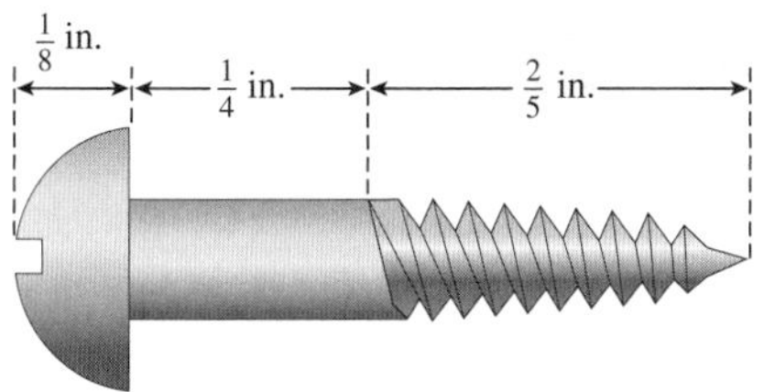

42. When installing a computer chassis, Bonnie Bottorff must be certain that the proper type and size of bolt is used. Find the total length of the bolt shown.

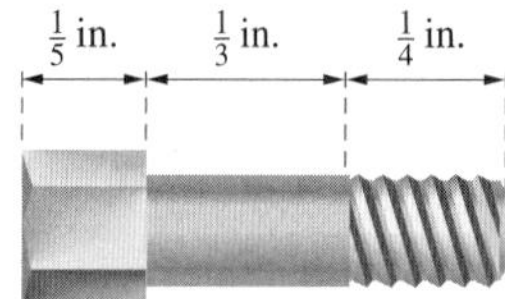

43. Bill Newton is a general contractor. He began a job with $\frac{3}{4}$ of a tank of fuel in his backhoe. He used $\frac{1}{3}$ of the tank in the morning and $\frac{3}{8}$ of the tank in the afternoon. What fraction of the tank of fuel remains?

44. Cliff Dinsmore is coordinating the refurbishing of the community swimming pool. The pool was $\frac{7}{8}$ full when workers began draining it. By noon, another $\frac{3}{16}$ of the pool had been drained. An additional $\frac{1}{3}$ of the pool was drained in the afternoon. Find the fraction of the pool water remaining.

45. Step 1 in adding or subtracting unlike fractions is to rewrite the fractions so they have the least common multiple as a denominator. Explain in your own words why this is necessary.

46. Briefly list the three steps used for addition and subtraction of unlike fractions.

A survey of 1200 users of social networking sites showed that honesty was not always practiced. Refer to the circle graph to answer Exercises 47–50.

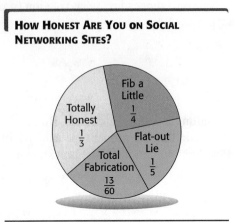

How Honest Are You on Social Networking Sites?

Fib a Little $\frac{1}{4}$

Totally Honest $\frac{1}{3}$

Flat-out Lie $\frac{1}{5}$

Total Fabrication $\frac{13}{60}$

Source: *USA Today.*

47. What fraction of those surveyed are totally honest?

48. What fraction of those surveyed fib a little?

49. Which response was given most often? How many people gave this response? What fraction of the users gave this response and the "fib a little" response?

50. Which response was given least often? How many people gave this response? What fraction of the users gave this response and the "total fabrication" response?

51. Find the diameter of the hole in the mounting bracket shown. (The diameter is the distance across the center of the hole.)

Diameter

$\frac{3}{8}$ in. $\frac{3}{8}$ in.

$\frac{15}{16}$ in.

52. Chakotay is fitting a turquoise stone into a bear claw pendant. Find the diameter of the hole in the pendant. (The diameter is the distance across the center of the hole.)

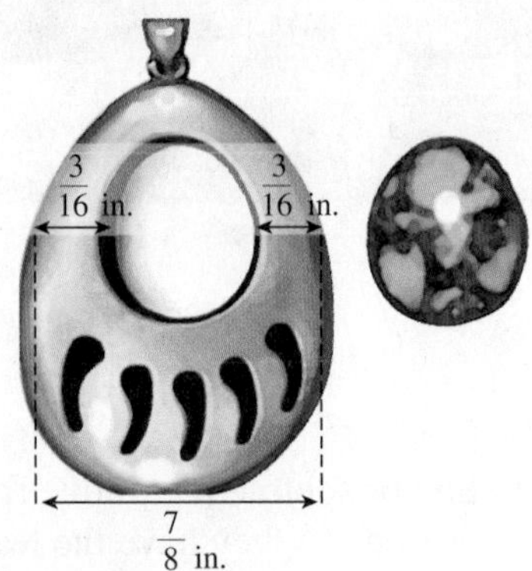

3.4 Adding and Subtracting Mixed Numbers

Recall that a mixed number is the sum of a whole number and a fraction. For example,

$$3\frac{2}{5} \quad \text{means} \quad 3 + \frac{2}{5}.$$

OBJECTIVE 1 Estimate an answer, then add or subtract mixed numbers. Add or subtract mixed numbers by adding or subtracting the fraction parts and then the whole number parts. It is a good idea to estimate the answer first, as we did when multiplying and dividing mixed numbers in **Chapter 2.**

Work Problem 1 at the Side. ▶

EXAMPLE 1 Adding and Subtracting Mixed Numbers

First estimate the answer. Then add or subtract to find the exact answer.

(a) $16\frac{1}{8} + 5\frac{5}{8}$

Estimate:

16 ← Rounds to — $16\frac{1}{8}$

$+\ 6$ ← Rounds to — $+\ 5\frac{5}{8}$

First, estimate the answer. → 22

Exact:

$$16\frac{1}{8} + 5\frac{5}{8} = 21\frac{6}{8} = 21\frac{3}{4} \leftarrow \text{Lowest terms}$$

Sum of whole numbers; Sum of fractions. The exact answer is close to the estimate.

In lowest terms $\frac{6}{8}$ is $\frac{3}{4}$, so the exact answer of $21\frac{3}{4}$ is in lowest terms. The exact answer is *reasonable* because it is close to the estimate of 22.

(b) $8\frac{5}{8} - 3\frac{1}{12}$

Estimate:

9 ← Rounds to — $8\frac{5}{8}$

$-\ 3$ ← Rounds to — $-3\frac{1}{12}$

6

Exact:

$$8\frac{5}{8} = 8\frac{15}{24}$$

$$-3\frac{1}{12} = -3\frac{2}{24}$$

24 is the least common denominator.

$$5\frac{13}{24} \leftarrow \text{Lowest terms}$$

Subtract whole numbers. Subtract fractions.

The exact answer of $5\frac{13}{24}$ is *reasonable* because it is close to the estimated answer of 6. Check by adding $5\frac{13}{24}$ and $3\frac{1}{12}$; the sum should be $8\frac{5}{8}$.

Continued on Next Page

OBJECTIVES

1. Estimate an answer, then add or subtract mixed numbers.
2. Estimate an answer, then subtract mixed numbers by regrouping.
3. Add or subtract mixed numbers using an alternative method.

1 As a review of mixed numbers, write each mixed number as an improper fraction and each improper fraction as a mixed number.

(a) $\frac{9}{2}$

(b) $\frac{8}{3}$

(c) $4\frac{3}{4}$

(d) $3\frac{7}{8}$

Answers

1. (a) $4\frac{1}{2}$ (b) $2\frac{2}{3}$ (c) $\frac{19}{4}$ (d) $\frac{31}{8}$

2 First estimate, and then add or subtract to find the exact answer.

GS **(a)** *Estimate:* *Exact:*

$7 \leftarrow$ Rounds to $\left\{ 6\frac{7}{8} = 6\frac{7}{8} \right.$

$+2 \leftarrow$ Rounds to $\left\{ +2\frac{1}{4} = 2\frac{2}{8} \right.$

(b) *Estimate:* *Exact:*

$\leftarrow$ Rounds to $\left\{ 4\frac{7}{9} \right.$

$- \quad \leftarrow$ Rounds to $\left\{ -2\frac{2}{3} \right.$

3 First estimate, and then add to find the exact answer.

GS **(a)** *Estimate:* *Exact:*

$10 \leftarrow$ Rounds to $\left\{ 9\frac{3}{4} \right.$

$+\ 8 \leftarrow$ Rounds to $\left\{ +7\frac{1}{2} \right.$

(b) *Estimate:* *Exact:*

$\leftarrow$ Rounds to $\left\{ 15\frac{4}{5} \right.$

$+ \quad \leftarrow$ Rounds to $\left\{ +12\frac{2}{3} \right.$

Answers

2. (a) $7 + 2 = 9;\ 9\frac{1}{8}$

(b) $5 - 3 = 2;\ 2\frac{1}{9}$

3. (a) $10 + 8 = 18;\ 17\frac{1}{4}$

(b) $16 + 13 = 29;\ 28\frac{7}{15}$

Note

When estimating, if the numerator is *half* of the denominator or *more*, round up the whole number part. If the numerator is *less* than *half* the denominator, leave the whole number part as it is.

◀ **Work Problem 2 at the Side.**

When you add the fraction parts of mixed numbers, the sum may be greater than 1. If this happens, simplify the fraction and regroup in the whole number column.

EXAMPLE 2 Simplify and Regroup When Adding Mixed Numbers

First estimate, and then add $9\frac{5}{8} + 13\frac{7}{8}$.

Estimate: *Exact:*

$10 \leftarrow$ Rounds to $\left\{ 9\frac{5}{8} \right.$

$+\ 14 \leftarrow$ Rounds to $\left\{ +13\frac{7}{8} \right.$

$24 \qquad 22\frac{12}{8}$

22: Sum of whole numbers; $\frac{12}{8}$: Sum of fractions

First, add the fractions, then the whole numbers.

The improper fraction $\frac{12}{8}$ can be written in lowest terms as $\frac{3}{2}$. Then $\frac{3}{2} = 1\frac{1}{2}$, so the simplified sum is

$$22\frac{12}{8} = 22 + \frac{12}{8} = 22 + \frac{3}{2} = 22 + 1\frac{1}{2} = 23\frac{1}{2}.$$

($\frac{12}{8}$ Becomes $\frac{3}{2}$; $\frac{3}{2}$ Becomes $1\frac{1}{2}$)

The estimate was 24, so the exact answer of $23\frac{1}{2}$ is reasonable.

Note

When adding mixed numbers, first add the fraction parts, then add the whole number parts. Finally, combine the two answers and simplify.

◀ **Work Problem 3 at the Side.**

OBJECTIVE 2 Estimate an answer, then subtract mixed numbers by regrouping. When subtracting mixed numbers, **regrouping** is necessary when the fraction part of the first number is less than the fraction part of the second number.

EXAMPLE 3 Regroup When Subtracting Mixed Numbers

First estimate, and then subtract to find the exact answer.

(a) $7 - 2\frac{5}{6}$

Continued on Next Page

Estimate:

$$\begin{array}{r} 7 \\ -\ 3 \\ \hline 4 \end{array}$$

(7 Rounds to 7; $2\frac{5}{6}$ Rounds to 3)

Exact:

$$\begin{array}{r} 7 \\ -\ 2\frac{5}{6} \\ \hline \end{array}$$

There is no fraction here from which to subtract $\frac{5}{6}$.

It is **not** possible to subtract $\frac{5}{6}$ without regrouping the whole number **7** first.

Regroup 7 as 6 + 1.

$$7 = 6 + 1$$

$1 = \frac{6}{6}$

$$= 6 + \frac{6}{6}$$

$$= 6\frac{6}{6}$$

Now you can subtract.

$$\begin{array}{rr} 7 = & 6\frac{6}{6} \\ -\ 2\frac{5}{6} = & -\ 2\frac{5}{6} \\ \hline & 4\frac{1}{6} \end{array}$$

7 was rewritten as $6\frac{6}{6}$.

Exact answer is close to the estimate.

The estimate was 4, so the exact answer of $4\frac{1}{6}$ is reasonable.

(b) $8\frac{1}{3} - 4\frac{3}{5}$

Estimate:

$$\begin{array}{r} 8 \\ -\ 5 \\ \hline 3 \end{array}$$

($8\frac{1}{3}$ Rounds to 8; $4\frac{3}{5}$ Rounds to 5)

Exact:

$$\begin{array}{rr} 8\frac{1}{3} = & 8\frac{5}{15} \\ -\ 4\frac{3}{5} = & -\ 4\frac{9}{15} \\ \hline \end{array}$$

15 is the least common denominator.

It is **not** possible to subtract $\frac{9}{15}$ from $\frac{5}{15}$, so regroup the whole number **8.**

Regroup 8 as 7 + 1.

$$8\frac{5}{15} = 8 + \frac{5}{15} = 7 + 1 + \frac{5}{15}$$

$1 = \frac{15}{15}$

$$= 7 + \frac{15}{15} + \frac{5}{15}$$

$$= 7 + \frac{20}{15} \leftarrow \frac{15}{15} + \frac{5}{15}$$

$$= 7\frac{20}{15}$$

Continued on Next Page

4 First estimate and then subtract to find the exact answer.

GS **(a)** *Estimate:* *Exact:*

$7 \xleftarrow{\text{Rounds to}} \left\{ 7\frac{1}{3} = 7\frac{2}{6} \right.$

$-5 \xleftarrow{\text{Rounds to}} \left\{ -4\frac{5}{6} = 4\frac{5}{6} \right.$

(b) *Estimate:* *Exact:*

$\xleftarrow{\text{Rounds to}} \left\{ 4\frac{5}{8} \right.$

$- \xleftarrow{\text{Rounds to}} \left\{ -2\frac{15}{16} \right.$

(c) *Estimate:* *Exact:*

$\xleftarrow{\text{Rounds to}} \left\{ 15 \right.$

$- \xleftarrow{\text{Rounds to}} \left\{ -6\frac{4}{9} \right.$

5 Add or subtract by changing mixed numbers to improper fractions. Simplify answers.

GS **(a)** $3\frac{3}{8} = \frac{27}{8} = \frac{27}{8}$

$+2\frac{1}{2} = \frac{5}{2} = \frac{}{8}$ 8 is LCD.

(b) $6\frac{3}{4}$

$-4\frac{2}{3}$

Answers

4. (a) $7 - 5 = 2;\ 2\frac{1}{2}$

(b) $5 - 3 = 2;\ 1\frac{11}{16}$

(c) $15 - 6 = 9;\ 8\frac{5}{9}$

5. (a) $\frac{27}{8} + \frac{20}{8} = \frac{47}{8} = 5\frac{7}{8}$

(b) $\frac{25}{12} = 2\frac{1}{12}$

Now you can subtract.

$$8\frac{1}{3} = 8\frac{5}{15} = 7\frac{20}{15}$$
$$-4\frac{3}{5} = 4\frac{9}{15} = 4\frac{9}{15}$$
$$3\frac{11}{15}$$

The exact answer is $3\frac{11}{15}$ (lowest terms), which is reasonable because it is close to the estimate of 3.

◀ **Work Problem 4 at the Side.**

OBJECTIVE 3 Add or subtract mixed numbers using an alternative method. An alternative method for adding or subtracting mixed numbers is to first change the mixed numbers to improper fractions. Then rewrite the unlike fractions as like fractions. Finally, add or subtract the numerators and write the answer in lowest terms.

EXAMPLE 4 Adding or Subtracting Mixed Numbers

Add or subtract.

(a)
$$2\frac{3}{8} = \frac{19}{8} = \frac{19}{8}$$
$$+3\frac{3}{4} = \frac{15}{4} = \frac{30}{8}$$
$$\frac{49}{8} = 6\frac{1}{8}$$

8 is the least common denominator.

Rewrite $2\frac{3}{8}$ as $\frac{19}{8}$ and $3\frac{3}{4}$ as $\frac{15}{4}$.

Answer as mixed number

(b)
$$4\frac{2}{3} = \frac{14}{3} = \frac{70}{15}$$
$$-2\frac{1}{5} = \frac{11}{5} = \frac{33}{15}$$
$$\frac{37}{15} = 2\frac{7}{15}$$

15 is the least common denominator.

Improper fractions

Simplify the answer by writing it as a mixed number.

◀ **Work Problem 5 at the Side.**

Note

The advantage of this alternative method of adding or subtracting mixed numbers is that it eliminates the need to regroup. It is also the most useful method for working with algebraic fractions. (See **Chapter 9.**) However, if the mixed numbers are large, then the numerators of the improper fractions may become so large that they are difficult to work with. In such cases, you may want to keep the numbers as mixed numbers.

3.4 Exercises

FOR EXTRA HELP

Download the MyDashBoard App

MyMathLab®

CONCEPT CHECK *Round each of the following mixed numbers to the nearest whole number.*

1. $5\frac{1}{3}$ **2.** $6\frac{3}{10}$ **3.** $8\frac{4}{5}$ **4.** $12\frac{1}{2}$

5. $15\frac{7}{15}$ **6.** $20\frac{5}{8}$ **7.** $16\frac{2}{3}$ **8.** $3\frac{5}{12}$

First estimate the answer. Then add to find the exact answer. Write answers as mixed numbers. ***See Examples 1 and 2.***

9. *Estimate:* GS
6 ← Rounds to $5\frac{1}{2} = 5\frac{3}{6}$
+ 3 ← Rounds to $+\,3\frac{1}{3} = 3\frac{2}{6}$
9

Exact: $5\frac{1}{2} = 5\frac{3}{6}$, $+\,3\frac{1}{3} = 3\frac{2}{6}$

10. *Estimate:* GS
7 ← $6\frac{3}{5} = 6\frac{6}{10}$
+ 7 ← $+\,7\frac{1}{10} = 7\frac{1}{10}$
14

Exact: $6\frac{3}{5} = 6\frac{6}{10}$, $+\,7\frac{1}{10} = 7\frac{1}{10}$

11. *Estimate:* + *Exact:* $7\frac{1}{3}$ $+\,4\frac{1}{6}$

12. *Estimate:* + *Exact:* $10\frac{1}{4}$ $+\,5\frac{5}{8}$

13. *Estimate:* + *Exact:* $\frac{5}{8}$ $+\,3\frac{7}{12}$

14. *Estimate:* + *Exact:* $12\frac{4}{5}$ $+\,\frac{7}{10}$

15. *Estimate:* + *Exact:* $24\frac{5}{6}$ $+\,18\frac{5}{6}$

16. *Estimate:* + *Exact:* $14\frac{6}{7}$ $+\,15\frac{1}{2}$

17. *Estimate:* + *Exact:* $33\frac{3}{5}$ $+\,18\frac{1}{2}$

18. *Estimate:* + *Exact:* $18\frac{5}{8}$ $+\,6\frac{2}{3}$

19. *Estimate:* + *Exact:* $22\frac{3}{4}$ $+\,15\frac{3}{7}$

20. *Estimate:* + *Exact:* $7\frac{1}{4}$ $+\,25\frac{7}{8}$

21. *Estimate:* *Exact:*

$$12\frac{8}{15}$$
$$18\frac{3}{5}$$
$$+\ 14\frac{7}{10}$$

22. *Estimate:* *Exact:*

$$14\frac{9}{10}$$
$$8\frac{1}{4}$$
$$+\ 13\frac{3}{5}$$

First estimate the answer. Then subtract to find the exact answer. Simplify all answers.
See Examples 1 and 3.

23. GS *Estimate:* *Exact:*

$$15 \leftarrow 14\frac{7}{8} = 14\frac{7}{8}$$
$$-\ 12 \leftarrow -\ 12\frac{1}{4} = 12\frac{2}{8}$$
$$3$$

24. GS *Estimate:* *Exact:*

$$15 \leftarrow 14\frac{3}{4} = 14\frac{6}{8}$$
$$-\ 11 \leftarrow -\ 11\frac{3}{8} = 11\frac{3}{8}$$
$$4$$

25. *Estimate:* *Exact:*

$$12\frac{2}{3}$$
$$-\ 1\frac{1}{5}$$

26. *Estimate:* *Exact:*

$$11\frac{9}{20}$$
$$-\ 4\frac{3}{5}$$

27. *Estimate:* *Exact:*

$$28\frac{3}{10}$$
$$-\ 6\frac{1}{15}$$

28. *Estimate:* *Exact:*

$$15\frac{7}{20}$$
$$-\ 6\frac{1}{8}$$

29. GS *Estimate:* *Exact:*

$$17 = 16\frac{8}{8}$$
$$-\ 6\frac{5}{8} = 6\frac{5}{8}$$

30. GS *Estimate:* *Exact:*

$$22 = 21\frac{6}{6}$$
$$-\ 4\frac{5}{6} = 4\frac{5}{6}$$

31. *Estimate:* *Exact:*

$$18\frac{3}{4}$$
$$-\ 5\frac{4}{5}$$

32. *Estimate:* $-$ *Exact:* $14\frac{5}{8} - 3\frac{2}{3}$

33. *Estimate:* $-$ *Exact:* $19\frac{2}{3} - 11\frac{3}{4}$

34. *Estimate:* $-$ *Exact:* $20\frac{3}{5} - 12\frac{7}{15}$

CONCEPT CHECK *Write each mixed number as an improper fraction and each improper fraction as a mixed number.*

35. $3\frac{3}{4}$ **36.** $7\frac{7}{8}$ **37.** $\frac{12}{5}$ **38.** $\frac{24}{7}$

39. $5\frac{3}{8}$ **40.** $6\frac{2}{15}$ **41.** $\frac{56}{3}$ **42.** $\frac{29}{8}$

Add or subtract by changing mixed numbers to improper fractions. Write answers as mixed numbers when possible. ***See Example 4.***

43. GS
$$7\frac{5}{8} = \frac{61}{8} = \frac{61}{8}$$
$$+\,1\frac{1}{2} = \frac{3}{2} = \frac{12}{8}$$
$$\frac{73}{8} =$$

44. GS
$$8\frac{3}{4} = \frac{35}{4} = \frac{70}{8}$$
$$+\,1\frac{5}{8} = \frac{13}{8} = \frac{13}{8}$$
$$\frac{83}{8} =$$

45. $4\frac{2}{3} + 6\frac{5}{6}$

46. $3\frac{3}{5} + 6\frac{1}{2}$

47. $2\frac{2}{3} + 1\frac{1}{6}$

48. $4\frac{1}{2} + 2\frac{3}{4}$

49. $3\frac{1}{4} + 3\frac{2}{3}$

50. $2\frac{4}{5} + 5\frac{1}{3}$

51. $1\frac{3}{8} + 6\frac{3}{4}$

52. $1\frac{5}{12} + 1\frac{7}{8}$

53. $3\frac{1}{2} - 2\frac{2}{3}$

54. $4\frac{1}{4} - 3\frac{7}{12}$

55. $8\frac{3}{4} - 5\frac{7}{8}$

56. $5 - 4\frac{7}{8}$

57. $7\frac{1}{4} - 4\frac{2}{3}$

58. $4\frac{1}{10} - 3\frac{7}{8}$

59. $9\frac{1}{5} - 3\frac{3}{4}$

60. $9 - 7\frac{5}{6}$

61. $6\frac{3}{7} - 2\frac{2}{3}$

62. $8\frac{2}{15} - 6\frac{1}{2}$

63. In your own words, explain the steps you would take to add two large mixed numbers.

64. When subtracting mixed numbers, explain when you need to regroup. Explain how to regroup using your own example.

First estimate the answer. Then solve each application problem.

65. At the beginning of this chapter you read about Bryan Berg, who builds houses of cards. While in high school, Bryan built a house of cards $14\frac{1}{2}$ ft tall, setting his first world record. Today his current world record is $25\frac{3}{4}$ ft tall. How much taller is his current world record than his first world record? (*Source: Guinness World Records.*)

Estimate:

Exact:

66. The average age of a fast-food worker has increased over the past 12 years from $21\frac{3}{4}$ years of age to $29\frac{1}{2}$. How much older is the average fast-food worker today than 12 years ago? (*Source:* U.S. Census Bureau.)

Estimate:

Exact:

Use the newspaper advertisement for these industrial quality professional pliers to answer Exercises 67–70. The ″ symbol is for inches and 2-15/16″ means $2\frac{15}{16}$ *inches.* (*Source:* Harbor Freight Tools.)

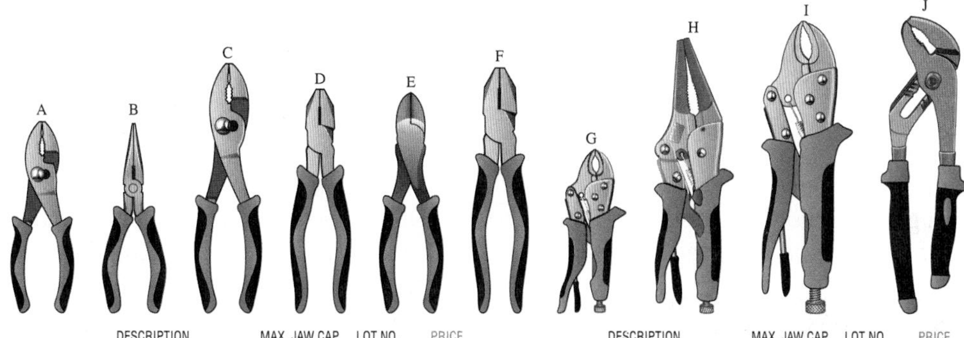

	DESCRIPTION	MAX. JAW CAP.	LOT NO.	PRICE
A	6" SLIP JOINT	2-15/16"	94380	$2.99
B	6" LONG NOSE	1-1/2"	94378	$4.29
C	8" SLIP JOINT	3-3/8"	94381	$4.29
D	7" LINEMAN'S	1-1/2"	94382	$5.29
E	7" DIAGONAL	7/8"	94383	$5.29
F	8" LINEMAN'S	1-7/16"	94385	$6.29
G	5" CURVED JAW LOCKING	1-3/8"	94289	$3.29
H	12" LONG NOSE LOCKING	2-1/4"	94285	$6.29
I	10" CURVED JAW LOCKING	2-1/4"	94286	$7.29
J	12" GROOVE JOINT	2"	94288	$8.29

67. How much larger is the maximum jaw capacity of the 8″ slip joint pliers than of the 6″ slip joint pliers?

Estimate:

Exact:

68. How much larger is the maximum jaw capacity of the 8″ lineman's pliers than of the 5″ curver jaw locking pliers?

Estimate:

Exact:

69. What is the difference in the maximum jaw capacity of the smallest pliers and that of the largest pliers?

Estimate:

Exact:

70. What is the difference in the maximum jaw capacity of the smallest pliers and that of the third smallest pliers?

Estimate:

Exact:

Storehouse
LOT NO. 1420
34 PC. GEAR HOSE CLAMP ASSORTMENT
Includes: two 2-3/4", two 2-1/2", four 2-1/4", four 2", six 1-3/4", six 1-1/2", six 1-1/4", four 9/16" through 1-1/16"
SALE!
$5.97
SAVE 40%
REGULAR PRICE $9.99

A mechanic buys the 34-piece hose clamp assortment shown at the left. Use the advertisement to answer Exercises 71–72. The ″ symbol is for inches and 2-3/4″ means $2\frac{3}{4}$ *inches.* (*Source:* Harbor Freight Tools.)

71. Find the difference in size between the largest hose clamp and the smallest hose clamp.

Estimate:

Exact:

72. Find the difference in size between the second to largest hose clamp and the smallest hose clamp.

Estimate:

Exact:

73. The four sides of Pam Prentiss' vegetable garden are $15\frac{1}{2}$ feet, $18\frac{3}{4}$ feet, $24\frac{1}{4}$ feet, and $30\frac{1}{2}$ feet. How many feet of fencing are needed to go around the garden?

Estimate:

Exact:

74. On a recent vacation to Canada, Janeen Cartmill drove for $7\frac{3}{4}$ hours on the first day, $5\frac{1}{4}$ hours on the second day, $6\frac{1}{2}$ hours on the third day, and 9 hours on the fourth day. How many hours did she drive altogether?

Estimate:

Exact:

75. A craftsperson must attach a lead strip around all four sides of a stained glass window before it is installed. Find the length of lead stripping needed.

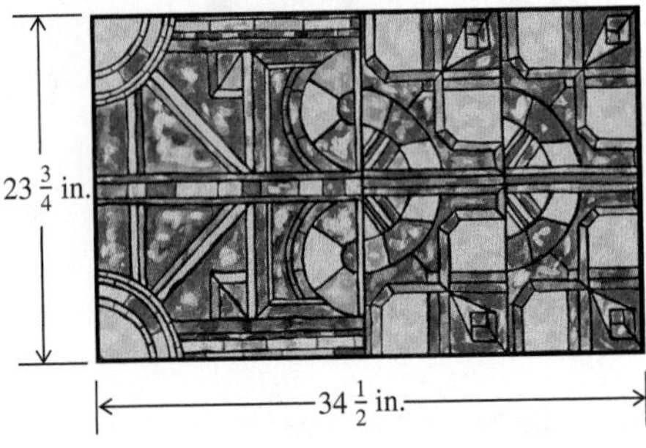

Estimate:

Exact:

76. To complete a custom order, Zak Morten of Home Depot must find the number of inches of brass trim needed to go around the four sides of the lamp base plate shown. Find the length of brass trim needed.

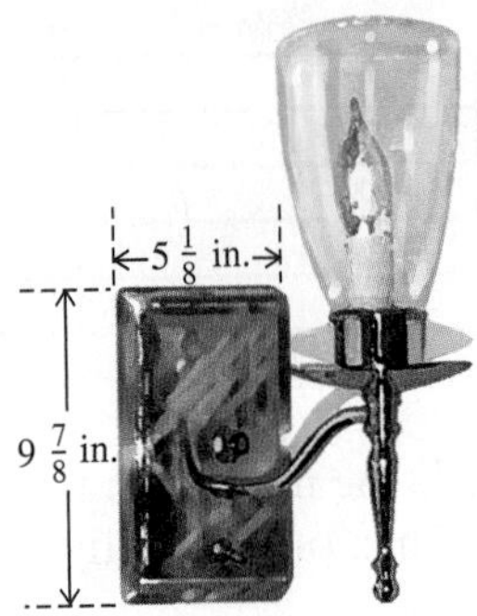

Estimate:

Exact:

77. A museum humidifier contains 100 gallons of water. The system uses $10\frac{1}{4}$ gallons of water on Monday, $13\frac{1}{2}$ gallons on Tuesday, $8\frac{7}{8}$ gallons on Wednesday, $18\frac{3}{4}$ gallons on Thursday, $12\frac{3}{8}$ gallons on Friday, $9\frac{1}{2}$ gallons on Saturday, and $14\frac{1}{8}$ gallons on Sunday. Find the total number of gallons of water remaining.

Estimate:

Exact:

78. Scott Alamo had 16 cubic yards of mulch delivered to the home he recently purchased. Using his wheelbarrow, he spread $3\frac{3}{4}$ yards of the mulch in his front yard, $4\frac{7}{8}$ yards in his back yard, $2\frac{1}{2}$ yards in the side yard, and used the remainder in the children's outdoor play area. Find the number of cubic yards of mulch remaining for the play area.

Estimate:

Exact:

79. The exercise yard at the correction center has four sides and is surrounded by $527\frac{1}{24}$ ft of security fencing. Three sides of the yard measure $107\frac{2}{3}$ ft, $150\frac{3}{4}$ ft, and $138\frac{5}{8}$ ft. Find the length of the fourth side.

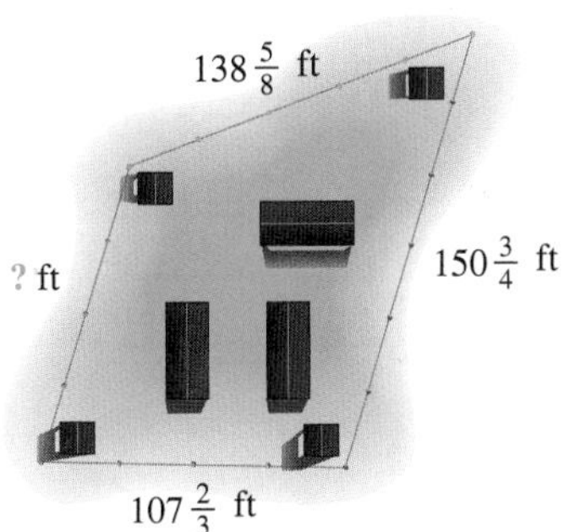

Estimate:

Exact:

80. Three sides of a parking lot are $108\frac{1}{4}$ ft, $162\frac{3}{8}$ ft, and $143\frac{1}{2}$ ft. The total distance around the lot is $518\frac{3}{4}$ ft. What is the length of the fourth side?

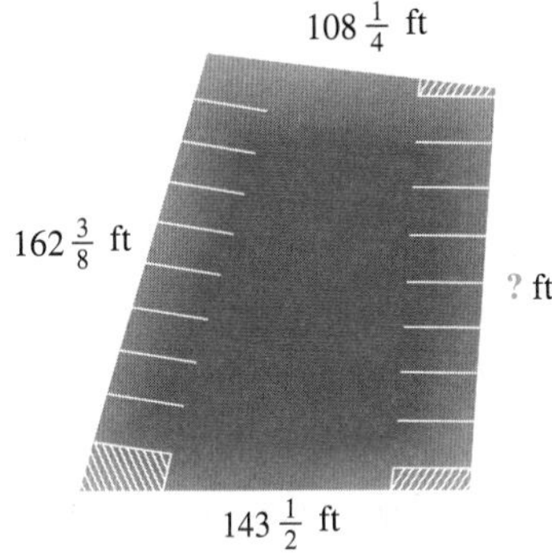

Estimate:

Exact:

81. A freight car is loaded with Morton Salt products consisting of $58\frac{1}{2}$ tons of coarse rock salt, $23\frac{5}{8}$ tons of medium rock salt, $16\frac{5}{6}$ tons of table salt, and $29\frac{1}{4}$ tons of animal salt lick blocks. The weight of the unloaded freight car is $58\frac{1}{3}$ tons. Find the weight of the loaded freight car.

Estimate:

Exact:

82. Bryan Berg, from the opening page of this chapter, built houses of cards reaching heights of $14\frac{1}{2}$ ft, $19\frac{3}{8}$ ft, $23\frac{5}{12}$ ft, and $25\frac{3}{4}$ ft (his current world record). Find the total height of these four houses of cards. (*Source: Reader's Digest.*)

Estimate:

Exact:

Find the unknown length, labeled with a question mark, in each figure.

83.

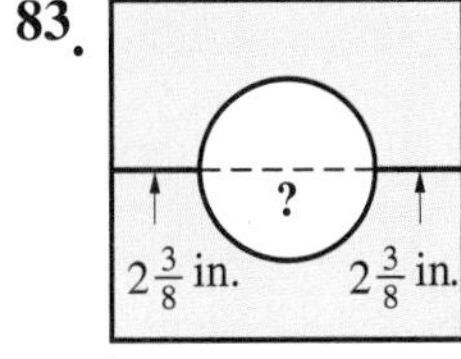

84.

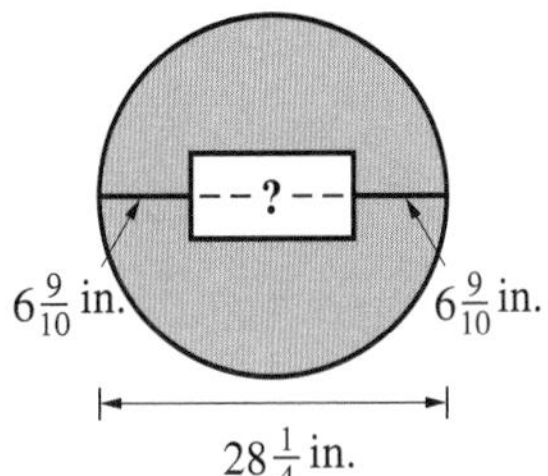

85.

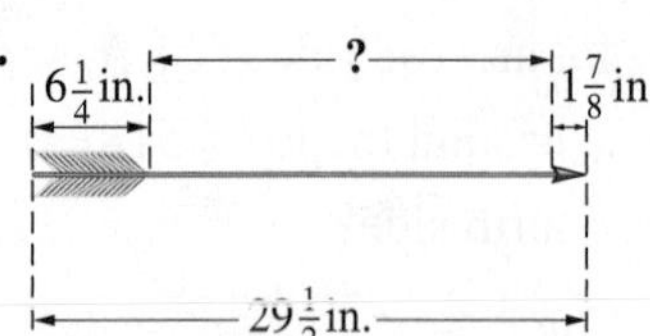

86.

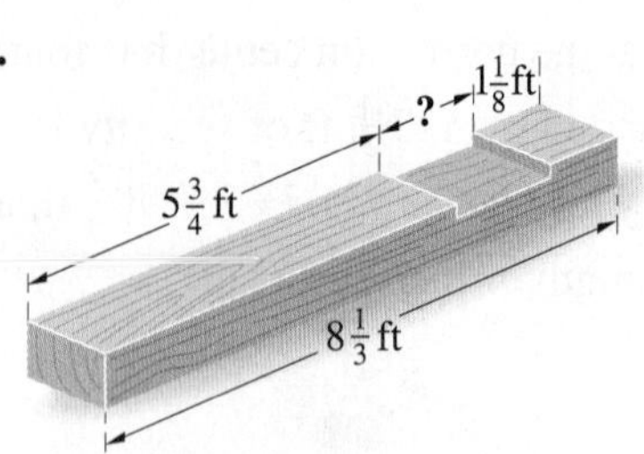

Relating Concepts (Exercises 87–92) For Individual or Group Work

Most fraction problems include fractions with different denominators.
Work Exercises 87–92 in order.

87. To add or subtract fractions, we must first rewrite them as like fractions. Rewrite each fraction with the indicated denominator.

(a) $\frac{5}{9} = \frac{}{54}$ **(b)** $\frac{7}{12} = \frac{}{48}$

(c) $\frac{5}{8} = \frac{}{40}$ **(d)** $\frac{11}{5} = \frac{}{120}$

88. When rewriting unlike fractions as like fractions with the least common multiple as a denominator, the new denominator is called the __________ __________ __________, or LCD.

89. Add or subtract as indicated. Write answers in lowest terms.

(a) $\frac{5}{8} + \frac{1}{3}$ **(b)** $\frac{19}{20} - \frac{5}{12}$

(c) $\frac{7}{12} + \frac{3}{16} + \frac{3}{24}$ **(d)** $\frac{6}{7} - \frac{2}{3}$

90. A common method for adding or subtracting mixed numbers is to add or subtract the __________ __________ and then add or subtract the whole number parts.

91. Another method for adding or subtracting mixed numbers is to first change the mixed numbers to __________ fractions. After adding or subtracting, write the answer in lowest terms and as a mixed number when possible. This method is difficult to use if the mixed numbers are (*large/small*).

92. Add or subtract these fractions as indicated. First use the method where you add or subtract fraction parts and then whole number parts. Then use the method where you change each mixed number to an improper fraction before adding or subtracting. Do you get the same answer using both methods? Which method do you prefer?

(a) $4\frac{5}{8} + 3\frac{3}{4}$ **(b)** $12\frac{2}{5} - 8\frac{7}{8}$

Summary Exercises *Operations with Fractions*

CONCEPT CHECK *Write* proper *or* improper *for each fraction.*

1. $\frac{3}{4}$

2. $\frac{4}{3}$

3. $\frac{10}{10}$

4. $\frac{11}{12}$

Write each fraction in lowest terms.

5. $\frac{30}{36}$

6. $\frac{175}{200}$

7. $\frac{15}{35}$

8. $\frac{115}{235}$

Add, subtract, multiply, or divide as indicated. Simplify all answers.

9. $\frac{3}{4} \cdot \frac{2}{3}$

10. $\frac{7}{12} \cdot \frac{9}{14}$

11. $56 \cdot \frac{5}{8}$

12. $\frac{5}{8} \div \frac{3}{4}$

13. $\frac{35}{45} \div \frac{10}{15}$

14. $21 \div \frac{3}{8}$

15. $\frac{7}{8} + \frac{2}{3}$

16. $\frac{5}{8} + \frac{3}{4} + \frac{7}{16}$

17. $\frac{7}{12} + \frac{5}{6} + \frac{2}{3}$

18. $\frac{5}{6} - \frac{3}{4}$

19. $\frac{7}{8} - \frac{5}{12}$

20. $\frac{4}{5} - \frac{2}{3}$

First estimate the answer. Then add, subtract, multiply, or divide to find the exact answer.

21. $3\frac{1}{2} \cdot 2\frac{1}{4}$

Estimate:

____ • ____ = ____

Exact:

22. $5\frac{3}{8} \cdot 3\frac{1}{4}$

Estimate:

____ • ____ = ____

Exact:

23. $8 \cdot 5\frac{2}{3} \cdot 2\frac{3}{8}$

Estimate:

____ • ____ • ____ = ____

Exact:

24. $4\frac{3}{8} \div 3\frac{3}{4}$

Estimate:

____ ÷ ____ = ____

Exact:

25. $6\frac{7}{8} \div 2$

Estimate:

____ ÷ ____ = ____

Exact:

26. $4\frac{5}{8} \div \frac{3}{4}$

Estimate:

____ ÷ ____ = ____

Exact:

27. *Estimate:* Rounds to $\square$ + Rounds to $\square$ *Exact:*

$$\begin{array}{r} 5\frac{2}{3} \\ +\ 4\frac{1}{4} \\ \hline \end{array}$$

28. *Estimate:* *Exact:*

$$\begin{array}{r} 18\frac{5}{12} \\ +\ 9\frac{3}{4} \\ \hline \end{array}$$

29. *Estimate:* *Exact:*

$$\begin{array}{r} 14\frac{3}{5} \\ +\ 10\frac{2}{3} \\ \hline \end{array}$$

30. *Estimate:* *Exact:*

$$\begin{array}{r} 8\frac{3}{4} \\ -\ 3\frac{4}{5} \\ \hline \end{array}$$

31. *Estimate:* *Exact:*

$$\begin{array}{r} 14 \\ -\ 7\frac{3}{8} \\ \hline \end{array}$$

32. *Estimate:* *Exact:*

$$\begin{array}{r} 31\frac{5}{6} \\ -\ 22\frac{7}{12} \\ \hline \end{array}$$

Find the least common multiple of each set of numbers.

33. 8, 10

34. 4, 15

35. 3, 5, 10

36. 6, 8, 16

37. 9, 18, 24

38. 4, 12, 21

Rewrite each fraction with the indicated denominator.

39. $\frac{5}{6} = \frac{\quad}{42}$

40. $\frac{3}{4} = \frac{\quad}{16}$

41. $\frac{3}{7} = \frac{\quad}{28}$

42. $\frac{5}{8} = \frac{\quad}{40}$

43. $\frac{3}{9} = \frac{\quad}{45}$

44. $\frac{11}{12} = \frac{\quad}{60}$

Study Skills

MAKING A MIND MAP

OBJECTIVES

1. **Create mind maps for appropriate concepts.**
2. **Visually show how concepts relate to each other using arrows or lines.**

Mind mapping is a visual way to show information that you have learned. It is an excellent way to review. Mapping is flexible and can be personalized, which is helpful for your memory. Your brain likes to see things that are **pleasing** to look at, **colorful**, and that **show connections** between ideas. Take advantage of that by creating maps that

- are easy to read,
- use color in a systematic way, and
- clearly show you how different concepts are related (using arrows or dotted lines, for example).

Why Is Mapping Brain Friendly?

Remember that your brain grows dendrites when you are **actively thinking** about and working with information. Making a map requires you to think hard about ***how to place the information***, ***how to show connections*** between parts of the map, and ***how color will be useful***. It also takes a lot of thinking to fill in all related details and **show how those details connect to the larger concept.** All that thinking will let your brain grow a complex, many-branched neural network of interconnected dendrites. It is time well spent.

Directions for Making a Mind Map

Here are some general directions for making a map. After you read them, work on completing the map that has been started for you on the next page. It is from **Chapters 2 and 3: Fractions.**

- To begin a mind map, write the concept in the center of a piece of paper and either circle it or draw a box around it.
- Make a line out from the center concept, and draw a box large enough to write the definition of the concept.
- Think of the other aspects (subpoints) of the concept that you have learned, such as procedures to follow or formulas. Make a separate line and box connecting each subpoint to the center.
- From each of the new boxes, add the information you've learned. You can continue making new lines and boxes or circles, or you can list items below the new information.
- Use color to highlight the major points. For example, everything related to one subpoint might be the same color. That way you can easily see related ideas.
- You may also use arrows, underlining, or small drawings to help yourself remember.

Study Skills

Continued from page 237

Try This Fractions Mind Map

On a separate paper, make a map that summarizes Computations with Fractions. Follow the directions and use the starter map below.

- The longest rectangles are *instructions* for all four operations. (The first one starts "Rewrite all numbers as fractions. . ." and the second one is at the bottom of the map.)
- Notice the wavy dividing lines that separate the map into two sides.
- Your job is to complete the map by writing the steps used in multiplying and dividing fractions and the steps used in adding and subtracting fractions.

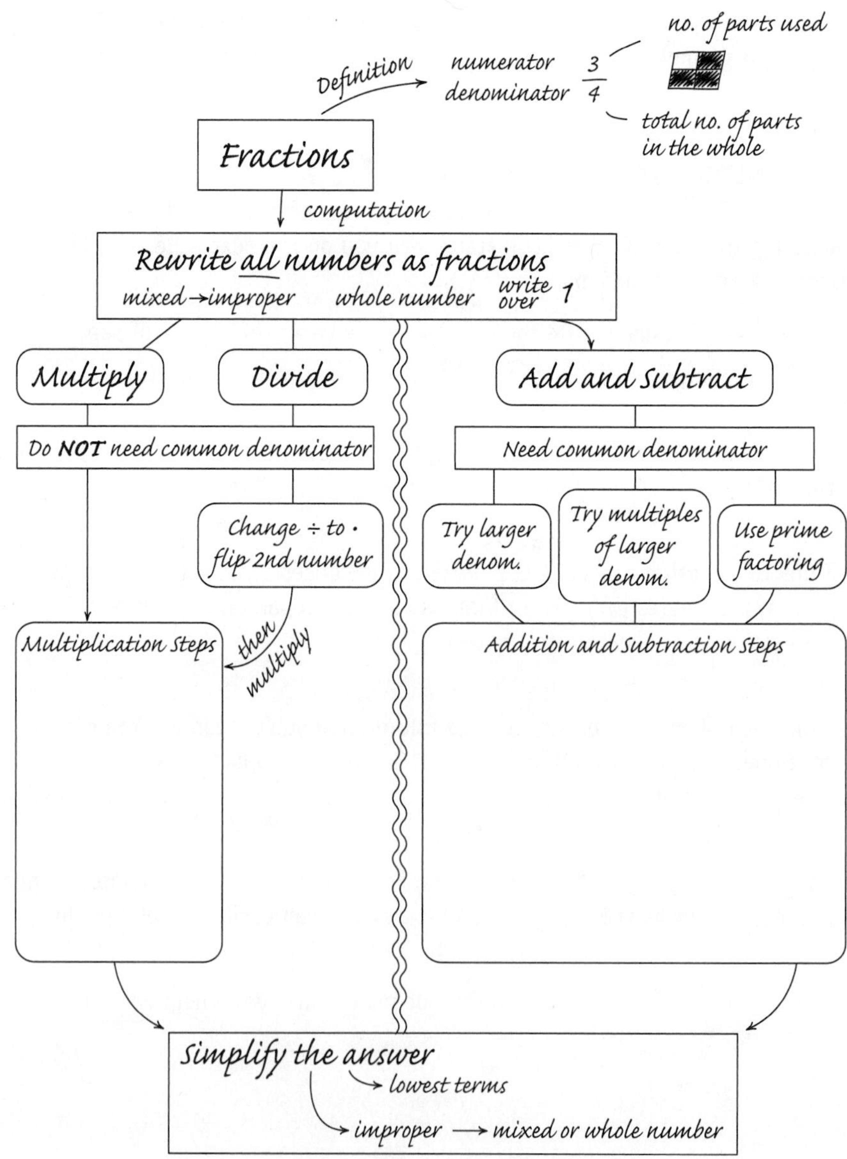

3.5 Order Relations and the Order of Operations

OBJECTIVES

1. Identify the greater or lesser of two fractions.
2. Use exponents with fractions.
3. Use the order of operations with fractions.

There are times when we want to compare the size of two numbers. For example, we might want to know which is the greater amount, the larger size, or the longer distance.

Fractions, like whole numbers, can be graphed on a number line. Fractions divide the space between whole numbers into equal parts.

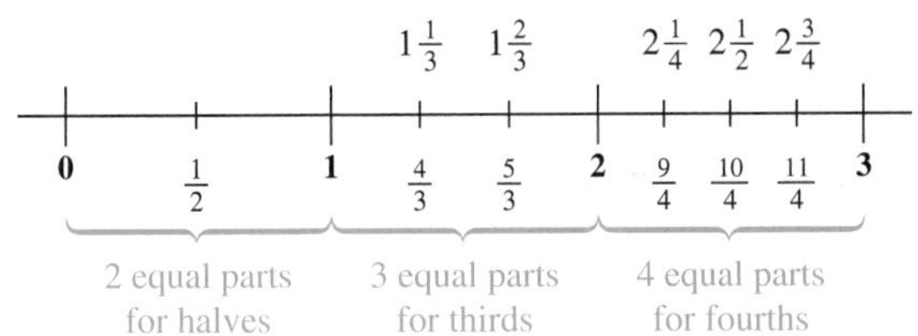

OBJECTIVE 1 Identify the greater or lesser of two fractions. To compare the size of two numbers, place the two numbers on a number line and use the following rule.

Comparing the Size of Two Numbers

The number farther to the *left* on the number line is always *less*, and the number farther to the *right* on the number line is always *greater.*

For example, on the number line above, $\frac{1}{2}$ is to the *left* of $\frac{4}{3}$, so $\frac{1}{2}$ is *less than* $\frac{4}{3}$.

Work Problem 1 at the Side. ▶

Write *order relations* using the symbols shown below.

Symbols Used to Show Order Relations

$<$ is less than $>$ is greater than

EXAMPLE 1 Using Less-Than and Greater-Than Symbols

Rewrite the following using $<$ and $>$ symbols.

(a) $\frac{1}{2}$ is less than $\frac{4}{3}$.

$\frac{1}{2}$ is less than $\frac{4}{3}$ is written as $\frac{1}{2} < \frac{4}{3}$.

$\frac{1}{2}$ is farther to the *left* on the number line, so it is *less* than $\frac{4}{3}$.

(b) $\frac{9}{4}$ is greater than 1.

$\frac{9}{4}$ is greater than 1 is written as $\frac{9}{4} > 1$.

$\frac{9}{4}$ is farther to the *right* on the number line, so it is *greater* than 1.

(c) $\frac{5}{3}$ is less than $\frac{11}{4}$.

$\frac{5}{3}$ is less than $\frac{11}{4}$ is written as $\frac{5}{3} < \frac{11}{4}$.

Continued on Next Page

1 Locate each fraction on the number line.

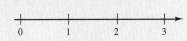

GS **(a)** $\frac{2}{3}$ Put a dot on the number line between 0 and 1.

(b) $1\frac{1}{2}$

(c) $2\frac{3}{4}$

Answer

1.
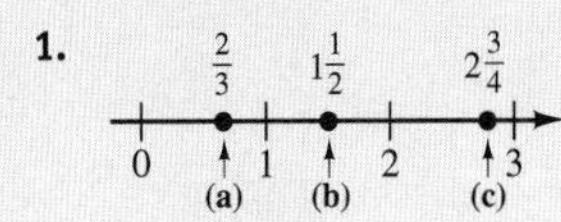

❷ Use the number line on the previous page to help you write < or > in each blank to make a true statement.

GS (a) $1 \_\_\_ \frac{5}{4}$ The point on the symbol points to the lesser value.

(b) $\frac{8}{3} \_\_\_ \frac{3}{2}$

(c) $0 \_\_\_ 1$

(d) $\frac{17}{8} \_\_\_ \frac{8}{4}$

❸ Write < or > in each blank to make a true statement.

GS (a) $\frac{7}{8} \_\_\_ \frac{3}{4}$ The point on the symbol points to the lesser value.

(b) $\frac{13}{8} \_\_\_ \frac{15}{9}$

(c) $\frac{9}{4} \_\_\_ \frac{7}{3}$

(d) $\frac{9}{10} \_\_\_ \frac{14}{15}$

Answers

2. (a) < (b) > (c) < (d) >
3. (a) > (b) < (c) < (d) <

Note

A number line is a very useful tool when working with order relations.

◀ **Work Problem ❷ at the Side.**

The fraction $\frac{7}{8}$ represents 7 of 8 equal parts, while $\frac{3}{8}$ means 3 of 8 equal parts. Because $\frac{7}{8}$ represents more of the equal parts, $\frac{7}{8}$ is greater than $\frac{3}{8}$.

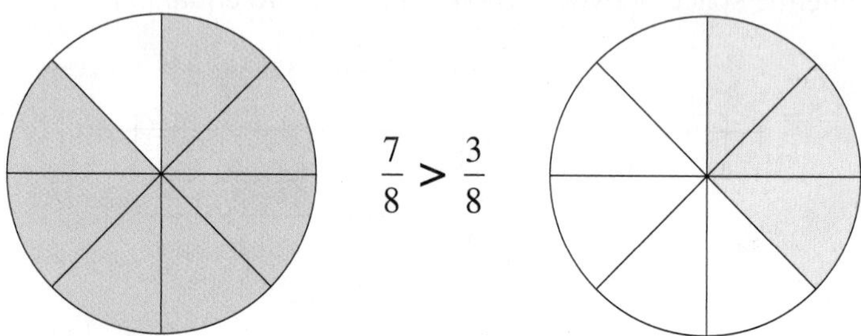

To identify the greater fraction, use the following steps.

Identifying the Greater Fraction

Step 1 Write the fractions as like fractions (same denominators).

Step 2 Compare the numerators. The fraction with the greater numerator is the greater fraction.

EXAMPLE 2 Identifying the Greater Fraction

Determine which fraction in each pair is greater.

(a) $\frac{7}{8}, \frac{9}{10}$

First, write the fractions as like fractions. The least common multiple for 8 and 10 is 40.

$$\frac{7}{8} = \frac{7 \cdot 5}{8 \cdot 5} = \frac{35}{40} \quad \text{and} \quad \frac{9}{10} = \frac{9 \cdot 4}{10 \cdot 4} = \frac{36}{40}$$

Rewrite both fractions with 40 as the denominator.

Look at the numerators. Because 36 is greater than 35, $\frac{36}{40}$ is greater than $\frac{35}{40}$. Then, because $\frac{36}{40}$ is equivalent to $\frac{9}{10}$,

$$\frac{9}{10} > \frac{7}{8} \quad \text{or} \quad \frac{7}{8} < \frac{9}{10}.$$

The greater fraction is $\frac{9}{10}$.

(b) $\frac{8}{5}, \frac{23}{15}$

The least common multiple of 5 and 15 is 15.

$$\frac{8}{5} = \frac{8 \cdot 3}{5 \cdot 3} = \frac{24}{15} \quad \text{and} \quad \frac{23}{15} = \frac{23}{15}$$

This shows that $\frac{8}{5}$ is greater than $\frac{23}{15}$, or

$$\frac{8}{5} > \frac{23}{15}.$$

◀ **Work Problem ❸ at the Side.**

OBJECTIVE 2 Use exponents with fractions. Exponents were used in **Chapter 1** to write repeated multiplication.

$$3^2 = 3 \cdot 3 = 9 \quad \text{and} \quad 5^3 = 5 \cdot 5 \cdot 5 = 125$$

(Exponent; Two factors of 3) (Exponent; Three factors of 5)

The next example shows exponents used with fractions.

EXAMPLE 3 Using Exponents with Fractions

Simplify.

(a) $\left(\frac{1}{2}\right)^3$

Three factors of $\frac{1}{2}$ — $\frac{1}{2}$ is multiplied by itself three times.

$$\left(\frac{1}{2}\right)^3 = \frac{1}{2} \cdot \frac{1}{2} \cdot \frac{1}{2} = \frac{1}{8}$$

(b) $\left(\frac{5}{8}\right)^2$

Two factors of $\frac{5}{8}$

$$\left(\frac{5}{8}\right)^2 = \frac{5}{8} \cdot \frac{5}{8} = \frac{25}{64}$$

(c) $\left(\frac{3}{4}\right)^2 \cdot \left(\frac{2}{3}\right)^3$

Two factors of $\frac{3}{4}$; Three factors of $\frac{2}{3}$

$$\left(\frac{3}{4}\right)^2 \cdot \left(\frac{2}{3}\right)^3 = \left(\frac{3}{4} \cdot \frac{3}{4}\right) \cdot \left(\frac{2}{3} \cdot \frac{2}{3} \cdot \frac{2}{3}\right)$$

$$= \frac{\overset{1}{\cancel{3}} \cdot \overset{1}{\cancel{3}} \cdot \overset{1}{\cancel{2}} \cdot \overset{1}{\cancel{2}} \cdot \overset{1}{\cancel{2}}}{\underset{2}{\cancel{4}} \cdot \underset{\underset{1}{\cancel{2}}}{\cancel{4}} \cdot \underset{1}{\cancel{3}} \cdot \underset{1}{\cancel{3}} \cdot 3} \quad \text{Divide out all the common factors.}$$

$$= \frac{1}{6}$$

The fraction is in lowest terms after all common factors are divided out.

Work Problem 4 at the Side. ▶

OBJECTIVE 3 Use the order of operations with fractions. Recall the *order of operations* from **Chapter 1.**

Order of Operations

1. Do all operations inside *parentheses or other grouping symbols.*
2. Simplify any expressions with *exponents* and find any *square roots.*
3. *Multiply* or *divide,* proceeding from left to right.
4. *Add* or *subtract,* proceeding from left to right.

4 Simplify.

GS **(a)** $\left(\frac{1}{2}\right)^4$

Four factors of $\frac{1}{2}$

$$\frac{1}{2} \cdot \frac{1}{2} \cdot \frac{1}{2} \cdot \frac{\_\_\_}{\_\_\_} = \frac{\_\_\_}{\_\_\_}$$

(b) $\left(\frac{3}{4}\right)^2$

(c) $\left(\frac{1}{2}\right)^3 \cdot \left(\frac{2}{3}\right)^2$

(d) $\left(\frac{1}{5}\right)^2 \cdot \left(\frac{5}{3}\right)^2$

Answers

4. (a) 1; 2; $\frac{1}{16}$ **(b)** $\frac{9}{16}$ **(c)** $\frac{1}{18}$ **(d)** $\frac{1}{9}$

5 Simplify by using the order of operations.

(a) $\frac{5}{9} - \frac{3}{4}\left(\frac{2}{3}\right)$

$\frac{5}{9} - \frac{\overset{1}{\cancel{3}}}{\underset{2}{\cancel{4}}}\left(\frac{\overset{1}{\cancel{2}}}{\underset{1}{\cancel{3}}}\right)$ Multiply first.

$\frac{5}{9} - \frac{1}{2} = \frac{10}{18} - \frac{9}{\_\_\_\_} =$

(b) $\frac{3}{4}\left(\frac{2}{3} \cdot \frac{3}{5}\right)$

(c) $\frac{7}{8}\left(\frac{2}{3}\right) - \left(\frac{1}{2}\right)^2$

(d) $\frac{\left(\frac{5}{6}\right)^2}{\frac{4}{3}}$

Answers

5. (a) 18; $\frac{1}{18}$ (b) $\frac{3}{10}$ (c) $\frac{1}{3}$ (d) $\frac{25}{48}$

The next example shows how to apply the order of operations with fractions.

EXAMPLE 4 Using the Order of Operations with Fractions

Simplify by using the order of operations.

(a) $\frac{1}{3} + \frac{1}{2}\left(\frac{4}{5}\right)$

Multiply $\frac{1}{2}\left(\frac{4}{5}\right)$ first because multiplication and division are done *before* adding.

$$\frac{1}{3} + \frac{1}{\underset{1}{\cancel{2}}}\left(\frac{\overset{2}{\cancel{4}}}{5}\right) = \frac{1}{3} + \frac{2}{5}$$

Do **not** add $\frac{1}{3} + \frac{1}{2}$ as the first step.

Now add. The least common denominator of 3 and 5 is 15.

$$\frac{1}{3} + \frac{2}{5} = \frac{5}{15} + \frac{6}{15} = \frac{11}{15} \leftarrow \text{Lowest terms}$$

(b) $\frac{3}{8}\left(\frac{1}{2} + \frac{1}{3}\right)$

$$\frac{3}{8}\left(\frac{1}{2} + \frac{1}{3}\right) = \frac{3}{8}\left(\frac{3}{6} + \frac{2}{6}\right)$$ Work inside parentheses first.

$$= \frac{3}{8}\left(\frac{5}{6}\right)$$

$$= \frac{\overset{1}{\cancel{3}}}{8}\left(\frac{5}{\underset{2}{\cancel{6}}}\right)$$ Divide numerator and denominator by 3. Then multiply.

$$= \frac{5}{16} \leftarrow \text{Lowest terms}$$

(c) $\left(\frac{2}{3}\right)^2 - \frac{4}{5}\left(\frac{1}{2}\right)$

$$\left(\frac{2}{3}\right)^2 - \frac{4}{5}\left(\frac{1}{2}\right) = \frac{4}{9} - \frac{4}{5}\left(\frac{1}{2}\right)$$ Simplify the expression with the exponent. $\frac{2}{3} \cdot \frac{2}{3}$ is $\frac{4}{9}$.

No work inside the parentheses, so applying the exponent is next.

$$= \frac{4}{9} - \frac{\overset{2}{\cancel{4}}}{5}\left(\frac{1}{\underset{1}{\cancel{2}}}\right)$$ Multiply next.

$$= \frac{4}{9} - \frac{2}{5}$$

$$= \frac{20}{45} - \frac{18}{45}$$ Subtract last. (Least common denominator is 45.)

$$= \frac{2}{45} \leftarrow \text{Lowest terms}$$

◀ Work Problem **5** at the Side.

3.5 Exercises

FOR EXTRA HELP

MyMathLab®

CONCEPT CHECK *Locate each fraction in Exercises 1–12 on the following number line.*

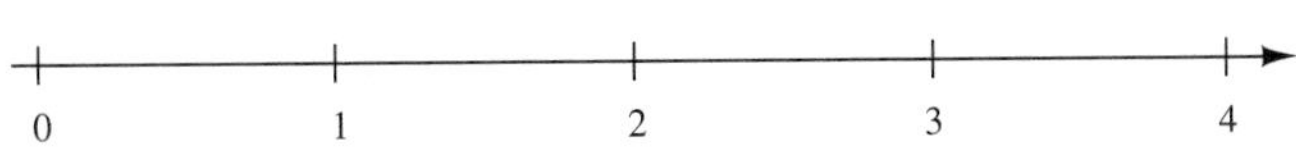

1. $\frac{1}{2}$ **2.** $\frac{1}{4}$ **3.** $\frac{3}{2}$ **4.** $\frac{5}{4}$ **5.** $\frac{7}{3}$ **6.** $\frac{11}{4}$

7. $2\frac{1}{6}$ **8.** $3\frac{4}{5}$ **9.** $\frac{7}{2}$ **10.** $\frac{7}{8}$ **11.** $3\frac{1}{4}$ **12.** $1\frac{7}{8}$

Write < or > to make a true statement. **See Examples 1 and 2.**

13. $\frac{1}{2}$ ____ $\frac{3}{8}$ **14.** $\frac{5}{8}$ ____ $\frac{3}{4}$ **15.** $\frac{5}{6}$ ____ $\frac{11}{12}$ **16.** $\frac{13}{18}$ ____ $\frac{5}{6}$

17. $\frac{5}{12}$ ____ $\frac{3}{8}$ **18.** $\frac{17}{24}$ ____ $\frac{5}{6}$ **19.** $\frac{11}{18}$ ____ $\frac{5}{9}$ **20.** $\frac{7}{12}$ ____ $\frac{11}{20}$

CONCEPT CHECK *Write* true *or* false *for each statement, Also show how to complete each solution.*

21. $\left(\frac{1}{2}\right)^2 = \frac{1}{4}$

$\left(\frac{1}{2}\right)^2 = \frac{1}{2} \cdot \frac{1}{\_\_\_} = \frac{1}{\_\_\_}$

22. $\left(\frac{3}{8}\right)^2 = \frac{6}{16}$

$\left(\frac{3}{8}\right)^2 = \frac{3}{8} \cdot \frac{\_\_\_}{8} = \frac{\_\_\_}{64}$

23. $\left(\frac{2}{5}\right)^3 = \frac{6}{15}$

24. $\left(\frac{5}{6}\right)^3 = \frac{125}{216}$

Simplify. **See Example 3.**

25. $\left(\frac{1}{3}\right)^2$ **26.** $\left(\frac{2}{3}\right)^2$ **27.** $\left(\frac{5}{8}\right)^2$

28. $\left(\frac{7}{8}\right)^2$ **29.** $\left(\frac{3}{4}\right)^2$ **30.** $\left(\frac{3}{5}\right)^3$

31. $\left(\frac{4}{5}\right)^3$ **32.** $\left(\frac{4}{7}\right)^3$ **33.** $\left(\frac{3}{2}\right)^4 = \frac{3}{2} \cdot \frac{3}{2} \cdot \frac{3}{2} \cdot \frac{3}{2} =$

34. $\left(\frac{4}{3}\right)^4 = \frac{4}{3} \cdot \frac{4}{3} \cdot \frac{4}{3} \cdot \frac{4}{3} =$ **35.** $\left(\frac{3}{4}\right)^4$ **36.** $\left(\frac{2}{3}\right)^5$

37. Describe in your own words what a number line is, and draw a picture of one. Be sure to include how it works and how it can be used.

38. You have used the order of operations with whole numbers and again with fractions. List from memory the steps in the order of operations.

CONCEPT CHECK *Use the order of operations to simplify each expression. Circle the correct answer.*

39. $2^4 - 4(3)$

30 4 12 2

40. $3^2 + 4(1)$

9 13 11 10

41. $3 \cdot 2^2 - \frac{6}{3}$

9 4 2 10

42. $5 \cdot 2^3 - \frac{6}{2}$

37 12 27 22

Use the order of operations to simplify each expression. ***See Example 4.***

43. $\left(\frac{1}{2}\right)^2 \cdot 4$

44. $\left(\frac{1}{4}\right)^2 \cdot 4$

45. $\left(\frac{3}{4}\right)^2 \cdot \left(\frac{1}{3}\right)$

46. $\left(\frac{2}{3}\right)^3 \cdot \left(\frac{1}{2}\right)$

47. $\left(\frac{4}{5}\right)^2 \cdot \left(\frac{5}{6}\right)^2$

48. $\left(\frac{5}{8}\right)^2 \cdot \left(\frac{4}{25}\right)^2$

49. $6\left(\frac{2}{3}\right)^2\left(\frac{1}{2}\right)^3$

50. $9\left(\frac{1}{3}\right)^3\left(\frac{4}{3}\right)^2$

51. $\frac{3}{5}\left(\frac{1}{3}\right) + \frac{2}{5}\left(\frac{3}{4}\right)$

$\underbrace{\frac{3}{5} \cdot \frac{1}{3}} + \underbrace{\frac{2}{5} \cdot \frac{3}{4}}$

$\frac{3}{15} + \frac{6}{20}$

$\frac{12}{60} + \frac{18}{60} = \frac{30}{60} =$

52. $\frac{1}{4}\left(\frac{3}{4}\right) + \frac{3}{8}\left(\frac{4}{3}\right)$

$\underbrace{\frac{1}{4} \cdot \frac{3}{4}} + \underbrace{\frac{3}{8} \cdot \frac{4}{3}}$

$\frac{3}{16} + \frac{12}{24}$

$\frac{9}{48} + \frac{24}{48} = \frac{33}{48} =$

53. $\frac{1}{2} + \left(\frac{1}{2}\right)^2 - \frac{3}{8}$

54. $\frac{2}{3} + \left(\frac{1}{3}\right)^2 - \frac{5}{9}$

55. $\left(\frac{1}{3}+\frac{1}{6}\right) \cdot \frac{1}{2}$

56. $\left(\frac{3}{5}-\frac{3}{20}\right) \cdot \frac{4}{3}$

57. $\frac{9}{8} \div \left(\frac{2}{3}+\frac{1}{12}\right)$

58. $\frac{6}{5} \div \left(\frac{3}{5}-\frac{3}{10}\right)$

59. $\left(\frac{7}{8}-\frac{3}{4}\right) \div \frac{3}{2}$

GS

$\left(\frac{7}{8}-\frac{6}{8}\right) \div \frac{3}{2}$

$\frac{1}{8} \div \frac{3}{2}$

$\frac{1}{8} \cdot \frac{2}{3} = \frac{2}{24} =$

60. $\left(\frac{4}{5}-\frac{3}{10}\right) \div \frac{4}{5}$

GS

$\left(\frac{8}{10}-\frac{3}{10}\right) \div \frac{4}{5}$

$\frac{5}{10} \div \frac{4}{5}$

$\frac{5}{10} \cdot \frac{5}{4} = \frac{25}{40} =$

61. $\frac{3}{8}\left(\frac{1}{4}+\frac{1}{2}\right) \cdot \frac{32}{3}$

62. $\frac{1}{3}\left(\frac{4}{5}-\frac{3}{10}\right) \cdot \frac{4}{2}$

63. $\left(\frac{3}{4}\right)^2 - \left(\frac{1}{2}-\frac{1}{6}\right) \div \frac{4}{3}$

64. $\left(\frac{2}{3}\right)^2 - \left(\frac{5}{8}-\frac{1}{2}\right) \div \frac{3}{2}$

65. $\left(\frac{7}{8}-\frac{1}{4}\right) - \frac{2}{3}\left(\frac{3}{4}\right)^2$

66. $\left(\frac{5}{6}-\frac{7}{12}\right) - \frac{3}{4}\left(\frac{1}{3}\right)^2$

67. $\left(\frac{3}{4}\right)^2\left(\frac{2}{3}-\frac{5}{9}\right) - \frac{1}{4}\left(\frac{1}{8}\right)$

68. $\left(\frac{2}{3}\right)^2\left(\frac{1}{2}-\frac{1}{8}\right) - \frac{2}{3}\left(\frac{1}{8}\right)$

Solve each application problem.

69. The population of Las Vegas, Nevada has had an increase of $\frac{11}{50}$ since the turn of the century. During this same period, the population in Atlanta, Georgia has had an increase of $\frac{5}{30}$. Which city has had a higher rate of population growth? (*Source:* Census Bureau estimates.)

70. Jacob Evan weighs $22\frac{11}{16}$ pounds and his stroller weighs $22\frac{5}{8}$ pounds. Which weighs more, Jacob or the stroller?

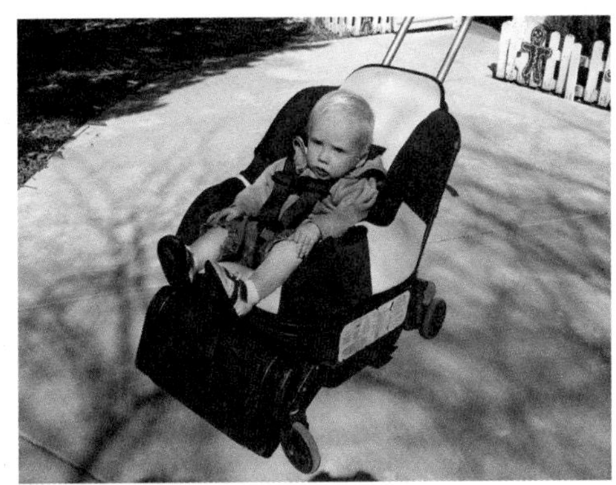

Relating Concepts (Exercises 71–80) For Individual or Group Work

You often need to use order relations and the order of operations when solving problems. ***Work Exercises 71–80 in order.***

71. When comparing the size of two numbers, the symbol ___________ means **is less than** and the symbol ___________ means **is greater than.**

72. (a) To identify the greater of two fractions, we must first write the fractions as ___________ fractions and then compare the ___________. The fraction with the greater ___________ is the greater fraction.

(b) Write four pairs of fractions, all with different denominators. Write the symbol for **less than** or for **greater than** between each pair.

Use the order of operations to simplify each expression.

73. $\left(\frac{2}{3}\right)^2 - \left(\frac{4}{5} - \frac{3}{10}\right) \div \frac{5}{4}$

74. $\left(\frac{1}{2}\right)^2\left(\frac{3}{8} - \frac{1}{4}\right) + \frac{2}{3} \div \left(\frac{1}{3}\right)$

Simplify, then place the results on the number line.

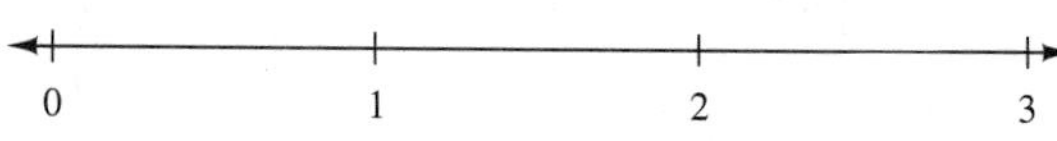

75. $\left(\frac{2}{3}\right)^2$

76. $\left(\frac{3}{2}\right)^2$

77. $\left(\frac{1}{2}\right)^3$

78. $\left(\frac{5}{3}\right)^2$

79. $4 + 2 - 2^2$

80. $\left(\frac{3}{4}\right)^2 + \left(\frac{5}{8} - \frac{1}{4}\right) \div \frac{2}{3}$

Study Skills

PREPARING FOR TESTS

OBJECTIVES

1. **Restate the importance of sleep and good nutrition as it affects learning.**
2. **Explain the effect of anxiety and stress on learning.**

Many things besides studying can improve your test scores. You may not realize that eating the right foods, and getting enough exercise and sleep, can also improve your scores. Your brain (and therefore your ability to think) is affected by the condition of your whole body. So, part of your preparation for tests includes keeping yourself in good physical shape as well as spending time on the actual course material. Try these suggestions and see the difference.

Performance Health Tips

To Improve Your Test Score	Explanation
Get ***seven to eight hours of sleep*** the night before the exam. (It's helpful to get that much sleep *every* night.)	***Fatigue and exhaustion*** reduce efficiency. They also cause poor memory and recall. If you didn't sleep much the night before a test, 20 minutes of relaxation or meditation can help.
Eat a ***small, high-energy meal*** about two hours before the test. Start the meal with a small amount of protein such as fish, chicken, or nonfat yogurt. Include carbohydrates if you like, but no high-fat foods.	Just 3 to 4 ounces of protein increases the amount of a chemical in the brain called tyrosine, which ***improves your alertness, accuracy, and motivation.*** High-fat foods dull your mind and slow down your brain.
Drink plenty of water. Don't wait until you feel thirsty; your body is already dehydrated by the time you feel it.	Research suggests that staying well hydrated improves the electrochemical communications in your brain.
Give your brain the time it needs to grow dendrites!	***Cramming doesn't work;*** your brain cannot grow dendrites that quickly. ***Studying every day*** is the way to give your brain the time it needs.

Anxiety Prevention Tips

To Prevent Anxiety	Explanation
Practice slow, deep breathing for five minutes each day. Then do a minute or two of deep breathing right before the test. Also, if you feel your anxiety building during the test, stop for a minute, close your eyes, and do some deep breathing.	When ***test anxiety*** hits, you breathe more quickly and shallowly, which causes hyperventilation. Symptoms may be confusion, inability to concentrate, shaking, dizziness, and more. Slow, deep breathing will ***calm you and prevent panic.***

Study Skills

Continued from page 247

Anxiety Prevention Tips *(Continued)*

To Prevent Anxiety	Explanation
Do 15 to 20 minutes of ***moderate exercise*** (like walking) shortly before the test. Daily exercise is even better!	***Exercise reduces stress*** and will help prevent "blanking out" on a test. Exercise also increases your alertness, clear thinking, and energy.
To help you sleep the night before the test, or any time you need to calm down, ***eat high carbohydrate foods*** such as popcorn, bread, rice, crackers, muffins, bagels, pasta, corn, baked potatoes (not fries or chips), and cereals.	Carbohydrates increase the level of a chemical in the brain called serotonin, which has a *calming effect on the mind*. It reduces feelings of tension and stress and improves your ability to concentrate. You only need to eat a small amount, like half a bagel, to get this effect.
Before the test, ***go easy on caffeinated beverages*** such as coffee, tea, and soft drinks. Do not eat candy bars or other sugary snacks.	Extra caffeine can ***make you jittery,*** "hyper," and shaky for the test. It can increase the tendency to panic. Too much sugar causes negative emotional reactions in some people.

Now Try This

What will you do to improve your next test score?
List the three or four tips you think will help you the most.

1. ______________________________
2. ______________________________
3. ______________________________
4. ______________________________

What changes will you have to make in order to try the tips you chose?

____

____

____

____

See *Tips for Taking Math Tests* and *Preparing for Your Final Exam* for more ideas about managing anxiety. (Check the Table of Contents to find their locations.)

Chapter 3 Summary

Key Terms

3.1

like fractions Fractions with the same denominator are called *like fractions.*

unlike fractions Fractions with different denominators are called *unlike fractions.*

3.2

least common multiple Given two or more whole numbers, the least common multiple is the smallest whole number that is divisible by all the numbers.

LCM The abbreviation for *least common multiple* is LCM.

3.3

least common denominator When unlike fractions are rewritten as like fractions with the least common multiple as their denominator, the new denominator is the least common denominator.

LCD The abbreviation for *least common denominator* is LCD.

3.4

regrouping when adding fractions Regrouping is used in the addition of mixed numbers when the sum of the fractions is greater than 1.

regrouping when subtracting fractions Regrouping is used in the subtraction of mixed numbers when the fraction part of the first number is less than the fraction part of the second number.

New Symbols

$<$ is less than $(2 < 5)$

$>$ is greater than $(4 > 2)$

Test Your Word Power

See how well you have learned the vocabulary in this chapter.

1. **Like fractions** are
   A. fractions that are equivalent
   B. fractions that have the same numerator
   C. fractions that have the same denominator.

2. Two or more fractions are **unlike fractions** if
   A. they are not equivalent
   B. they have different numerators
   C. they have different denominators.

3. The abbreviation **LCM** stands for
   A. the largest common multiple
   B. the longest common multiple
   C. the least common multiple.

4. The **least common multiple** is
   A. the smallest whole number that is divisible by each of two or more numbers
   B. the smallest numerator
   C. the smallest denominator.

5. The abbreviation **LCD** stands for
   A. the largest common denominator
   B. the least common denominator
   C. the least common divisor.

6. The **least common denominator** is
   A. needed when multiplying fractions
   B. needed when dividing fractions
   C. the least common multiple of the denominators in a fraction problem.

Answers to Test Your Word Power

1. C; *Example:* Because the fractions $\frac{3}{8}$ and $\frac{10}{8}$ both have 8 as a denominator, they are like fractions.

2. C; *Example:* The fractions $\frac{2}{3}$ and $\frac{3}{4}$ are unlike fractions because they have different denominators.

3. C; *Example:* LCM is the abbreviation for least common multiple.

4. A; *Example:* The least common multiple of 4 and 5 is 20 because 20 is the smallest number into which both 4 and 5 will divide evenly.

5. B; *Example:* LCD is the abbreviation for least common denominator.

6. C; *Example:* The least common denominator of the fractions $\frac{2}{3}$ and $\frac{1}{2}$ is 6 because 6 is the least common multiple of 3 and 2. When written using the least common denominator, $\frac{2}{3}$ and $\frac{1}{2}$ become $\frac{4}{6}$ and $\frac{3}{6}$, respectively.

Quick Review

Concepts	Examples
3.1 Adding Like Fractions Add numerators and keep the same denominator. Simplify the answer.	$\frac{3}{4}+\frac{1}{4}+\frac{5}{4}=\frac{3+1+5}{4}=\frac{9}{4}=2\frac{1}{4}$
3.1 Subtracting Like Fractions Subtract numerators and keep the same denominator. Simplify the answer.	$\frac{7}{8}-\frac{5}{8}=\frac{7-5}{8}=\frac{2}{8}=\frac{2\div 2}{8\div 2}=\frac{1}{4}$
3.2 Finding the Least Common Multiple (LCM) Method of using multiples of the larger number: List the first few multiples of the larger number. Then find the first multiple that is divisible by the smaller number.	$\frac{1}{3}+\frac{1}{4}$ 4, 8, 12, 16, … ← Multiples of 4 12: First multiple divisible by 3 ($12 \div 3 = 4$) The least common multiple (LCM) of 3 and 4 is 12.
3.2 Finding the Least Common Multiple (LCM) Method of prime numbers: First find the prime factorization of each number. Then use the prime factors to build the least common multiple.	$9 = ③ \cdot ③$ $15 = 3 \cdot ⑤$ $\text{LCM} = 3 \cdot 3 \cdot 5 = 45$ (Factors of 9: $3 \cdot 3$; Factors of 15: $3 \cdot 5$) The least common multiple (LCM) of 9 and 15 is 45.
3.2 Finding the Least Common Multiple (LCM) Alternative method: Start by trying to divide the numbers by the first prime number. Continue dividing by prime numbers until all quotients are 1. The product of the prime numbers is the least common multiple.	4 and 6 2 | 4 6 2 | 2 3 3 | 1 3 1 1 The least common multiple (LCM) of 4 and 6 = $2 \cdot 2 \cdot 3 = 12$.
3.3 Adding Unlike Fractions *Step 1* Find the least common multiple (LCM). *Step 2* Rewrite the fractions with the least common multiple as the denominator. *Step 3* Add the numerators, placing the sum over the common denominator, and simplify the answer.	$\frac{1}{3} + \frac{1}{4} + \frac{1}{10}$ LCM of 3, 4, and 10 = 60 $\frac{1}{3}=\frac{20}{60}$ $\frac{1}{4}=\frac{15}{60}$ $\frac{1}{10}=\frac{6}{60}$ $\frac{20}{60}+\frac{15}{60}+\frac{6}{60}=\frac{41}{60}$ Lowest terms

Concepts	Examples
3.3 Subtracting Unlike Fractions *Step 1* Find the least common multiple (LCM). *Step 2* Rewrite the fractions with the least common multiple as the denominator. *Step 3* Subtract the numerators, place the difference over the common denominator, and simplify the answer.	$\frac{5}{8} - \frac{1}{3}$ LCM of 8 and 3 = 24 $\frac{5}{8} = \frac{15}{24}$ $\frac{1}{3} = \frac{8}{24}$ $\frac{15}{24} - \frac{8}{24} = \frac{7}{24}$ Lowest terms
3.4 Adding Mixed Numbers Round the numbers and estimate the answer. Then find the exact answer using these steps. *Step 1* Add the fractions, using a common denominator. *Step 2* Add the whole numbers. *Step 3* Combine the sums of the whole numbers and the fractions. Simplify the fraction part when necessary. Compare the exact answer to the estimate to see if it is reasonable.	*Estimate:* *Exact:* 10 ← Rounds to $9\frac{2}{3} = 9\frac{8}{12}$ $+\ 7$ ← Rounds to $+\ 6\frac{3}{4} = 6\frac{9}{12}$ 17 $15\frac{17}{12} = 16\frac{5}{12}$ The exact answer of $16\frac{5}{12}$ is reasonable because it is close to the estimate of 17.
3.4 Subtracting Mixed Numbers Round the numbers and estimate the answer. Then find the exact answer using these steps. *Step 1* Subtract the fractions, regrouping if necessary. *Step 2* Subtract the whole numbers. *Step 3* Combine the differences of the whole numbers and the fractions. Simplify the fraction part when necessary. Compare the exact answer to the estimate to see if it is reasonable.	*Estimate:* *Exact:* 9 ← Rounds to $8\frac{5}{8} = 8\frac{15}{24} = 7\frac{39}{24}$ $-\ 4$ ← Rounds to $-\ 3\frac{11}{12} = 3\frac{22}{24} = 3\frac{22}{24}$ 5 $4\frac{17}{24}$ The exact answer of $4\frac{17}{24}$ is reasonable because it is close to the estimate of 5.

Concepts	Examples

3.4 Adding or Subtracting Mixed Numbers Using an Alternative Method

Step 1 Change the mixed numbers to improper fractions.

Step 2 Rewrite the unlike fractions as like fractions.

Step 3 Add or subtract the numerators and simplify the answer.

Add.

$$2\frac{2}{3} = \frac{8}{3} = \frac{64}{24}$$

$$+1\frac{3}{8} = \frac{11}{8} = +\frac{33}{24}$$

24 is the least common denominator.

$$\frac{97}{24} = 4\frac{1}{24}$$ Answer as mixed number

Improper fractions

Subtract.

$$8\frac{2}{3} = \frac{26}{3} = \frac{104}{12}$$

$$-5\frac{3}{4} = \frac{23}{4} = -\frac{69}{12}$$

12 is the least common denominator.

$$\frac{35}{12} = 2\frac{11}{12}$$ Answer as mixed number

Improper fractions

3.5 Identifying the Greater of Two Fractions

With unlike fractions, change to like fractions first. The fraction with the greater numerator is the greater fraction. Use these symbols:

$<$ is less than

$>$ is greater than

Identify the greater fraction.

$$\frac{7}{8} \quad \frac{9}{10}$$

$$\frac{7}{8} = \frac{7 \cdot 5}{8 \cdot 5} = \frac{35}{40}$$

$$\frac{9}{10} = \frac{9 \cdot 4}{10 \cdot 4} = \frac{36}{40}$$

$\frac{36}{40}$ is greater than $\frac{35}{40}$, so $\frac{9}{10} > \frac{7}{8}$.

$\frac{9}{10}$ is greater.

3.5 Using the Order of Operations with Fractions

Follow the order of operations.

1. Do all operations inside parentheses or other grouping symbols.
2. Simplify any expressions with exponents and find any square roots.
3. Multiply or divide, proceeding from left to right.
4. Add or subtract, proceeding from left to right.

Simplify by using the order of operations.

$$\frac{1}{2}\left(\frac{2}{3}\right) - \left(\frac{1}{4}\right)^2$$ Apply the exponent first.

$$= \frac{1}{2}\left(\frac{2}{3}\right) - \frac{1}{16}$$ Multiply next.

$$= \frac{1}{3} - \frac{1}{16}$$ Subtract last.

$$= \frac{16}{48} - \frac{3}{48}$$ Rewrite fractions so they have a common denominator. Subtract numerators.

$$= \frac{13}{48}$$ Simplest form

Chapter 3 Review Exercises

3.1 *Add or subtract. Write answers in lowest terms.*

1. $\frac{5}{7}+\frac{1}{7}$

2. $\frac{4}{9}+\frac{3}{9}$

3. $\frac{1}{8}+\frac{3}{8}+\frac{2}{8}$

4. $\frac{5}{16}-\frac{3}{16}$

5. $\frac{5}{10}+\frac{3}{10}$

6. $\frac{5}{12}-\frac{3}{12}$

7. $\frac{36}{62}-\frac{10}{62}$

8. $\frac{68}{75}-\frac{43}{75}$

Solve each application problem. Write answers in lowest terms.

9. Jaime Villagranna earns $\frac{7}{12}$ of his income installing kitchen cabinets for Home Depot and $\frac{4}{12}$ of his income by operating his own cabinet business. What fraction of his total income comes from the two jobs?

10. At the annual Cub Scout Pinewood Derby competition, $\frac{3}{8}$ of the events were completed before lunch and $\frac{5}{8}$ after lunch. How much more was completed after lunch than before lunch?

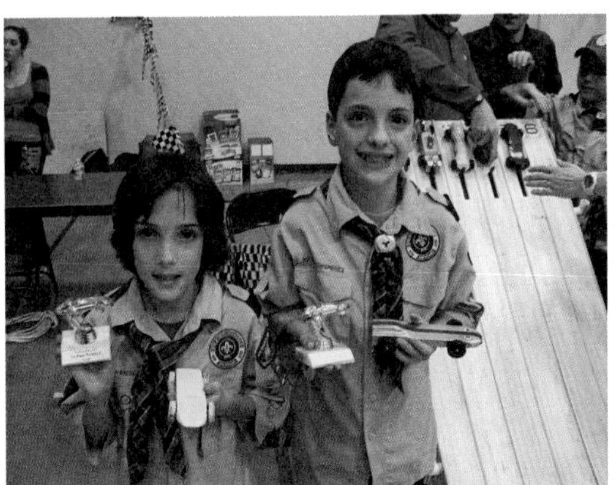

3.2 *Find the least common multiple of each set of numbers.*

11. 5, 2

12. 3, 4

13. 10, 12, 20

14. 3, 8, 4

15. 6, 8, 5, 15

16. 15, 9, 20

Rewrite each fraction using the indicated denominator.

17. $\frac{2}{3} = \frac{}{12}$

18. $\frac{3}{8} = \frac{}{56}$

19. $\frac{2}{5} = \frac{}{25}$

20. $\frac{5}{9} = \frac{}{81}$

21. $\frac{4}{5} = \frac{}{40}$

22. $\frac{5}{16} = \frac{}{64}$

3.1–3.3 *Add or subtract. Write answers in lowest terms.*

23. $\frac{1}{2} + \frac{1}{3}$

24. $\frac{1}{5} + \frac{3}{10} + \frac{3}{8}$

25. $\frac{5}{12} + \frac{5}{24}$

26. $\frac{2}{3} - \frac{1}{4}$

27. $\frac{7}{8} - \frac{1}{3}$

28. $\frac{11}{12} - \frac{4}{9}$

Solve each application problem.

29. The San Juan School District operates an after school program for students. This year $\frac{2}{5}$ of the students played after school sports, $\frac{1}{6}$ participated in arts and crafts, and $\frac{1}{3}$ spent their time in tutoring and study hall. What fraction of the total students participated in these activities?

30. When budgeting for her wedding, Kara Salzie plans to spend $\frac{1}{3}$ of her total budget on the wedding site, $\frac{3}{8}$ on food and beverages, $\frac{3}{16}$ on photos and video, and $\frac{1}{16}$ on entertainment. What portion of her budget will be spent on these four categories?

3.4 *First estimate the answer. Then add or subtract to find the exact answer. Simplify all exact answers.*

31. *Estimate:* $\underline{\quad} + \underline{\quad}$ (Rounds to) *Exact:* $18\frac{5}{8} + 13\frac{3}{4}$

32. *Estimate:* $\underline{\quad} + \underline{\quad}$ *Exact:* $22\frac{2}{3} + 15\frac{4}{9}$

33. *Estimate:* $\underline{\quad} + \underline{\quad} + \underline{\quad}$ *Exact:* $12\frac{3}{5} + 8\frac{5}{8} + 10\frac{5}{16}$

34. *Estimate:* $\underline{\quad} - \underline{\quad}$ *Exact:* $31\frac{3}{4} - 14\frac{2}{3}$

35. *Estimate:* $\underline{\quad} - \underline{\quad}$ *Exact:* $34 - 15\frac{2}{3}$

36. *Estimate:* $\underline{\quad} - \underline{\quad}$ *Exact:* $215\frac{7}{16} - 136$

Add or subtract by changing mixed numbers to improper fractions. Simplify all answers.

37. $5\frac{2}{5} + 3\frac{7}{10}$

38. $4\frac{3}{4} + 5\frac{2}{3}$

39. $5 - 1\frac{3}{4}$

40. $6\frac{1}{2} - 4\frac{5}{6}$

41. $8\frac{1}{3} - 2\frac{5}{6}$

42. $5\frac{5}{12} - 2\frac{5}{8}$

First estimate the answer and then find the exact answer for each application problem. Simplify all exact answers.

43. Two long-distance runners began an $18\frac{3}{4}$ mile run. They ran uphill $5\frac{5}{8}$ miles, downhill $7\frac{1}{3}$ miles, and the rest of the course was level. Find the distance of the level portion of the course.

Estimate:

Exact:

44. The Boys Scouts had a paper drive. They collected $28\frac{2}{3}$ tons of newspapers on Saturday and $24\frac{3}{4}$ tons on Sunday. Find the total weight of the newspapers collected.

Estimate:

Exact:

45. On a recent fishing trip to Kemmerer, Wyoming, Roy Abriani caught four fish. One fish was a German brown trout weighing $7\frac{1}{2}$ pounds, while the other three were rainbow trout weighing $2\frac{3}{4}$ pounds, $4\frac{7}{8}$ pounds, and $3\frac{3}{8}$ pounds. Find the total weight of the four fish.

Estimate:

Exact:

46. The Safeway World Championship Pumpkin Weigh-off awarded first place to the grower of a pumpkin weighing $1535\frac{3}{8}$ pounds. The second place pumpkin weighed $1475\frac{11}{16}$ pounds. How much more did the first place pumpkin weigh? (*Source:* Sacramento Bee.)

Estimate:

Exact:

3.5 *Locate each fraction in Exercises 47–50 on the number line.*

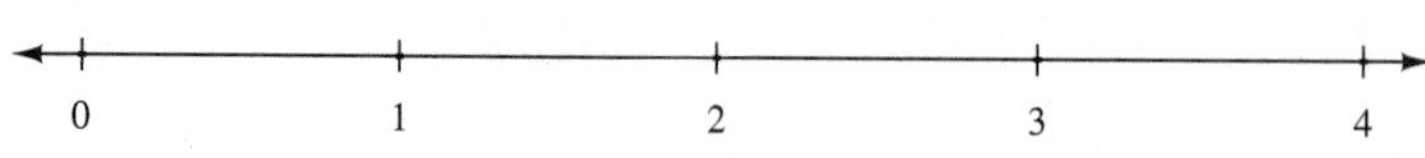

47. $\frac{3}{8}$

48. $\frac{7}{4}$

49. $\frac{8}{3}$

50. $3\frac{1}{5}$

Write < or > in each blank to make a true statement.

51. $\frac{2}{3}$ ___ $\frac{3}{4}$

52. $\frac{3}{4}$ ___ $\frac{7}{8}$

53. $\frac{1}{2}$ ___ $\frac{7}{15}$

54. $\frac{7}{10}$ ___ $\frac{8}{15}$

55. $\frac{9}{16}$ ___ $\frac{5}{8}$

56. $\frac{7}{20}$ ___ $\frac{8}{25}$

57. $\frac{19}{36}$ ___ $\frac{29}{54}$

58. $\frac{19}{132}$ ___ $\frac{7}{55}$

Simplify each expression.

59. $\left(\frac{1}{2}\right)^2$

60. $\left(\frac{2}{3}\right)^2$

61. $\left(\frac{3}{10}\right)^3$

62. $\left(\frac{3}{8}\right)^4$

Simplify by using the order of operations.

63. $8\left(\frac{1}{4}\right)^2$

64. $12\left(\frac{3}{4}\right)^2$

65. $\left(\frac{2}{3}\right)^2 \cdot \left(\frac{3}{8}\right)^2$

66. $\frac{7}{8} \div \left(\frac{1}{8} + \frac{3}{4}\right)$

67. $\left(\frac{1}{2}\right)^2 \cdot \left(\frac{1}{4} + \frac{1}{2}\right)$

68. $\left(\frac{1}{4}\right)^3 + \left(\frac{5}{8} + \frac{3}{4}\right)$

Mixed Review Exercises

Simplify. Use the order of operations as necessary.

69. $\frac{7}{8} - \frac{1}{8}$

70. $\frac{7}{10} - \frac{3}{10}$

71. $\frac{29}{32} - \frac{5}{16}$

72. $\frac{1}{4} + \frac{1}{8} + \frac{5}{16}$

73. $\begin{array}{r} 6\frac{2}{3} \\ -\ 4\frac{1}{2} \\ \hline \end{array}$

74. $\begin{array}{r} 9\frac{1}{2} \\ +\ 16\frac{3}{4} \\ \hline \end{array}$

75. $\begin{array}{r} 7 \\ -\ 1\frac{5}{8} \\ \hline \end{array}$

76. $\begin{array}{r} 2\frac{3}{5} \\ 8\frac{5}{8} \\ +\ \frac{5}{16} \\ \hline \end{array}$

77. $\begin{array}{r} 32\frac{5}{12} \\ -\ 17 \\ \hline \end{array}$

78. $\frac{7}{22} + \frac{3}{22} + \frac{3}{11}$

79. $\left(\frac{1}{4}\right)^2 \cdot \left(\frac{2}{5}\right)^3$

80. $\frac{3}{8} \div \left(\frac{1}{2} + \frac{1}{4}\right)$

81. $\left(\frac{2}{3}\right)^2 \cdot \left(\frac{1}{3} + \frac{1}{6}\right)$

82. $\left(\frac{2}{3}\right)^3 + \left(\frac{2}{3} - \frac{5}{9}\right)$

Write $<$ or $>$ in each blank to make a true statement.

83. $\frac{2}{3}$ ____ $\frac{7}{12}$

84. $\frac{8}{9}$ ____ $\frac{15}{8}$

85. $\frac{17}{30}$ ____ $\frac{36}{60}$

86. $\frac{5}{8}$ ____ $\frac{17}{30}$

Find the least common multiple of each set of numbers.

87. 12, 18

88. 6, 8, 10, 12

89. 9, 14, 21

Rewrite each fraction using the indicated denominator.

90. $\frac{2}{3} = \frac{}{27}$

91. $\frac{9}{12} = \frac{}{144}$

92. $\frac{4}{5} = \frac{}{75}$

First estimate the answer and then find the exact answer for each application problem. Simplify exact answers.

93. A cement contractor needs $13\frac{1}{2}$ ft of wire mesh for a concrete walkway and $22\frac{3}{8}$ ft of wire mesh for a driveway. If the contractor starts with a roll of wire that is $92\frac{3}{4}$ ft long, find the number of feet remaining after the two jobs have been completed.

Estimate:

Exact:

94. A baker had four 50-pound bags of sugar. She used $68\frac{1}{2}$ pounds of sugar to bake cakes, $76\frac{5}{8}$ pounds for baking pies, and $33\frac{1}{4}$ pounds for baking cookies. How many pounds of sugar remain?

Estimate:

Exact:

Chapter 3 Test

The Chapter Test Prep Videos with test solutions are available on DVD, in MyMathLab, and on YouTube — search "LialBasicCollegeMath" and click on "Channels."

Add or subtract. Write answers in lowest terms.

1. $\frac{5}{8}+\frac{1}{8}$

2. $\frac{1}{16}+\frac{7}{16}$

3. $\frac{7}{10}-\frac{3}{10}$

4. $\frac{7}{12}-\frac{5}{12}$

Find the least common multiple of each set of numbers.

5. 2, 3, 4

6. 6, 3, 5, 15

7. 6, 9, 27, 36

Add or subtract. Write answers in lowest terms.

8. $\frac{3}{8}+\frac{1}{4}$

9. $\frac{2}{9}+\frac{5}{12}$

10. $\frac{7}{8}-\frac{2}{3}$

11. $\frac{2}{5}-\frac{3}{8}$

First estimate the answer. Then add or subtract to find the exact answer. Simplify exact answers.

12. $7\frac{2}{3}+4\frac{5}{6}$

Estimate:

Exact:

13. $16\frac{2}{5}-11\frac{2}{3}$

Estimate:

Exact:

14. $18\frac{3}{4}+9\frac{2}{5}+12\frac{1}{3}$

Estimate:

Exact:

15. $24-18\frac{3}{8}$

Estimate:

Exact:

16. Most students say that "addition and subtraction of fractions are more difficult than multiplication and division of fractions." Why do you think they say this? Do you agree with these students?

17. Explain a method of estimating an answer to addition and subtraction problems involving mixed numbers. Could your estimated answer vary from the exact answer? If it did, what would the estimation accomplish?

First estimate the answer and then find the exact answer for each application problem. Simplify exact answers.

18. In one week, a kennel owner used $10\frac{3}{8}$ pounds of puppy chow, $84\frac{1}{2}$ pounds of dry kibble, $36\frac{5}{6}$ pounds of high-protein mature dog mix, and $8\frac{1}{3}$ pounds of fresh ground meat products. Find the total number of pounds used.

Estimate:

Exact:

19. A painting contractor arrived at a 6-unit apartment complex with $147\frac{1}{2}$ gallons of exterior paint. If his crew sprayed $68\frac{1}{2}$ gallons on the wood siding and rolled $37\frac{3}{8}$ gallons on the masonry exterior, find the number of gallons of paint remaining.

Estimate:

Exact:

Write $<$ or $>$ between each pair of fractions to make a true statement.

20. $\frac{3}{4}$ ___ $\frac{17}{24}$

21. $\frac{19}{24}$ ___ $\frac{17}{36}$

Simplify. Use the order of operations as needed.

22. $\left(\frac{1}{3}\right)^3 \cdot 54$

23. $\left(\frac{3}{4}\right)^2 - \left(\frac{7}{8} \cdot \frac{1}{3}\right)$

24. $4\left(\frac{7}{8} - \frac{7}{16}\right)$

25. $\frac{5}{6} + \frac{4}{3}\left(\frac{3}{8}\right)$

Chapters 1–3 Cumulative Review Exercises

1. In this number, name the digit that has the given place value. 5,639,428

millions ten-thousands

thousands hundreds

2. Round 59,803 to the nearest ten, then to the nearest hundred, and finally to the nearest thousand.

Use front end rounding to estimate each answer. Then find the exact answer.

3. *Estimate:* *Exact:* $24{,}276 - 9\,887$

4. *Estimate:* *Exact:* $35\overline{)112{,}385}$

Add, subtract, multiply, or divide as indicated.

5. $375{,}899 + 521{,}742 + 357{,}968$

6. $3{,}896{,}502 - 1{,}094{,}807$

7. $5(8)(4)$

8. 962×384

9. $8\overline{)1080}$

10. $13{,}467 \div 5$

Use front end rounding to estimate the answer to each application problem. Then find the exact answer.

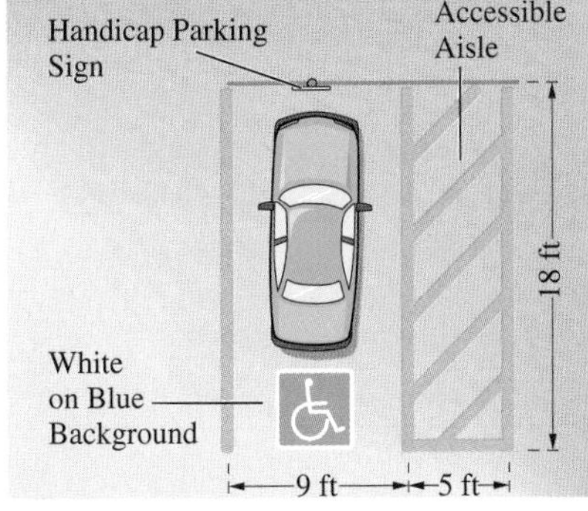

11. The Americans with Disabilities Act provides the single parking space design shown at the left. Find the perimeter of (distance around) this parking space, including the accessible aisle.

Estimate:

Exact:

12. The single parking space design in **Exercise 11** measures 18 ft by 14 ft. Find its area.

Estimate:

Exact:

Round the mixed numbers in each problem to the nearest whole number and estimate the answer. Then find the exact answer. Simplify exact answers.

13. The first car to be marketed specifically to women was the 1953 Nash Metropolitan. Ads said, "Metropolitan is a good thing in a small package." The car could get $47\frac{3}{10}$ miles per gallon and the gas tank capacity was $8\frac{3}{4}$ gallons. How many miles could the car go on a full tank of gasoline? (*Source:* uniquecarsandparts.com and Wikipedia.)

Estimate:

Exact:

14. Fleas can jump 130 times higher than their own height. In human terms, how far could a $70\frac{1}{4}$ inch person jump? (*Source:* High Tech Productions Science and Technology Center.)

Estimate:

Exact:

Simplify.

15. $2^4 \cdot 3^2$

16. $5^2 - 2(8)$

17. $\sqrt{25} + 5 \cdot 9 - 6$

18. $\frac{2}{3}\left(\frac{4}{5} - \frac{2}{3}\right)$

19. $\frac{3}{4} \div \left(\frac{1}{3} + \frac{1}{2}\right)$

20. $\frac{7}{8} + \left(\frac{3}{4}\right)^2 - \frac{3}{8}$

21. $\frac{3}{4} \cdot \frac{2}{3}$

22. $42 \cdot \frac{7}{8}$

23. $\frac{25}{40} \div \frac{10}{35}$

24. $9 \div \frac{2}{3}$

First estimate the answer and then add or subtract to find the exact answer. Write exact answers as mixed numbers.

25. *Estimate:* *Exact:*

$$\begin{array}{r} ___ \\ + ___ \\ \hline \end{array} \quad \begin{array}{l} \xleftarrow{\text{Rounds to}} \\ \xleftarrow{\text{Rounds to}} \end{array} \quad \begin{array}{r} 3\frac{3}{8} \\ + 4\frac{1}{2} \\ \hline \end{array}$$

26. *Estimate:* *Exact:*

$$\begin{array}{r} \\ + \\ \hline \end{array} \qquad \begin{array}{r} 21\frac{7}{8} \\ + 4\frac{5}{12} \\ \hline \end{array}$$

27. *Estimate:* *Exact:*

$$\begin{array}{r} \\ - \\ \hline \end{array} \qquad \begin{array}{r} 5 \\ - 2\frac{3}{8} \\ \hline \end{array}$$

Locate each fraction in Exercises 28–31 on the number line at the right.

28. $2\frac{3}{4}$

29. $\frac{1}{9}$

30. $\frac{5}{3}$

31. $\frac{10}{3}$

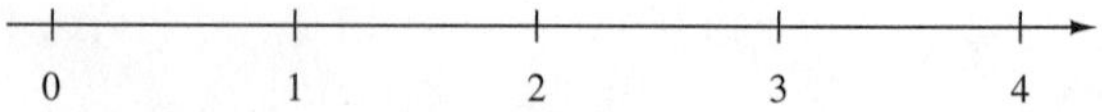

Write < or > in each blank to make a true statement.

32. $\frac{3}{5}$ ____ $\frac{5}{8}$

33. $\frac{17}{20}$ ____ $\frac{3}{4}$

34. $\frac{7}{12}$ ____ $\frac{11}{18}$

4 Decimals

Over 42 million Americans go fishing at least once a year, spending more than $2 billion on gear and tackle. This grandfather and grandson will use decimal numbers when paying for new fishing equipment. But will decimals help them catch their limit?

4.1 Reading and Writing Decimal Numbers

OBJECTIVES

1. Write parts of a whole using decimals.
2. Identify the place value of a digit.
3. Read and write decimals in words.
4. Write decimals as fractions or mixed numbers.

Fractions are used to represent parts of a whole. In this chapter, **decimals** are used as another way to show parts of a whole. For example, our money system is based on decimals. One dollar is divided into 100 equal parts. One cent ($0.01) is one of the parts, and a dime ($0.10) is 10 of the parts. Metric measurement (see **Chapter 7**) is also based on decimals.

OBJECTIVE 1 Write parts of a whole using decimals. Decimals are used when a whole is divided into 10 equal parts, or into 100 or 1000 or 10,000 equal parts. In other words, decimals are fractions with denominators that are a power of 10. For example, the square below is cut into 10 equal parts. Written as a fraction, each part is $\frac{1}{10}$ of the whole. Written as a decimal, each part is 0.1. Both $\frac{1}{10}$ and 0.1 are read as "*one tenth.*"

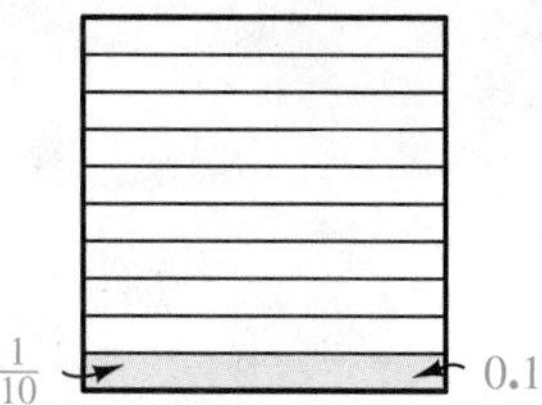

One-tenth of the square is shaded.

The dot in 0.1 is called the **decimal point.**

0.1

Decimal point

The square at the right has 7 of its 10 parts shaded.

Written as a *fraction,* $\frac{7}{10}$ of the square is shaded.

Written as a *decimal,* 0.7 of the square is shaded.

Both $\frac{7}{10}$ and 0.7 are read as "*seven tenths.*"

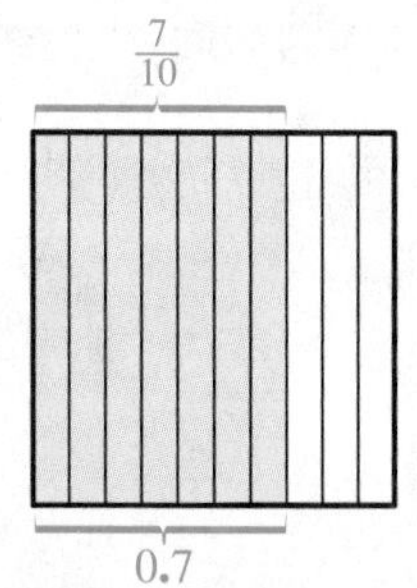

Seven-tenths of the square is shaded.

◀ **Work Problem 1 at the Side.**

The square below is cut into 100 equal parts.

Written as a *fraction,* each part is $\frac{1}{100}$ of the whole.

Written as a decimal, each part is **0.01** of the whole.

Both $\frac{1}{100}$ and 0.01 are read as "*one hundredth.*"

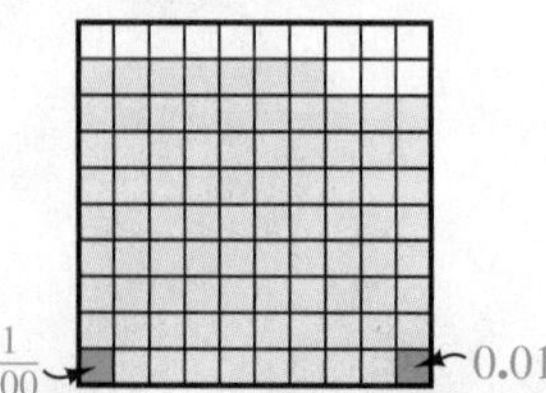

This square has 87 parts shaded.

Written as a fraction, $\frac{87}{100}$ of the total area is shaded.

Written as a decimal, **0.87** of the total area is shaded.

Both $\frac{87}{100}$ and 0.87 are read as "*eighty-seven hundredths.*"

VOCABULARY TIP

Deci The prefix **deci** means **tenth.** For example, a **decimeter** is one **tenth** of a meter.

1 There are 10 dimes in one dollar. Each dime is $\frac{1}{10}$ of a dollar. Label the yellow shaded portion of each dollar as a fraction, as a decimal, and in words.

(a)

(b)

(c)

Answers

1. **(a)** $\frac{1}{10}$; 0.1; one tenth

(b) $\frac{3}{10}$; 0.3; three tenths

(c) $\frac{9}{10}$; 0.9; nine tenths

Work Problem 2 at the Side. ▶

Example 1 below shows several numbers written as fractions, as decimals, and in words.

EXAMPLE 1 Using the Decimal Forms of Fractions

	Fraction	Decimal	Read As
(a)	$\frac{4}{10}$	0.4	four tenths
(b)	$\frac{9}{100}$	0.09	nine hundredths
(c)	$\frac{71}{100}$	0.71	seventy-one hundredths
(d)	$\frac{8}{1000}$	0.008	eight thousandths
(e)	$\frac{45}{1000}$	0.045	forty-five thousandths
(f)	$\frac{832}{1000}$	0.832	eight hundred thirty-two thousandths

Work Problem 3 at the Side. ▶

OBJECTIVE 2 **Identify the place value of a digit.** The decimal point separates the *whole number part* from the *fractional part* in a decimal number. In the chart below, you see that the **place value** names for fractional parts are similar to those on the whole number side, but end in "*ths*."

Decimal Place Value Chart

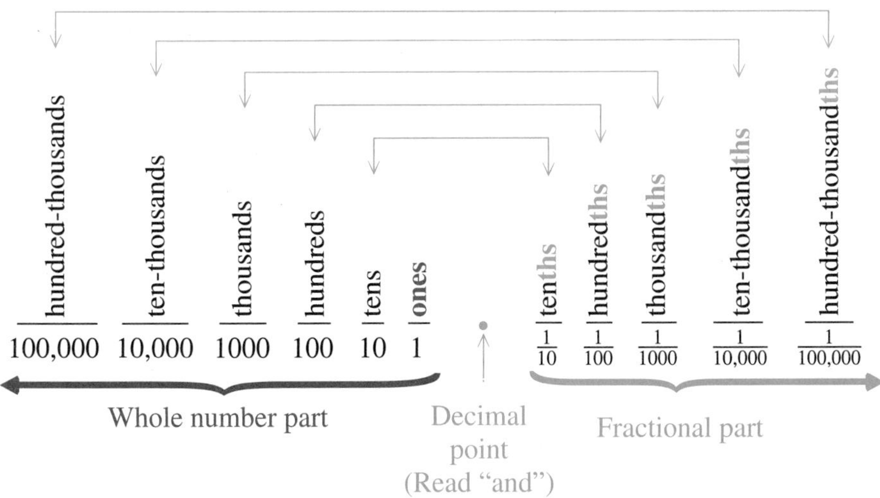

Note

Notice that the **ones** place is at the center of the place value chart. There is no "oneths" place.

Also notice that each place is 10 times the value of the place to its right.

Finally, be sure to write a **hyphen** (dash) in ten–thousandths and hundred–thousandths.

2 Write the portion of each square that is shaded as a fraction, as a decimal, and in words.

GS **(a)**

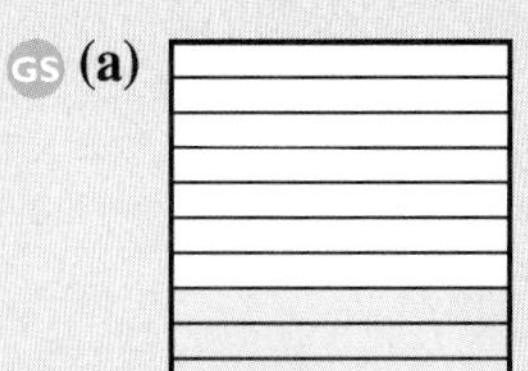

$\frac{3}{10}$ = 0.____ =

_______ tenths

(b)

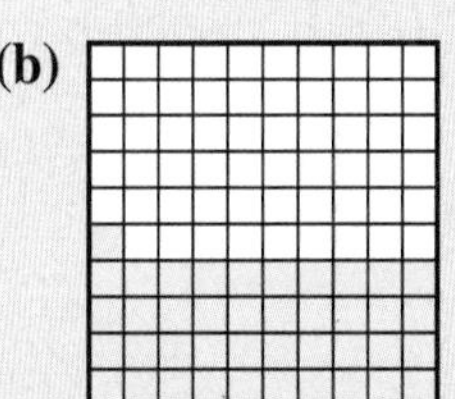

3 Write each decimal as a fraction.

GS **(a)** $0.7 = \frac{7}{\_\_\_}$

(b) 0.2

(c) 0.03

(d) 0.69

(e) 0.047

(f) 0.351

Answers

2. **(a)** 0.3; three tenths

(b) $\frac{41}{100}$; 0.41; forty-one hundredths

3. **(a)** 10; $\frac{7}{10}$ **(b)** $\frac{2}{10}$ **(c)** $\frac{3}{100}$

(d) $\frac{69}{100}$ **(e)** $\frac{47}{1000}$ **(f)** $\frac{351}{1000}$

4 Identify the place value of each digit.

(a) 971.54

(b) 0.4

(c) 5.60

(d) 0.0835

5 Tell how to read each decimal in words.

GS (a) 0.6 six ______

GS (b) 0.46 forty-six ______

(c) 0.05 (d) 0.409

(e) 0.0003 (f) 0.2703

Answers

4. (a) 9 (hundreds) 7 (tens) 1 (ones) . 5 (tenths) 4 (hundredths) (b) 0 (ones) . 4 (tenths)

(c) 5 (ones) . 6 (tenths) 0 (hundredths) (d) 0 (ones) . 0 (tenths) 8 (hundredths) 3 (thousandths) 5 (ten-thousandths)

5. (a) six tenths
(b) forty-six hundredths
(c) five hundredths
(d) four hundred nine thousandths
(e) three ten-thousandths
(f) two thousand seven hundred three ten-thousandths

CAUTION

If a number does *not* have a decimal point, it is a *whole number.* A whole number has no fractional part. If you want to show the decimal point in a whole number, it is just to the ***right*** of the digit in the ones place. Here are two examples.

8 = 8. (Decimal point) **306 = 306.** (Decimal point)

EXAMPLE 2 Identifying the Place Value of a Digit

Identify the place value of each digit.

(a) 178.36 (b) 0.00935

Notice the hyphen (dash) in ten-thousandths and in hundred-thousandths.

hundreds	tens	ones	.	tenths	hundredths
1	7	8	.	3	6

ones	.	tenths	hundredths	thousandths	ten-thousandths	hundred-thousandths
0	.	0	0	9	3	5

Notice in **Example 2(b)** that we do *not* use commas on the right side of the decimal point.

◄ Work Problem 4 at the Side.

OBJECTIVE 3 Read and write decimals in words. A decimal is read according to its form as a fraction.

ones	.	tenths
0	.	9

We read 0.9 as "nine tenths" because 0.9 is the same as $\frac{9}{10}$. Notice that 0.9 ends in the tenths place.

ones	.	tenths	hundredths
0	.	0	2

We read 0.02 as "two hundredths" because 0.02 is the same as $\frac{2}{100}$. Notice that 0.02 ends in the hundredths place.

EXAMPLE 3 Reading Decimal Numbers

Tell how to read each decimal in words.

(a) 0.3

Because $0.3 = \frac{3}{10}$, read the decimal as three tenths.

(b) 0.49 Read it as: forty-nine hundredths.

(c) 0.08 Read it as: eight hundredths.

Think: $0.08 = \frac{8}{100}$ so write *hundredths*.

(d) 0.918 Read it as: nine hundred eighteen thousandths.

(e) 0.0106 Read it as: one hundred six ten-thousandths.

Think: $0.0106 = \frac{106}{10{,}000}$

◄ Work Problem 5 at the Side.

Reading Decimal Numbers

Step 1 Read any whole number part to the ***left*** of the decimal point as you normally would.

Step 2 Read the decimal point as "***and.***"

Step 3 Read the part of the number to the ***right*** of the decimal point as if it were an ordinary whole number.

Step 4 Finish with the place value name of the rightmost digit; these names all end in "***ths.***"

Note

If there is *no whole number part,* you will use only Steps 3 and 4.

EXAMPLE 4 Reading Decimals

Read each decimal.

(a) 9 is in tenths place.

16.9

sixteen **and** nine **tenths**

Remember to say or write "and" *only* when you see a decimal point.

16.9 is read "sixteen **and** nine tenths."

(b) 5 is in hundredths place.

482.35

four hundred eighty-two **and** thirty-five **hundredths**

482.35 is read "four hundred eighty-two and thirty-five hundredths."

3 is in thousandths place.

(c) 0.063 is "sixty-three **thousandths**." (no whole number part)

(d) 11.1085 is "eleven **and** one thousand eighty-five **ten-thousandths**."

CAUTION

Use "and" when reading a decimal point. A common mistake is to read the whole number 405 as "four hundred *and* five." But there is ***no decimal point*** shown in 405, so it is read "four hundred five."

Work Problem 6 at the Side. ▶

OBJECTIVE 4 Write decimals as fractions or mixed numbers. Knowing how to read decimals will help you when writing decimals as fractions.

Writing Decimals as Fractions or Mixed Numbers

Step 1 The digits to the right of the decimal point are the numerator of the fraction.

Step 2 The denominator is 10 for tenths, 100 for hundredths, 1000 for thousandths, 10,000 for ten-thousandths, and so on.

Step 3 If the decimal has a whole number part, it will be written as a mixed number with the same whole number part.

6 Tell how to read each decimal in words.

GS **(a)** 3.8 is read

three and eight ________.

(b) 15.001

(c) 0.0073

(d) 64.309

Answers

6. (a) three and eight tenths
(b) fifteen and one thousandth
(c) seventy-three ten-thousandths
(d) sixty-four and three hundred nine thousandths

7 Write each decimal as a fraction or mixed number.

(a) 0.7

(b) 12.21

(c) 0.101

(d) 0.007

(e) 1.3717

8 Write each decimal as a fraction or mixed number in lowest terms.

GS **(a)** 0.5 Write in lowest terms.

$$\frac{5}{10} = \frac{1}{\_\_\_}$$

GS **(b)** 12.6 Write in lowest terms.

$$12.6 = 12\frac{6}{10} = 12\frac{\_\_\_}{\_\_\_}$$

(c) 0.85

(d) 3.05

(e) 0.225

(f) 420.0802

Answers

7. (a) $\frac{7}{10}$ **(b)** $12\frac{21}{100}$ **(c)** $\frac{101}{1000}$
(d) $\frac{7}{1000}$ **(e)** $1\frac{3717}{10,000}$

8. (a) $\frac{1}{2}$ **(b)** $12\frac{3}{5}$ **(c)** $\frac{17}{20}$ **(d)** $3\frac{1}{20}$
(e) $\frac{9}{40}$ **(f)** $420\frac{401}{5000}$

EXAMPLE 5 Writing Decimals as Fractions or Mixed Numbers

Write each decimal as a fraction or mixed number.

(a) 0.19

The digits to the right of the decimal point, 19, are the numerator of the fraction. The denominator is 100 for hundredths because the rightmost digit is in the hundredths place.

$$0.19 = \frac{19}{100} \leftarrow \text{100 for hundredths}$$

Hundredths place (points to the 9 in 0.19)

(b) 0.863

$$0.863 = \frac{863}{1000} \leftarrow \text{1000 for thousandths}$$

Thousandths place (points to the 3 in 0.863)

(c) 4.0099

The whole number part stays the same.

$$4.0099 = 4\frac{99}{10,000} \leftarrow \text{10,000 for ten-thousandths}$$

Ten-thousandths place (points to the last 9 in 4.0099)

◀ Work Problem **7** at the Side.

CAUTION

After you write a decimal as a fraction or a mixed number, make sure the fraction is in lowest terms.

EXAMPLE 6 Writing Decimals as Fractions or Mixed Numbers in Lowest Terms

Write each decimal as a fraction or mixed number in lowest terms.

(a) $0.4 = \frac{4}{10} \leftarrow$ 10 for tenths

Write $\frac{4}{10}$ in lowest terms.

$$\frac{4}{10} = \frac{4 \div 2}{10 \div 2} = \frac{2}{5} \leftarrow \text{Lowest terms}$$

(b) $0.75 = \frac{75}{100} = \frac{75 \div 25}{100 \div 25} = \frac{3}{4} \leftarrow$ Lowest terms

The whole number part stays the same.

(c) $18.105 = 18\frac{105}{1000} = 18\frac{105 \div 5}{1000 \div 5} = 18\frac{21}{200} \leftarrow$ Lowest terms

(d) $42.8085 = 42\frac{8085}{10,000} = 42\frac{8085 \div 5}{10,000 \div 5} = 42\frac{1617}{2000} \leftarrow$ Lowest terms

◀ Work Problem **8** at the Side.

Calculator Tip

In this book we will write a 0 in the ones place for decimal fractions. We write 0.45 instead of just .45, to emphasize that there is no whole number. Many calculators show these zeros also. Try entering [.] [4] [5]; the display probably shows 0.45 even though you did not press 0.

4.1 Exercises

FOR EXTRA HELP

Download the MyDashBoard App

MyMathLab®

CONCEPT CHECK *Circle the correct answer in Exercises 1–4.*

1. The number 11.084 has how many decimal places?
2 3 5

2. The number 0.7185 has how many decimal places?
4 5 1

3. In 22.85, the 8 means what?
8 tenths
8 hundredths
8 ones

4. In 57.213, the 3 means what?
3 thousandths
3 tenths
3 hundredths

Identify the digit that has the given place value. ***See Example 2.***

5. 70.489
tens
ones
tenths

6. 135.296
ones
tenths
tens

7. 0.83472
thousandths
ten-thousandths
tenths

8. 0.51968
tenths
ten-thousandths
hundredths

9. 149.0832
hundreds
hundredths
ones

10. 3458.712
hundreds
hundredths
tenths

11. 6285.7125
thousands
thousandths
hundredths

12. 5417.6832
thousands
thousandths
ones

Write the decimal number that has the specified place values. ***See Example 2.***

13. Fill in the blanks for this decimal number:
GS 0 ones, 5 hundredths, 1 ten, 4 hundreds, 2 tenths
410. ____ ____

14. Fill in the blanks for this decimal number:
GS 7 tens, 9 tenths, 3 ones, 6 hundredths, 8 hundreds
____ ____ ____.96

15. 3 thousandths, 4 hundredths, 6 ones, 2 ten-thousandths, 5 tenths

16. 8 ten-thousandths, 4 hundredths, 0 ones, 2 tenths, 6 thousandths

17. 4 hundredths, 4 hundreds, 0 tens, 0 tenths, 5 thousandths, 5 thousands, 6 ones

18. 7 tens, 7 tenths, 6 thousands, 6 thousandths, 3 hundreds, 3 hundredths, 2 ones

Write each decimal as a fraction or mixed number in lowest terms.
See Examples 5 and 6.

19. 0.7

20. 0.1

21. 13.4

22. 9.8

23. 0.25

24. 0.55

25. 0.66

26. 0.33

27. 10.17

28. 31.99

29. 0.06

30. 0.08

31. 0.205

32. 0.805

33. 5.002

34. 4.008

35. 0.686

36. 0.492

Tell how to read each decimal in words. ***See Examples 3 and 4.***

37. 0.5

38. 0.2

39. 0.78

40. 0.55

41. 0.105

42. 0.609

43. 12.04

44. 86.09

45. 1.075

46. 4.025

For each exercise, rewrite the words as a decimal number. ***See Examples 3 and 4.***

47. Fill in the blanks to write six and seven tenths as a decimal number.

____ . ____

48. Fill in the blanks to write eight and twelve hundredths as a decimal number.

____ . ____ ____

49. thirty-two hundredths

50. one hundred eleven thousandths

51. four hundred twenty and eight thousandths

52. two hundred and twenty-four thousandths

53. seven hundred three ten-thousandths

54. eight hundred and six hundredths

55. seventy-five and thirty thousandths

56. sixty and fifty hundredths

57. CONCEPT CHECK Anne read the number 4302 as "four thousand three hundred and two." Explain what is wrong with the way Anne read the number.

58. CONCEPT CHECK Jerry read the number 9.0106 as "nine and one hundred and six ten-thousandths." Explain the error he made.

The grandfather on the first page of this chapter needs to select the correct fishing line for his grandson's reel. Fishing line is sold according to how many pounds of "pull" the line can withstand before breaking. Use the table to answer Exercises 59–62. Write all fractions in lowest terms. (Note: The diameter of the fishing line is its thickness.)

RELATING FISHING LINE DIAMETER TO TEST STRENGTH

Test Strength (pounds)	Average Diameter (inches)
4	0.008
8	0.010
12	0.013
14	0.014
17	0.015
20	0.016

Source: Berkley Outdoor Technologies Group.

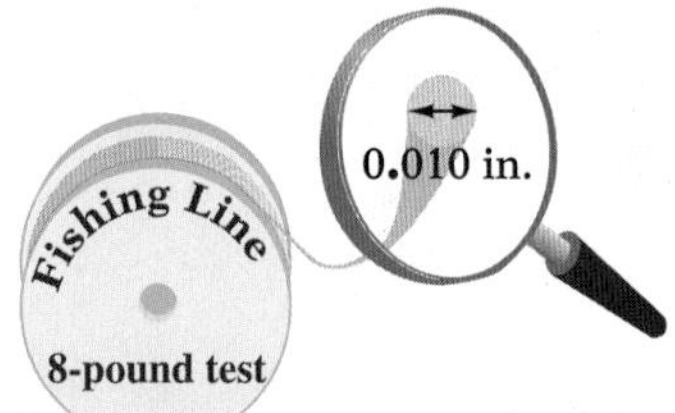

The diameter is the distance across the end of the line (or its thickness).

59. Write the diameter of 8-pound test line in words and as a fraction.

60. Write the diameter of 17-pound test line in words and as a fraction.

61. What is the test strength of the line with a diameter of $\frac{13}{1000}$ inch?

62. What is the test strength of the line with a diameter of sixteen thousandths inch?

CONCEPT CHECK *Suppose your job is to take phone orders for precision parts. Use the table, and in Exercises 63–66, write the correct part number that matches what you hear the customer say over the phone. In Exercises 67–68, write the words you would say to the customer.*

Part Number	Size in Centimeters
3-A	0.06
3-B	0.26
3-C	0.6
3-D	0.86
4-A	1.006
4-B	1.026
4-C	1.06
4-D	1.6
4-E	1.602

63. "Please send the six-tenths centimeter bolt."

Part number ______.

64. "The part missing from our order was the one and six hundredths size."

Part number ______.

65. "The size we need is one and six thousandths centimeters."

Part number ______.

66. "Do you still stock the twenty-six hundredths centimeter bolt?"

Part number ______.

67. "What size is part number 4-E?" Write your answer in words.

68. "What size is part number 4-B?" Write your answer in words.

Relating Concepts (Exercises 69–76) For Individual or Group Work

Use your knowledge of place value to ***work Exercises 69–76 in order.***

69. Look back at the decimal place value chart early in this section. What do you think would be the names of the next four places to the *right* of hundred-thousandths? What information did you use to come up with these names?

70. A common mistake is to think that the first place to the right of the decimal point is "oneths" and the second place is "tenths." Why might someone make that mistake? How would you explain why there is no "oneths" place?

71. Use your answer to **Exercise 69** to write 0.72436955 in words.

72. Use your answer to **Exercise 69** to write 0.000678554 in words.

73. Write 8006.500001 in words.

74. Write 20,060.000505 in words.

75. Write this decimal in numbers.

three hundred two thousand forty ten-millionths

76. Write this decimal in numbers.

nine billion, eight hundred seventy-six million, five hundred forty-three thousand, two hundred ten and one hundred million two hundred thousand three hundred billionths

4.2 Rounding Decimal Numbers

OBJECTIVES

1. Learn the rules for rounding decimals.
2. Round decimals to any given place.
3. Round money amounts to the nearest cent or nearest dollar.

In **Chapter 1,** you learned how to round whole numbers. For example, 89 rounded to the nearest ten is 90, and 8512 rounded to the nearest hundred is 8500.

OBJECTIVE 1 **Learn the rules for rounding decimals.** It is also important to be able to **round** decimals. For example, a store is selling 2 candy mints for $0.75 but you want only one mint. The price of each mint is $0.75 ÷ 2, which is $0.375, but you cannot pay part of a cent. Is $0.375 closer to $0.37 or to $0.38? Actually, it's exactly halfway between. When this happens in everyday situations, the rule is to round *up.* The store will charge you $0.38 for the mint.

Rounding Decimals

Step 1 Find the place to which the rounding is being done. Draw a "cut-off" line ***after*** that place to show that you are cutting off and dropping the rest of the digits.

Step 2 Look ***only*** at the ***first*** digit you are cutting off.

Step 3(a) If this digit is ***4 or less,*** the part of the number you are keeping ***stays the same.***

Step 3(b) If this digit is ***5 or more,*** you must ***round up*** the part of the number you are keeping.

Step 4 You can use the $\approx$ symbol or the $\doteq$ symbol to indicate that the rounded number is now an approximation (close, but not exact). Both symbols mean "is approximately equal to." In this book we will use the $\approx$ symbol.

CAUTION

Do ***not*** move the decimal point when rounding.

OBJECTIVE 2 **Round decimals to any given place.** The following examples show you how to round decimals.

EXAMPLE 1 Rounding a Decimal Number

Round 14.39652 to the nearest thousandth.

Step 1 Draw a "cut-off" line after the thousandths place.

14 . 3 9 6 | 5 2 — You are cutting off the 5 and 2. They will be dropped.

Thousandths (points to 6)

Step 2 Look *only* at the *first* digit you are cutting off. Ignore the other digits you are cutting off.

14 . 3 9 6 | 5 2 — Look *only* at the 5. Ignore the 2.

Continued on Next Page

Step 3 If the first digit you are cutting off is *5 or more*, round up the part of the number you are keeping.

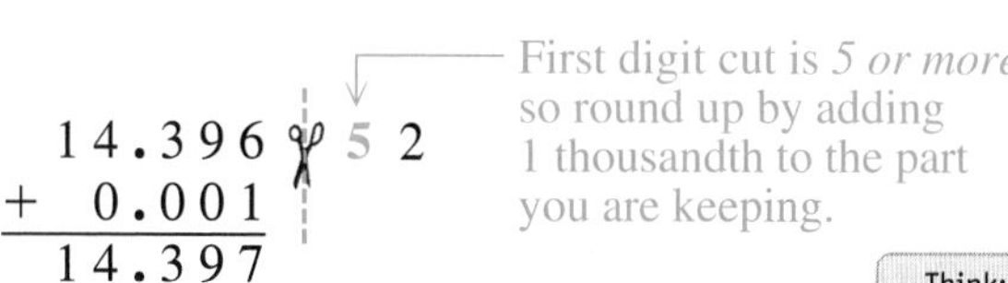
14.396 | 5 2
+ 0.001
14.397
First digit cut is *5 or more*, so round up by adding 1 thousandth to the part you are keeping.

Think: Rounding to *thousandths* means *three* decimal places.

So, 14.39652 rounded to the nearest thousandth is 14.397
We can write 14.39652 ≈ 14.397

CAUTION

When rounding whole numbers in **Chapter 1**, you kept all the digits but changed some to zeros. With decimals, you cut off and *drop the extra digits*. In **Example 1** above, 14.39652 rounds to 14.397, ***not*** 14.39700.

Work Problem 1 at the Side. ▶

In **Example 1**, the rounded number 14.397 had *three decimal places*. **Decimal places** are the number of digits to the *right* of the decimal point. The first decimal place is tenths, the second is hundredths, the third is thousandths, and so on.

EXAMPLE 2 Rounding Decimals to Different Places

Round to the place indicated.

(a) 5.3496 to the nearest tenth

Tenths is one decimal place.

Step 1 Draw a cut-off line after the tenths place.

5 . 3 | 4 9 6 — You are cutting off the 4, 9, and 6. They will be dropped.
Tenths ↑

Step 2 5 . 3 | 4 9 6 — Look *only* at the 4. Ignore these digits (9 6).

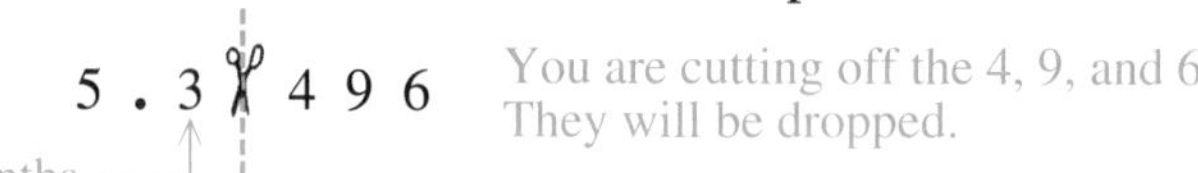

Step 3 5 . 3 | 4 9 6 — First digit cut is *4 or less*, so the part you are keeping stays the same.
5 . 3 ← Stays the same

5.3496 rounded to the nearest tenth is 5.3 (*one* decimal place for *tenths*).
We can write 5.3496 ≈ 5.3
Notice: 5.3496 does ***not*** round to 5.3000, which would be ten-thousandths.

(b) 0.69738 to the nearest hundredth

Step 1 0 . 69 | 7 3 8 — Draw a cut-off line after the hundredths place.
Hundredths ↑

Step 2 0 . 69 | 7 3 8 — Look *only* at the 7.

Continued on Next Page

1 Round to the nearest thousandth.

(a) 0.33492

0.334 | 92 — First digit cut is *5 or more*.
Thousandths ↑

(b) 8.00851

(c) 265.42038

(d) 10.70180

Answers

1. (a) 0.335 (b) 8.009 (c) 265.420 (d) 10.702

2 Round to the place indicated.

GS **(a)** 0.8988 to the nearest hundredth

0 . 8 9 | 88 — First digit cut is *5 or more,* so round up.

0 . 8 9 ← Keep this part.
+0 . 0 1 ← To round up, add 1 hundredth.
0 . _ _

(b) 5.8903 to the nearest hundredth

(c) 11.0299 to the nearest thousandth

(d) 0.545 to the nearest tenth

Answers

2. **(a)** 0.90 **(b)** 5.89 **(c)** 11.030 **(d)** 0.5

Step 3 0 . 6 9 | 7 3 8 — First digit cut is *5 or more*, so round up by adding 1 hundredth to the part you are keeping.

0 . 6 9 ← Keep this part.
+0 . 0 1 ← To round up, add 1 hundredth.
0 . 7 0 ← 9 + 1 is 10; write 0 and regroup 1 to the tenths place.

0.69738 rounded to the nearest hundredth is 0.70. Hundredths is *two* decimal places so you *must* write the 0 in the hundredths place.
We can write 0.69738 ≈ 0.70

Think: Rounding to *hundredths* means *two* decimal places.

CAUTION

If a *rounded* number has a 0 in the rightmost place, you *must* keep the 0. As shown above, 0.69738 rounded to the nearest hundredth is 0.70. Do ***not*** write 0.7, which is rounded to tenths instead of hundredths.

(c) 0.01806 to the nearest thousandth

0 . 0 1 8 | 0 6 — First digit cut is *4 or less*, so the part you are keeping stays the same.

0 . 0 1 8 ← Stays the same

Rounding to *thousandths* means *three* decimal places.

0.01806 rounded to the nearest thousandth is 0.018
We can write 0.01806 ≈ 0.018

(d) 57.976 to the nearest tenth

57.9 | 76 — First digit cut is *5 or more*, so round up by adding 1 tenth to the part you are keeping.

57.9 ← Keep this part.
+ 0.1 ← To round up, add 1 tenth.
58.0 ← 9 + 1 is 10; write the 0 and regroup the 1 to the ones place.

Be sure to write the 0 in the tenths place.

57.976 rounded to the nearest tenth is 58.0. We can write 57.976 ≈ 58.0 You *must* write the 0 in the tenths place to show that the number was rounded to the nearest tenth.

CAUTION

Check that your rounded answer shows ***exactly*** the number of decimal places asked for in the problem. Be sure your answer shows *one* decimal place if you rounded to *tenths, two* decimal places for *hundredths, three* decimal places for *thousandths,* and so on.

◀ **Work Problem 2 at the Side.**

OBJECTIVE 3 Round money amounts to the nearest cent or nearest dollar. In many everyday situations, such as shopping in a store, money amounts are rounded to the nearest cent. There are 100 cents in a dollar.

$$\text{Each cent is } \frac{1}{100} \text{ of a dollar.}$$

Another way to write $\frac{1}{100}$ is 0.01. So rounding to the *nearest cent* is the same as rounding to the *nearest hundredth of a dollar.*

EXAMPLE 3 Rounding to the Nearest Cent

How much will you pay in each shopping situation? Round each money amount to the nearest cent.

(a) $2.4238 (Is it closer to $2.42 or to $2.43?)

$2.42 | 38 ← First digit cut is *4 or less*, so the part you are keeping stays the same.

$2.42 ← You pay. (Rounding to the *nearest cent* is rounding to *hundredths*.)

You pay $2.42 because $2.4238 is closer to $2.42 than to $2.43.

(b) $0.695 (Is it closer to $0.69 or to $0.70?)

$0.69 | 5 ← *5 or more;* round up.

$0.69
+ $0.01 ← To round up, add 1 hundredth (1 cent).
$0.70 ← You pay.

Work Problem 3 at the Side. ▶

Note

Some stores round *all* money amounts up to the next higher cent, even if the next digit is *4 or less*. In **Example 3(a)** above, some stores would round $2.4238 *up* to $2.43, even though it is closer to $2.42.

It is also common to round money amounts to the nearest dollar. For example, you can do that on your federal and state income tax returns to make the calculations easier. Rounding to the nearest dollar is rounding to the ones place, that is, to the nearest whole number.

EXAMPLE 4 Rounding to the Nearest Dollar

Round to the nearest dollar.

(a) $48.69 (Is it closer to $48 or to $49?)

Draw a cut-off line after the ones place. First digit cut is *5 or more*, so round up by adding $1.

$48 | .69

$48
+ 1
$49

$48.69 is closer to $49 than to $48.
So $48.69 rounded to the nearest dollar is $49 (Write $49 *not* $49.00)

CAUTION

$48.69 rounded to the nearest dollar is $49. Write the answer as **$49** to show that the rounding is to the *nearest dollar*. Writing $49.00 would show rounding to the *nearest cent*.

Continued on Next Page

3 Round each money amount to the nearest cent.

(a) $14.595

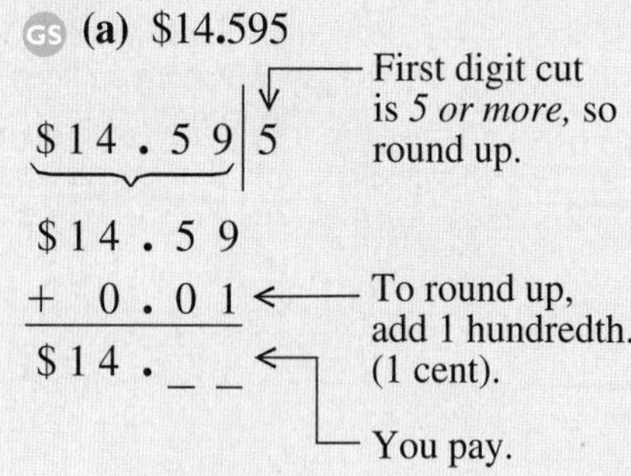

$14.59 | 5
First digit cut is *5 or more*, so round up.
$14.59
+ 0.01
$14.__
To round up, add 1 hundredth. (1 cent).
You pay.

(b) $578.0663

(c) $0.849

(d) $0.0548

Answers

3. (a) $14.60 (b) $578.07 (c) $0.85 (d) $0.05

4 Round to the nearest dollar.

GS (a) $29.10

$29 | .10 — First digit cut is *4 or less,* so the part you keep stays the same.

(b) $136.49

(c) $990.91

(d) $5949.88

(e) $49.60

(f) $0.55

(g) $1.08

Answers

4. (a) $29 (b) $136 (c) $991 (d) $5950 (e) $50 (f) $1 (g) $1

(b) $594.36 (Is it closer to $594 or $595?)

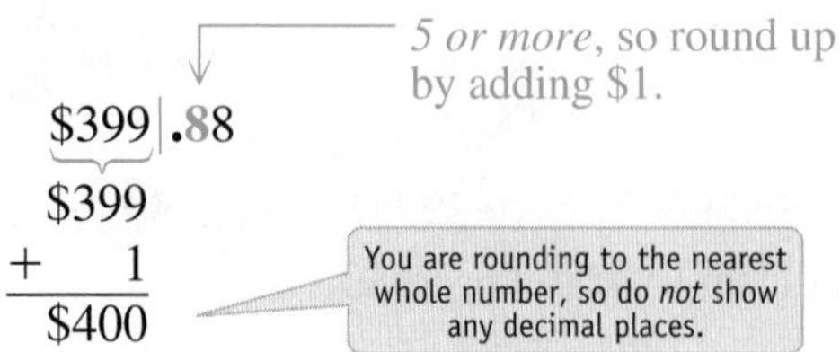

$594.36 rounded to the nearest dollar is $594

(c) $399.88 (Is it closer to $399 or to $400?)

5 or more, so round up by adding $1.
$399 | .88
$399
+ 1
$400
You are rounding to the nearest whole number, so do *not* show any decimal places.

$399.88 rounded to the nearest dollar is $400

(d) $2689.50 (Is it closer to $2689 or $2690?)

5 or more, so round up by adding $1.

$2689 | .50

$2689
+ 1
$2690

Careful! Write $2690 (*not* $2690.00)

$2689.50 rounded to the nearest dollar is $2690

Note

When rounding $2689.50 to the nearest dollar, above, notice that it is exactly halfway between $2689 and $2690. When this happens in every-day situations, the rule is to round *up.* (Scientists working with technical data may use a more complicated rule when rounding numbers that are exactly in the middle.)

(e) $0.61 (Is it closer to $0 or to $1?)

5 or more, so round up.
$0 | .61
Write the rounded amount as $1, *not* $1.00

$0.61 rounded to the nearest dollar is $1

Calculator Tip

Accountants and other people who work with money amounts often set their calculators to automatically round to two decimal places (nearest cent) or to round to zero decimal places (nearest dollar). Your calculator may have this feature.

◀ Work Problem 4 at the Side.

4.2 Exercises

FOR EXTRA HELP

MyMathLab®

1. CONCEPT CHECK Which digit would you look at when deciding how to round 4.8073 to the nearest tenth?
____

2. CONCEPT CHECK Which digit would you look at when deciding how to round 875.639 to the nearest hundredth?
____

3. CONCEPT CHECK Explain how to round 5.70961 to the nearest thousandth.

4. CONCEPT CHECK Explain how to round 10.028 to the nearest tenth.

Round each number to the place indicated. ***See Examples 1 and 2.***

5. 16.8|974 to the nearest tenth (First digit cut is *5 or more.*)

6. 193.84|5 to the nearest hundredth (First digit cut is *5 or more.*)

7. 0.956|47 to the nearest thousandth (First digit cut is *4 or less.*)

8. 96.8158|4 to the nearest ten-thousandth (First digit cut is *4 or less.*)

9. 0.799 to the nearest hundredth

10. 0.952 to the nearest tenth

11. 3.66062 to the nearest thousandth

12. 1.5074 to the nearest hundredth

13. 793.988 to the nearest tenth

14. 476.1196 to the nearest thousandth

15. 0.09804 to the nearest ten-thousandth

16. 176.004 to the nearest tenth

17. 9.0906 to the nearest hundredth

18. 30.1290 to the nearest thousandth

19. 82.000151 to the nearest ten-thousandth

20. 0.400594 to the nearest ten-thousandth

Nardos is grocery shopping. The store will round the amount she pays for each item to the nearest cent. Write the rounded amounts. ***See Example 3.***

21. Soup is three cans for \$2.45, so one can is \$0.81666. Nardos pays _______.

22. Orange juice is two cartons for \$3.89, so one carton is \$1.945. Nardos pays _______.

23. Facial tissue is four boxes for \$4.89, so one box is \$1.2225. Nardos pays _______.

24. Muffin mix is three packages for \$1.99, so one package is \$0.66333. Nardos pays _______.

25. Candy bars are six for \$4.19, so one bar is \$0.6983. Nardos pays _______.

26. Spaghetti is four boxes for \$4.39, so one box is \$1.0975. Nardos pays _______.

As she gets ready to do her income tax return, Ms. Chen rounds each amount to the nearest dollar. Write the rounded amounts. ***See Example 4.***

27. Income from job, \$48,649.60

28. Income from interest on bank account, \$69.58

29. Donations to charity, \$840.08

30. Federal withholding, \$6064.49

Round each money amount as indicated. ***See Examples 3 and 4.***

31. \$499.98 to the nearest dollar

32. \$9899.59 to the nearest dollar

33. \$0.996 to the nearest cent

34. \$0.09929 to the nearest cent

35. \$999.73 to the nearest dollar

36. \$9999.80 to the nearest dollar

The table lists speed records for various types of transportation. Use the table to answer Exercises 37–40.

Record	Speed (miles per hour)
Land speed record (specially built car)	763.04
Motorcycle speed record (conventional motorcycle)	252.662
Fastest roller coaster	134.8
Fastest military jet	2193.167
Boeing 757-300 airplane (regular passenger service)	509
Indianapolis 500 auto race (fastest average winning speed)	185.981
Daytona 500 auto race (fastest average winning speed)	177.602

Source: *Guinness World Records* and *World Almanac and Book of Facts.*

37. Round these speed records to the nearest whole number.

(a) Motorcycle

(b) Roller coaster

38. Round these speed records to the nearest hundredth.

(a) Daytona 500 average winning speed

(b) Indianapolis 500 average winning speed

39. Round these speed records to the nearest tenth.

(a) Indianapolis 500 average winning speed

(b) Land speed record

40. Round these speed records to the nearest hundred.

(a) military jet

(b) Boeing 757-300 airplane

Relating Concepts (Exercises 41–44) For Individual or Group Work

Use your knowledge about rounding money amounts to ***work Exercises 41–44 in order.***

41. Explain what happens when you round \$0.499 to the nearest dollar. Why does this happen?

42. Look again at **Exercise 41**. How else could you round \$0.499 that would be more helpful? What kind of guideline does this suggest about rounding to the nearest dollar?

43. Explain what happens when you round \$0.0015 to the nearest cent. Why does this happen?

44. Suppose you want to know which of these amounts is less, so you round them both to the nearest cent.

\$0.5968 \$0.6014

Explain what happens. Describe what you could do instead of rounding to the nearest cent.

4.3 Adding and Subtracting Decimal Numbers

OBJECTIVES

1. Add decimals.
2. Subtract decimals.
3. Estimate the answer when adding or subtracting decimals.

OBJECTIVE 1 Add decimals. When adding or subtracting *whole* numbers in **Chapter 1,** you lined up the numbers in columns so that you were adding ones to ones, tens to tens, and so on. A similar idea applies to adding or subtracting *decimal* numbers. With decimals, you line up the decimal points to make sure you are adding tenths to tenths, hundredths to hundredths, and so on.

Adding and Subtracting Decimals

Step 1 Write the numbers in columns with the decimal points lined up.

Step 2 If necessary, write in zeros so both numbers have the same number of decimal places. Then add or subtract as if they were whole numbers.

Step 3 Line up the decimal point in the answer directly below the decimal points in the problem.

EXAMPLE 1 Adding Decimal Numbers

Find each sum.

(a) 16.92 and 48.34

Step 1 Write the numbers in columns with the decimal points lined up.

```
tens ones . tenths hundredths
  1 6 . 9 2
+ 4 8 . 3 4
```

Decimal points are lined up.

Step 2 Add as if these were whole numbers.

```
   1 1
   16.92
 + 48.34
 -------
   65.26
```

Step 3 Decimal point in answer is lined up under decimal points in problem.

(b) 5.897 + 4.632 + 12.174

Write the numbers vertically with decimal points lined up. Then add.

```
  11 21
    5.897
    4.632
 + 12.174
 --------
   22.703
```

When you rewrite the numbers in columns, and then add, be careful to line up the decimal points.

Work Problem 1 at the Side. ▶

In **Example 1(a)** above, both numbers had *two decimal places* (two digits to the right of the decimal point). In **Example 1(b),** all the numbers had *three decimal places* (three digits to the right of the decimal point). That made it easy to add tenths to tenths, hundredths to hundredths, and so on.

1 Find each sum.

GS **(a)** 2.86 + 7.09

```
   2.86
 + 7.09
 ------
    .
```

Decimal points are lined up.

(b) 13.761 + 8.325

(c) 0.319 + 56.007 + 8.252

(d) 39.4 + 0.4 + 177.2

Answers

1. (a) 9.95 (b) 22.086 (c) 64.578 (d) 217.0

2 Find each sum.

GS **(a)** 6.54 + 9.8

$$\begin{array}{r} 6.54 \\ +\ 9.80 \\ \hline \_\_.\_\_ \end{array}$$

(b) 0.831 + 222.2 + 10

(c) 8.64 + 39.115 + 3.0076

(d) 5 + 429.823 + 0.76

Answers

2. **(a)** 16.34 **(b)** 233.031 **(c)** 50.7626 **(d)** 435.583

If the number of decimal places does *not* match, you can write in zeros as placeholders to make them match. This is shown in **Example 2**.

EXAMPLE 2 Writing Zeros as Placeholders before Adding

Find each sum.

(a) 7.3 + 0.85

There are two decimal places in 0.85 (tenths and hundredths), so write zero in the hundredths place in 7.3 so that it has two decimal places also.

$$\begin{array}{r} 7.30 \\ +\ 0.85 \\ \hline 8.15 \end{array}$$ ← One 0 is written in.

7.30 is equivalent to 7.3 because $7\frac{30}{100}$ in lowest terms is $7\frac{3}{10}$

(b) 6.42 + 9 + 2.576

Write in zeros so that all the addends have three decimal places. Notice how the whole number 9 is written with the decimal point at the *far right* side. (If you put the decimal point on the *left* side of the 9, you would turn it into the decimal fraction 0.9.)

Write the decimal point in 9 on the *right* side.

$$\begin{array}{r} 6.420 \\ 9.000 \\ +\ 2.576 \\ \hline 17.996 \end{array}$$

6.420 ← One 0 is written in.
9.000 ← 9 is a whole number; decimal point and three zeros are written in.
+2.576 ← No zeros are needed.

Decimal points are lined up.

Note

Writing zeros to the right of a *decimal* number does *not* change the value of the number, as shown in **Example 2(a)** above.

◀ **Work Problem 2 at the Side.**

OBJECTIVE 2 Subtract decimals. Subtraction of decimals is done in much the same way as addition of decimals. You can check the answers to subtraction problems using addition, as you did with whole numbers.

EXAMPLE 3 Subtracting Decimal Numbers

Find each difference. Check your answers using addition.

(a) 15.82 from 28.93

Watch the order of the numbers when you see "from." Subtraction is **not** commutative like addition. So the number you are subtracting "from" goes first.

Step 1

$$\begin{array}{r} 28.93 \\ -\ 15.82 \\ \hline \end{array}$$

Line up decimal points. Then you will be subtracting hundredths from hundredths and tenths from tenths.

Continued on Next Page

Step 2

```
  28.93     Both numbers have two decimal places:
− 15.82     no need to write in zeros.
  13.11  ← Subtract as if they were whole numbers.
    ↑
    Decimal point in answer is lined up.
```

Step 3

Check the answer by adding 13.11 and 15.82. If the subtraction is done correctly, the sum will be 28.93.

(b) 146.35 minus 58.98

Regrouping is needed here.

```
  0 13 15   12 15
  1  4  6 .  3  5
−    5  8 .  9  8
     8  7 .  3  7
            ↑
            Line up decimal points.
```

Check the answer by adding 87.37 and 58.98. If you did the subtraction correctly, the sum will be 146.35. (If it *isn't,* rework the problem.)

Work Problem 3 at the Side. ▶

EXAMPLE 4 Writing Zeros as Placeholders before Subtracting

Find each difference.

(a) 16.5 from 28.362

Use the same steps as in **Example 3** above. Remember to write in zeros so both numbers have three decimal places.

```
                       Line up decimal points.
             28.362
16.500 is  − 16.500 ← Write two zeros.
equivalent   11.862 ← Subtract as usual.
to 16.5
```

Check the answer by adding.

```
  16.500
+ 11.862
  28.362 ← Matches minuend in original problem.
```

(b) 59.7 − 38.914

```
  59.700 ← Write two zeros.
− 38.914
  20.786 ← Subtract as usual.
```

(c) 12 less 5.83

```
             12.00 ← Write a decimal point and two zeros.
12.00 is   −  5.83
equivalent    6.17 ← Subtract as usual.
to 12
```

Work Problem 4 at the Side. ▶

OBJECTIVE 3 Estimate the answer when adding or subtracting decimals.

A common error in working decimal problems by hand is to misplace the decimal point in the answer. Or, when using a calculator, you may accidentally press the wrong key. **Estimating** the answer will help you avoid these mistakes. Start by using *front end rounding* on each number (as you did with whole numbers). Here are several examples. Notice that in the rounded numbers, only the leftmost digit is something other than 0.

3.25	rounds to	3	0.812	rounds to	1
532.6	rounds to	500	26.397	rounds to	30
7094.2	rounds to	7000	351.24	rounds to	400

---

3 Find each difference. Check your answers using addition.

GS **(a)** 22.7 from 72.9

```
  7 2 . 9
− 2 2 . 7
  _ _ . _
```

(b) 6.425 from 11.813

(c) 20.15 − 19.67

4 Find each difference. Check your answers using addition.

GS **(a)** 18.651 from 25.3

```
  25.300
− 18.651
```

(b) 5.816 − 4.98

(c) 40 less 3.66

(d) 1 − 0.325

Answers

3. (a) 50.2; 50.2 + 22.7 = 72.9
(b) 5.388; 5.388 + 6.425 = 11.813
(c) 0.48; 0.48 + 19.67 = 20.15

4. (a) 6.649; 6.649 + 18.651 = 25.3
(b) 0.836; 0.836 + 4.98 = 5.816
(c) 36.34; 36.34 + 3.66 = 40
(d) 0.675; 0.675 + 0.325 = 1

5 First, use front end rounding and estimate each answer. Then add or subtract to find the exact answer.

GS **(a)** 2.83 + 5.009 + 76.1

Estimate:		*Exact:*
3	⟵	2.830
5	⟵	5.009
+ 80	⟵	76.100

(b) 11.365 from 58

Estimate: *Exact:*

(c) 398.81 + 47.658 + 4158.7

Estimate: *Exact:*

(d) Find the difference between 12.837 meters and 46.091 meters.

Estimate: *Exact:*

(e) $19.28 plus $1.53

Estimate: *Exact:*

EXAMPLE 5 Estimating Decimal Answers

Use front end rounding to round each number. Then add or subtract the rounded numbers to get an estimated answer. Finally, find the exact answer.

(a) Find the sum of 194.2 and 6.825.

Estimate:		*Exact:*
200	⟵ Rounds to	194.200
+ 7	⟵ Rounds to	+ 6.825
207		201.025

The estimate goes out to the hundreds place (three places to the *left* of the decimal point), and so does the exact answer. Therefore, the decimal point is probably in the correct place in the exact answer.

(b) $69.42 + $13.78

Estimate:		*Exact:*	
$70	⟵ Rounds to	$69.42	
+ 10	⟵ Rounds to	+ 13.78	
$80		$83.20	⟵ Exact answer is close to estimate, so it is reasonable.

(c) Find the difference between 0.92 ft and 8 ft.

Use subtraction to find the difference between two numbers. The larger number, 8, is written on top.

Estimate:		*Exact:*	
8	⟵ Rounds to	8.00	⟵ Write a decimal point and two zeros.
− 1	⟵ Rounds to	− 0.92	
7		7.08 ft	0.92 has a digit in the ones place, so round to the ones place (nearest whole number).

(d) Subtract 1.8614 from 7.3.

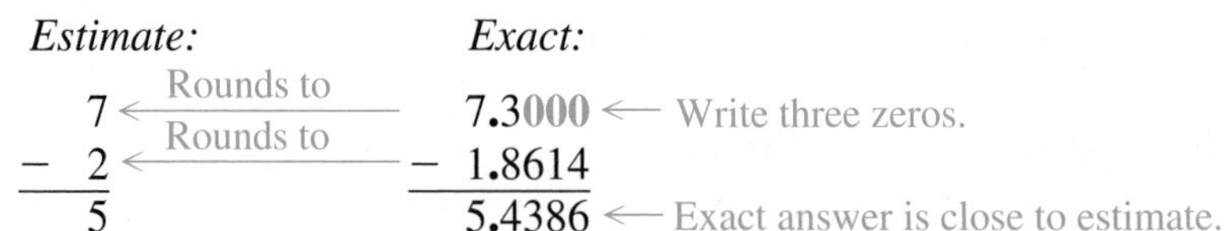

Estimate: *Exact:*
7 ⟵ Rounds to 7.3000 ⟵ Write three zeros.
− 2 ⟵ Rounds to − 1.8614
5 5.4386 ⟵ Exact answer is close to estimate.

◀ **Work Problem 5 at the Side.**

Calculator Tip

If you are *adding* numbers, you can enter them in any order on your calculator. Try these; jot down the answers.

9.82 (+) 1.86 (=) ________ 1.86 (+) 9.82 (=) ________

The answers are the same because addition is *commutative.* (See **Chapter 1.**) But subtraction is *not* commutative. It *does* matter which number you enter first. Try these:

9.82 (−) 1.86 (=) ________ 1.86 (−) 9.82 (=) ________

The second answer has a negative sign (−) next to it. A negative number is *less* than 0. If it was shown on your bank statement, you'd be "in the hole" by $7.96. (See **Chapter 9** for more about negative numbers.)

Answers

5. (a) *Estimate:* 3 + 5 + 80 = 88; *Exact:* 83.939
   (b) *Estimate:* 60 − 10 = 50; *Exact:* 46.635
   (c) *Estimate:* 400 + 50 + 4000 = 4450; *Exact:* 4605.168
   (d) *Estimate:* 50 − 10 = 40; *Exact:* 33.254 meters
   (e) *Estimate:* $20 + $2 = $22; *Exact:* $20.81

4.3 Exercises

FOR EXTRA HELP

Download the MyDashBoard App

MyMathLab®

CONCEPT CHECK *Rewrite each addition or subtraction in columns. Write in any decimal points or zeros, as needed. Do not complete the calculation.*

1. 6.42 + 10.163

2. 7 + 9.204

3. 20 − 9.1263

4. 137.06 − 12

Find each sum or difference. **See Examples 1–4.**

5. 5.69 + 11.79

6. 0.7759 + 9.8883

7. 8.263 − 0.5

8. 47.658 − 20.9

9. 76.5 + 0.506

10. 1.87 + 9.749

11. 21 − 0.896

12. 9 − 1.183

13. Subtract 0.291 from 0.4

14. Subtract 0.088 from 0.35

15. 39.76005 + 182 + 4.799 + 98.31 + 5.9999

16. 489.76 + 0.9993 + 38 + 8.55087 + 80.697

This drawing of a human skeleton shows the average length of the longest bones, in inches. Use the drawing to answer Exercises 17–20. (Source: The Human Body.)

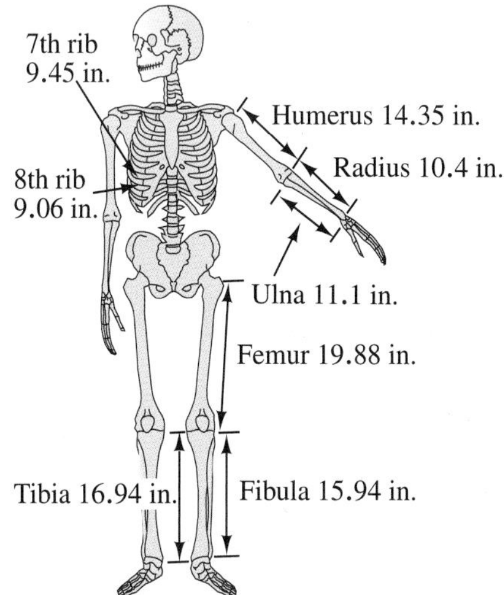

17. (a) What is the combined length of the humerus and radius bones?

(b) What is the difference in the lengths of these two bones?

18. (a) What is the total length of the femur and tibia bones?

(b) How much longer is the femur than the tibia?

19. (a) Find the sum of the lengths of the humerus, ulna, femur, and tibia.

(b) How much shorter is the 8th rib than the 7th rib?

20. (a) What is the difference in the lengths of the two bones in the lower arm?

(b) What is the difference in the lengths of the two bones in the lower leg?

21. CONCEPT CHECK Explain and correct the error that a student made when he added 0.72 + 6 + 39.5 this way:

```
   0.72
   6
+ 39.50
  40.28
```

22. CONCEPT CHECK Explain the difference between saying "subtract 2.9 from 8" and saying "2.9 minus 8."

Use front end rounding to round each number. Then add or subtract the rounded numbers to get an estimated answer. Finally, find the exact answer. ***See Example 5.***

23. The *Estimate* is already done. Now find the *Exact* answer.

Estimate:	*Exact:*
\$20 ←	\$19.74
− 7 ←	− 6.58
\$13	

24. The *Estimate* is already done. Now find the *Exact* answer.

Estimate:	*Exact:*
\$30 ←	\$27.96
− 8 ←	− 8.39
\$22	

25. *Estimate:* *Exact:*

```
  392.7
    0.865
+  21.08
```

26. *Estimate:* *Exact:*

```
  38.55
   7.716
+  0.6
```

27. What is 8.6 less 3.751?

Estimate: *Exact:*

28. What is 31.7 less 4.271?

Estimate: *Exact:*

29. *Estimate:* *Exact:*

```
   62.8173
  539.99
+   5.629
```

30. *Estimate:* *Exact:*

```
  332.607
   12.5
+ 823.3949
```

Use your estimation skills to pick the most reasonable answer for each example. Do **not** *solve the problems. Circle your choice.*

31. 12 − 11.725

2.75 0.275 27.5

32. 20 − 1.37

0.1863 1.863 18. 63

33. 6.5 + 0.007

6.507 0.6507 65.07

34. 9.67 + 0.09

0.976 9.76 0.00976

35. 456.71 − 454.9

18.1 181 1.81

36. 803.25 − 0.6

802.65 0.80265 8.0265

37. 6004.003 + 52.7172

60.567202 605.67202 6056.7202

38. 128.35 + 97.0093

2253.593 225.3593 0.2253593

First use front end rounding to round each number and estimate the answer. Then find the exact answer. Use the information in the table below for Exercises 39–42.

INTERNET USERS IN SELECTED COUNTRIES

Country	Number of Users
China	420 million
United States	239.9 million
Japan	99.14 million
India	81 million
Nigeria	43.98 million
Mexico	30.6 million
Canada	26.2 million
World total	1966.5 million

Source**:** www.internetworldstats.com

39. How many fewer Internet users are there in Mexico than in India?

Estimate:

Exact:

40. How many more users are there in China than in Nigeria?

Estimate:

Exact:

41. How many Internet users are there in all the countries listed in the table?

Estimate:

Exact:

42. Using the exact answer from **Exercise 41,** calculate the number of worldwide Internet users in countries other than the ones in the table.

Estimate:

Exact:

43. The tallest known land mammal, a prehistoric ancestor of the rhino, was 6.4 meters tall. Compare the rhino's height to the combined heights of these three NBA basketball players: Dirk Nowitzki at 2.13 meters, Kobe Bryant at 1.98 meters, and Kevin Love at 2.08 meters. Is their combined height greater or less than the rhino's height? By how much?
(*Source:* www.NBA.com/players)

Estimate:

Exact:

44. At a bakery, Sue Chee bought $7.42 worth of muffins and $10.09 worth of croissants for a staff party and a $0.69 cookie for herself. How much change did she receive from two $10 bills?

Estimate:

Exact:

45. Namiko is comparing two boxes of chicken nuggets. One box weighs 9.85 ounces and the other weighs 10.5 ounces. What is the difference in the weight of the two boxes?

Estimate:

Exact:

46. Sammy works in a veterinarian's office. He weighed two young kittens. One was 3.9 ounces and the other was 4.05 ounces. What was the difference in the weight of the two kittens?

Estimate:

Exact:

Find the perimeter of (distance around) each figure by adding the lengths of the sides.

47.

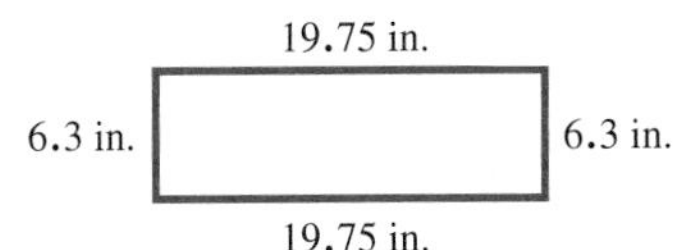

Estimate:

Exact:

48.

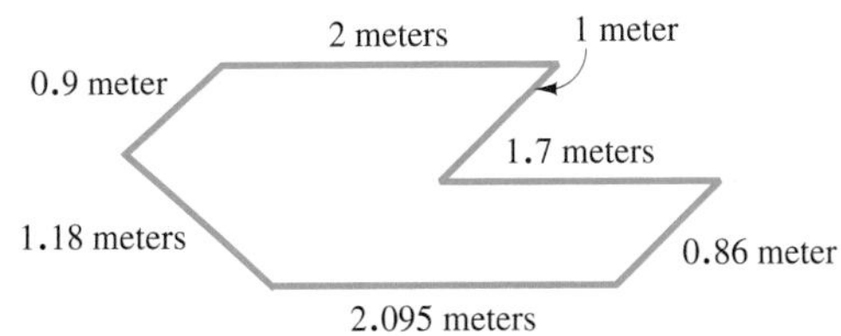

Estimate:

Exact:

The grandfather and grandson on the first page of this chapter are buying fishing equipment. They brought along the store's sale insert from the Sunday paper. Use the information below on sale prices to answer Exercises 49–52. When estimating, round to the nearest whole number.

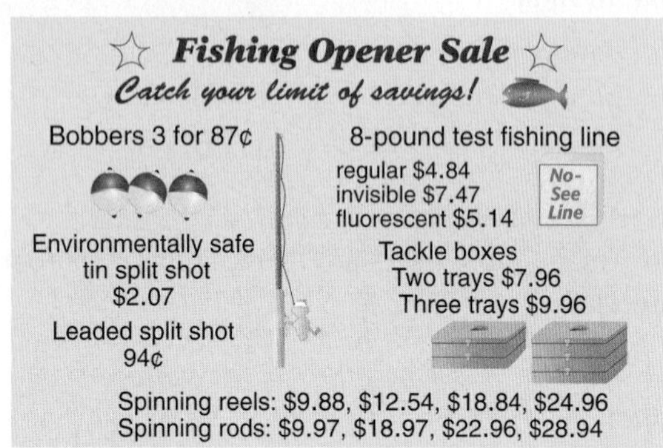

Fishing Opener Sale
Catch your limit of savings!
Bobbers 3 for 87¢
8-pound test fishing line
regular $4.84
invisible $7.47
fluorescent $5.14
No-See Line
Environmentally safe tin split shot $2.07
Leaded split shot 94¢
Tackle boxes
Two trays $7.96
Three trays $9.96
Spinning reels: $9.88, $12.54, $18.84, $24.96
Spinning rods: $9.97, $18.97, $22.96, $28.94

Source: Walmart.

49. What is the difference in price between the fluorescent and regular fishing line?

Estimate:

Exact:

50. How much more does the least expensive spinning rod cost than the least expensive spinning reel?

Estimate:

Exact:

51. Find the total cost of the second highest priced spinning reel, two packages of tin split shot, and a three-tray tackle box. Sales tax for all the items was $2.31.

Estimate:

Exact:

52. The grandfather bought three bobbers on sale. He also bought some SPF15 sunscreen for $7.53 and a flotation vest for $44.96. Sales tax was $3.74. How much did he spend in all?

Estimate:

Exact:

Olivia Sanchez kept track of her expenses for one month. Use her list to answer Exercises 53–58.

MONTHLY EXPENSES

Rent	$994
Car payment	$290.78
Car repairs, gas	$205
Cable TV	$49.95
Internet access	$29.95
Electricity	$40.80
Cell phone	$57.32
Groceries	$186.81
Entertainment	$97.75
Clothing, laundry	$107

53. What were Olivia's total expenses for the month?

54. How much did Olivia pay for cell phone, cable TV, and Internet access?

55. What was the difference in the amounts spent for groceries and for the car payment?

56. Compare the amount Olivia spent on entertainment to the amount spent on car repairs and gas. What is the difference?

57. How much more did Olivia spend on rent than on all her car expenses?

58. How much less did Olivia spend on clothing and laundry than on all her car expenses?

Find the length of the dashed line in each rectangle or circle.

59.

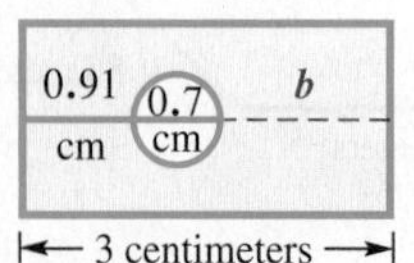

60.

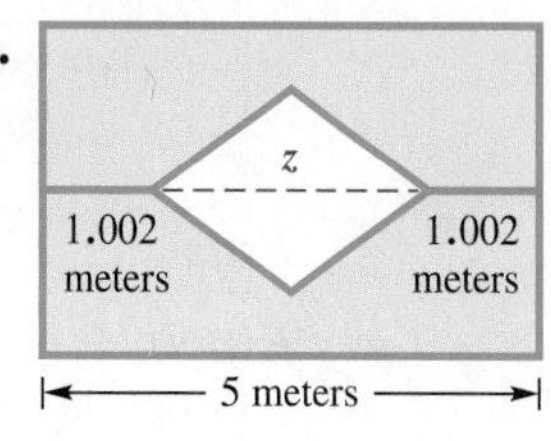

61.

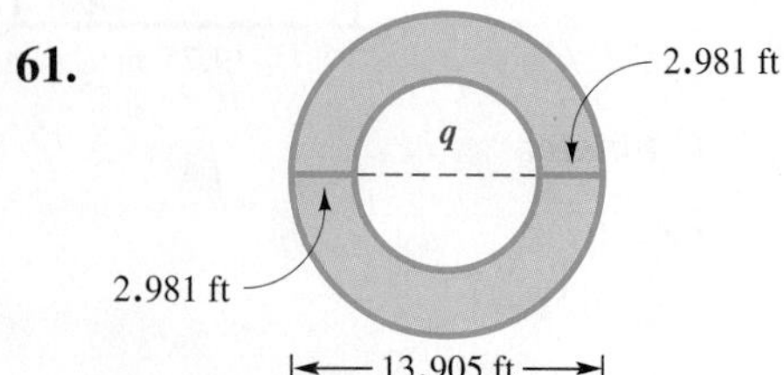

4.4 Multiplying Decimal Numbers

OBJECTIVES

1. Multiply decimals.
2. Estimate the answer when multiplying decimals.

OBJECTIVE 1 **Multiply decimals.** The decimals 0.3 and 0.07 can be multiplied by writing them as fractions.

$$0.3 \times 0.07 = \frac{3}{10} \times \frac{7}{100} = \frac{3 \times 7}{10 \times 100} = \frac{21}{1000} = 0.021$$

1 decimal place + 2 decimal places → 3 decimal places

Can you see a way to multiply decimals without writing them as fractions? Use these steps. Remember that each number in a multiplication problem is called a *factor*, and the answer is called the *product*.

Multiplying Decimal Numbers

Step 1 Multiply the numbers (the factors) as if they were whole numbers.

Step 2 Find the *total* number of decimal places in *both* factors.

Step 3 Write the decimal point in the product (the answer) so it has the same number of decimal places as the total from Step 2. You may need to write in extra zeros on the left side of the product to get the correct number of decimal places.

Note

When multiplying decimals, you do ***not*** need to line up decimal points. (You ***do*** need to line up decimal points when adding or subtracting.)

EXAMPLE 1 **Multiplying Decimal Numbers**

Find the product of 8.34 and 4.2

Step 1 Multiply the numbers as if they were whole numbers.

```
   8.34
 ×  4.2
  1668
 3336
 35028
```

You do *not* have to line up decimal points when multiplying.

Step 2 Count the total number of decimal places in both factors.

```
   8.34   ← 2 decimal places
 ×  4.2   ← 1 decimal place
  1668      3 total decimal places
 3336
 35028
```

Step 3 Count over 3 places in the product and write the decimal point.

```
   8.34   ← 2 decimal places
 ×  4.2   ← 1 decimal place
  1668      3 total decimal places
 3336
 35.028   ← 3 decimal places in product
```

Count over 3 places; count from *right* to *left*.

Work Problem 1 at the Side. ▶

1 Find each product.

(a)
```
   2.6   ← 1 decimal place
 × 0.4   ← 1 decimal place
         ← 2 decimal places in the product
```

(b)
```
   45.2
 × 0.25
```

(c)
```
   0.104 ← 3 decimal places
 ×     7 ← 0 decimal places
         ← 3 decimal places in the product
```

(d)
```
   3.18
 × 2.23
```

(e)
```
   611
 × 3.7
```

Answers

1. (a) 1.04 (b) 11.300 (c) 0.728 (d) 7.0914 (e) 2260.7

2 Find each product.

(a) 0.04×0.09

(b) $(0.2)(0.008)$

(c) $(0.003)^2$ *Hint:* Recall that the 2 is an exponent, so multiply $(0.003)(0.003)$.

(d) $(0.0081)(0.003)$

(e) $(0.11)(0.0005)$

3 First use front end rounding and estimate the answer. Then find the exact answer.

(a) $(11.62)(4.01)$

(b) $(5.986)(33)$

(c) $8.31(4.2)$

(d) 58.6×17.4

Answers

2. (a) 0.0036 (b) 0.0016 (c) 0.000009 (d) 0.0000243 (e) 0.000055

3. (a) $(10)(4) = 40$; 46.5962 (b) $(6)(30) = 180$; 197.538 (c) $8(4) = 32$; 34.902 (d) $60 \times 20 = 1200$; 1019.64

EXAMPLE 2 Writing Zeros as Placeholders in the Product

Find the product: $(0.042)(0.03)$.

Start by multiplying, then count decimal places.

$$\begin{array}{r l} 0.0\,4\,2 & \leftarrow \text{3 decimal places} \\ \times \quad 0.0\,3 & \leftarrow \text{2 decimal places} \\ \hline 1\,2\,6 & \leftarrow \text{5 decimal places needed in product} \end{array}$$

After multiplying, the answer has only three decimal places, but five are needed, so write two zeros on the *left* side of the answer.

$$\begin{array}{r} 0.0\,4\,2 \\ \times \quad 0.0\,3 \\ \hline 0\,0\,1\,2\,6 \end{array}$$

Write two zeros on *left* side of answer.

$$\begin{array}{r l} 0.0\,4\,2 & \leftarrow \text{3 decimal places} \\ \times \quad 0.0\,3 & \leftarrow \text{2 decimal places} \\ \hline .0\,0\,1\,2\,6 & \leftarrow \text{5 decimal places} \end{array}$$

Now count over 5 places and write in the decimal point.

The final product is 0.00126, which has five decimal places.

◀ Work Problem **2** at the Side.

OBJECTIVE 2 Estimate the answer when multiplying decimals. If you are doing multiplication problems by hand, estimating the answer helps you check that the decimal point is in the right place. When you are using a calculator, estimating helps you catch an error like pressing the ÷ key instead of the × key.

EXAMPLE 3 Estimating before Multiplying

First estimate the answer to $(76.34)(12.5)$ using front end rounding. Then find the exact answer.

Estimate:

$$\begin{array}{r l} 80 & \leftarrow \text{Rounds to} \\ \times\, 10 & \leftarrow \text{Rounds to} \\ \hline 800 & \end{array}$$

Exact:

$$\begin{array}{r l} 7\,6.3\,4 & \leftarrow \text{2 decimal places} \\ \times \quad 1\,2.5 & \leftarrow \text{1 decimal place} \\ \hline 3\,8\,1\,7\,0 & \text{3 decimal places needed in product} \\ 1\,5\,2\,6\,8 & \\ 7\,6\,3\,4 & \\ \hline 9\,5\,4.2\,5\,0 & \end{array}$$

Both the estimate and the exact answer go out to the hundreds place, so the decimal point in 954.250 is probably in the correct place.

◀ Work Problem **3** at the Side.

Calculator Tip

When working with money amounts, you may need to write a 0 in your answer. For example, try multiplying $3.54 × 5 on your calculator. Write down the result.

3.54 × 5 = ________

Notice that the result is 17.7, which is *not* the way to write a money amount. You have to write the 0 in the hundredths place: $17.70 is correct. The calculator does not show the "extra" 0 because:

$$17.70 \text{ or } 17\frac{70}{100} \quad \text{simplifies to} \quad 17\frac{7}{10} \text{ or } 17.7$$

So keep an eye on your calculator—it doesn't know when you're working with money amounts.

4.4 Exercises

FOR EXTRA HELP

Download the MyDashBoard App

MyMathLab®

CONCEPT CHECK *Fill in the blanks for Exercises 1–4.*

1. In 4.2×3.46, how many decimal places will the product have? ______

2. In 8.071×2.79, how many decimal places will the product have? ______

3. In $(0.12)(0.03)$, how many zeros do you have to write in the product as placeholders? ______

4. In $(0.0006)(0.07)$, how many zeros do you have to write in the product as placeholders? ______

Find each product. ***See Example 1.***

5. $\begin{array}{r} 0.042 \\ \times\ 3.2 \\ \hline \end{array}$

6. $\begin{array}{r} 0.571 \\ \times\ 2.9 \\ \hline \end{array}$

7. $\begin{array}{r} 21.5 \\ \times\ 7.4 \\ \hline \end{array}$

8. $\begin{array}{r} 85.4 \\ \times\ 3.5 \\ \hline \end{array}$

9. $(0.666)(23.4)$

10. $(0.799)(0.896)$

11. $\begin{array}{r} \$51.88 \\ \times\ 665 \\ \hline \end{array}$

12. $\begin{array}{r} \$736.75 \\ \times\ 118 \\ \hline \end{array}$

CONCEPT CHECK *Use the fact that $72 \times 6 = 432$ to solve Exercises 13–18 by simply counting decimal places and writing the decimal point in the correct location.*

13. $72 \times 0.6 =$ 4 3 2

14. $7.2 \times 6 =$ 4 3 2

15. $(7.2)(0.06) =$ 4 3 2

16. $(0.72)(0.6) =$ 4 3 2

17. $0.72(0.06) =$ 4 3 2

18. $72(0.0006) =$ 4 3 2

Find each product. ***See Example 2.***

19. $(0.006)(0.0052)$

20. $(0.0052)(0.009)$

21. $(0.005)^2$

22. $(0.03)^2$

First use front end rounding to round each number and estimate the answer. Then find the exact answer. ***See Example 3.***

23. *Estimate:* $\begin{array}{r} 40 \\ \times\ 5 \\ \hline 200 \end{array}$ ← Rounds to ← Rounds to *Exact:* $\begin{array}{r} 39.6 \\ \times\ 4.8 \\ \hline \end{array}$

24. *Estimate:* $\begin{array}{r} 20 \\ \times\ 2 \\ \hline 40 \end{array}$ ← Rounds to ← Rounds to *Exact:* $\begin{array}{r} 18.7 \\ \times\ 2.3 \\ \hline \end{array}$

25. *Estimate:* $\begin{array}{r} \\ \times\ \\ \hline \end{array}$ *Exact:* $\begin{array}{r} 37.1 \\ \times\ 42 \\ \hline \end{array}$

26. *Estimate:* $\begin{array}{r} \\ \times\ \\ \hline \end{array}$ *Exact:* $\begin{array}{r} 5.08 \\ \times\ 71 \\ \hline \end{array}$

27. *Estimate:* $\begin{array}{r} \\ \times\ \\ \hline \end{array}$ *Exact:* $\begin{array}{r} 6.53 \\ \times\ 4.6 \\ \hline \end{array}$

28. *Estimate:* $\begin{array}{r} \\ \times\ \\ \hline \end{array}$ *Exact:* $\begin{array}{r} 7.51 \\ \times\ 8.2 \\ \hline \end{array}$

29. *Estimate:* (____)(____) = ____ *Exact:* $(2.809)(6.85) =$

30. *Estimate:* (____)(____) = ____ *Exact:* $(73.52)(22.34) =$

Even with most of the problem missing, you can tell whether or not these answers are reasonable. Circle reasonable *or* unreasonable. *If the answer is unreasonable, move the decimal point, or insert a decimal point, to make the answer reasonable.*

31. How much was his car payment? $28.90
reasonable
unreasonable, should be ____________

32. How many hours did she work today? 25 hours
reasonable
unreasonable, should be ____________

33. How tall is her son? 60.5 in.
reasonable
unreasonable, should be ____________

34. How much does he pay for rent now? $6.92
reasonable
unreasonable, should be ____________

35. What is the price of one gallon of milk? $419
reasonable
unreasonable, should be ____________

36. How long is the living room? 16.8 feet
reasonable
unreasonable, should be ____________

37. How much did Mrs. Brown's baby weigh? 0.095 pound
reasonable
unreasonable, should be ____________

38. What was the sale price of the jacket? $1.49
reasonable
unreasonable, should be ____________

Solve each application problem. Round money answers to the nearest cent when necessary.

39. LaTasha worked 50.5 hours over the last two weeks. She earns $18.73 per hour. How much did she make?

40. Michael's time card shows 42.2 hours at $10.03 per hour. What are his earnings?

41. Sid needs 0.6 meter of canvas material to make a carry-all bag that fits on his wheelchair. If canvas is $4.09 per meter, how much will Sid spend? (*Note:* $4.09 *per* meter means $4.09 for *one* meter.)

42. How much will Mrs. Nguyen pay for 3.5 yards of lace trim that costs $0.87 per yard?

43. Michelle filled the tank of her pickup truck with regular unleaded gas. Use the information shown on the pump to find how much she paid for gas.

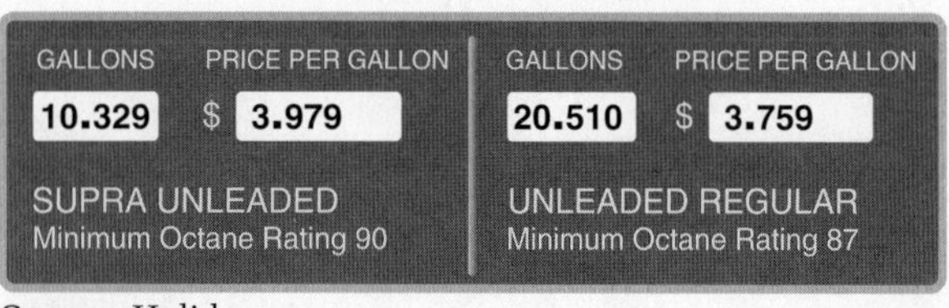

Source: Holiday.

44. Ground beef and chicken legs are on sale. Juma bought 1.7 pounds of legs. Use the information in the ad to find the amount she paid.

45. Ms. Rolack is a real estate broker who helps people sell their homes. Her fee is 0.07 times the price of the home. What was her fee for selling a $289,500 home?

46. Josh Hamilton of the Texas Rangers baseball team had a batting average of 0.359 in the 2010 season. He went to bat 518 times. How many hits did he make? (*Hint:* Multiply the number of times at bat by his batting average.) Round to the nearest whole number. (*Source: World Almanac and Book of Facts.*)

Paper money in the United States has not always been the same size. Shown below are the measurements of bills printed before 1929 and the measurements from 1929 on. Use this information to answer Exercises 47–50. (Source: www.moneyfactory.com*)*

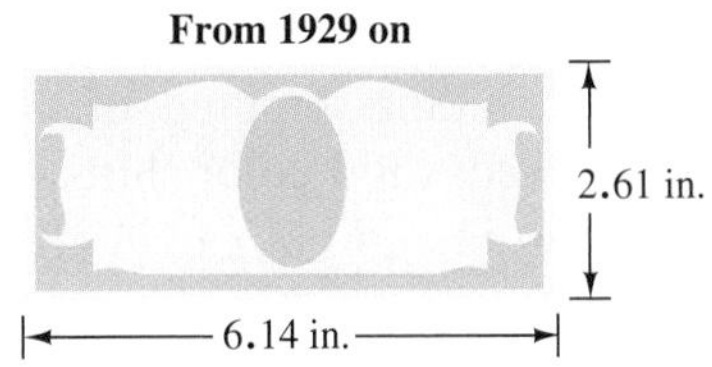

47. (a) Find the area of each bill, rounded to the nearest tenth. (*Hint:* Multiply to find area.)

(b) What is the difference in the rounded areas?

48. (a) Find the perimeter of each bill, to the nearest hundredth. (*Hint:* Add to find perimeter.)

(b) How much less is the perimeter of today's bills than the bills printed before 1929?

49. The thickness of one piece of today's paper money is 0.0043 inch.

(a) If you had a pile of 100 bills, how high would the pile be?

(b) How high would a pile of 1000 bills be?

50. (a) Use your answers from **Exercise 49** to find the number of bills in a pile that is 43 inches high.

(b) How much money would you have if the pile is all $20 bills?

51. Judy Lewis pays $48.96 per month for basic cable TV. The one-time installation fee was $89. How much will she pay for cable over two years? How much would she pay in two years for the deluxe cable package that costs $109.78 per month?

52. Chuck's car payment is $420.27 per month for four years. He also made a down payment of $5000 at the time he bought the car. How much will he pay altogether?

53. Barry bought 16.5 meters of rope at $0.47 per meter and three meters of wire at $1.05 per meter. How much change did he get from three $5 bills?

54. Susan bought a 46-inch LED Smart HDTV that cost $1249.97. She paid $66.59 per month for 24 months. How much could she have saved by paying for the HDTV when she bought it?

Use the information from the Look Smart online catalog to answer Exercises 55–56. Disregard sales tax.

Knit Shirt Ordering Information		
43-2A	Short-sleeved, solid colors	$14.75 each
43-2B	Short-sleeved, stripes	$16.75 each
43-3A	Long-sleeved, solid colors	$18.95 each
43-3B	Long-sleeved, stripes	$21.95 each
XXL size, add $2 per shirt.		
Monogram, $4.95 each. Gift box, $5 each.		

Total Price of All Items (excluding monograms and gift boxes)	Shipping, Packing, and Handling
$0–25.00	$5.50
$25.01–75.00	$7.95
$75.01–125.00	$9.95
$125.01 +	$11.95
Shipping to each additional address, add $4.25.	

55. **(a)** What is the total cost, including shipping, of sending three short-sleeved, solid-color shirts, size M, with monograms, in a gift box to your aunt for her birthday?

(b) How much did the monograms, gift box, and shipping add to the cost of your gift?

56. **(a)** Suppose you order one of each type of shirt for yourself, adding a monogram to each of the solid-color shirts. At the same time, you order three long-sleeved striped shirts, in the XXL size, shipped to your dad in a gift box. Find the total cost of your order.

(b) What is the difference in total cost (excluding shipping) between the shirts for yourself and the gift for your dad?

Relating Concepts (Exercises 57–58) For Individual or Group Work

Look for patterns in the multiplications as you ***work Exercises 57 and 58 in order.***

57. Do these multiplications:

$(5.96)(10) =$ ______ $(3.2)(10) =$ ______

$(0.476)(10) =$ ______ $(80.35)(10) =$ ______

$(722.6)(10) =$ ______ $(0.9)(10) =$ ______

What pattern do you see? Write a "rule" for multiplying by 10. What do you think the rule is for multiplying by 100? by 1000? Write the rules and try them out on the numbers above.

58. Do these multiplications:

$(59.6)(0.1) =$ ______ $(3.2)(0.1) =$ ______

$(0.476)(0.1) =$ ______ $(80.35)(0.1) =$ ______

$(65)(0.1) =$ ______ $(523)(0.1) =$ ______

What pattern do you see? Write a "rule" for multiplying by 0.1. What do you think the rule is for multiplying by 0.01? by 0.001? Write the rules and try them out on the numbers above.

Summary Exercises *Adding, Subtracting, and Multiplying Decimal Numbers*

CONCEPT CHECK *Write each decimal as a fraction or mixed number in lowest terms.*

1. 0.8

2. 6.004

3. 0.35

Write each decimal in words.

4. 94.5

5. 2.0003

6. 0.706

Write each decimal in numbers.

7. five hundredths

8. three hundred nine ten-thousandths

9. ten and seven tenths

Round to the place indicated.

10. 6.1873 to the nearest hundredth

11. 0.95 to the nearest tenth

12. 0.42025 to the nearest thousandth

13. $0.893 to the nearest cent

14. $3.0017 to the nearest cent

15. $99.64 to the nearest dollar

Simplify.

16. 0.27(3.5)

17. 50 − 0.3801

18. $\begin{array}{r} 0.205 \\ \times \quad 9 \\ \hline \end{array}$

19. 25($3.74)

20. (0.004)(1.22)

21. 3.7 − 1.55

22. 0.95 + 10.005

23. 3.6 + 0.718 + 9 + 5.0829

24. 32.305 − 28 − 0.0007

25. 8.9 − 0.4 − 0.03 − 7.6

26. $18 + $2.09 + $90 + $0.75

27. (0.99)(83.672) Round your answer to the nearest hundredth.

28. Find the perimeter and area of this rectangular computer chip. Round your answers to the nearest tenth.

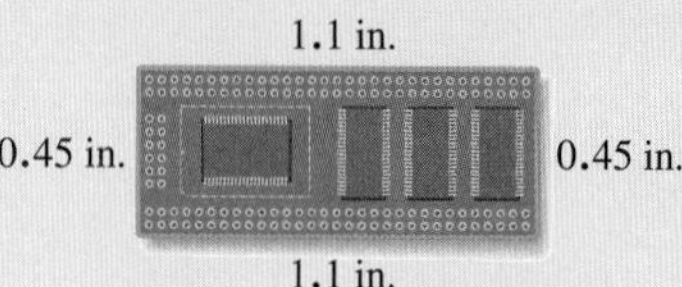

29. Find the perimeter and area of this rectangular watch face. Round your answers to the nearest hundredth.

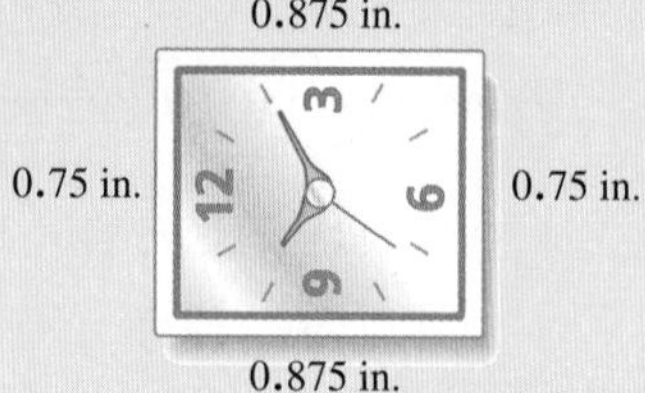

Use the table on heaviest fruits and vegetables to answer Exercises 30–34.

HEAVIEST FRUITS AND VEGETABLES

Fruit/Vegetable	Weight (in pounds)	Location
Apple	4.0625	Japan
Grapefruit	6.75	Australia
Onion	16.52	England
Peach	1.6	Michigan, USA
Radish	68.563	Japan
Pumpkin	1810	Wisconsin, USA
Strawberry	0.5	England
Tomato	7.75	Oklahoma, USA

Source: *Guinness World Records*, www.funonthenet.inc, Associated Press.

30. **(a)** Identify the heaviest and lightest weight items in the table.

(b) Find the difference in weight between the heaviest and lightest items.

31. If you used the radish, tomato, and onion in a salad, what would be their combined weight to the nearest tenth of a pound?

32. **(a)** How much would six grapefruits weigh?

(b) Find the total weight of a dozen strawberries.

33. **(a)** How much heavier is the tomato than the apple?

(b) How much less does the peach weigh than the apple?

34. **(a)** Round the weight of each fruit or vegetable to the nearest whole number.

(b) Use the rounded numbers to find an estimated total weight if you had one of each item in the table. Then find the exact total weight.

35. A craft cooperative paid \$60.32 for enough fabric to make eight baby blankets for their store. They sold the blankets for \$18.95 each. How much profit was made on the blankets?

4.5 Dividing Decimal Numbers

There are two kinds of decimal division problems: those in which a decimal is divided by a whole number, and those in which a number is divided by a decimal. First recall the parts of a division problem from **Chapter 1.**

```
                 8 ← Quotient
Divisor →     4)33 ← Dividend
                32
                 1 ← Remainder
```

OBJECTIVES

1. Divide a decimal by a whole number.
2. Divide a number by a decimal.
3. Estimate the answer when dividing decimals.
4. Use the order of operations with decimals.

OBJECTIVE ▶ 1 Divide a decimal by a whole number. When the divisor is a whole number, use these steps.

Dividing Decimals by Whole Numbers

Step 1 Write the decimal point in the quotient (answer) directly above the decimal point in the dividend.

Step 2 Divide as if both numbers were whole numbers.

EXAMPLE 1 Dividing Decimals by Whole Numbers

Find each quotient. Check the quotients by multiplying.

(a) 21.93 by 3

Dividend — Divisor

Rewrite the division problem. $3\overline{)21.93}$

Step 1 Write the decimal point in the quotient directly above the decimal point in the dividend. $3\overline{)21.93}$ (Decimal points lined up)

Step 2 Divide as if the numbers were whole numbers.

```
   7.31
3)21.93
```

Check by multiplying the quotient times the divisor.

CHECK

```
  7.31
×    3
 21.93
```

Matches, so 7.31 is correct.

The quotient (answer) is 7.31.

(b) $9\overline{)470.7}$

Divisor — Dividend

Write the decimal point in the quotient above the decimal point in the dividend. Then divide as if the numbers were whole numbers.

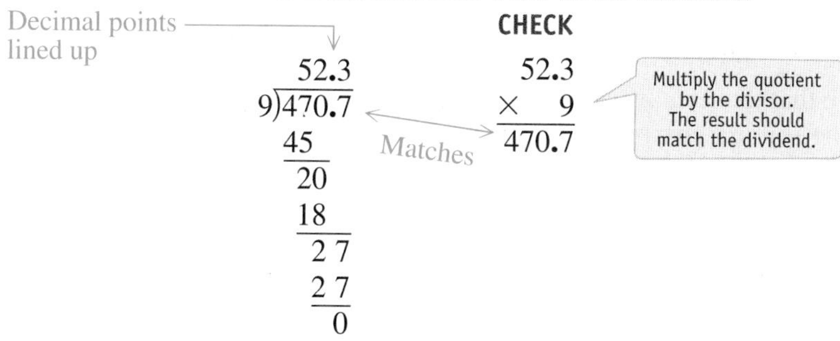

Decimal points lined up
52.3
9)470.7
45
20
18
27
27
0
Matches
CHECK
52.3
× 9
470.7
Multiply the quotient by the divisor. The result should match the dividend.

The quotient is 52.3.

Work Problem **1** at the Side. ▶

1 Find each quotient. Check the quotients by multiplying.

GS **(a)**

```
   2 3 . _
4)9 3 . 6
  8
  1 3
```

Finish the division. Then multiply the quotient by 4. You should get 93.6 if the quotient is correct.

(b) $6\overline{)6.804}$

(c) $11\overline{)278.3}$

(d) $0.51835 \div 5$

(e) $213.45 \div 15$

Answers

1. (a) 23.4; (23.4)(4) = 93.6
   (b) 1.134; (1.134)(6) = 6.804
   (c) 25.3; (25.3)(11) = 278.3
   (d) 0.10367; (0.10367)(5) = 0.51835
   (e) 14.23; (14.23)(15) = 213.45

2 Divide. Check each quotient by multiplying.

GS **(a)** $5\overline{)6.4} \rightarrow 5\overline{)6.40}$

Finish the division.

(b) $30.87 \div 14$

(c) $\dfrac{259.5}{30}$

(d) $0.3 \div 8$

Answers

2. **(a)** 1.28; (1.28)(5) = 6.40 or 6.4
(b) 2.205; (2.205)(14) = 30.870 or 30.87
(c) 8.65; (8.65)(30) = 259.50 or 259.5
(d) 0.0375; (0.0375)(8) = 0.3000 or 0.3

EXAMPLE 2 **Writing Extra Zeros to Complete a Division**

Divide 1.5 by 8. Check the quotient by multiplying.

Keep dividing until the remainder is 0, or until the digits in the quotient begin to repeat in a pattern. In **Example 1(b),** you ended up with a remainder of 0. But sometimes you run out of digits in the dividend before that happens. If so, write extra zeros on the right side of the dividend so you can continue dividing.

$$8\overline{)1.5} = 0.1$$

1.5 ← All digits have been used.
8
7 ← Remainder is not yet 0.

Write a 0 after the 5 in the dividend so you can continue dividing. Keep writing more zeros in the dividend, if needed. Recall that writing zeros to the *right* of a decimal number does *not* change its value.

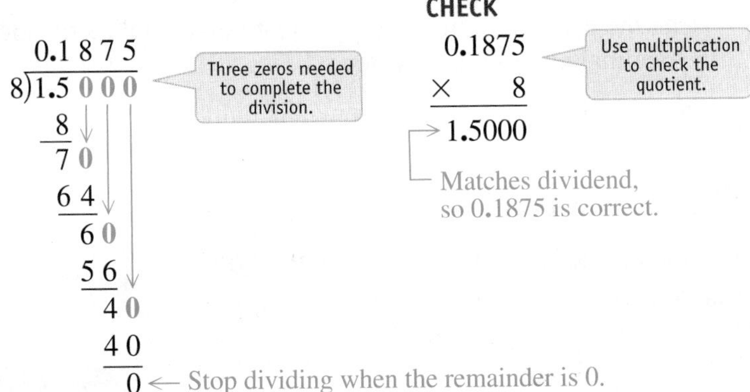
0.1 8 7 5
8)1.5 0 0 0
8
7 0
6 4
6 0
5 6
4 0
4 0
0 ← Stop dividing when the remainder is 0.
Three zeros needed to complete the division.
CHECK
0.1875
× 8
1.5000
Use multiplication to check the quotient.
Matches dividend, so 0.1875 is correct.

CAUTION

When dividing decimals, notice that the dividend might *not* be the greater number. In **Example 2** above, the dividend is 1.5, which is *less* than the divisor 8.

◀ **Work Problem 2 at the Side.**

Calculator Tip

In **Chapter 1,** you learned that when *multiplying* numbers, you can enter them in any order because multiplication is commutative. But division is *not* commutative. It *does* matter which number you enter first. Try **Example 2** both ways; jot down your answers.

1.5 (÷) 8 (=) ________ 8 (÷) 1.5 (=) ________

Notice that the first answer, 0.1875, matches the result from **Example 2.** But the second answer is much different: 5.333333333. Be careful to enter the dividend first.

The next example shows a quotient (answer) that must be rounded because you will never get a remainder of 0.

EXAMPLE 3 **Rounding a Decimal Quotient**

Divide 4.7 by 3. Round the quotient to the nearest thousandth. Write extra zeros in the dividend so you can continue dividing.

$$\begin{array}{r} 1.5666 \\ 3\overline{)4.7000} \\ \underline{3} \\ 17 \\ \underline{15} \\ 20 \\ \underline{18} \\ 20 \\ \underline{18} \\ 20 \\ \underline{18} \\ 2 \end{array}$$

← Three zeros added so far

← Remainder is still not 0.

Notice that the digit 6 in the answer is repeating. It will continue to do so. The remainder will *never be 0.* There are two ways to show that the answer is a **repeating decimal** that goes on forever. You can write three dots after the answer, or you can write a bar above the digits that repeat (in this case, the 6).

$1.5666\ldots$ (Three dots) or $1.5\overline{6}$ ← Bar above repeating digit

When repeating decimals occur, round the quotient according to the directions in the problem. In this example, to round to thousandths, divide out one *more* place, to ten-thousandths.

$4.7 \div 3 = 1.5666\ldots$ rounds to 1.567 — Nearest thousandth is *three* decimal places.

Check the answer by multiplying 1.567 by 3. Because 1.567 is a rounded answer, the check will not give exactly 4.7, but it should be very close.

$(1.567)(3) = 4.701$ ← Does not equal exactly 4.7 because 1.567 was rounded.

CAUTION

When checking quotients that you've rounded, the check will *not* match the dividend exactly, but it should be very close.

Work Problem 3 at the Side. ▶

OBJECTIVE 2 **Divide a number by a decimal.** To divide by a *decimal* divisor, first change the divisor to a whole number. Then divide as before. To see how this is done, write the problem in fraction form. Here is an example.

$1.2\overline{)6.36}$ can be written $\dfrac{6.36}{1.2}$

In **Chapter 3** you learned that multiplying the numerator and denominator by the same number gives an equivalent fraction. We want the divisor (1.2) to be a whole number. Multiplying by 10 will accomplish that.

$$\frac{6.36}{1.2} = \frac{(6.36)(10)}{(1.2)(10)} = \frac{63.6}{12}$$

Decimal divisor → 1.2; 12 ← Whole number divisor

VOCABULARY TIP

Repeating decimals To show a repeating decimal, use three dots ***or*** write a bar above the digits that repeat. Don't use both! The bar and three dots mean the same thing.

3 Divide. Round quotients to the nearest thousandth. If it is a repeating decimal, also write the answer using a bar. Check your quotients by multiplying.

(a) $13\overline{)267.010}$

Use your calculator.

$267.01 : 13 \approx$ ____________

Is the quotient a repeating decimal? ________

So round your answer to the nearest thousandth.

____

(b) $6\overline{)20.5}$

(c) $\dfrac{10.22}{9}$

(d) $16.15 \div 3$

(e) $116.3 \div 11$

Answers

3. **(a)** 20.53923; no repeating digits visible on calculator: 20.539 (rounded); $(20.539)(13) = 267.007$
(b) 3.417 (rounded); $3.41\overline{6}$; $(3.417)(6) = 20.502$
(c) 1.136 (rounded); $1.13\overline{5}$; $(1.136)(9) = 10.224$
(d) 5.383 (rounded); $5.38\overline{3}$; $(5.383)(3) = 16.149$
(e) 10.573 (rounded); $10.5\overline{72}$; $(10.573)(11) = 116.303$

4 Divide. If the quotient does not come out even, round to the nearest hundredth.

GS **(a)** $0.2\overline{)1.04}$

(b) $0.06\overline{)1.8072}$

(c) $0.005\overline{)32}$

(d) $8.1 \div 0.025$

(e) $\dfrac{7}{1.3}$

(f) $5.3091 \div 6.2$

Answers

4. **(a)** 5.2 **(b)** 30.12 **(c)** 6400 **(d)** 324 **(e)** 5.38 (rounded) **(f)** 0.86 (rounded)

The short way to multiply by 10 is to move the decimal point *one place* to the *right* in both the divisor and the dividend.

$$1.2\overline{)6.36} \quad \text{is equivalent to} \quad 12\overline{)63.6}$$

Note

Moving the decimal points the ***same*** number of places in ***both*** the divisor and dividend will ***not*** change the answer.

Dividing by a Decimal Number

Step 1 Count the number of decimal places in the divisor and move the decimal point that many places to the *right*. (This changes the divisor to a whole number.)

Step 2 Move the decimal point in the dividend the *same* number of places to the *right*. (Write in extra zeros if needed.)

Step 3 Write the decimal point in the quotient directly above the decimal point in the dividend. Then divide as usual.

EXAMPLE 4 Dividing by Decimal Numbers

(a) $0.003\overline{)27.69}$

Move the decimal point in the divisor *three* places to the *right* so 0.003 becomes the whole number 3. To move the decimal point in the dividend the same number of places, write in an extra 0.

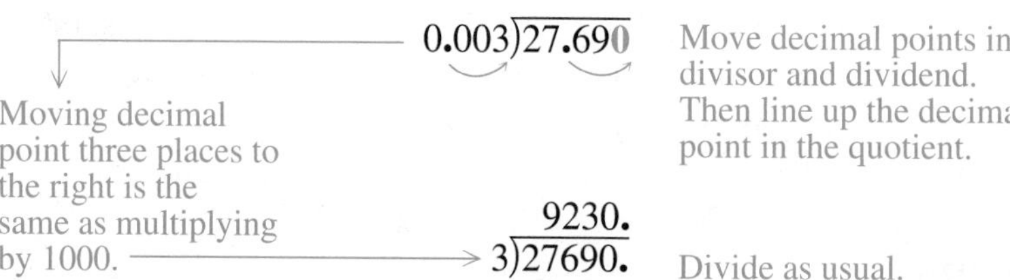

(b) Divide 5 by 4.2. Round to the nearest hundredth.

Move the decimal point in the divisor one place to the right so 4.2 becomes the whole number 42. The decimal point in the dividend starts on the right side of 5 and is also moved one place to the right.

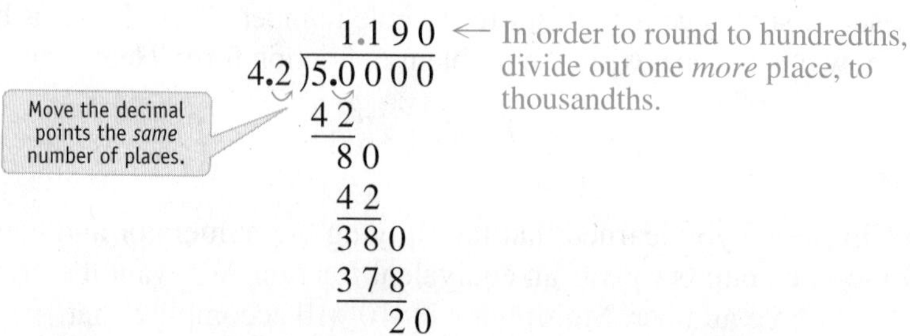

1.190
In order to round to hundredths, divide out one *more* place, to thousandths.
4.2)5.0000
Move the decimal points the *same* number of places.
42
80
42
380
378
20

Round the quotient. It is 1.19 (rounded to the nearest hundredth).

◀ **Work Problem 4 at the Side.**

OBJECTIVE 3 Estimate the answer when dividing decimals. Estimating answers helps you catch errors. Compare the estimate to your exact answer. If they are very different, work the problem again.

EXAMPLE 5 Estimating before Dividing

First use front end rounding to round each number and estimate the answer. Then divide to find the exact answer.

$$580.44 \div 2.8$$

Here is how one student solved this problem. She rounded 580.44 to 600 and rounded 2.8 to 3 to estimate the answer.

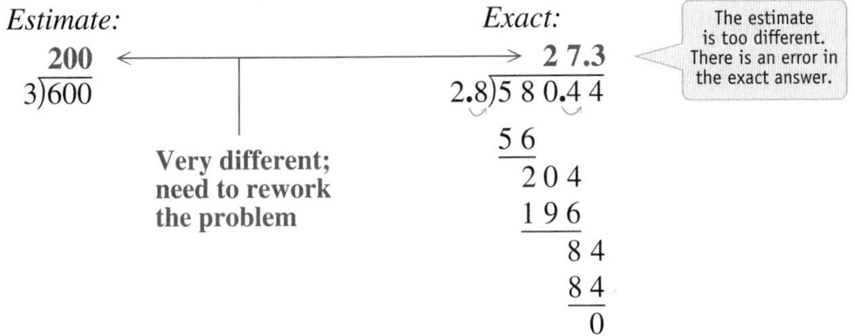

Estimate:
200
3)600
Very different; need to rework the problem
Exact:
27.3
2.8)580.44
56
204
196
84
84
0
The estimate is too different. There is an error in the exact answer.

Notice that the estimate, which is in the hundreds, is very different from the exact answer, which is only in the tens. This tells the student that she needs to rework the problem. Can you find the error?
(The exact answer should be 207.3, which fits with the estimate of 200.)

Work Problem 5 at the Side. ▶

OBJECTIVE 4 Use the order of operations with decimals. Use the order of operations when a decimal problem involves more than one operation, as you did with whole numbers in **Chapter 1.**

Order of Operations

1. Do all operations inside ***parentheses*** or ***other grouping symbols.***
2. Simplify any expressions with ***exponents*** and find any ***square roots.***
3. ***Multiply*** or ***divide,*** proceeding from left to right.
4. ***Add*** or ***subtract,*** proceeding from left to right.

EXAMPLE 6 Using the Order of Operations

Use the order of operations to simplify each expression.

(a) $2.5 + 6.3^2 + 9.62$ Apply the exponent: (6.3)(6.3) is 39.69

$2.5 + 39.69 + 9.62$ Add from left to right.

$42.19 + 9.62$

51.81

(b) $1.82 + (6.7 - 5.2)(5.8)$ Work inside parentheses.

$1.82 + (1.5)(5.8)$ Multiply next.

$1.82 + 8.7$ Add last.

10.52

Continued on Next Page

5 Decide whether each answer is reasonable by using front end rounding to estimate the answer. If the exact answer is *not* reasonable, find and correct the error.

GS **(a)** $42.75 \div 3.8 = 1.125$

Estimate: $40 \div 4 = 10$

Answer of 1.125 is **not** reasonable. Rework.

3.8)42.750

(b) $807.1 \div 1.76 = 458.580$ to nearest thousandth

Estimate:

(c) $48.63 \div 52 = 93.519$ to nearest thousandth

Estimate:

(d) $9.0584 \div 2.68 = 0.338$

Estimate:

Answers

5. (a) Exact answer should be 11.25.
(b) Estimate is $800 \div 2 = 400$; answer is reasonable.
(c) Estimate is $50 \div 50 = 1$; answer is not reasonable, should be 0.935 (rounded).
(d) Estimate is $9 \div 3 = 3$; answer is not reasonable, should be 3.38.

6 Use the order of operations to simplify each expression. The black brace shows you where to start.

(a) $4.6 - 0.79 + \underbrace{1.5^2}$

(b) $\underbrace{3.64 \div 1.3} \times 3.6$

(c) $0.08 + 0.6\underbrace{(3 - 2.99)}$

(d) $10.85 - \underbrace{2.3(5.2)} \div 3.2$

Answers

6. (a) 6.06 (b) 10.08 (c) 0.086 (d) 7.1125

(c) $\underbrace{3.7^2} - 1.8 \div 5(1.5)$ Apply the exponent.

CAREFUL! Do **not** subtract yet.

$13.69 - \underbrace{1.8 \div 5}(1.5)$ Multiply and divide from left to right, so first divide 1.8 by 5 to get 0.36

$13.69 - \underbrace{0.36(1.5)}$ Then multiply 0.36 by 1.5

$\underbrace{13.69 - 0.54}$ Subtract last.

13.15

◀ Work Problem **6** at the Side.

Calculator Tip

You may want to use a calculator to check your work. Most scientific calculators that have parentheses keys ( ) can handle calculations like those in **Example 6** if you just enter the numbers in the order given. For example, the keystrokes for **Example 6(b)** on the previous page are:

Parentheses

1.82 (+) (( ) 6.7 (−) 5.2 ( )) (×) 5.8 (=) Answer is 10.52

Standard, four-function calculators generally do not have parentheses keys and will *not* give the correct answer if you simply enter the numbers in the order given.

Check the instruction manual that came with your calculator for information on "order of calculations" to see if your model has the rules for order of operations built into it. For a quick check, try entering this problem.

2 (+) 2 (×) 2 (=)

If the result is 6, the calculator follows the order of operations. If the result is 8, it does *not* have the rules built into it. To see why this test works, do the calculations by hand.

Follow the order of operations.	Work from left to right.
$2 + \underbrace{2 \times 2}$ Multiply before adding.	$\underbrace{2 + 2} \times 2$
$\underbrace{2 + 4}$	$\underbrace{4 \times 2}$
6 ← Correct	8 ← **Incorrect**

4.5 Exercises

FOR EXTRA HELP

Download the MyDashBoard App

MyMathLab®

1. CONCEPT CHECK Which problem below has a whole number for the divisor? Find it and then rewrite the problem using the $\overline{)\ \ }$ symbol.
   (a) Divide 0.25 by 0.05 (b) $25 \div 0.5$ (c) $25.5 \div 5$

2. CONCEPT CHECK Which problem below has a whole number for the divisor? Find it and then rewrite the problem using the $\overline{)\ \ }$ symbol.
   (a) $0.423 \div 0.07$ (b) $42.3 \div 7$ (c) Divide 423 by 0.7

3. CONCEPT CHECK In $2.2\overline{)8.24}$ how do you make 2.2 a whole number, and how does that change 8.24? Use arrows to show how to move the decimal points.

4. CONCEPT CHECK In $5.1\overline{)10.5}$ how do you make 5.1 a whole number, and how does that change 10.5? Use arrows to show how to move the decimal points.

Find each quotient. ***See Examples 1 and 4.***

5. $7\overline{)27.3}$
6. $8\overline{)50.4}$
7. $\dfrac{4.23}{9}$
8. $\dfrac{1.62}{6}$
9. $0.05\overline{)20.01}$
10. $0.08\overline{)16.04}$
11. $1.5\overline{)54.0}$
12. $2.4\overline{)132.0}$

CONCEPT CHECK *Use the fact that* $108 \div 18 = 6$ *to work Exercises 13–16 simply by moving decimal points.*

13. $0.108 \div 1.8$
14. $10.8 \div 18$
15. $0.018\overline{)108}$
16. $0.18\overline{)1.08}$

Divide. Round quotients to the nearest hundredth if necessary. ***See Examples 3 and 4.***

17. $4.6\overline{)116.38}$
18. $2.6\overline{)4.992}$
19. $\dfrac{3.1}{0.006}$
20. $\dfrac{1.7}{0.09}$

Divide. Round quotients to the nearest thousandth.

21. $240.8 \div 9$
22. $76.43 \div 7$
23. $0.034\overline{)342.81}$
24. $0.043\overline{)1748.4}$

Decide whether each answer is reasonable *or* unreasonable *by rounding the numbers and estimating the answer. If the exact answer is not reasonable, find the correct answer.* ***See Example 5.***

25. $37.8 \div 8 = 47.25$
    Estimate:

26. $345.6 \div 3 = 11.52$
    Estimate:

27. $54.6 \div 48.1 = 1.135$
    Estimate:

28. $2428.8 \div 4.8 = 56$
    Estimate:

29. $307.02 \div 5.1 = 6.2$
    Estimate:

30. $395.415 \div 5.05 = 78.3$
    Estimate:

Solve each application problem. Round money answers to the nearest cent, if necessary.

31. Rob discovered that his daughter's favorite brand of tights is on sale. He decided to buy one pair as a surprise for her. How much did he pay?

Special Purchase!
Girls'
Tights
6 pairs
for $23.98
Stock up now!

32. The bookstore has a special price on notepads. How much did Randall pay for one notepad?

33. It will take 21 equal monthly payments for Aimee to pay off her credit card balance of $1408.68. How much is she paying each month?

34. Marcella Anderson bought 2.6 meters of microfiber woven suede fabric for $33.77. How much did she pay per meter? (*Hint:* Cost *per* meter means the cost for *one* meter.)

35. Darren Jackson earned $476.80 for 40 hours of work. Find his earnings per hour.

36. Adrian Webb bought 108 patio blocks to build a backyard patio. He paid $237.60. Find the cost per block. (*Hint:* Cost *per* block means the cost for *one* block.)

37. It took 12.3 gallons of gas to fill the gas tank of Kim's car. She had driven 344.1 miles since her last fill-up. How many miles per gallon did her car get? Round to the nearest tenth.

38. Mr. Rodriquez pays $53.19 each month to Household Finance. How many months will it take him to pay off $1436.13?

39. Soup is on sale at six cans for $3.25, or you may purchase individual cans for $0.57. How much will you save per can if you buy six cans? Round to the nearest cent.

40. Nadia's diet allows her to eat 3.5 ounces of chicken nuggets. The package weighs 10.5 ounces and contains 15 nuggets. How many nuggets can Nadia eat?

Use the table of world records for the women's long jump (through the year 2010) to answer Exercises 41–46. To find an average, add up the values you are interested in and then divide the sum by the number of values. Round your answers to the nearest hundredth.

Athlete	Country	Year	Length (meters)
Galina Christyakova	USSR	1988	7.52
Jackie Joyner-Kersee	U.S.	1994	7.49
Heike Drechsler	Germany	1992	7.48
Jackie Joyner-Kersee	U.S.	1987	7.45
Jackie Joyner-Kersee	U.S.	1988	7.40
Jackie Joyner-Kersee	U.S.	1991	7.32
Jackie Joyner-Kersee	U.S.	1996	7.20
Chioma Ajunwa	Nigeria	1996	7.12
Fiona May	Italy	2000	7.09
Tatyana Lebedeva	Russia	2004	7.07

Source: CNNSI.com

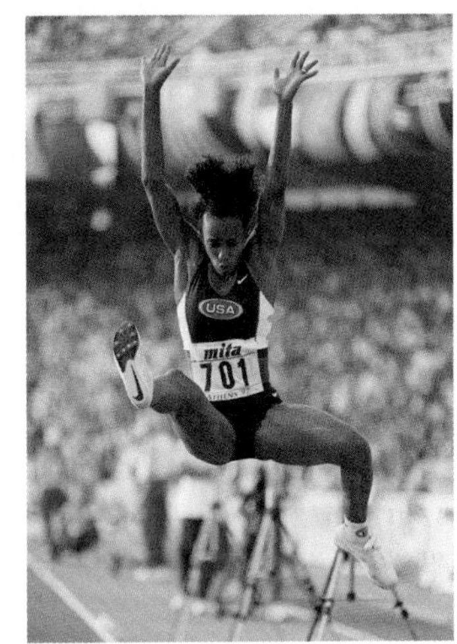

41. Find the average length of the long jumps made by Jackie Joyner-Kersee.

42. Find the average length of all the long jumps listed in the table.

43. How much longer was the fifth-longest jump than the sixth-longest jump?

44. If the first athlete in the table made five jumps of the same length, what would be the total distance jumped?

45. What was the total length jumped by the top three athletes in the table?

46. How much less was the shortest jump than the next-to-shortest jump?

47. For $2.2^2 + (9.5 - 3.1)$ list what you would do, in the correct order, to simplify the expression.
(a) First

(b) Second

(c) Third

48. For $60.41 - (0.4 + 5.07)(3)$ list what you would do, in the correct order, to simplify the expression.
(a) First

(b) Second

(c) Third

Use the order of operations to simplify each expression. ***See Example 6.***

49. $7.2 - 5.2 + 3.5^2$

50. $6.2 + 4.3^2 - 9.72$

51. $38.6 + 11.6(13.4 - 10.4)$

52. $2.25 - 1.06(4.85 - 3.95)$

53. $8.68 - 4.6(10.4) \div 6.4$

54. $25.1 + 11.4 \div 7.5(3.75)$

55. $33 - 3.2(0.68 + 9) - 1.3^2$

56. $0.6 + (1.89 + 0.11) \div 0.004(0.5)$

57. In 2010, the U.S. Treasury printed about 26,000,000 pieces of paper money each day. The printing presses run 24 hours a day. How many pieces of money are printed, to the nearest whole number:
(a) each hour?

(b) each minute?

(c) each second?

(*Source:* www.moneyfactory.com)

26,000,000 pieces of paper money are printed each day.

58. Mach 1 is the speed of sound. Dividing a vehicle's speed by the speed of sound gives its speed on the Mach scale. In 1997, a specially built car with two 110,000-horsepower engines broke the world land speed record by traveling 763.035 miles per hour. The speed of sound that day was 748.11 miles per hour. What was the car's Mach speed, to the nearest hundredth? (*Source:* Associated Press.)

General Mills will give a school 10¢ for each box top logo from its cereals and other products. A school can earn up to $10,000 per year. Use this information to answer Exercises 59–62. Round your answers to the nearest whole number when necessary. (Source: General Mills.*)*

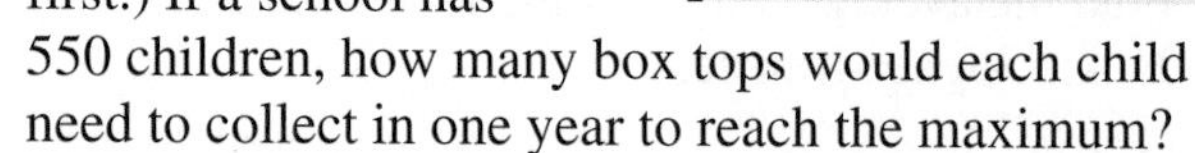

59. How many box tops would a school need to collect in one year to earn the maximum amount?

60. (Complete **Exercise 59** first.) If a school has 550 children, how many box tops would each child need to collect in one year to reach the maximum?

61. How many box tops would need to be collected during each of the 38 weeks in the school year to reach the maximum amount?

62. How many box tops would each of the 550 children need to collect during each of the 38 weeks of school to reach the maximum amount?

Relating Concepts (Exercises 63–64) For Individual or Group Work

Look for patterns as you ***work Exercises 63 and 64 in order.***

63. Do these division problems:

$3.77 \div 10 =$ ______ $9.1 \div 10 =$ ______

$0.886 \div 10 =$ ______ $30.19 \div 10 =$ ______

$406.5 \div 10 =$ ______ $6625.7 \div 10 =$ ______

(a) What pattern do you see? Write a "rule" for dividing by 10. What do you think the rule is for dividing by 100? by 1000? Write the rules and try them out on the numbers above.

(b) Compare your rules to the ones you wrote in the previous section of this chapter for **Exercise 57**. How are they different?

64. Do these division problems:

$40.2 \div 0.1 =$ ______ $7.1 \div 0.1 =$ ______

$0.339 \div 0.1 =$ ______ $15.77 \div 0.1 =$ ______

$46 \div 0.1 =$ ______ $873 \div 0.1 =$ ______

(a) What pattern do you see? Write a "rule" for dividing by 0.1. What do you think the rule is for dividing by 0.01? by 0.001? Write the rules and try them out on the numbers above.

(b) Compare your rules to the ones you wrote in the previous section of this chapter for **Exercise 58**. How are they different?

4.6 Fractions and Decimals

Writing fractions as equivalent decimals can help you do calculations more easily or compare the size of two numbers.

OBJECTIVE 1 Write fractions as equivalent decimals. Recall from **Chapter 1** that a fraction is one way to show division. For example, $\frac{3}{4}$ means $3 \div 4$. If you are doing the division by hand, write it as $4\overline{)3}$. When you do the division, the result is 0.75, the decimal equivalent of $\frac{3}{4}$.

Writing a Fraction as an Equivalent Decimal

Step 1 Divide the numerator of the fraction by the denominator.

Step 2 If necessary, round the answer to the place indicated.

Work Problem 1 at the Side. ▶

EXAMPLE 1 Writing Fractions or Mixed Numbers as Decimals

(a) Write $\frac{1}{8}$ as a decimal.

$\frac{1}{8}$ means $1 \div 8$. Write it as $8\overline{)1}$. The decimal point in the dividend is on the *right* side of the 1. Write extra zeros in the dividend so you can continue dividing until the remainder is 0.

$$\frac{1}{8} \rightarrow 1 \div 8 \rightarrow 8\overline{)1} \rightarrow \begin{array}{r} 0.125 \\ 8\overline{)1.000} \\ \underline{8} \\ 20 \\ \underline{16} \\ 40 \\ \underline{40} \\ 0 \end{array}$$

Decimal points lined up

Three extra zeros needed

Remainder is 0.

Therefore, $\frac{1}{8} = 0.125$.

To check this, write 0.125 as a fraction, then change it to lowest terms.

$$0.125 = \frac{125}{1000} \quad \text{In lowest terms} \quad \frac{125 \div 125}{1000 \div 125} = \frac{1}{8} \leftarrow \text{Original fraction}$$

Calculator Tip

When using your calculator to write fractions as decimals, enter the numbers from the top down. Remember that the order in which you enter the numbers *does* matter in division. **Example 1(a)** above works like this:

$\frac{1}{8}$ Top down Enter 1 (÷) 8 (=) Answer is 0.125

What happens if you enter 8 (÷) 1 (=) ? Do you see why that cannot possibly be correct? (Answer: $8 \div 1 = 8$. A proper fraction like $\frac{1}{8}$ *cannot* be equivalent to a whole number.)

Continued on Next Page

OBJECTIVES

1. Write fractions as equivalent decimals.
2. Compare the size of fractions and decimals.

1 Rewrite each fraction so you could do the division by hand. Do *not* complete the division.

GS **(a)** $\frac{1}{9}$ is written $9\overline{)}$

(b) $\frac{2}{3}$ is written $\overline{)}$

(c) $\frac{5}{4}$ is written $\overline{)}$

(d) $\frac{3}{10}$ is written $\overline{)}$

(e) $\frac{21}{16}$ is written $\overline{)}$

(f) $\frac{1}{50}$ is written $\overline{)}$

Answers

1. (a) $9\overline{)1}$ **(b)** $3\overline{)2}$ **(c)** $4\overline{)5}$ **(d)** $10\overline{)3}$ **(e)** $16\overline{)21}$ **(f)** $50\overline{)1}$

2 Write each fraction or mixed number as a decimal.

GS **(a)** $\frac{1}{4} \rightarrow 4\overline{)1.00}$ (quotient: 0. _ _)

GS **(b)** $2\frac{1}{2} = \frac{5}{2} \rightarrow 2\overline{)5.0}$

(c) $\frac{5}{8}$

(d) $4\frac{3}{5}$

(e) $\frac{7}{8}$

Answers

2. **(a)** 0.25 **(b)** 2.5 **(c)** 0.625 **(d)** 4.6 **(e)** 0.875

(b) Write $2\frac{3}{4}$ as a decimal.

One method is to divide 3 by 4 to get 0.75 for the fraction part. Then add the whole number part to 0.75.

$$\frac{3}{4} \rightarrow \begin{array}{r} 0.75 \\ 4\overline{)3.00} \\ \underline{2\,8} \\ 20 \\ \underline{20} \\ 0 \end{array}$$

Fraction part →

$$\begin{array}{r} 2.00 \\ +\,0.75 \\ \hline 2.75 \end{array}$$

2.00 ← Whole number part

So, $2\frac{3}{4} = 2.75$ **CHECK** $2.75 = 2\frac{75}{100} = 2\frac{3}{4}$ ← Lowest terms

Whole number parts match.

A second method is to first write $2\frac{3}{4}$ as an improper fraction and then divide numerator by denominator.

$$2\frac{3}{4} = \frac{11}{4}$$

$$\frac{11}{4} \rightarrow 11 \div 4 \rightarrow 4\overline{)11} \rightarrow \begin{array}{r} 2.75 \\ 4\overline{)11.00} \\ \underline{8} \\ 3\,0 \\ \underline{2\,8} \\ 20 \\ \underline{20} \\ 0 \end{array}$$

← Two extra zeros needed

Whole number parts match.

So, $2\frac{3}{4} = 2.75$

$\frac{3}{4}$ **is equivalent to** $\frac{75}{100}$ **or 0.75**

◄ **Work Problem 2 at the Side.**

EXAMPLE 2 Writing a Fraction as a Decimal with Rounding

Write $\frac{2}{3}$ as a decimal and round to the nearest thousandth.

$\frac{2}{3}$ means $2 \div 3$. To round to thousandths, divide out one *more* place, to ten-thousandths.

$$\frac{2}{3} \rightarrow 2 \div 3 \rightarrow 3\overline{)2} \rightarrow \begin{array}{r} 0.6666 \\ 3\overline{)2.0000} \\ \underline{1\,8} \\ 20 \\ \underline{18} \\ 20 \\ \underline{18} \\ 20 \\ \underline{18} \\ 2 \end{array}$$

← Four zeros needed for ten-thousandths

Be careful to divide in the correct order!

Written as a repeating decimal, $\frac{2}{3} = 0.\overline{6}$ ← Bar above repeating digit

Rounded to the nearest thousandth, $\frac{2}{3} \approx 0.667$

Continued on Next Page

Calculator Tip

Try **Example 2** on your calculator. Enter 2 ÷ 3. Which answer do you get?

0.666666667 or ***0.6666666***

Many scientific calculators will show a 7 as the last digit. Because the sixes keep on repeating forever, the calculator automatically rounds in the last decimal place it has room to show. If you have a 10-digit display space, the calculator is rounding as shown below.

0.6666666666 (11 digits) rounds to 0.666666667

Next digit is 5 or more, so 6 rounds to 7.

Other calculators, especially standard, four-function ones, may *not* round. They just cut off, or *truncate,* the extra digits. Such a calculator would show 0.6666666 in the display.

Would this difference in calculators show up when changing $\frac{1}{3}$ to a decimal? Why not? (Answer: The repeating digit is 3, which is *4 or less,* so it stays as 3 whether it's rounded or not.)

Work Problem **3** at the Side. ▶

OBJECTIVE 2 Compare the size of fractions and decimals. You can use a number line to compare fractions and decimals. For example, the number line below shows the space between 0 and 1. The locations of some commonly used fractions are marked, along with their decimal equivalents.

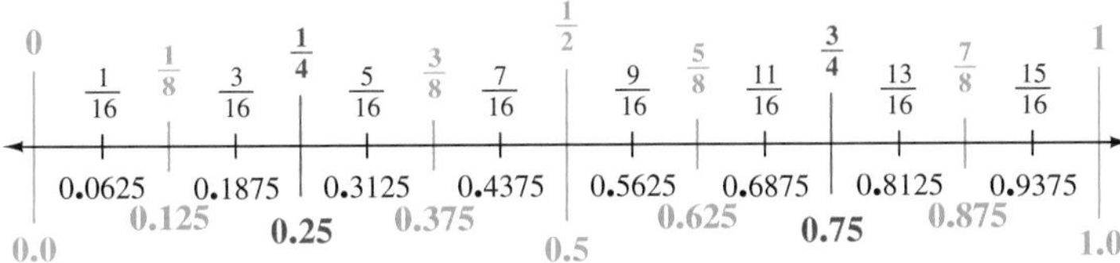

The next number line shows the locations of some commonly used fractions between 0 and 1 that are equivalent to *repeating* decimals. The decimal equivalents use a bar above repeating digits.

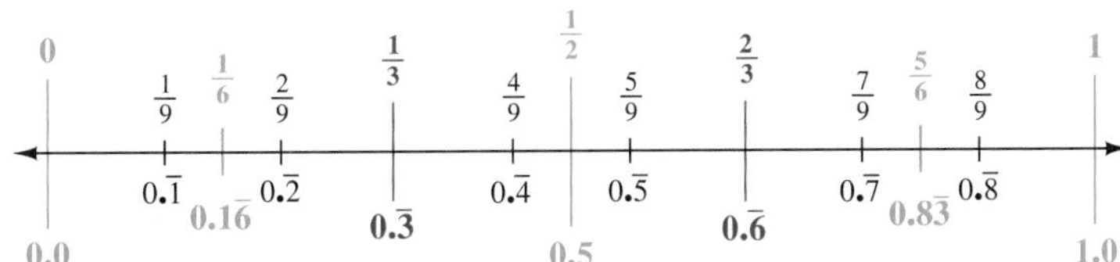

EXAMPLE 3 Using a Number Line to Compare Numbers

Use the number lines above to decide whether to write >, <, or = in the blank between each pair of numbers.

(a) 0.6875 ________ 0.625

You learned in **Chapter 3** that the number farther to the right on the number line is the greater number. Look at the first number line above. Because 0.6875 is to the *right* of 0.625, use the > symbol.

0.6875 is greater than 0.625 can be written as 0.6875 > 0.625

The larger end of the > symbol faces the greater number.

Continued on Next Page

3 Write as decimals. Round to the nearest thousandth.

(a) $\frac{1}{3}$

(b) $2\frac{7}{9} = \frac{25}{9} = 25 \div 9$

Now use your calculator to do the division and round the answer to the nearest thousandth.

(c) $\frac{10}{11}$

(d) $\frac{3}{7}$

(e) $3\frac{5}{6}$

Answers

3. All answers are rounded.
(a) 0.333 **(b)** 2.778 **(c)** 0.909
(d) 0.429 **(e)** 3.833

4 Use the number lines on the previous page to help you decide whether to write <, >, or = in each blank.

(a) 0.4375 ____ 0.5

(b) 0.75 ____ 0.6875

(c) 0.625 ____ 0.0625

(d) $\frac{2}{8}$ ____ 0.375

(e) $0.8\overline{3}$ ____ $\frac{5}{6}$

(f) $\frac{1}{2}$ ____ $0.\overline{5}$

(g) $0.\overline{1}$ ____ $0.1\overline{6}$

(h) $\frac{8}{9}$ ____ $0.\overline{8}$

(i) $0.\overline{7}$ ____ $\frac{4}{6}$

(j) $\frac{1}{4}$ ____ 0.25

5 Arrange each group in order from least to greatest.

(a) 0.7, 0.703, 0.7029

(b) 6.39, 6.309, 6.401, 6.4

(c) 1.085, $1\frac{3}{4}$, 0.9

(d) $\frac{1}{4}, \frac{2}{5}, \frac{3}{7}$, 0.428

Answers

4. (a) < **(b)** > **(c)** > **(d)** < **(e)** = **(f)** < **(g)** < **(h)** = **(i)** > **(j)** =

5. (a) 0.7, 0.7029, 0.703
(b) 6.309, 6.39, 6.4, 6.401
(c) 0.9, 1.085, $1\frac{3}{4}$ **(d)** $\frac{1}{4}, \frac{2}{5}$, 0.428, $\frac{3}{7}$

(b) $\frac{3}{4}$ ________ 0.75

On the first number line, $\frac{3}{4}$ and 0.75 are at the same point on the number line. They are equivalent.

$$\frac{3}{4} = 0.75$$

(c) 0.5 ________ $0.\overline{5}$

On the second number line, 0.5 is to the *left* of $0.\overline{5}$ (which is actually 0.555 . . .) so use the < symbol.

0.5 is less than $0.\overline{5}$ can be written as $0.5 < 0.\overline{5}$

The smaller end of the < symbol points to the lesser number.

(d) $\frac{2}{6}$ ________ $0.\overline{3}$

Write $\frac{2}{6}$ in lowest terms as $\frac{1}{3}$.

On the second number line you can see that $\frac{1}{3} = 0.\overline{3}$.

◀ **Work Problem 4 at the Side.**

Fractions can also be compared by first writing each one as a decimal. The decimals can then be compared by writing each one with the same number of decimal places.

EXAMPLE 4 Arranging Numbers in Order

Write each group of numbers in order, from least to greatest.

(a) 0.49 0.487 0.4903

It is easier to compare decimals if they are all tenths, or all hundredths, and so on. Because 0.4903 has four decimal places (ten-thousandths), write zeros to the right of 0.49 and 0.487 so they also have four decimal places. Recall that writing zeros to the right of a decimal number does *not* change its value (see the section on adding decimals). Then find the least and greatest number of ten-thousandths.

0.49 = 0.4900 = **4900** ten-thousandths ← 4900 is in the middle.
0.487 = 0.4870 = **4870** ten-thousandths ← 4870 is the least.
0.4903 = **4903** ten-thousandths ← 4903 is the greatest.

From least to greatest, the correct order is: 0.487 0.49 0.4903

(b) $2\frac{5}{8}$ 2.63 2.6

Write $2\frac{5}{8}$ as $\frac{21}{8}$ and divide $8\overline{)21}$ to get the decimal form, 2.625. Then, because 2.625 has three decimal places, write zeros so all the numbers have three decimal places.

$2\frac{5}{8}$ = 2.625 = 2 and **625** thousandths ← 625 is in the middle.
2.63 = 2.630 = 2 and **630** thousandths ← 630 is the greatest.
2.6 = 2.600 = 2 and **600** thousandths ← 600 is the least.

From least to greatest, the correct order is: 2.6 $2\frac{5}{8}$ 2.63

◀ **Work Problem 5 at the Side.**

4.6 Exercises

FOR EXTRA HELP

Download the MyDashBoard App

MyMathLab®

CONCEPT CHECK *In Exercises 1 and 2, circle all the divisions that are correctly set up to convert the given fraction to a decimal. If a division is* **not** *set up correctly, fix it.*

1. **(a)** $\frac{2}{5} \rightarrow 2\overline{)5}$ **(b)** $\frac{1}{7} \rightarrow 7\overline{)1}$ **(c)** $\frac{3}{8} \rightarrow 8\overline{)3}$

2. **(a)** $\frac{4}{9} \rightarrow 9\overline{)4}$ **(b)** $\frac{1}{3} \rightarrow 1\overline{)3}$ **(c)** $\frac{5}{6} \rightarrow 6\overline{)5}$

CONCEPT CHECK *In Exercises 3 and 4, show the correct division set up for converting the given fraction to a decimal. Place the decimal points in the dividend and quotient. Then write the first digit in the quotient. Finally, explain what you will do next in order to continue dividing.*

3. $\frac{3}{4} \rightarrow$

4. $\frac{1}{8} \rightarrow$

CONCEPT CHECK *In Exercises 5 and 6, write* greater than *or* less than *between each pair of numbers.*

5. **(a)** $\frac{2}{5}$ is __________ 0.5

(b) $\frac{3}{4}$ is __________ 0.6

(c) 0.2 is __________ $\frac{5}{8}$

6. **(a)** 0.1 is __________ $\frac{1}{5}$

(b) $\frac{2}{3}$ is __________ 0.5

(c) 0.7 is __________ $\frac{1}{2}$

Write each fraction or mixed number as a decimal. Round to the nearest thousandth if necessary. ***See Examples 1 and 2.***

7. Finish the division. $\frac{1}{2} \rightarrow 2\overline{)1.0}$ (quotient $0.$)

8. Finish the division. $\frac{1}{4} \rightarrow 4\overline{)1.00}$ (quotient $0.$)

9. $\frac{3}{4}$

10. $\frac{1}{10}$

11. $\frac{3}{10}$

12. $\frac{7}{10}$

13. $\frac{9}{10}$

14. $\frac{4}{5}$

15. $\frac{3}{5}$

16. $\frac{2}{5}$

17. $\frac{7}{8}$

18. $\frac{3}{8}$

19. $2\frac{1}{4}$

20. $1\frac{1}{2}$

21. $14\frac{7}{10}$

22. $23\frac{3}{5}$

23. $3\frac{5}{8}$

24. $2\frac{7}{8}$

25. $6\frac{1}{3}$

26. $5\frac{2}{3}$

27. $\frac{5}{6}$

28. $\frac{1}{6}$

29. $1\frac{8}{9}$

30. $5\frac{4}{7}$

Find each decimal or fraction equivalent. Write fractions in lowest terms.

	Fraction	Decimal		Fraction	Decimal
31.	____	0.4	**32.**	____	0.75
33.	____	0.625	**34.**	____	0.111
35.	____	0.35	**36.**	____	0.9
37.	$\frac{7}{20}$	____	**38.**	$\frac{1}{40}$	____
39.	____	0.04	**40.**	____	0.52
41.	$\frac{1}{5}$	____	**42.**	$\frac{1}{8}$	____
43.	____	0.09	**44.**	____	0.02

Solve each application problem.

45. The average length of a newborn baby is 20.8 inches. Charlene's baby is 20.08 inches long. Is her baby longer or shorter than the average? By how much?

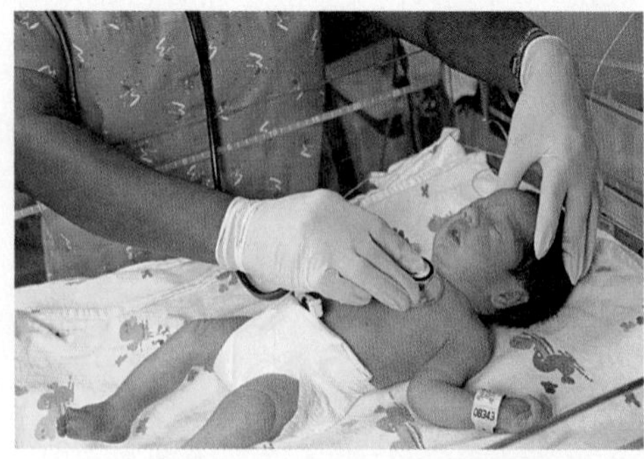

46. The patient in room 830 is supposed to get 8.3 milligrams of medicine. She was actually given 8.03 milligrams. Did she get too much or too little medicine? What was the difference?

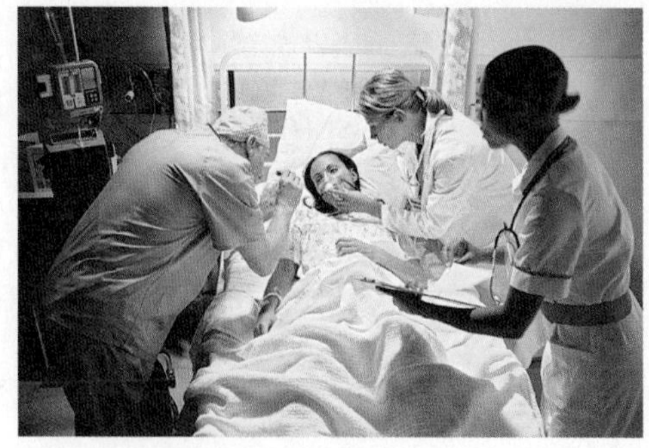

47. Ginny Brown hoped her crops would get $3\frac{3}{4}$ inches of rain this month. The newspaper said the area received 3.8 inches of rain. Was that more or less than Ginny had hoped for? By how much?

48. The rats in a medical experiment gained $\frac{3}{8}$ ounce. They were expected to gain 0.3 ounce. Was their actual gain more or less than expected? By how much?

49. The label on the bottle of vitamins says that each capsule contains 0.5 gram of calcium. When checked, each capsule had 0.505 gram of calcium. Was there too much or too little calcium? What was the difference?

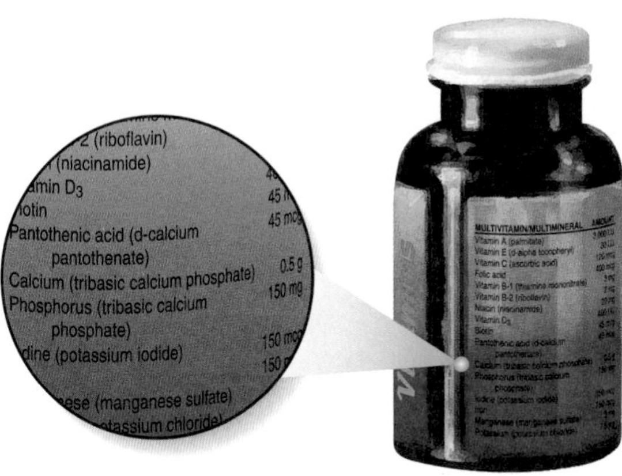

50. The glass mirror of the Hubble telescope had to be repaired in space because it would not focus properly. The problem was that the mirror's outer edge had a thickness of 0.6248 centimeter when it was supposed to be 0.625 centimeter. Was the edge too thick or too thin? By how much? (*Source:* NASA.)

51. CONCEPT CHECK Precision Medical Parts makes an artificial heart valve that must measure between 0.998 centimeter and 1.002 centimeters. Circle the lengths that are acceptable.

1.01 cm   0.9991 cm   1.0007 cm   0.99 cm

52. CONCEPT CHECK The white rats in a medical experiment must start out weighing between 2.95 ounces and 3.05 ounces. Circle the weights that can be used.

3.0 ounces   2.995 ounces   3.055 ounces

3.005 ounces

Arrange each group of numbers in order, from least to greatest. ***See Example 4.***

53. 0.54, 0.5455, 0.5399

54. 0.76, 0.7, 0.7006

55. 5.8, 5.79, 5.0079, 5.804

56. 12.99, 12.5, 13.0001, 12.77

57. 0.628, 0.62812, 0.609, 0.6009

58. 0.27, 0.281, 0.296, 0.3

59. 5.8751, 4.876, 2.8902, 3.88

60. 0.98, 0.89, 0.904, 0.9

61. 0.043, 0.051, 0.006, $\frac{1}{20}$

62. 0.629, $\frac{5}{8}$, 0.65, $\frac{7}{10}$

63. $\frac{3}{8}$, $\frac{2}{5}$, 0.37, 0.4001

64. 0.1501, 0.25, $\frac{1}{10}$, $\frac{1}{5}$

Four boxes of fishing line are in the sale bin. The thicker the line, the stronger it is. The diameter of the fishing line is its thickness. Use the information on the boxes to answer Exercises 65–68.

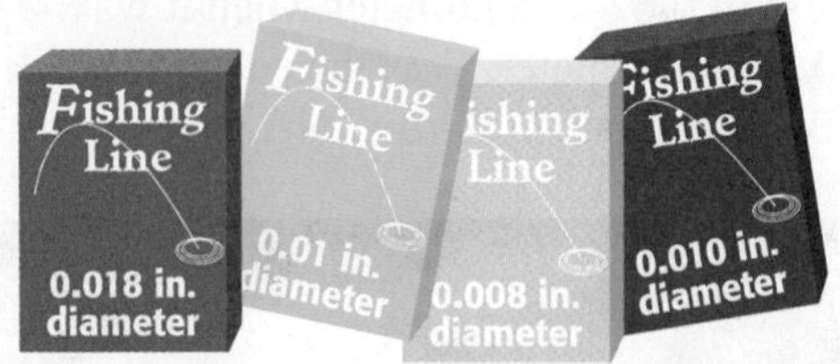

65. Which color box has the strongest line?

66. Which color box has the line with the least strength?

67. Which color box has the line that is $\frac{1}{125}$ inch in diameter?

68. What is the difference in line diameter between the blue and purple boxes?

Some rulers for technical occupations show each inch divided into tenths. Use this scale drawing for Exercises 69–74. Change the measurements on the drawing to decimals and round them to the nearest tenth of an inch.

69. Length **(a)** is ____________

70. Length **(b)** is ____________

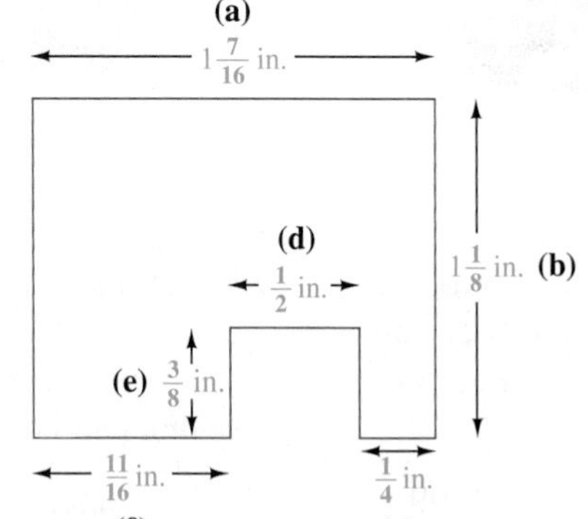

71. Length **(c)** is ____________

72. Length **(d)** is ____________

73. Length **(e)** is ____________

74. Length **(f)** is ____________

Relating Concepts (Exercises 75–78) For Individual or Group Work

Use your knowledge of fractions and decimals to ***work Exercises 75–78 in order.***

75. (a) Explain how you can tell that Keith made an error *just by looking at his final answer*. Here is his work.

$$\frac{5}{9} = 5\overline{)9.0}\ (1.8) \quad \text{so} \quad \frac{5}{9} = 1.8$$

(b) Show the correct way to change $\frac{5}{9}$ to a decimal. Explain why your answer makes sense.

76. (a) How can you prove to Sandra that $2\frac{7}{20}$ is *not* equivalent to 2.035? Here is her work.

$$2\frac{7}{20} = 20\overline{)7.00}\ (0.35) \quad \text{so} \quad 2\frac{7}{20} = 2.035$$

(b) What is the correct answer? Show how to prove that it is correct.

77. Ving knows that $\frac{3}{8} = 0.375$. How can he write $1\frac{3}{8}$ as a decimal *without* having to do a division? How can he write $3\frac{3}{8}$ as a decimal? $295\frac{3}{8}$? Explain your answer.

78. Iris has found a shortcut for writing mixed numbers as decimals.

$$2\frac{7}{10} = 2.7 \qquad 1\frac{13}{100} = 1.13$$

Does her shortcut work for all mixed numbers? Explain when it works and why it works.

Study Skills
ANALYZING YOUR TEST RESULTS

OBJECTIVES

1. Determine the reason for errors.
2. Develop a plan to avoid test-taking errors.
3. Review material to correct misunderstandings.

After taking a test, many students heave a big sigh of relief and try to forget it ever happened. Don't fall into this trap! An exam is a learning opportunity. It gives you clues about *what your instructor thinks is important,* what *concepts and skills are valued* in mathematics, and *if you are on the right track.*

Immediately After the Test

Jot down problems that caused you trouble. Find out how to solve them by checking your textbook, looking at your notes, or asking your instructor or tutor (if available). You might see those same problems again on a final exam.

After the Test Is Returned

Find out what you got wrong and why you had points deducted. Write down the problem so you can learn how to do it correctly. Sometimes you only have a short time in class to review your test. *If you need more time,* ask your instructor if you can look at the test in his or her office.

Find Out Why You Made the Errors You Made

Here is a list of typical reasons for making errors on math tests.

1. You read the directions wrong.
2. You read the question wrong or skipped over something.
3. You made a computation error (maybe even an easy one).
4. Your answer is not accurate.
5. Your answer is not complete.
6. You labeled your answer wrong. For example, you labeled it "feet" and it should have been "feet^2."
7. You didn't show your work.
8. *You didn't understand the concept.
9. *You were unable to go from words (in a word problem) to setting up the problem.
10. *You were unable to apply a procedure to a new situation.
11. You were so anxious that you made errors even when you knew the material.

The first seven errors are **test-taking errors.** They are easy to correct if you decide to carefully read test questions and directions, proofread or rework your problems, show all your work, and double check units and labels every time.

The three starred errors (*) are **test preparation errors.** Remember that to grow a complex neural network, you need to practice the kinds of problems that you will see on the tests. So, for example, if application problems are difficult for you, you must *do more application problems!* If you have practiced the study skills techniques, however, you are less likely to make these kinds of errors on tests because you will have a deeper understanding of course concepts and you will be able to remember them better.

The last error isn't really an error. **Anxiety** can play a big part in your test results. Go back to the *Preparing for Tests* activity and read the suggestions about exercise and deep breathing. Recall from the *Your Brain* **Can** *Learn Mathematics* activity that when you are anxious, your body produces adrenaline. The presence of *adrenaline in the brain blocks connections* between dendrites. If you can

Study Skills *Continued from page 315*

reduce the adrenaline in your system, you will be able to *think more clearly* during your test. Just five minutes of brisk walking right before your test can help do that. Also, *practicing a relaxation technique while you do your homework* will make it more likely that you can benefit from using the technique during a test. *Deep breathing* is helpful because it gets *oxygen into your brain*. When you are anxious you tend to breathe more shallowly, which can make you feel confused and easily distracted.

Make a Plan for the Next Test

Make a plan for your next test based on your results from this test. You might review the Chapter Summary and work the problems in the Chapter Review Exercises or the Chapter Test. Ask your instructor or a tutor (if available) for more help if you are confused about any of the problems.

Now Try This

Below is a record sheet to track your progress in test-taking. Use it to find out if you make particular kinds of errors. Then you can work specifically on correcting them. Just place a check in the box when you made one of the errors. If you take more than four tests, make your own grid on separate paper.

Test-taking Errors

Test #	Read directions wrong	Read question wrong	Computation error	Not exact or accurate	Not complete	Labeled wrong	Didn't show work
1							
2							
3							
4							

Test Preparation Errors

Test #	Didn't understand concept	Didn't set up problem correctly	Couldn't apply concept to new situation
1			
2			
3			
4			

Anxiety

Test #	Felt anxious *before* the exam	Felt anxious *during* the exam	Blanked out on questions	Got questions wrong that I knew how to do
1				
2				
3				
4				

What will you do to avoid test-taking errors?

____

____

____

What will you do to avoid test preparation errors?

____

____

____

What will you do to reduce anxiety?

____

____

____

Chapter 4 Summary

Key Terms

4.1

decimals Decimals, like fractions, are used to show parts of a whole.

decimal point A decimal point is the dot that is used to separate the whole number part from the fractional part of a decimal number.

place value A place value is assigned to each place to the right or left of the decimal point. Whole numbers, such as ones and tens, are to the *left* of the decimal point. Fractional parts, such as tenths and hundredths, are to the *right* of the decimal point.

4.2

rounding Rounding is "cutting off" a number after a certain place, such as rounding to the nearest hundredth. The rounded number is less accurate than the original number. You can use the symbol "≈" to mean "is approximately equal to."

decimal places Decimal places are the number of digits to the *right* of the decimal point. For example, 6.37 has two decimal places, and 4.706 has three decimal places.

4.3

estimating Estimating is the process of rounding the numbers in a problem and getting an approximate answer. This helps you check that the decimal point is in the correct place in the exact answer.

4.5

repeating decimal A repeating decimal is a decimal number with one or more digits that repeat forever; it never ends. For example, in 0.1666 . . . , the digit 6 continues to repeat. Use three dots to indicate that it is a repeating decimal. Or, write the number with a bar above the repeating digits, as in $0.1\overline{6}$. (Use the dots or the bar, but not both.)

New Symbols

$3.8\overline{6}$ ← Bar above repeating digit(s) in a decimal number

3.866 . . . Three dots indicate a repeating decimal.

Test Your Word Power

See how well you have learned the vocabulary in this chapter.

1. **Decimal numbers** are like fractions in that they both
   A. need common denominators
   B. have decimal points
   C. represent parts of a whole.

2. **Decimal places** refer to
   A. the digits from 0 to 9
   B. digits to the left of the decimal point
   C. digits to the right of the decimal point.

3. When a decimal number is **rounded,** it
   A. always ends in 0
   B. is less accurate than the original number
   C. is less than one whole.

4. The **decimal point**
   A. separates the whole number part from the fractional part
   B. separates tenths from hundredths
   C. is at the far left side of a whole number.

5. The number $\mathbf{0.\overline{3}}$ is an example of
   A. an estimate
   B. a repeating decimal
   C. a rounded number.

6. The **place value** names on the right side of the decimal point are
   A. ones, tens, hundreds, and so on
   B. ones, tenths, hundredths, and so on
   C. tenths, hundredths, thousandths, and so on.

Answers To Test Your Word Power

1. C; *Example:* For 0.7, the whole is cut into ten equal parts, and you are interested in 7 of the parts.

2. C; *Examples:* The number 6.87 has two decimal places; 0.309 has three decimal places.

3. B; *Example:* When 0.815 is rounded to 0.8 it is accurate only to the nearest tenth, while the original number was accurate to the nearest thousandth.

4. A; *Example:* In 5.42, the whole number part is 5 ones, and the decimal part is 42 hundredths.

5. B; *Example:* The bar above the 3 in $0.\overline{3}$ indicates that the 3 repeats forever.

6. C; *Example:* In 6.219, the 2 is in the tenths place, the 1 is in the hundredths place, and the 9 is in the thousandths place.

Quick Review

Concepts	Examples
4.1 Reading and Writing Decimal Numbers 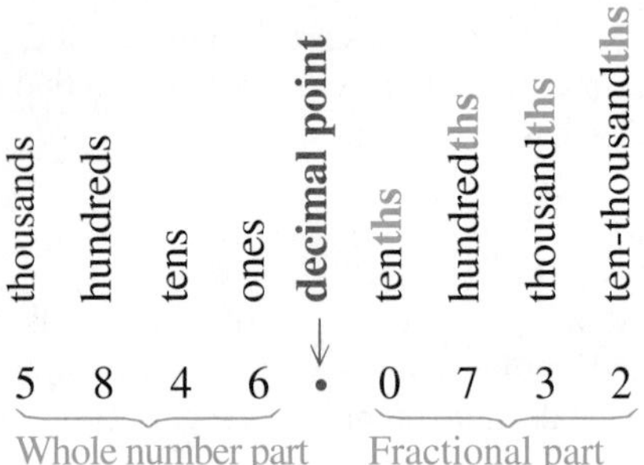	Write each decimal in words. 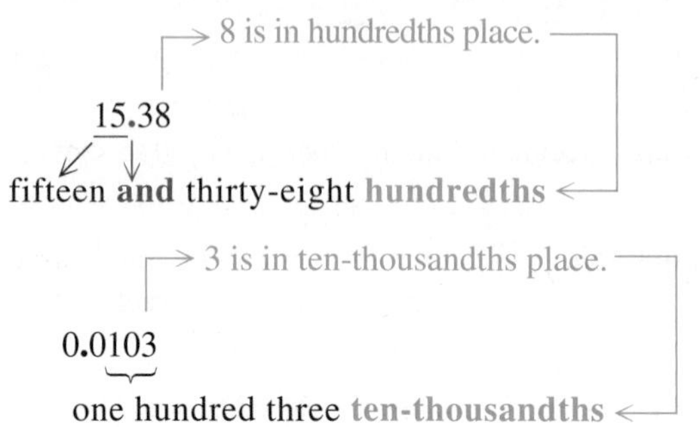
4.1 Writing Decimal Numbers as Fractions The digits to the right of the decimal point are the numerator. The place value of the rightmost digit determines the denominator. Always write the fractions in lowest terms.	Write 0.45 as a fraction in lowest terms. The numerator is 45. The rightmost digit, 5, is in the hundredths place, so the denominator is 100. Then write the fraction in lowest terms. $\frac{45}{100} = \frac{45 \div 5}{100 \div 5} = \frac{9}{20}$ ← Lowest terms
4.2 Rounding Decimal Numbers Find the place to which you are rounding. Draw a cut-off line to the right of that place; the rest of the digits will be dropped. Look *only* at the first digit being cut. If it is *4 or less,* the part you are keeping stays the same. If it is *5 or more,* the part you are keeping rounds up. Do not move the decimal point when rounding. Write "≈" to mean "is approximately equal to."	Round 0.17952 to the nearest thousandth. 0.179|52 ← First digit cut is *5 or more,* so round up. 0.179 ← Keep this part. + 0.001 ← To round up, add 1 thousandth. = 0.180. 0.17952 rounds to 0.180. Write 0.17952 ≈ 0.180
4.3 Adding and Subtracting Decimal Numbers Use front end rounding to round each number and estimate the answer. To find the exact answer, line up the decimal points. If needed, write in zeros as placeholders. Add or subtract as if they were whole numbers. Line up the decimal point in the answer directly below the decimal points in the problem.	Add 5.68 + 785.3 + 12 + 2.007. *Estimate:* 6 + 800 + 10 + 2 = 818. *Exact:* 5.680 + 785.300 + 12.000 + 2.007 = 804.987. Use zeros as placeholders so that all numbers have three decimal places. Line up decimal points. The estimate and exact answer are both in the hundreds, so the decimal point is probably in the correct place.
4.4 Multiplying Decimal Numbers *Step 1* Multiply as you would for whole numbers. *Step 2* Count the total number of decimal places in both factors. *Step 3* Write the decimal point in the product so it has the same number of decimal places as the total from Step 2. You may need to write extra zeros on the left side of the product to get enough decimal places.	Multiply 0.169 × 0.21. 0.169 ← 3 decimal places; × 0.21 ← 2 decimal places; 5 total decimal places. Partial products 169 and 338. .03549 ← 5 decimal places in product. Write a 0 in the product so you can count over 5 decimal places. The final answer is 0.03549.

Concepts	Examples
4.5 Dividing by Decimal Numbers *Step 1* Change the divisor to a whole number by moving the decimal point to the right. *Step 2* Move the decimal point in the dividend the same number of places to the right. *Step 3* Write the decimal point in the quotient directly above the decimal point in the dividend. *Step 4* Divide as with whole numbers.	Divide 52.8 by 0.75. $$0.75\overline{)52.800}$$ quotient 70.4; 525; 300; 300; 0 Move the decimal point two places to the right in the divisor and dividend. Write zeros in the dividend so you can move the decimal point and continue dividing until the remainder is 0. To check your answer, multiply 70.4 times 0.75. If the result matches the dividend (52.8), you solved the problem correctly.
4.6 Writing Fractions as Decimal Numbers Divide the numerator by the denominator. If necessary, round to the place indicated.	Write $\frac{1}{8}$ as a decimal. $\frac{1}{8}$ means $1 \div 8$. Write it as $8\overline{)1}$. The decimal point is on the right side of 1. $$8\overline{)1.000}$$ quotient 0.125; 8; 20; 16; 40; 40; 0 ← Write a decimal point and three zeros so you can continue dividing. Therefore, $\frac{1}{8}$ is equivalent to 0.125
4.6 Comparing the Size of Fractions and Decimal Numbers *Step 1* Write any fractions as decimals. *Step 2* Write zeros so that all the numbers being compared have the same number of decimal places. *Step 3* Use $<$ to mean "is less than," $>$ to mean "is greater than," or list the numbers from least to greatest.	Arrange in order from least to greatest. $0.505 \quad \frac{1}{2} \quad 0.55$ $0.505 = 505$ thousandths ← 505 is in the middle. $\frac{1}{2} = 0.5 = 0.500 = 500$ thousandths ← 500 is least. $0.55 = 0.550 = 550$ thousandths ← 550 is greatest. (least) $\frac{1}{2}$ 0.505 0.55 (greatest)

Chapter 4 Review Exercises

4.1 *Name the digit that has the given place value.*

1. 243.059
tenths
hundredths

2. 0.6817
ones
tenths

3. $5824.39
hundreds
hundredths

4. 896.503
tenths
tens

5. 20.73861
tenths
ten-thousandths

Write each decimal as a fraction or mixed number in lowest terms.

6. 0.5

7. 0.75

8. 4.05

9. 0.875

10. 0.027

11. 27.8

Write each decimal in words.

12. 0.8

13. 400.29

14. 12.007

15. 0.0306

Write each decimal in numbers.

16. eight and three tenths

17. two hundred five thousandths

18. seventy and sixty-six ten-thousandths

19. thirty hundredths

4.2 *Round each decimal to the place indicated.*

20. 275.635 to the nearest tenth

21. 72.789 to the nearest hundredth

22. 0.1604 to the nearest thousandth

23. 0.0905 to the nearest thousandth

24. 0.98 to the nearest tenth

Round each money amount to the nearest cent.

25. $15.8333

26. $0.698

27. $17,625.7906

Round each income or expense item to the nearest dollar.

28. Income from pancake breakfast was $350.48.

29. Members paid $129.50 in dues.

30. Refreshments cost $99.61.

31. Bank charges were $29.37.

4.3 *First use front end rounding to round each number and estimate the answer. Then find the exact answer.*

32. *Estimate:* $+$ ____ *Exact:*

$$\begin{array}{r} 5.81 \\ 423.96 \\ +\ 15.09 \\ \hline \end{array}$$

33. *Estimate:* $+$ ____ *Exact:*

$$\begin{array}{r} 75.6 \\ 1.29 \\ 122.045 \\ 0.88 \\ +\ 33.7 \\ \hline \end{array}$$

34. *Estimate:* $-$ ____ *Exact:*

$$\begin{array}{r} 308.5 \\ -\ 17.8 \\ \hline \end{array}$$

35. *Estimate:* $-$ ____ *Exact:*

$$\begin{array}{r} 9.2 \\ -\ 7.9316 \\ \hline \end{array}$$

36. Americans' favorite household pet is a cat. There are about 93.6 million pet cats, 77.5 million pet dogs, and 16.6 million pet birds. How many more pet cats are there than pet birds? (*Source:* American Pet Products Association.)

Estimate:

Exact:

37. Jasmin started with $406 in her bank account. She paid $315.53 to the day care center and $74.67 at the grocery store. What is the new amount in her account?

Estimate:

Exact:

38. Joey spent $1.59 for toothpaste, $5.33 for vitamins, and $18.94 for a toaster. He gave the clerk three $10 bills. How much change did he get?

Estimate:

Exact:

39. Roseanne is training for a wheelchair race. She raced 2.3 kilometers on Monday, 4 kilometers on Wednesday, and 5.25 kilometers on Friday. How far did she race altogether?

Estimate:

Exact:

4.4 *First use front end rounding to round each number and estimate the answer. Then find the exact answer.*

40. *Estimate:* $\times$ ____ *Exact:*

$$\begin{array}{r} 6.138 \\ \times\ \ 3.7 \\ \hline \end{array}$$

41. *Estimate:* $\times$ ____ *Exact:*

$$\begin{array}{r} 42.9 \\ \times\ 3.3 \\ \hline \end{array}$$

Find each product.

42. $(5.6)(0.002)$

43. $0.071(0.005)$

4.5 *Decide whether each answer is reasonable by rounding the numbers and estimating the answer. If the exact answer is not reasonable, find and correct the error.*

44. $706.2 \div 12 = 58.85$

Estimate:

45. $26.6 \div 2.8 = 0.95$

Estimate:

Divide. Round each quotient to the nearest thousandth if necessary.

46. $3\overline{)43.4}$

47. $\dfrac{72}{0.06}$

48. $0.00048 \div 0.0012$

4.4–4.5 *Solve each application problem.*

49. Adrienne worked 46.5 hours this week. Her hourly wage is \$14.24 for the first 40 hours and 1.5 times that rate over 40 hours. Find her total earnings to the nearest dollar.

50. A book of 12 tickets costs \$35.89 at the State Fair midway. What is the cost per ticket, to the nearest cent?

51. Stock in MathTronic sells for \$3.75 per share. Kenneth is thinking of investing \$500. How many whole shares could he buy?

52. Grapes are on sale at \$0.99 per pound. How much will Ms. Lee pay for 3.5 pounds of grapes, to the nearest cent?

Simplify each expression.

53. $3.5^2 + 8.7(1.95)$

54. $11 - 3.06 \div (3.95 - 0.35)$

4.6 *Write each fraction or mixed number as a decimal. Round to the nearest thousandth when necessary.*

55. $3\frac{4}{5}$

56. $\dfrac{16}{25}$

57. $1\frac{7}{8}$

58. $\dfrac{1}{9}$

Arrange each group of numbers in order from least to greatest.

59. 3.68, 3.806, 3.6008

60. 0.215, 0.22, 0.209, 0.2102

61. $0.17, \frac{3}{20}, \frac{1}{8}, 0.159$

Mixed Review Exercises

Add, subtract, multiply, or divide as indicated.

62. $89.19 + 0.075 + 310.6 + 5$

63. 72.8×3.5

64. $1648.3 \div 0.46$ Round to the nearest thousandth.

65. $30 - 0.9102$

66. $4.38(0.007)$

67. $0.005\overline{)0.047}$

68. $72.105 + 8.2 + 95.37$

69. $81.36 \div 9$

70. $(5.6 - 1.22) + 4.8(3.15)$

71. 0.455×18

72. $(1.6)(0.58)$

73. $0.218\overline{)7.63}$

74. $21.059 - 20.8$

75. $18.3 - 3^2 \div 0.5$

Use the information in the ad to answer Exercises 76–80. Round money answers to the nearest cent. (Disregard any sales tax.)

Grand Opening Sale!
Save on Clothing for the Entire Family
Jeans for Teens
only $19.95 each
women's sizes $24.99
Athletic Shoes
regularly priced
$89.99 to $149.50
NOW just $71 to $119.60
Men's socks NOW 3 pairs for $8.99
Children's socks 6 pairs for $5
Hurry in — TWO DAYS ONLY

76. How much would one pair of men's socks cost?

77. How much more would one pair of men's socks cost than one pair of children's socks?

78. How much would Fernando pay for a dozen pair of men's socks?

79. How much would Akiko pay for five pairs of teen jeans and four pairs of women's jeans?

80. What is the difference between the cheapest sale price for athletic shoes and the highest regular price?

To decrease your risk of clogged arteries, it is recommended that you get at least 2 milligrams of vitamin B-6 each day. Use the information in the table to answer Exercises 81–82.

SOURCES OF VITAMIN B-6
(AMOUNTS IN MILLIGRAMS)

Food	Amount
$\frac{1}{2}$ cup green peas	0.11
1 banana	0.68
1 baked potato with skin	0.7
$\frac{1}{2}$ cup strawberries	0.45
3 ounces skinless chicken	0.5
3 ounces water-packed tuna	0.2
$\frac{1}{2}$ cup chickpeas	0.57

Source: National Institutes of Health.

81. **(a)** Which food item has the highest amount of vitamin B-6?

(b) Which food item has the lowest amount?

(c) What is the difference in the amount of vitamin B-6 between the food items with the highest and lowest amounts?

82. **(a)** Suppose you ate a banana, 3 ounces of skinless chicken, and 1 cup of strawberries. How many milligrams of vitamin B-6 would you get?

(b) Did you get more or less than the recommended daily amount? By how much?

Chapter 4 Test

The Chapter Test Prep Videos with test solutions are available on DVD, in MyMathLab, and on YouTube—search "LialBasicCollegeMath" and click on "Channels."

Write each decimal as a fraction or mixed number in lowest terms.

1. 18.4

2. 0.075

Write each decimal in words.

3. 60.007

4. 0.0208

Round each decimal to the place indicated.

5. 725.6089 to the nearest tenth

6. 0.62951 to the nearest thousandth

7. \$1.4945 to the nearest cent

8. \$7859.51 to the nearest dollar

First use front end rounding to round each number and estimate the answer. Then find the exact answer.

9. $7.6 + 82.0128 + 39.59$
Estimate:
Exact:

10. $79.1 - 3.602$
Estimate:
Exact:

11. $5.79(1.2)$
Estimate:
Exact:

12. $20.04 \div 4.8$
Estimate:
Exact:

Find the exact answer.

13. $53.1 + 4.631 + 782 + 0.031$

14. $670 - 0.996$

15. $(0.0069)(0.007)$

16. $0.15\overline{)72}$

17. Write $2\frac{5}{8}$ as a decimal. Round to the nearest thousandth, if necessary.

18. Arrange in order from least to greatest.

$0.44, 0.451, \frac{9}{20}, 0.4506$

19. Simplify this expression.

$6.3^2 - 5.9 + 3.4(0.5)$

Solve each application problem.

20. Jennifer bought a 51-inch HDTV with 3D technology. The TV had an original price of $1299.99, and she bought two pairs of 3D glasses at $49.99 each. The store gave her a $200 discount, but she had to pay $94.50 in sales tax. How much did Jennifer pay?

21. Three types of ducks that are hunted in the United States are gadwalls, wigeons, and pintails. The estimated populations of these ducks are 2.5 million gadwalls, 2.551 million wigeons, and 2.56 million pintails. List the ducks in order from the greatest number to the least. (*Source:* U.S. Fish and Wildlife Service.)

22. Mr. Yamamoto bought 1.85 pounds of cheese at $2.89 per pound. What was the total amount he paid for the cheese, to the nearest cent?

23. Loren's baby had a temperature of 102.7 degrees. Later in the day it was 99.9 degrees. How much had the baby's temperature dropped?

24. Pat bought 6.5 ft of decorative gold chain to hang a light over her dining table. She paid $24.64. What was the cost per foot, to the nearest cent?

25. Write your own application problem using decimals. Make it different from **Problems 20–24.** Then show how to solve your problem.

Chapters 1–4 Cumulative Review Exercises

Round each number as indicated.

1. 499,501 to the nearest thousand

2. 602.4937 to the nearest hundredth

3. \$709.60 to the nearest dollar

4. \$0.0528 to the nearest cent

Simplify each expression in Exercises 5–21.

5. $10 - 0.329$

6. $2\frac{3}{5} \cdot \frac{5}{9}$

7. $11\frac{1}{5} \div 8$

8. $5006 - 92$

9. $0.7 + 85 + 7.903$

10. Write your answer using R for the remainder.

$7\overline{)2831}$

11. $\frac{5}{6} + \frac{7}{8}$

12. 332×704

13. $(0.006)(5.44)$

14. $3.2(2.5)$

15. $25.2 \div 0.56$

16. $\frac{2}{3} \div 5\frac{1}{6}$

17. $5\frac{1}{4} - 4\frac{7}{12}$

18. $4.7 \div 9.3$
Round to the nearest hundredth.

19. $10 - 4 \div 2 \cdot 3$

20. $\sqrt{36} + 3(8) - 4^2$

21. Write 40.035 in words.

22. Write three hundred six ten-thousandths in numbers.

Arrange each group of numbers in order, from least to greatest.

23. 7.005, 7.5005, 7.5, 7.505

24. $\frac{7}{8}$, 0.8, $\frac{21}{25}$, 0.8015

First round the numbers and estimate the answer to each application problem. Then find the exact answer.

25. Lameck had three \$20 bills. He spent \$47.96 on gasoline and \$0.87 for a candy bar at the convenience store. How much money does he have left?

Estimate:

Exact:

26. Manuela's daughter is 50 inches tall. Last year she was $46\frac{5}{8}$ inches tall. How much has she grown? (When estimating, round to the nearest whole number.)

Estimate:

Exact:

27. Sharon records textbooks on CDs for students who are blind. Her hourly wage is \$11.63. How much did she earn working 16.5 hours last week, to the nearest cent?

Estimate:

Exact:

28. The Farnsworth Elementary School has eight classrooms with 22 students in each one and 12 classrooms with 26 students in each one. How many students attend the school?

Estimate:

Exact:

29. Toshihiro bought $2\frac{1}{3}$ yd of cotton fabric and $3\frac{7}{8}$ yd of wool fabric. How many yards did he buy in all?

Estimate:

Exact:

30. Kimberly had \$29.44 in her bank account. She paid \$40 at the gas station and deposited a \$220.06 paycheck into her account, but not in time to prevent a \$35 charge for insufficient funds. What is the new balance in her account?

Estimate:

Exact:

31. Paulette bought 2.7 pounds of apples for \$6.18. What was the cost per pound, to the nearest cent?

Estimate:

Exact:

32. Carter Community College received a \$78,000 grant from a local hospital to help students pay tuition for nursing classes. How much money could be given to each of 107 students? Round to the nearest dollar.

Estimate:

Exact:

Use the information in the table to answer Exercises 33–36. All measurements are in inches.

Children's Hats
Measure around head above eyebrow ridges.

Head Size	Order Size:
16½"–18"	XXS
18¼"–19"	XS
19¼"–20"	S
20¼"–21⅛"	M
21½"–22¼"	L

Source: Lands' End.

33. If the distance around your child's head is $21\frac{1}{16}$ in., which hat size should you order?

34. Find the difference between the smaller and larger measurements for each of the hat sizes.

35. Change the measurements for the medium (M) size hat to decimals. Then find the difference in the measurements. Show why this answer is equivalent to your answer for the medium (M) size in **Exercise 34.**

36. Did you prefer doing the subtraction using fractions (as in **Exercise 34**) or using decimals (as in **Exercise 35**)? Explain your reasoning.

Math in the Media

LAWN FERTILIZER

Gotta Be Green

A lot's being written about personal responsibility these days, and the idea seems to be ending up on the front lawn—literally! Each spring, homeowners across the country gear up to green up their lawns, and the increased use of fertilizer has a lot of environmentalists concerned about the potential effects of chemical runoff into nearby rivers and streams.

Every year, according to a study conducted by the University of Minnesota's Department of Agriculture, each household in the Minneapolis/St. Paul metro area uses an average of 36 pounds of lawn fertilizer. That adds up to 25,529,295 pounds, or 12,765 tons. Add to that another 193,000 pounds of weed killer and you're looking at the total picture for keeping it green in the Twin Cities.

Source: *Minneapolis Star Tribune.*

1. Refer to the article.

(a) How many pounds of lawn fertilizer are used each year in the *entire metro area?*

(b) Do a division on your calculator to find the number of *households* in the metro area.

(c) Why does it make sense to round your answer to part (b)? How would you round it?

2. There are 2000 pounds in one ton.

(a) Find the number of tons equivalent to 25,529,295 pounds of fertilizer.

(b) Does your answer match the figure given in the article? If not, what did the author of the article do to get 12,765 tons?

(c) Is the author's figure accurate? Why or why not?

3. (a) When the average amount of lawn fertilizer per household was calculated, the answer was probably not exactly 36 pounds. List six different values that are less than 36 that would round to 36. List two values with one decimal place; two values with two decimal places; and two values with three decimal places.

(b) List six different values that are greater than 36 that would round to 36. List two values each with one, two, and three decimal places.

5 Ratio and Proportion

Nine out of every ten American teens and adults have a cell phone. Worldwide, about 2 out of 3 people use a cell phone! (*Source:* International Telecommunication Union.) In this chapter, you will use unit rates to find the best deal on cell phone service plans and on international calling cards.

5.1 Ratios

OBJECTIVES

1. Write ratios as fractions.
2. Solve ratio problems involving decimals or mixed numbers.
3. Solve ratio problems after converting units.

A **ratio** compares two quantities. You can compare two numbers, such as 8 and 4, or two measurements that have the *same* type of units, such as 3 *days* and 12 *days*. (*Rates* compare measurements with different types of units and are covered in the next section.)

Ratios can help you see important relationships. For example, if the ratio of your monthly expenses to your monthly income is 10 to 9, then you are spending $10 for every $9 you earn and going deeper into debt.

OBJECTIVE 1 **Write ratios as fractions.** A ratio can be written in three ways.

Writing a Ratio

The ratio of $7 **to** $3 can be written:

$$7 \textbf{ to } 3 \quad \text{or} \quad 7{:}3 \quad \text{or} \quad \frac{7}{3} \leftarrow \text{Fraction bar indicates "to."}$$

↑ ":" indicates "**to**"

Writing a ratio as a fraction is the most common method, and the one we will use here. All three ways are read, "the ratio of 7 to 3." The word to separates the quantities being compared.

Writing a Ratio as a Fraction

Order is important when writing a ratio. The quantity mentioned **first** is the **numerator**. The quantity mentioned **second** is the **denominator**. For example:

$$\text{The ratio of } \mathbf{5} \text{ to } \mathbf{12} \text{ is written } \frac{5}{12}$$

EXAMPLE 1 Writing Ratios

Ancestors of the Pueblo Indians built multistory apartment towns in New Mexico about 1100 years ago. A room might measure 14 ft long, 11 ft wide, and 15 ft high.

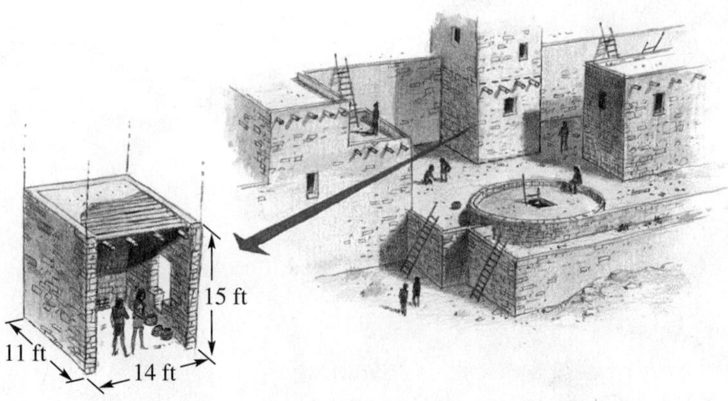

Continued on Next Page

Write each ratio as a fraction, using the room measurements.

(a) Ratio of length to width

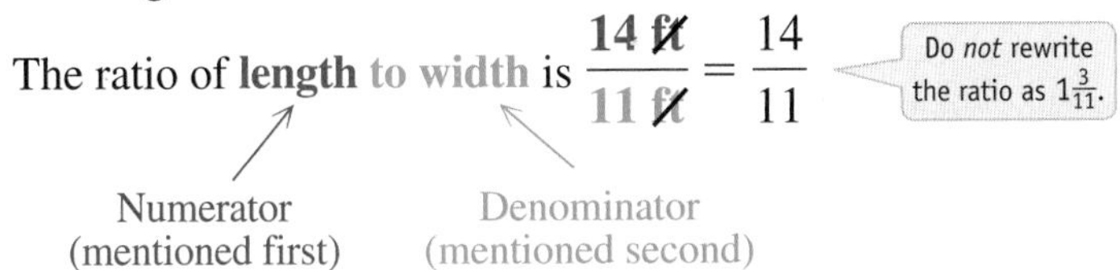

The ratio of length to width is $\frac{14 \text{ ft}}{11 \text{ ft}} = \frac{14}{11}$
Do *not* rewrite the ratio as $1\frac{3}{11}$.
Numerator (mentioned first)
Denominator (mentioned second)

You can divide out common *units* just like you divided out common *factors* when writing fractions in lowest terms. (See **Chapter 2.**) However, do *not* rewrite the fraction as a mixed number. Keep it as the ratio of 14 to 11.

(b) Ratio of width to height

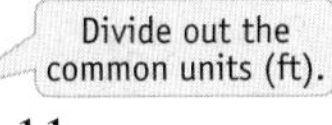

The ratio of width to height is $\frac{11 \text{ ft}}{15 \text{ ft}} = \frac{11}{15}$

CAUTION

Remember, the order of the numbers is important in a ratio. Look for the words "ratio of *a* to *b*." Write the ratio as $\frac{a}{b}$, ***not*** $\frac{b}{a}$. The quantity mentioned first is the numerator.

Work Problem 1 at the Side. ▶

Any ratio can be written as a fraction. Therefore, you can write a ratio in *lowest terms,* just as you do with any fraction.

EXAMPLE 2 Writing Ratios in Lowest Terms

Write each ratio in lowest terms.

(a) 60 days of sun to 20 days of rain

The ratio is $\frac{60 \text{ days}}{20 \text{ days}}$. Divide out the common units. Then write this ratio in lowest terms by dividing the numerator and denominator by 20.

$$\frac{60 \text{ days}}{20 \text{ days}} = \frac{60}{20} = \frac{60 \div 20}{20 \div 20} = \frac{3}{1} \leftarrow \text{Ratio in lowest terms}$$

So, the ratio of 60 days to 20 days is 3 to 1 or, written as a fraction, $\frac{3}{1}$. In other words, for every 3 days of sun, there is 1 day of rain.

CAUTION

In the fractions chapters you would have rewritten $\frac{3}{1}$ as 3. But a *ratio* compares *two* quantities, so you need to keep both parts of the ratio and write it as $\frac{3}{1}$.

(b) 50 ounces of medicine to 120 ounces of water

The ratio is $\frac{50 \text{ ounces}}{120 \text{ ounces}}$. Divide out the common units. Then divide the numerator and denominator by 10.

$$\frac{50 \text{ ounces}}{120 \text{ ounces}} = \frac{50}{120} = \frac{50 \div 10}{120 \div 10} = \frac{5}{12} \leftarrow \text{Ratio in lowest terms}$$

So, the ratio of 50 ounces to 120 ounces is $\frac{5}{12}$. In other words, for every 5 ounces of medicine, there are 12 ounces of water.

Continued on Next Page

VOCABULARY TIP

Ratio A **ratio** compares two quantities that have the **same** units. For example, if you are comparing two lengths, both measurements must be in feet, or both in yards, and so on. In a ratio, the common units divide out.

1 Shane spent \$14 on meat, \$5 on milk, and \$7 on fresh fruit. Write each ratio as a fraction.

GS **(a)** The ratio of amount spent on fruit to amount spent on milk.

Spent on fruit → ____
Spent on milk → 5

(b) The ratio of amount spent on milk to amount spent on meat.

(c) The ratio of amount spent on meat to amount spent on milk.

Answers

1. (a) 7; $\frac{7}{5}$ **(b)** $\frac{5}{14}$ **(c)** $\frac{14}{5}$

2 Write each ratio as a fraction in lowest terms.

GS (a) 9 hours to 12 hours

$$\frac{9 \text{ hours}}{12 \text{ hours}} = \frac{9}{12} = \frac{9 \div 3}{12 \div 3} = \frac{\_\_\_}{\_\_\_}$$

(b) 100 meters to 50 meters

(c) Write the ratio of width to length for this rectangle.

Length 48 ft

Width 24 ft

3 Write each ratio as a ratio of whole numbers in lowest terms.

(a) The price of Tamar's favorite brand of lipstick increased from \$5.50 to \$7.00. Find the ratio of the increase in price to the original price.

(b) Last week, Lance worked 4.5 hours each day. This week he cut back to 3 hours each day. Find the ratio of the decrease in hours to the original number of hours.

Answers

2. (a) $\frac{3}{4}$ (b) $\frac{2}{1}$ (c) $\frac{1}{2}$

3. (a) $\frac{1.50 \times 100}{5.50 \times 100} = \frac{150 \div 50}{550 \div 50} = \frac{3}{11}$

(b) $\frac{1.5 \times 10}{4.5 \times 10} = \frac{15 \div 15}{45 \div 15} = \frac{1}{3}$

(c) 15 people in a large van to 6 people in a small van

$$\text{The ratio is } \frac{15 \text{ people}}{6 \text{ people}} = \frac{15}{6} = \frac{15 \div 3}{6 \div 3} = \frac{5}{2} \leftarrow \text{Ratio in lowest terms}$$

Note

Although $\frac{5}{2} = 2\frac{1}{2}$, ratios are *not* written as mixed numbers. Nevertheless, in **Example 2(c)** above, the ratio $\frac{5}{2}$ does mean the large van holds $2\frac{1}{2}$ times as many people as the small van.

◀ **Work Problem 2 at the Side.**

OBJECTIVE 2 Solve ratio problems involving decimals or mixed numbers. Sometimes a ratio compares two decimal numbers or two fractions. It is easier to understand if we rewrite the ratio as a ratio of two whole numbers.

EXAMPLE 3 Using Decimal Numbers in a Ratio

The price of a Sunday newspaper increased from \$1.50 to \$1.75. Find the ratio of the increase in price to the original price.

The words increase in price are mentioned first, so the increase will be the numerator. How much did the price go up? Use subtraction.

$$\text{new price} - \text{original price} = \text{increase}$$
$$\$1.75 - \$1.50 = \$0.25$$

Subtract first to find how much the price went up.

The words the original price are mentioned second, so the original price of \$1.50 is the denominator.

The ratio of increase in price to original price is shown below.

$$\frac{0.25}{1.50} \begin{matrix} \leftarrow \text{increase in price} \\ \leftarrow \text{original price} \end{matrix}$$

Now rewrite the ratio as a ratio of whole numbers. Recall that if you multiply both the numerator and denominator of a fraction by the same number, you get an equivalent fraction. The decimals in this example are hundredths, so multiply by 100 to get whole numbers. (If the decimals are tenths, multiply by 10. If thousandths, multiply by 1000.) Then write the ratio in lowest terms.

$$\frac{0.25}{1.50} = \frac{0.25 \times 100}{1.50 \times 100} = \frac{25}{150} = \frac{25 \div 25}{150 \div 25} = \frac{1}{6} \leftarrow \text{Ratio in lowest terms}$$

Ratio as two whole numbers

◀ **Work Problem 3 at the Side.**

EXAMPLE 4 Using Mixed Numbers in Ratios

Write each ratio as a comparison of whole numbers in lowest terms.

(a) 2 days to $2\frac{1}{4}$ days

Write the ratio as follows. Divide out the common units.

$$\frac{2 \text{ days}}{2\frac{1}{4} \text{ days}} = \frac{2}{2\frac{1}{4}}$$

Continued on Next Page

Next, write 2 as $\frac{2}{1}$ and $2\frac{1}{4}$ as the improper fraction $\frac{9}{4}$.

$$\frac{2}{2\frac{1}{4}} = \frac{\frac{2}{1}}{\frac{9}{4}}$$

Think: 2 • 4 = 8 and 8 + 1 = 9 so $2\frac{1}{4} = \frac{9}{4}$.

Rewrite the problem in horizontal format, using the "÷" symbol for division. Finally, multiply by the reciprocal of the divisor, as you did in **Chapter 2.**

$$\frac{\frac{2}{1}}{\frac{9}{4}} = \frac{2}{1} \div \frac{9}{4} = \frac{2}{1} \cdot \frac{4}{9} = \frac{8}{9}$$

Reciprocals

The ratio, in lowest terms, is $\frac{8}{9}$.

(b) $3\frac{1}{4}$ to $1\frac{1}{2}$

Write the ratio as $\frac{3\frac{1}{4}}{1\frac{1}{2}}$. Then write $3\frac{1}{4}$ and $1\frac{1}{2}$ as improper fractions.

Think: 3 • 4 = 12 and 12 + 1 = 13 so $3\frac{1}{4} = \frac{13}{4}$.

$$3\frac{1}{4} = \frac{13}{4} \quad \text{and} \quad 1\frac{1}{2} = \frac{3}{2}$$

Think: 1 • 2 = 2 and 2 + 1 = 3 so $1\frac{1}{2} = \frac{3}{2}$.

The ratio is shown here.

$$\frac{3\frac{1}{4}}{1\frac{1}{2}} = \frac{\frac{13}{4}}{\frac{3}{2}}$$

Rewrite as a division problem in horizontal format, using the "÷" symbol. Then multiply by the reciprocal of the divisor.

$$\frac{13}{4} \div \frac{3}{2} = \frac{13}{\underset{2}{\cancel{4}}} \cdot \frac{\overset{1}{\cancel{2}}}{3} = \frac{13}{6}$$

Ratio in lowest terms

Work Problem 4 at the Side. ▶

OBJECTIVE 3 Solve ratio problems after converting units. When a ratio compares measurements, both measurements must be in the *same* units. For example, *feet* must be compared to *feet, hours* to *hours, pints* to *pints*, and *inches* to *inches*.

EXAMPLE 5 Ratio Applications Using Measurement

(a) Write the ratio of the length of the shorter board on the left to the length of the longer board on the right. Compare in inches.

2 ft 30 in.

First, express 2 ft in inches. Because 1 ft has 12 in., 2 ft is

$$2 \cdot 12 \text{ in.} = 24 \text{ in.}$$

Continued on Next Page

4 Write each ratio as a ratio of whole numbers in lowest terms.

GS **(a)** $3\frac{1}{2}$ to 4

$$\frac{3\frac{1}{2}}{4} = \frac{\frac{7}{2}}{\frac{4}{1}}$$

$$= \frac{7}{2} \div \frac{4}{1} = \frac{7}{2} \cdot \frac{1}{4}$$

Reciprocals

$$= \frac{\quad}{\quad}$$

(b) $5\frac{5}{8}$ pounds to $3\frac{3}{4}$ pounds

(c) $3\frac{1}{2}$ in. to $\frac{7}{8}$ in.

Answers

4. (a) $\frac{7}{8}$ **(b)** $\frac{3}{2}$ **(c)** $\frac{4}{1}$

5 Write each ratio as a fraction in lowest terms. (*Hint:* Recall that it is usually easier to write the ratio using the smaller measurement unit.)

GS **(a)** 9 in. to 6 ft

Change 6 ft to inches.

6 • 12 in. = 72 in.

$$\frac{9 \text{ in.}}{6 \text{ ft}} = \frac{9 \text{ in.}}{\_\_\_ \text{ in.}}$$

$$= \frac{9 \div \_\_\_}{\_\_\_ \div \_\_\_} =$$

(b) 2 days to 8 hours

(c) 7 yd to 14 ft

(d) 3 quarts to 3 gallons

(e) 25 minutes to 2 hours

(f) 4 pounds to 12 ounces

On the previous page, the length of the board on the left is 24 in., so the ratio of the lengths is

$$\frac{2 \text{ ft}}{30 \text{ in.}} = \frac{24 \text{ in.}}{30 \text{ in.}} = \frac{24}{30}$$

Once the units match, you can divide them out.

Write the ratio in lowest terms.

$$\frac{24}{30} = \frac{24 \div 6}{30 \div 6} = \frac{4}{5} \leftarrow \text{Ratio in lowest terms}$$

The shorter board is $\frac{4}{5}$ the length of the longer board.

Note

Notice in the example above that we wrote the ratio using the smaller unit (inches are smaller than feet). Using the smaller unit will help you avoid working with fractions. If we wrote the ratio using feet, then

$$30 \text{ in.} = 2\frac{1}{2} \text{ ft}$$

So the ratio in feet is shown below.

$$\frac{2 \text{ ft}}{2\frac{1}{2} \text{ ft}} = \frac{2}{1} \div \frac{5}{2} = \frac{2}{1} \bullet \frac{2}{5} = \frac{4}{5} \leftarrow \text{Same result}$$

The ratio is the same, but it takes more steps to get the answer. Using the smaller unit is usually easier.

(b) Write the ratio of 28 days to 3 weeks.

Since it is easier to write the ratio using the smaller measurement unit, compare in *days* because days are shorter than weeks.

First express 3 weeks in days. Because 1 week has 7 days, 3 weeks is

$$3 \bullet 7 \text{ days} = 21 \text{ days.}$$

So the ratio in days is shown below.

$$\frac{28 \text{ days}}{3 \text{ weeks}} = \frac{28 \text{ days}}{21 \text{ days}} = \frac{28}{21} = \frac{28 \div 7}{21 \div 7} = \frac{4}{3} \leftarrow \text{Ratio in lowest terms}$$

The following table will help you set up ratios that compare measurements. You will work with these measurements again in **Chapter 7.**

Measurement Comparisons

Length	Capacity (Volume)
12 inches = 1 foot	2 cups = 1 pint
3 feet = 1 yard	2 pints = 1 quart
5280 feet = 1 mile	4 quarts = 1 gallon
Weight	**Time**
16 ounces = 1 pound	60 seconds = 1 minute
2000 pounds = 1 ton	60 minutes = 1 hour
	24 hours = 1 day
	7 days = 1 week

◀ Work Problem 5 at the Side.

Answers

5. (a) 72 in.; $\frac{9 \div 9}{72 \div 9} = \frac{1}{8}$ (b) $\frac{6}{1}$ (c) $\frac{3}{2}$
(d) $\frac{1}{4}$ (e) $\frac{5}{24}$ (f) $\frac{16}{3}$

5.1 Exercises

FOR EXTRA HELP

Download the MyDashBoard App

MyMathLab®

1. CONCEPT CHECK What is a ratio? Write an explanation. Then make up an example of a ratio that is different from the ones in the textbook.

2. CONCEPT CHECK Both quantities in a ratio have the same units. That means you can do what to the units?

3. CONCEPT CHECK To rewrite the ratio $\frac{80}{20}$ in lowest terms, divide both the numerator and the denominator by _______. The ratio in lowest terms is _______.

4. CONCEPT CHECK To rewrite the ratio $\frac{20}{75}$ in lowest terms, divide both the numerator and the denominator by _______. The ratio in lowest terms is _______.

Write each ratio as a fraction in lowest terms. ***See Examples 1 and 2.***

5. 8 days to 9 days

6. \$11 to \$15

7. \$100 to \$50

8. 35¢ to 7¢

9. 30 minutes to 90 minutes

10. 9 pounds to 36 pounds

11. 80 miles to 50 miles

12. 300 people to 450 people

13. 6 hours to 16 hours

14. 45 books to 35 books

Write each ratio as a ratio of whole numbers in lowest terms. ***See Examples 3 and 4.***

15. \$4.50 to \$3.50

16. \$0.08 to \$0.06

17. 15 to $2\frac{1}{2} = \frac{15}{2\frac{1}{2}} = \frac{\frac{15}{1}}{\frac{5}{2}}$
GS
$= \frac{15}{1} \div \frac{5}{2} = \frac{\overset{3}{\cancel{15}}}{1} \cdot \frac{2}{\underset{1}{\cancel{5}}} = \frac{\_\_\_}{\_\_\_}$

18. 5 to $1\frac{1}{4} = \frac{5}{1\frac{1}{4}} = \frac{\frac{5}{1}}{\frac{5}{4}}$
GS
$= \frac{5}{1} \div \frac{5}{4} = \frac{\overset{1}{\cancel{5}}}{1} \cdot \frac{4}{\underset{1}{\cancel{5}}} = \frac{\_\_\_}{\_\_\_}$

19. $1\frac{1}{4}$ to $1\frac{1}{2}$

20. $2\frac{1}{3}$ to $2\frac{2}{3}$

21. CONCEPT CHECK When writing the ratio of 20 ounces to 3 pounds as a fraction in lowest terms, would it be easier to use ounces or pounds? _______ Explain why.

22. CONCEPT CHECK When writing the ratio of 4 days to 36 hours as a fraction in lowest terms, would it be easier to use days or hours? _______ Explain why.

Write each ratio as a fraction in lowest terms. For help, use the table of measurement relationships in this section. ***See Example 5.***

23. 4 ft to 30 in.
Hint: First, convert 4 ft to inches.

24. 8 ft to 4 yd
Hint: First, convert 4 yd to feet.

25. 5 minutes to 1 hour

26. 8 quarts to 5 pints

27. 15 hours to 2 days

28. 3 pounds to 6 ounces

29. 5 gallons to 5 quarts

30. 3 cups to 3 pints

The table shows the number of greeting cards sold in the United States for various occasions. Use the information to answer Exercises 31–36. Write each ratio as a fraction in lowest terms.

Holiday/Event	Cards Sold
Valentine's Day	150 million
Mother's Day	130 million
Father's Day	95 million
Graduation	60 million
Halloween	20 million
Thanksgiving	10 million

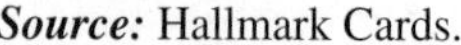
Source: Hallmark Cards.

31. Find the ratio of Thanksgiving cards to graduation cards.

32. Find the ratio of Halloween cards to Mother's Day cards.

33. Find the ratio of Valentine's Day cards to Thanksgiving cards.

34. Find the ratio of Mother's Day cards to Father's Day cards.

35. Explain how you might use the information in the table if you owned a shop selling gifts and greeting cards.

36. Why is the ratio of Valentine's Day cards to graduation cards $\frac{5}{2}$? Give two possible reasons.

The bar graph shows worldwide sales of the most popular songs of all time. Use the graph to complete Exercises 37–40. Write each ratio as a fraction in lowest terms.

Source: The Music Information Database.

37. Write a ratio that compares the top-selling song to the second best seller, and a ratio that compares the top-selling song to the third best seller.

38. Write two ratios that compare sales of Elvis Presley's song with sales of the songs just ahead and just behind it in the graph.

39. Sales of which two songs give a ratio of $\frac{3}{1}$? There may be more than one correct answer.

40. Sales of which two songs give a ratio of $\frac{5}{6}$? There may be more than one correct answer.

For each figure, find the ratio of the length of the longest side to the length of the shortest side. Write each ratio as a fraction in lowest terms.

41.

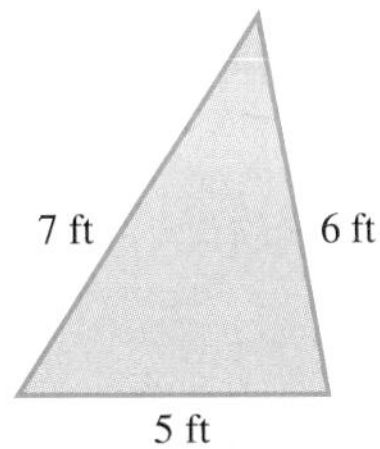

42.

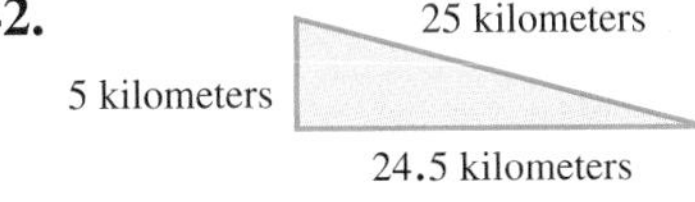

43.

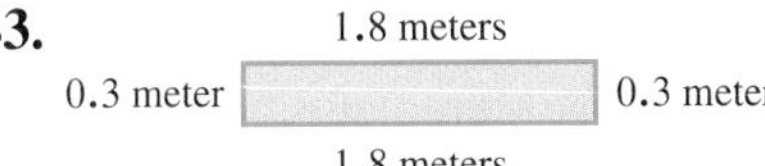

44.

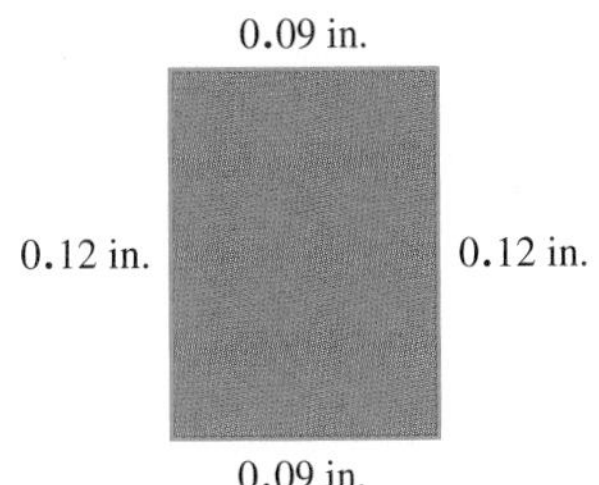

45.

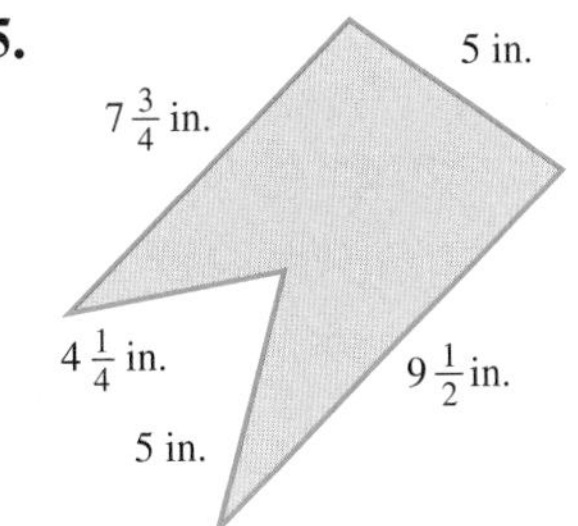

46.

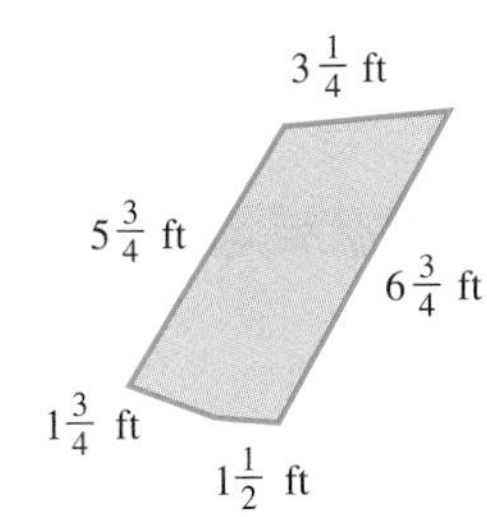

Write each ratio as a fraction in lowest terms.

47. The price of automobile engine oil has gone from \$10 to \$12.50 for the 5 quarts needed for an oil change. Find the ratio of the increase in price to the original price.

48. The price that a pharmacy pays for an antibiotic decreased from \$8.80 to \$5.60 for 10 tablets. Find the ratio of the decrease in price to the original price.

49. The first time a movie was made in Minnesota, the cast and crew spent $59\frac{1}{2}$ days filming winter scenes. The next year, another movie was filmed in $8\frac{3}{4}$ weeks. Find the ratio of the first movie's filming time to the second movie's time. Compare in weeks.

50. The percheron, a large draft horse, measures about $5\frac{3}{4}$ ft at the shoulder. The prehistoric ancestor of the horse measured only $15\frac{3}{4}$ in. at the shoulder. Find the ratio of the percheron's height to its prehistoric ancestor's height. Compare in inches. (*Source: Eyewitness Books: Horse.*)

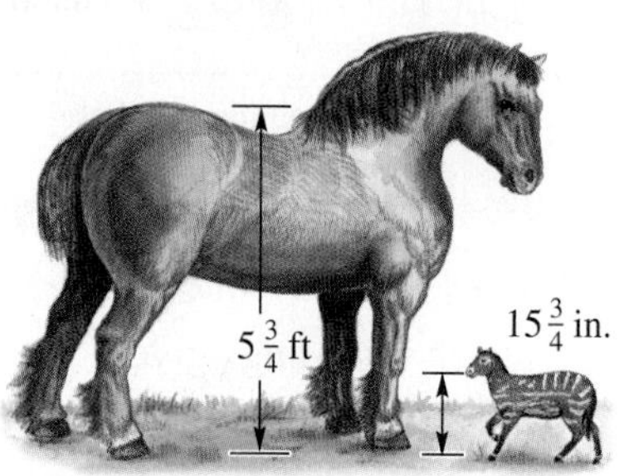

Relating Concepts (Exercises 51–54) For Individual or Group Work

Use your knowledge of ratios to ***work Exercises 51–54 in order.***

51. In this painting, what is the ratio of the length of the longest side to the length of the shortest side? What other measurements could the painting have and still maintain the same ratio?

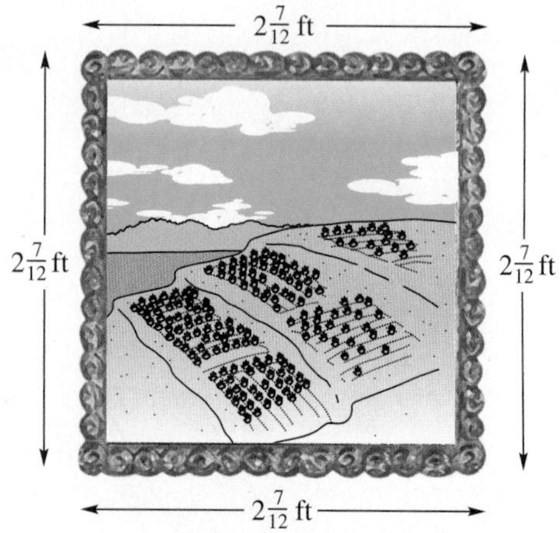

52. The ratio of my son's age to my daughter's age is 4 to 5. One possibility is that my son is 4 years old and my daughter is 5 years old. Find six other possibilities that fit the 4 to 5 ratio.

53. Amelia said that the ratio of her age to her mother's age is 5 to 3. Is this possible? Explain your answer.

54. Would you prefer that the ratio of your income to your friend's income be 1 to 3 or 3 to 1? Explain your answers.

5.2 Rates

A *ratio* compares two measurements with the **same** units, such as 9 feet to 12 feet (both length measurements). But many of the comparisons we make use measurements with **different** units.

160 dollars for 8 hours (money to time)

450 miles on 15 gallons (distance to capacity)

This type of comparison is called a **rate.**

OBJECTIVES

1. Write rates as fractions.
2. Find unit rates.
3. Find the best buy based on cost per unit.

OBJECTIVE 1 Write rates as fractions. Suppose you hiked 14 miles in 4 hours. The *rate* at which you hiked can be written as a fraction in lowest terms.

$$\frac{14 \text{ miles}}{4 \text{ hours}} = \frac{14 \text{ miles} \div 2}{4 \text{ hours} \div 2} = \frac{7 \text{ miles}}{2 \text{ hours}}$$ ← Rate in lowest terms

In a rate, you often find these words separating the quantities you are comparing.

in for on per from

VOCABULARY TIP

Ratio vs. Rate **Ratios** compare two quantities that have the **same** units, so the units divide out. However, **rates** compare two quantities with **different** units, so you must write the units in a rate.

CAUTION

When writing a rate, always include the units, such as miles, hours, dollars, and so on. Because the units in a rate are different, the units do *not* divide out.

EXAMPLE 1 Writing Rates in Lowest Terms

Write each rate as a fraction in lowest terms.

(a) 5 gallons of chemical for $60.

$$\frac{5 \text{ gallons} \div 5}{60 \text{ dollars} \div 5} = \frac{1 \text{ gallon}}{12 \text{ dollars}}$$ ← Write the units: gallons and dollars.

(b) $1500 wages in 10 weeks

$$\frac{1500 \text{ dollars} \div 10}{10 \text{ weeks} \div 10} = \frac{150 \text{ dollars}}{1 \text{ week}}$$

Be sure to write the units in a *rate*: dollars, miles, gallons, and so on.

(c) 2225 miles on 75 gallons of gas

$$\frac{2225 \text{ miles} \div 25}{75 \text{ gallons} \div 25} = \frac{89 \text{ miles}}{3 \text{ gallons}}$$

Work Problem 1 at the Side. ▶

1 Write each rate as a fraction in lowest terms.

GS **(a)** $6 for 30 packets

$$\frac{\$6}{30 \text{ packets}} = \frac{\$\_\_\_}{\_\_\_ \text{ packets}}$$

To write the rate in lowest terms, divide both numerator and denominator by ____.

(b) 500 miles in 10 hours

(c) 4 teachers for 90 students

OBJECTIVE 2 Find unit rates. When the *denominator* of a rate is 1, it is called a **unit rate.** We use unit rates frequently. For example, you earn $12.75 for *1 hour* of work. This unit rate is written:

$12.75 per hour or $12.75/hour.

Use per or a slash mark (/) when writing unit rates.

Answers

1. (a) $\frac{\$1}{5 \text{ packets}}$; divide by 6.
(b) $\frac{50 \text{ miles}}{1 \text{ hour}}$ (c) $\frac{2 \text{ teachers}}{45 \text{ students}}$

VOCABULARY TIP

Per means "for each." For example, if your car gets 30 miles **per** gallon, it gets 30 miles *for each gallon* of gas. Use **per** or a slash mark, like this: 30 miles/gallon.

2 Find each unit rate.

GS **(a)** \$4.35 for 3 pounds of apples

$\frac{\$4.35}{3 \text{ pounds}}$ ← Fraction bar indicates division.

$3\overline{)4.35}$ with 1. above — Finish the division.

(b) 304 miles on 9.5 gallons of gas

(c) \$850 in 5 days

(d) 24-pound turkey for 15 people

Answers

2. **(a)** \$1.45/pound **(b)** 32 miles/gallon **(c)** \$170/day **(d)** 1.6 pounds/person

EXAMPLE 2 Finding Unit Rates

Find each unit rate.

(a) 337.5 miles on 13.5 gallons of gas
Write the rate as a fraction.

$$\frac{337.5 \text{ miles}}{13.5 \text{ gallons}}$$ ← The fraction bar indicates division.

Divide 337.5 by 13.5 to find the unit rate.

$$13.5\overline{)337.5} = 25.$$

$$\frac{337.5 \text{ miles} \div 13.5}{13.5 \text{ gallons} \div 13.5} = \frac{25 \text{ miles}}{1 \text{ gallon}}$$

Be sure to write **miles** and **gallon** in your answer.

The unit rate is 25 miles **per** gallon, or 25 miles/gallon.

(b) 549 miles in 18 hours

$$\frac{549 \text{ miles}}{18 \text{ hours}} \qquad \text{Divide: } 18\overline{)549.0} = 30.5$$

The unit rate is 30.5 miles/hour.

(c) \$810 in 6 days

$$\frac{810 \text{ dollars}}{6 \text{ days}} \qquad \text{Divide: } 6\overline{)810} = 135$$

Use *per* or a slash mark to write unit rates.

The unit rate is \$135/day.

◀ **Work Problem 2 at the Side.**

OBJECTIVE 3 **Find the best buy based on cost per unit.** When shopping for groceries, household supplies, and health and beauty items, you will find many different brands and package sizes. You can save money by finding the lowest *cost per unit.*

Cost per Unit

Cost per unit is a rate that tells how much you pay for *one* item or *one* unit. Examples are \$3.75 per gallon, \$47 per shirt, and \$2.98 per pound.

EXAMPLE 3 Determining the Best Buy

The local store charges the following prices for pancake syrup. Find the best buy.

Continued on Next Page

The best buy is the container with the *lowest* cost per unit. All the containers are measured in *ounces* (oz), so you first need to find the *cost per ounce* for each one. Divide the price of the container by the number of ounces in it. Round to the nearest thousandth, if necessary.

Let the *order* of the *words* help you set up the rate.

$$\begin{array}{l} \text{cost} \longrightarrow \\ \text{per (means divide)} \longrightarrow \\ \text{ounce} \longrightarrow \end{array} \frac{\$1.28}{12 \text{ ounces}}$$

Size	Cost Per Unit (Rounded)
12 ounces	$\frac{\$1.28}{12 \text{ ounces}} \approx \$0.107$ per ounce (highest)
24 ounces	$\frac{\$1.81}{24 \text{ ounces}} \approx \$0.075$ per ounce (lowest)
36 ounces	$\frac{\$2.73}{36 \text{ ounces}} \approx \$0.076$ per ounce

The lowest cost per ounce is $0.075, so the 24-ounce container is the best buy.

Note

Earlier we rounded money amounts to the nearest hundredth (nearest cent). But when comparing unit costs, rounding to the nearest thousandth will help you see the difference between very similar unit costs. Notice that the 24-ounce and 36-ounce syrup containers above would both have rounded to $0.08 per ounce if we had rounded to hundredths.

Work Problem 3 at the Side. ▶

Calculator Tip

When using a calculator to find unit prices, remember that division is *not* commutative. In **Example 3** you wanted to find cost per ounce. Let the *order* of the *words* help you enter the numbers in the correct order.

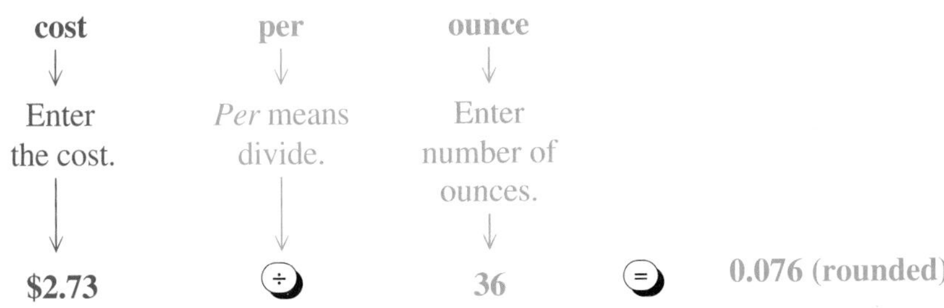

If you entered 36 ÷ 2.73 =, you'd get the number of *ounces* per *dollar*. How could you use that information to find the best buy? (*Answer:* The best buy would be to get the greatest number of ounces per dollar.)

Finding the best buy is sometimes a complicated process. Things that affect the cost per unit can include "cents off" coupons and differences in how much use you'll get out of each unit.

3 Find the best buy (lowest cost per unit) for each purchase.

GS **(a)** 2 quarts for $3.25
3 quarts for $4.95
4 quarts for $6.48

Divide to find each unit cost.

$$\begin{array}{l} \text{cost} \rightarrow \\ \text{per} \rightarrow \\ \text{quart} \rightarrow \end{array} \frac{\$3.25}{2 \text{ quarts}}$$

Unit cost is $ ____ per quart.

$$\begin{array}{l} \text{cost} \rightarrow \\ \text{per} \rightarrow \\ \text{quart} \rightarrow \end{array} \frac{\$\ \ \ \ }{3 \text{ quarts}}$$

Unit cost is $ ____ per ____.

$$\begin{array}{l} \text{cost} \rightarrow \\ \text{per} \rightarrow \\ \text{quart} \rightarrow \end{array} \frac{\ \ \ \ }{\ \ \ \ }$$

Unit cost is $ ____ per ____.

Now compare the three unit costs. The *lowest* one is the best buy. Circle it.

(b) 6 cans of cola for $1.99
12 cans of cola for $3.49
24 cans of cola for $7

Answers

3. (a) Unit cost is $1.625 per quart
$4.95; Unit cost is $1.65 per quart
$\frac{\$6.48}{4 \text{ quarts}}$; Unit cost is $1.62 per quart
lowest unit cost is 4 quarts, at $1.62 per quart

(b) 12 cans, at $0.291 per can (rounded)

❹ Solve each problem.

(a) Some batteries claim to last longer than others. If you believe these claims, which brand is the best buy?

Four-pack of AA-size batteries for $2.79

One AA-size battery for $1.19; lasts twice as long

(b) Which tube of toothpaste is the best buy? You have a coupon for 85¢ off Brand C and a coupon for 50¢ off Brand D.

Brand C is $3.89 for 6 ounces.

Brand D is $1.89 for 2.5 ounces.

Answers

4. **(a)** One battery that lasts twice as long (like getting two) is the better buy. The cost per unit is $0.595 per battery. The four-pack is $0.698 per battery (rounded).

(b) Brand C with the 85¢ coupon is the better buy at $0.507 per ounce (rounded). Brand D with the 50¢ coupon is $0.556 per ounce.

EXAMPLE 4 Solving Best Buy Applications

Solve each application problem.

(a) There are many brands of liquid laundry detergent. If you feel they all do a good job of cleaning your clothes, you can base your purchase on cost per unit. But some brands are "concentrated" so you can use less detergent for each load of clothes. The labels from two bottles are shown below. Which of the choices is the best buy?

To find Sudzy's unit cost, divide $3.99 by 64 ounces, not 50 ounces. Read the label. You're getting as many clothes washed as if you bought 64 ounces. Similarly, to find White-O's unit cost, divide $9.89 by 256 ounces (twice 128 ounces, or 2 • 128 ounces = 256 ounces).

$$\text{Sudzy} \quad \frac{\$3.99}{64 \text{ ounces}} \approx \$0.062 \text{ per ounce}$$

The best buy is the *lower* cost per ounce.

$$\text{White-O} \quad \frac{\$9.89}{256 \text{ ounces}} \approx \$0.039 \text{ per ounce}$$

White-O has the lower cost per ounce and is the better buy. (However, if you try it and it really doesn't get out all the stains, Sudzy may be worth the extra cost.)

(b) "Cents-off" coupons also affect the best buy. Suppose you are looking at these choices for "extra-strength" pain reliever. Both brands have the same amount of pain reliever in each tablet.

Brand X is $2.29 for 50 tablets.

Brand Y is $10.75 for 200 tablets.

You have a 40¢ coupon for Brand X and a 75¢ coupon for Brand Y. Which choice is the best buy?

To find the best buy, first subtract the coupon amounts, then divide to find the lower cost per ounce.

Brand X costs $2.29 − $0.40 = $1.89

$$\frac{\$1.89}{50 \text{ tablets}} \approx \$0.038 \text{ per tablet}$$

Brand Y costs $10.75 − $0.75 = $10.00

Look for the *lower* cost per tablet.

$$\frac{\$10.00}{200 \text{ tablets}} = \$0.05 \text{ per tablet}$$

Brand X has the lower cost per tablet and is the better buy.

◄ Work Problem ❹ at the Side.

5.2 Exercises

FOR EXTRA HELP

Download the MyDashBoard App

MyMathLab®

Write each rate as a fraction in lowest terms. ***See Example 1.***

1. 10 cups for 6 people

2. \$12 for 30 pens

3. 15 feet in 35 seconds

4. 100 miles in 30 hours

5. 72 miles on 4 gallons

6. 132 miles on 8 gallons

Find each unit rate. ***See Example 2.***

7. CONCEPT CHECK To find a unit rate, do you add, subtract, multiply, or divide? _________ Show the correct set-up to find this unit rate: \$5.85 for 3 boxes.

8. CONCEPT CHECK Circle the correct answer. When finding this unit rate, 15 hospital rooms for 3 nurses, the final answer will be:

5 to 1 5 rooms for 1 nurse 5 nurses for 1 room

9. \$60 in 5 hours

10. \$2500 in 20 days

11. 7.5 pounds for 6 people

12. 44 bushels from 8 trees

Earl kept the following record of the gas he bought for his car. For each entry, find the number of miles he traveled and the unit rate. Round your answers to the nearest tenth.

	Date	Odometer at Start	Odometer at End	Miles Traveled	Gallons Purchased	Miles per Gallon
13.	2/4	27,432.3	27,758.2		15.5	
14.	2/9	27,758.2	28,058.1		13.4	
15.	2/16	28,058.1	28,396.7		16.2	
16.	2/20	28,396.7	28,704.5		13.3	

Source: Author's car records.

Find the best buy (based on the cost per unit) for each item. Round to the nearest thousandth, if necessary. **See Example 3.** (*Source:* Cub Foods, Target, Rainbow Foods.)

17. Black pepper

Complete the divisions:

cost → $\dfrac{\$2.25}{2\text{ oz}} =$ ← per ounce

cost → $\dfrac{\$\quad}{4\text{ oz}} =$ ← per ounce

Lower unit cost (better buy) is __________, so the best buy is __________.

18. Shampoo

Complete the divisions:

cost → $\dfrac{\$3.59}{8\text{ oz}} =$ ← per ounce

cost → $\dfrac{\$\quad}{12\text{ oz}} =$ ← per ounce

Lower unit cost (better buy) is __________, so the best buy is __________.

19. Cereal

12 ounces for $2.49
14 ounces for $2.89
18 ounces for $3.96

20. Soup (same size cans)

2 cans for $2.18
3 cans for $3.57
5 cans for $5.29

21. Chunky peanut butter

12 ounces for $1.29
18 ounces for $1.79
28 ounces for $3.39
40 ounces for $4.39

22. Baked beans

8 ounces for $0.59
16 ounces for $0.99
21 ounces for $1.29
28 ounces for $1.89

23. Suppose you are choosing between two brands of chicken noodle soup. Brand A is $0.88 per can and Brand B is $0.98 per can. The cans are the same size but Brand B has more chunks of chicken in it. Which soup is the better buy? Explain your choice.

24. A small bag of potatoes costs $0.19 per pound. A large bag costs $0.15 per pound. But there are only two people in your family, so half the large bag would probably rot before you used it up. Which bag is the better buy? Explain.

Solve each application problem. **See Examples 2–4.**

25. Makesha lost 10.5 pounds in six weeks. What was her rate of loss in pounds per week?

26. Enrique's taco recipe uses three pounds of meat to feed 10 people. Give the rate in pounds per person.

27. Russ works 7 hours to earn $85.82. What is his pay rate per hour?

28. Find the cost of 1 gallon of Hawaiian Punch beverage if 18 gallons for a graduation party cost $55.62.

The table lists information about three long-distance calling cards. The connection fee is charged each time you make a call, no matter how long the call lasts. Use the table to answer Exercises 29–32. Round answers to the nearest thousandth when necessary.

LONG-DISTANCE CALLING CARDS (U.S.)

Card Name	Cost per Minute	Connection Fee
Penny Saver	$0.005	$0.39
Most Minutes	0.0025	0.49
USA Card	0.01	0.25

Source: www.noblecom.com

29. **(a)** Find the *actual* total cost, including the connection charge, for a five-minute call using each card.

(b) Find the cost per minute for this call using each card and select the best buy.

30. **(a)** Find the *actual* total cost, including the connection charge, for a 90-minute call using each card.

(b) Find the cost per minute for this call using each card and select the best buy.

31. Find the *actual* total cost and the per minute cost for a 30-minute call using each card. What do you see when you compare the unit rates for this call?

32. All the cards round calls up to the next full minute.

(a) Suppose you call the wrong number. How much would you pay for this 40-*second* call on each card?

(b) How much would you save on this call by using the USA Card instead of Most Minutes?

33. If you believe the claims that some batteries last longer, which is the better buy?

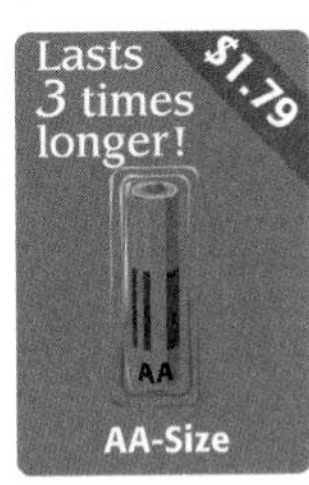

34. Which is the better buy, assuming these laundry detergents both clean equally well?

35. Three brands of cornflakes are available. Brand G is priced at $2.39 for 10 ounces. Brand K is $3.99 for 20.3 ounces and Brand P is $3.39 for 16.5 ounces. You have a coupon for 50¢ off Brand P and a coupon for 60¢ off Brand G. Which cereal is the best buy based on cost per unit?

36. Two brands of facial tissue are available. Brand K is priced at $5 for three boxes of 175 tissues each. Brand S is priced at $1.29 per box of 125 tissues. You have a coupon for 20¢ off one box of Brand S and a coupon for 45¢ off one box of Brand K. How can you get the best buy on one box of tissue?

Relating Concepts (Exercises 37–41) For Individual or Group Work

On the first page of this chapter, we said that unit rates can help you get the best deal on cell phone service. Use the information in the table to ***work Exercises 37–41 in order.*** *Round all money answers to the nearest cent.*

CELL PHONE PLANS

Plan	Anytime Minutes	Monthly Charge*	Cost per Minute (to nearest cent)	Overage†
A	450	\$39.95		45¢
B	900	\$59.95		40¢

*Does not include taxes or special fees.
†Cost per extra minute if you use more minutes than your plan covers.
Source: www.Verizon.com

37. Find the cost per minute for each plan. Round your answers to the nearest cent and write them in the table above. Which plan is the better buy based on cost per minute?

38. In a 30-day month, what is the average number of minutes you can use your phone each day on:
(a) Plan A
(b) Plan B

39. Suppose you are on Plan A. Last month you used 2 more minutes each day than the average number of minutes you calculated in **Exercise 38.** Find the new cost per minute for last month. (Assume 30 days in last month.)

40. Suppose you are on Plan B. Last month you used 3 more minutes each day than the average number of minutes you calculated in **Exercise 38.** Find the new cost per minute for last month. (Assume 30 days in last month.)

41. The company also offers Plan C, with unlimited minutes for \$69.95 per month. If you had Plan C, find the cost per minute for the situations described in **Exercises 39 and 40.**

5.3 Proportions

OBJECTIVE 1 Write proportions. A **proportion** states that two ratios (or rates) are equal. For example,

$$\frac{\$20}{4 \text{ hours}} = \frac{\$40}{8 \text{ hours}}$$

is a proportion that says the rate $\frac{\$20}{4 \text{ hours}}$ is equal to the rate $\frac{\$40}{8 \text{ hours}}$. As the amount of money doubles, the number of hours also doubles. This proportion is read:

20 dollars **is to** 4 hours **as** 40 dollars **is to** 8 hours.

EXAMPLE 1 Writing Proportions

Write each proportion.

(a) 6 ft is to 11 ft **as** 18 ft is to 33 ft.

$$\frac{6 \text{ ft}}{11 \text{ ft}} = \frac{18 \text{ ft}}{33 \text{ ft}} \quad \text{so} \quad \frac{6}{11} = \frac{18}{33}$$

The common units (ft) divide out and are not written.

(b) \$9 is to 6 liters **as** \$3 is to 2 liters.

$$\frac{\$9}{6 \text{ liters}} = \frac{\$3}{2 \text{ liters}}$$

The units do *not* match so you must write them in the proportion.

Work Problem **1** at the Side. ▶

OBJECTIVE 2 Determine whether proportions are true or false. There are two ways to see whether a proportion is true. One way is to *write both of the ratios in lowest terms.*

EXAMPLE 2 Writing Both Ratios in Lowest Terms

Determine whether each proportion is true or false by writing both ratios in lowest terms.

(a) $\frac{5}{9} = \frac{18}{27}$

Write each ratio in lowest terms.

$\frac{5}{9}$ ← Already in lowest terms $\qquad \frac{18 \div 9}{27 \div 9} = \frac{2}{3}$ ← Lowest terms

Because $\frac{5}{9}$ is *not* equivalent to $\frac{2}{3}$, the proportion is *false.*

(b) $\frac{16}{12} = \frac{28}{21}$

Write each ratio in lowest terms.

$$\frac{16 \div 4}{12 \div 4} = \frac{4}{3} \quad \text{and} \quad \frac{28 \div 7}{21 \div 7} = \frac{4}{3}$$

Both ratios in lowest terms.

Both ratios are equivalent to $\frac{4}{3}$, so the proportion is *true.*

Work Problem **2** at the Side. ▶

OBJECTIVES

1. Write proportions.
2. Determine whether proportions are true or false.
3. Find cross products.

1 Write each proportion.

GS **(a)** \$7 is to 3 cans as \$28 is to 12 cans

$\frac{\$7}{3 \text{ cans}} = \frac{\_\_\_}{\_\_\_}$ Complete the proportion.

(b) 9 meters is to 16 meters as 18 meters is to 32 meters

(c) 5 is to 7 as 35 is to 49

(d) 10 is to 30 as 60 is to 180

2 Determine whether each proportion is true or false by writing both ratios in lowest terms.

(a) $\frac{6}{12} = \frac{15}{30}$

(b) $\frac{20}{24} = \frac{3}{4}$

(c) $\frac{25}{40} = \frac{30}{48}$

(d) $\frac{35}{45} = \frac{12}{18}$

Answers

1. (a) $\frac{\$7}{3 \text{ cans}} = \frac{\$28}{12 \text{ cans}}$ **(b)** $\frac{9}{16} = \frac{18}{32}$
(c) $\frac{5}{7} = \frac{35}{49}$ **(d)** $\frac{10}{30} = \frac{60}{180}$

2. (a) $\frac{1}{2} = \frac{1}{2}$; true **(b)** $\frac{5}{6} \neq \frac{3}{4}$; false
(c) $\frac{5}{8} = \frac{5}{8}$; true **(d)** $\frac{7}{9} \neq \frac{2}{3}$; false

OBJECTIVE 3 **Find cross products.** Another way to test whether the ratios in a proportion are equivalent is to compare *cross products.*

Using Cross Products to Determine Whether a Proportion Is True

To see whether a proportion is true, first multiply along one diagonal, then multiply along the other diagonal, as shown here.

$$5 \cdot 4 = 20$$

$$\frac{2}{5} = \frac{4}{10}$$

$$2 \cdot 10 = 20$$

Cross products are equal.

In this case the **cross products** are both 20. When cross products are *equal,* the proportion is *true.* If the cross products are *unequal,* the proportion is *false.*

Note

The cross products test is based on rewriting both fractions with a common denominator of 5 • 10, or 50.

$$\frac{2 \cdot 10}{5 \cdot 10} = \frac{20}{50} \quad \text{and} \quad \frac{4 \cdot 5}{10 \cdot 5} = \frac{20}{50}$$

We see that $\frac{2}{5}$ and $\frac{4}{10}$ are equivalent because both can be rewritten as $\frac{20}{50}$. The cross products test takes a shortcut by comparing only the two numerators (20 = 20).

EXAMPLE 3 **Using Cross Products**

Use cross products to see whether each proportion is true or false.

(a) $\frac{3}{5} = \frac{12}{20}$

Multiply along one diagonal, then multiply along the other diagonal.

$$5 \cdot 12 = 60$$

$$\frac{3}{5} = \frac{12}{20}$$

$$3 \cdot 20 = 60$$

Equal cross products; proportion is *true.*

The cross products are *equal,* so the proportion is *true.*

CAUTION

Use cross products *only* when working with *proportions.* Do ***not*** use cross products when multiplying fractions, adding fractions, or writing fractions in lowest terms.

Continued on Next Page

(b) $\dfrac{2\frac{1}{3}}{3\frac{1}{3}} = \dfrac{9}{16}$

Find the cross products.

$$3\frac{1}{3} \cdot 9 = \frac{10}{\cancel{3}_{1}} \cdot \frac{\cancel{9}^{3}}{1} = \frac{30}{1} = \mathbf{30}$$

$$\dfrac{2\frac{1}{3}}{3\frac{1}{3}} = \dfrac{9}{16}$$

$$2\frac{1}{3} \cdot 16 = \frac{7}{3} \cdot \frac{16}{1} = \frac{112}{3} = \mathbf{37\frac{1}{3}}$$

Unequal cross products; proportion is *false*.

The cross products are *unequal,* so the proportion is *false.*

Note

The numbers in a proportion do *not* have to be whole numbers. They can be fractions, mixed numbers, decimal numbers, and so on.

Work Problem 3 at the Side. ▶

3 Find the cross products to see whether each proportion is true or false.

GS **(a)** $\dfrac{5}{9} = \dfrac{10}{18}$ $(9)(10) = \_\_\_\_$ $(5)(18) = \_\_\_\_$

(b) $\dfrac{32}{15} = \dfrac{16}{8}$

(c) $\dfrac{10}{17} = \dfrac{20}{34}$

GS **(d)** $\dfrac{2.4}{6} = \dfrac{5}{12}$ $(6)(5) = \_\_\_\_$ $(2.4)(12) = \_\_\_\_$

(e) $\dfrac{3}{4.25} = \dfrac{24}{34}$

(f) $\dfrac{1\frac{1}{6}}{2\frac{1}{3}} = \dfrac{4}{8}$

Answers

3. **(a)** $(9)(10) = 90$; $(5)(18) = 90$; true
(b) $240 \neq 256$; false
(c) $340 = 340$; true
(d) $(6)(5) = 30$; $(2.4)(12) = 28.8$; false
(e) $102 = 102$; true
(f) $9\frac{1}{3} = 9\frac{1}{3}$; true

5.3 Exercises

FOR EXTRA HELP

MyMathLab®

Write each proportion. **See Example 1.**

1. \$9 is to 12 cans as \$18 is to 24 cans.

2. 28 people is to 7 cars as 16 people is to 4 cars.

3. 200 adults is to 450 children as 4 adults is to 9 children.

4. 150 trees is to 1 acre as 1500 trees is to 10 acres.

5. 120 ft is to 150 ft as 8 ft is to 10 ft.

6. \$6 is to \$9 as \$10 is to \$15.

Determine whether each proportion is true or false by writing the ratios in lowest terms. Show the simplified ratios and then write true *or* false. **See Example 2.**

7. $\frac{6}{10} = \frac{3}{5}$

8. $\frac{1}{4} = \frac{9}{36}$

9. $\frac{5}{8} = \frac{25}{40}$

10. $\frac{2}{3} = \frac{20}{27}$

11. $\frac{150}{200} = \frac{200}{300}$

12. $\frac{100}{120} = \frac{75}{100}$

13. $\frac{42}{15} = \frac{28}{10}$

14. $\frac{18}{16} = \frac{36}{32}$

15. $\frac{32}{18} = \frac{48}{27}$

16. $\frac{15}{48} = \frac{10}{24}$

17. $\frac{7}{6} = \frac{54}{48}$

18. $\frac{28}{21} = \frac{44}{33}$

19. **CONCEPT CHECK** What is a proportion? Explain. Then draw arrows on the proportion below and show what numbers to multiply to find the cross products.

$$\frac{30}{6} = \frac{45}{9}$$

20. **CONCEPT CHECK** Explain how to tell if a proportion is true or false after you have found the cross products.

Use cross products to determine whether each proportion is true or false. Show the cross products and then circle true *or* false. **See Example 3.**

21. $\frac{2}{9} = \frac{6}{27}$

True False

22. $\frac{20}{25} = \frac{4}{5}$

True False

23. $\frac{20}{28} = \frac{12}{16}$

True False

24. $\dfrac{16}{40} = \dfrac{22}{55}$

True False

25. $\dfrac{110}{18} = \dfrac{160}{27}$

True False

26. $\dfrac{600}{420} = \dfrac{20}{14}$

True False

27. $\dfrac{3.5}{4} = \dfrac{7}{8}$

True False

28. $\dfrac{36}{23} = \dfrac{9}{5.75}$

True False

29. $\dfrac{18}{16} = \dfrac{2.8}{2.5}$

True False

30. $\dfrac{0.26}{0.39} = \dfrac{1.3}{1.9}$

True False

31. $\dfrac{6}{3\frac{2}{3}} = \dfrac{18}{11}$

True False

32. $\dfrac{16}{13} = \dfrac{2}{1\frac{5}{8}}$

True False

33. $\dfrac{2\frac{5}{8}}{3\frac{1}{4}} = \dfrac{21}{26}$

True False

34. $\dfrac{28}{17} = \dfrac{9\frac{1}{3}}{5\frac{2}{3}}$

True False

35. $\dfrac{\frac{2}{3}}{2} = \dfrac{2.7}{8}$ *Hint:* 2.7 is equivalent to $2\frac{7}{10}$.

True False

36. $\dfrac{3.75}{1\frac{1}{4}} = \dfrac{7.5}{2\frac{1}{2}}$

Hint: $1\frac{1}{4}$ is equivalent to 1.25 and $2\frac{1}{2}$ is equivalent to 2.5

True False

37. $\dfrac{2\frac{3}{10}}{8.05} = \dfrac{\frac{1}{4}}{0.9}$

True False

38. $\dfrac{\frac{3}{5}}{6} = \dfrac{1.5}{\frac{7}{12}}$

True False

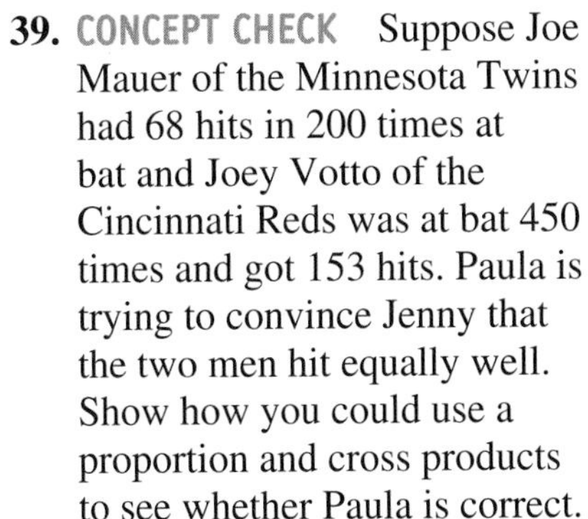

39. CONCEPT CHECK Suppose Joe Mauer of the Minnesota Twins had 68 hits in 200 times at bat and Joey Votto of the Cincinnati Reds was at bat 450 times and got 153 hits. Paula is trying to convince Jenny that the two men hit equally well. Show how you could use a proportion and cross products to see whether Paula is correct.

40. CONCEPT CHECK Jay worked 3.5 hours and packed 91 cartons. Craig packed 126 cartons in 5.25 hours. To see whether the men worked equally fast, Barry set up this proportion:

$$\frac{3.5}{91} = \frac{126}{5.25}$$

Explain what is wrong with Barry's proportion and write a correct one. Is the correct proportion true or false?

5.4 Solving Proportions

OBJECTIVES

1. Find the unknown number in a proportion.
2. Find the unknown number in a proportion with mixed numbers or decimals.

OBJECTIVE 1 Find the unknown number in a proportion. Four numbers are used in a proportion. If any three of these numbers are known, the fourth can be found. For example, find the unknown number that will make this proportion true.

$$\frac{3}{5} = \frac{x}{40}$$

The x represents the unknown number. Start by finding the cross products.

$$\frac{3}{5} = \frac{x}{40} \qquad 5 \cdot x \quad 3 \cdot 40$$

Cross products

To make the proportion true, the cross products must be equal.

$$5 \cdot x = \underbrace{3 \cdot 40}$$

$$5 \cdot x = 120$$

The equal sign says that $5 \cdot x$ and 120 are equal. If $5 \cdot x$ and 120 are *both* divided by 5, the results will still be equal.

$$\frac{5 \cdot x}{5} = \frac{120}{5} \leftarrow \text{Divide both sides by 5.}$$

On the left side, divide out the common factor of 5; slashes indicate the divisions.

$$\frac{\overset{1}{\cancel{5}} \cdot x}{\underset{1}{\cancel{5}}} = 24$$

On the right side, divide 120 by 5 to get 24.

Multiplying by 1 does *not* change a number, so in the numerator on the left side, $1 \cdot x$ is the same as x.

$$\frac{x}{1} = 24$$

Dividing by 1 does *not* change a number, so on the left side, $\frac{x}{1}$ is the same as x.

$$x = 24$$

The solution is 24.

The unknown number in the proportion is 24. The complete proportion is shown below.

$$\frac{3}{5} = \frac{24}{40}$$

x is 24.

Check by finding the cross products. If they are equal, you solved the problem correctly. If they are unequal, rework the problem.

$$\frac{3}{5} = \frac{24}{40} \qquad 5 \cdot 24 = 120 \quad 3 \cdot 40 = 120$$

Equal; proportion is true.

The cross products are equal, so the solution, $x = 24$, is correct.

CAUTION

The solution is 24, which is the unknown number in the proportion. 120 is ***not*** the solution; it is the cross product you get when *checking* the solution.

Solve a proportion for an unknown number by using the following steps.

Finding an Unknown Number in a Proportion

Step 1 Find the cross products.

Step 2 Show that the cross products are equal.

Step 3 Divide both products by the number multiplied by x (the number next to x).

Step 4 Check by writing the solution in the *original* proportion and finding the cross products.

EXAMPLE 1 Solving Proportions for Unknown Numbers

Find the unknown number in each proportion. Round answers to the nearest hundredth when necessary.

(a) $\frac{16}{x} = \frac{32}{20}$

Recall that ratios can be rewritten in lowest terms. If desired, you can do that *before* finding the cross products. In this example, write $\frac{32}{20}$ in lowest terms as $\frac{8}{5}$, which gives the proportion $\frac{16}{x} = \frac{8}{5}$.

Step 1 $\frac{\mathbf{16}}{\mathbf{x}} = \frac{\mathbf{8}}{\mathbf{5}}$ $x \cdot 8$ / $16 \cdot 5$ ← Find the cross products.

Step 2 $x \cdot 8 = 16 \cdot 5$ ← Show that cross products are equal.

$x \cdot 8 = 80$

Step 3 $\frac{x \cdot \overset{1}{\cancel{8}}}{\underset{1}{\cancel{8}}} = \frac{80}{8}$ ← Divide both sides by 8.

The solution is 10 → $x = 10$ ← Find x. (No rounding necessary)

The unknown number in the proportion is 10.

Step 4 Write the solution in the *original* proportion and check by finding cross products.

$10 \cdot 32 = 320$

x is 10. → $\frac{\mathbf{16}}{\mathbf{10}} = \frac{\mathbf{32}}{\mathbf{20}}$ Equal; proportion is true.

$16 \cdot 20 = 320$

The solution is 10, **not** 320.

The cross products are equal, so **10 is the correct solution.**

Note

It is not necessary to write the ratios in lowest terms before solving. However, if you do, you will work with smaller numbers.

Continued on Next Page

1 Find the unknown numbers. Round to hundredths when necessary. Check your solutions by finding the cross products.

GS **(a)**

$$\frac{1}{2} = \frac{x}{12} \quad 2 \cdot x \text{ and } 1 \cdot 12 = 12 \text{ (Cross products)}$$

$2 \cdot x = 12$ Now finish the solution. Divide both sides by 2.

(b) $\dfrac{6}{10} = \dfrac{15}{x}$

(c) $\dfrac{28}{x} = \dfrac{21}{9}$

(d) $\dfrac{x}{8} = \dfrac{3}{5}$

(e) $\dfrac{14}{11} = \dfrac{x}{3}$

(b) $\dfrac{7}{12} = \dfrac{15}{x}$

Step 1 $\dfrac{7}{12} = \dfrac{15}{x}$ $\quad 12 \cdot 15 = 180$ and $7 \cdot x$ — Find the cross products.

Step 2 $7 \cdot x = 180$ ← Show that cross products are equal.

Step 3 $\dfrac{\overset{1}{\cancel{7}} \cdot x}{\underset{1}{\cancel{7}}} = \dfrac{180}{7}$ ← Divide both sides by 7.

The rounded solution is 25.71 → $x \approx 25.71$ ← Rounded to nearest hundredth

When the division does not come out even, check for directions on how to round your answer. Divide out one more place, then round.

$$7\overline{)180.000} = 25.714$$ ← Divide out to thousandths so you can round to hundredths.

The unknown number in the proportion is 25.71 (rounded).

Step 4 Write the solution in the original proportion and check by finding the cross products.

$$\frac{7}{12} = \frac{15}{25.71} \quad 12 \cdot 15 = \mathbf{180} \qquad 7 \cdot 25.71 = \mathbf{179.97}$$

Very close, but *not* equal due to rounding the solution

The cross products are slightly different because you rounded the value of x. However, they are close enough to see that the problem was done correctly and that **25.71 is the approximate solution.**

◄ **Work Problem 1 at the Side.**

OBJECTIVE 2 **Find the unknown number in a proportion with mixed numbers or decimals.** The next example shows how to work with mixed numbers or decimals in a proportion.

EXAMPLE 2 Solving Proportions with Mixed Numbers and Decimals

Find the unknown number in each proportion.

(a) $\dfrac{2\frac{1}{5}}{6} = \dfrac{x}{10}$ $\qquad 6 \cdot x$ and $2\frac{1}{5} \cdot 10$ — Find the cross products.

Find $2\frac{1}{5} \cdot 10$.

$$2\frac{1}{5} \cdot 10 = \frac{11}{5} \cdot \frac{10}{1} = \frac{11}{\underset{1}{\cancel{5}}} \cdot \frac{\overset{2}{\cancel{10}}}{1} = \frac{22}{1} = 22$$

Changed to improper fraction

Continued on Next Page

Answers

1. (a) $\dfrac{\overset{1}{\cancel{2}} \cdot x}{\underset{1}{\cancel{2}}} = \dfrac{12}{2}$; $x = 6$ (b) $x = 25$
(c) $x = 12$ (d) $x = 4.8$
(e) $x \approx 3.82$ (rounded to nearest hundredth)

Show that the cross products are equal.

$$6 \cdot x = 22$$

Divide both sides by 6. $\dfrac{\overset{1}{\cancel{6}} \cdot x}{\underset{1}{\cancel{6}}} = \dfrac{22}{6}$

Write the solution as a mixed number in lowest terms.

$$x = \frac{22 \div 2}{6 \div 2} = \frac{11}{3} = 3\frac{2}{3}$$

The solution is $3\frac{2}{3}$.

The unknown number is $3\frac{2}{3}$.

Write the solution in the proportion and check by finding the cross products.

$$6 \cdot 3\frac{2}{3} = \frac{\overset{2}{\cancel{6}}}{1} \cdot \frac{11}{\underset{1}{\cancel{3}}} = \frac{22}{1} = 22$$

$$\frac{2\frac{1}{5}}{6} = \frac{3\frac{2}{3}}{10}$$

Equal

$$2\frac{1}{5} \cdot 10 = \frac{11}{\underset{1}{\cancel{5}}} \cdot \frac{\overset{2}{\cancel{10}}}{1} = \frac{22}{1} = 22$$

The solution is $3\frac{2}{3}$, **not** 22.

The cross products are equal, so $\mathbf{3\frac{2}{3}}$ is the correct solution.

(b) $\dfrac{1.5}{0.6} = \dfrac{2}{x}$

Show that cross products are equal.

$$(1.5)(x) = \underbrace{(0.6)(2)}$$

$$(1.5)(x) = 1.2$$

Divide both sides by 1.5 $\dfrac{(\overset{1}{\cancel{1.5}})(x)}{\underset{1}{\cancel{1.5}}} = \dfrac{1.2}{1.5}$

$$x = \frac{1.2}{1.5}$$

Complete the division.

The solution is 0.8

$x = 0.8$ $\qquad 1.5\overline{)1.20}$ quotient $.8$

So the unknown number is 0.8. Write the solution in the original proportion and check it by finding the cross products.

$$(0.6)(2) = 1.2$$

$$\frac{1.5}{0.6} = \frac{2}{0.8}$$

Equal

$$(1.5)(0.8) = 1.2$$

The solution is 0.8, **not** 1.2

The cross products are equal, so **0.8** is the correct solution.

Work Problem 2 at the Side. ▶

2 Find the unknown numbers. Round to hundredths on the decimal problems, if necessary. Check your solutions by finding the cross products.

GS **(a)** $\dfrac{3\frac{1}{4}}{2} = \dfrac{x}{8}$ $\qquad 2 \cdot x$

$$3\frac{1}{4} \cdot 8 = \frac{13}{\underset{1}{\cancel{4}}} \cdot \frac{\overset{2}{\cancel{8}}}{1} = \frac{26}{1}$$

$2 \cdot x = 26$ Finish the solution.

(b) $\dfrac{x}{3} = \dfrac{1\frac{2}{3}}{5}$

(c) $\dfrac{0.06}{x} = \dfrac{0.3}{0.4}$

(d) $\dfrac{2.2}{5} = \dfrac{13}{x}$

(e) $\dfrac{x}{6} = \dfrac{0.5}{1.2}$

(f) $\dfrac{0}{2} = \dfrac{x}{7.092}$

Answers

2. **(a)** $x = 13$ **(b)** $x = 1$ **(c)** $x = 0.08$
(d) $x \approx 29.55$ (rounded to nearest hundredth)
(e) $x = 2.5$ **(f)** $x = 0$

5.4 Exercises

FOR EXTRA HELP

Download the MyDashBoard App

MyMathLab®

1. CONCEPT CHECK In $\frac{7}{10} = \frac{5}{x}$, what is the first step when finding the unknown number? Show how to do this step.

2. CONCEPT CHECK In $\frac{7}{10} = \frac{5}{x}$, the cross products are $7 \cdot x$ and 50. What is the next step when finding the unknown number? Show how to do this step.

Find the unknown number in each proportion. Round your answers to hundredths, if necessary. Check your answers by finding the cross products. **See Examples 1 and 2.**

3. $\frac{1}{3} = \frac{x}{12}$ (cross products: $3 \cdot x$ and $1 \cdot 12$)

$3 \cdot x = \underbrace{1 \cdot 12}$

$3 \cdot x = \underline{\quad}$

$\frac{\overset{1}{\cancel{3}} \cdot x}{\underset{1}{\cancel{3}}} = \frac{\underline{\quad}}{3}$

$x = \underline{\quad}$

4. $\frac{x}{6} = \frac{15}{18}$ (cross products: $6 \cdot 15$ and $x \cdot 18$)

$x \cdot 18 = \underbrace{6 \cdot 15}$

$x \cdot 18 = \underline{\quad}$

$\frac{x \cdot \overset{1}{\cancel{18}}}{\underset{1}{\cancel{18}}} = \frac{\underline{\quad}}{18}$

$x = \underline{\quad}$

5. $\frac{15}{10} = \frac{3}{x}$

6. $\frac{5}{x} = \frac{20}{8}$

7. $\frac{x}{11} = \frac{32}{4}$

8. $\frac{12}{9} = \frac{8}{x}$

9. $\frac{42}{x} = \frac{18}{39}$

10. $\frac{49}{x} = \frac{14}{18}$

11. $\frac{x}{25} = \frac{4}{20}$

12. $\frac{6}{x} = \frac{4}{8}$

13. $\frac{8}{x} = \frac{24}{30}$

14. $\frac{32}{5} = \frac{x}{10}$

15. $\frac{99}{55} = \frac{44}{x}$

16. $\frac{x}{12} = \frac{101}{147}$

17. $\frac{0.7}{9.8} = \frac{3.6}{x}$

18. $\frac{x}{3.6} = \frac{4.5}{6}$

19. $\frac{250}{24.8} = \frac{x}{1.75}$

20. $\frac{4.75}{17} = \frac{43}{x}$

Find the unknown number in each proportion. Write your answers as whole or mixed numbers when possible. ***See Example 2.***

21. $\frac{15}{1\frac{2}{3}} = \frac{9}{x}$

22. $\frac{x}{\frac{3}{10}} = \frac{2\frac{2}{9}}{1}$

23. $\frac{2\frac{1}{3}}{1\frac{1}{2}} = \frac{x}{2\frac{1}{4}}$

24. $\frac{1\frac{5}{6}}{x} = \frac{\frac{3}{14}}{\frac{6}{7}}$

Solve each proportion two different ways. First change all the numbers to decimal form and solve. Then change all the numbers to fraction form and solve; write your answers in lowest terms.

25. $\frac{\frac{1}{2}}{x} = \frac{2}{0.8}$

26. $\frac{\frac{3}{20}}{0.1} = \frac{0.03}{x}$

27. $\frac{x}{\frac{3}{50}} = \frac{0.15}{1\frac{4}{5}}$

28. $\frac{8\frac{4}{5}}{1\frac{1}{10}} = \frac{x}{0.4}$

Relating Concepts (Exercises 29–30) For Individual or Group Work

Work Exercises 29–30 in order. *First prove that the proportions are* **not** *true. Then create four true proportions for each exercise by changing one number at a time.*

29. $\frac{10}{4} = \frac{5}{3}$

30. $\frac{6}{8} = \frac{24}{30}$

Summary Exercises *Ratios, Rates, and Proportions*

Use the bar graph of the countries with the most websites to complete Exercises 1–4. Write each ratio as a fraction in lowest terms.

1. Write the ratio of the number of websites in Japan to the number of websites in Germany.

2. What is the ratio of the number of websites in Brazil to the number of websites in Italy?

3. Find the ratio of the combined number of websites in Italy and Germany compared to the number of websites in Japan.

4. Write the ratio of the number of websites in the USA compared to the total number of websites in all the other countries shown in the graph.

COUNTRIES WITH THE MOST WEBSITES

Number of websites, rounded to the nearest 5 million.

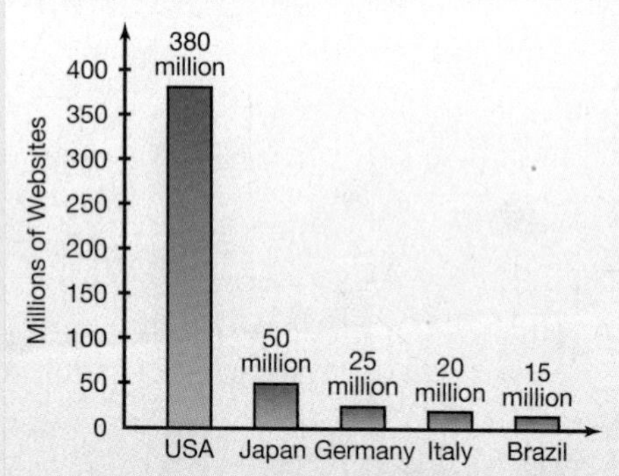

Source: *Scholastic Book of World Records.*

The bar graph shows the number of Americans who play various instruments. Use the graph to complete Exercises 5 and 6. Write each ratio as a fraction in lowest terms.

AMERICANS MAKE MUSIC BY THE MILLIONS

How many people play each type of instrument?

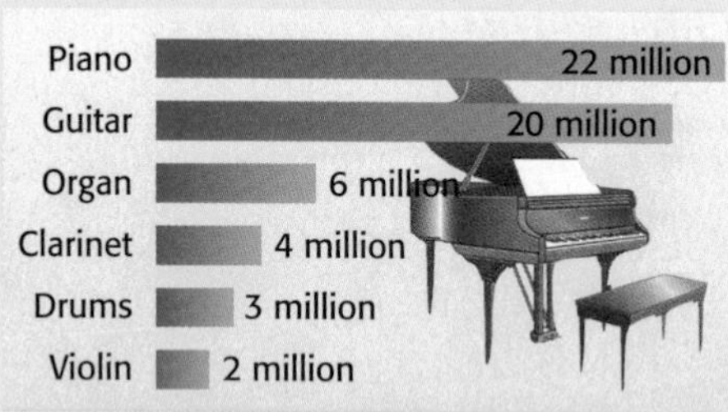

Source: *America by the Numbers.*

5. Write five ratios that compare the least popular instrument to each of the other instruments.

6. Which two instruments give each of these ratios?

   (a) $\frac{5}{1}$

   (b) $\frac{2}{1}$

   There may be more than one correct answer.

CONCEPT CHECK *The table lists data on the top three individual scoring NBA basketball games of all time. Use the data to answer Exercises 7 and 8. Round answers to the nearest tenth.*

7. What was Wilt Chamberlain's scoring rate in points per minute and in minutes per point?

8. Find Kobe Bryant's scoring rate in points per minute and minutes per point.

Player	Date	Points	Min.
Wilt Chamberlain	3/2/62	100	48
Kobe Bryant	1/22/06	81	48
David Thompson	9/4/78	73	43

Source: www.sportcity.com

Wilt Chamberlain (#13)

9. Lucinda's paycheck showed gross pay of \$652.80 for 40 hours of work and \$195.84 for 8 hours of overtime work. Find her regular hourly pay rate and her overtime rate.

10. Satellite TV service is being offered to new subscribers at \$29.99 per month for 150 channels, \$34.99 per month for 210 channels, or \$39.99 per month for 225 channels. What is the monthly cost per channel under each plan, to the nearest cent? (*Source:* DIRECTV advertisement.)

11. Find the best buy on gourmet coffee beans. Round to the nearest cent.

11 ounces for \$6.79

12 ounces for \$7.24

16 ounces for \$10.99

(*Source:* Cub Foods.)

12. Which brand of cat food is the best buy? You have a coupon for \$2 off Brand P, and another for \$1 off Brand N.

Brand N is \$3.75 for 3.5 pounds.

Brand P is \$5.99 for 7 pounds.

Brand R is \$10.79 for 18 pounds.

Use either the method of writing in lowest terms or the method of finding cross products to decide whether each proportion is true or false. Show your work and then write true *or* false.

13. $\frac{28}{21} = \frac{44}{33}$

14. $\frac{2.3}{8.05} = \frac{0.25}{0.9}$

15. $\frac{2\frac{5}{8}}{3\frac{1}{4}} = \frac{21}{26}$

Solve each proportion to find the unknown number. Round your answers to hundredths when necessary.

16. $\frac{7}{x} = \frac{25}{100}$

17. $\frac{15}{8} = \frac{6}{x}$

18. $\frac{x}{84} = \frac{78}{36}$

19. $\frac{10}{11} = \frac{x}{4}$

20. $\frac{x}{17} = \frac{3}{55}$

21. $\frac{2.6}{x} = \frac{13}{7.8}$

22. $\frac{0.14}{1.8} = \frac{x}{0.63}$

23. $\frac{\frac{1}{3}}{8} = \frac{x}{24}$

24. $\frac{6\frac{2}{3}}{4\frac{1}{6}} = \frac{\frac{6}{5}}{x}$

5.5 Solving Application Problems with Proportions

OBJECTIVE

1. **Use proportions to solve application problems.**

OBJECTIVE 1 Use proportions to solve application problems. Proportions can be used to solve a wide variety of problems. Watch for problems in which you are given a ratio or rate and then are asked to find part of a corresponding ratio or rate. Remember that a ratio or rate compares two quantities and often includes one of the following indicator words.

in for on per from to

Use the six problem-solving steps you learned in **Chapter 1.**

Step 1 **Read** the problem.

Step 2 **Work out a plan.**

Step 3 **Estimate** a reasonable answer.

Step 4 **Solve** the problem.

Step 5 **State the answer.**

Step 6 **Check** your work.

EXAMPLE 1 Solving a Proportion Application

Mike's car can travel 163 **miles on** 6.4 **gallons** of gas. How far can it travel on a full tank of 14 **gallons** of gas? Round to the nearest whole mile.

Step 1 **Read** the problem. The problem asks for the number of miles the car can travel on 14 gallons of gas.

Step 2 **Work out a plan.** Decide what is being compared. This example compares **miles** to **gallons**. Write a proportion using the two rates. Be sure that *both* rates compare miles to gallons in the same order. In other words, miles is in both numerators and gallons is in both denominators. Use a letter to represent the unknown number.

Matching units

This rate compares **miles** to **gallons**. → $\frac{163 \textbf{ miles}}{6.4 \textbf{ gallons}} = \frac{x \textbf{ miles}}{14 \textbf{ gallons}}$ ← This rate compares **miles** to **gallons**.

Matching units

Step 3 **Estimate** a reasonable answer. To estimate the answer, notice that 14 gallons is a little more than *twice as much* as 6.4 gallons, so the car should travel a little more than *twice as far.* So use $2 \cdot 163$ miles = 326 miles as the estimate.

Step 4 **Solve** the problem. Ignore the units while solving for x.

$$\frac{163 \text{ miles}}{6.4 \text{ gallons}} = \frac{x \text{ miles}}{14 \text{ gallons}}$$

$$(6.4)(x) = \underbrace{(163)(14)}$$ Show that cross products are equal.

$$(6.4)(x) = 2282$$

$$\frac{\overset{1}{\cancel{(6.4)}}(x)}{\underset{1}{\cancel{6.4}}} = \frac{2282}{6.4}$$ Divide both sides by 6.4

$$x = 356.5625$$ Round to 357.

Check the problem for rounding directions; this one asks for nearest whole number.

Continued on Next Page

Step 5 **State the answer.** Rounded to the nearest mile, the car can travel about **357 miles** on a full tank of gas.

Be sure to write **miles** in your answer.

Step 6 **Check** your work. The answer, 357 miles, is a little more than the estimate of 326 miles, so it is reasonable.

CAUTION

When setting up a proportion, do ***not*** mix up the units in the rates.

compares **miles** to **gallons** $\left\{ \frac{163 \text{ miles}}{6.4 \text{ gallons}} \neq \frac{14 \text{ gallons}}{x \text{ miles}} \right\}$ compares **gallons** to **miles**

These rates do ***not*** compare things in the same order and ***cannot*** be set up as a proportion.

Work Problem 1 at the Side. ▶

EXAMPLE 2 Solving a Proportion Application

A newspaper report says that 7 out of 10 people surveyed watch the news on TV. At that rate, how many of the 3200 people in town would you expect to watch the news?

Step 1 **Read** the problem. The problem asks how many of the 3200 people in town would be expected to watch TV news.

Step 2 **Work out a plan.** You are comparing people who watch the news to people surveyed. Set up a proportion using the two rates described in the example. Be sure that both rates make the same comparison. "People who watch the news" is mentioned first, so it should be in the numerator of *both* rates.

People who watch news → 7 / Total group (people surveyed) → 10 = x ← People who watch news / 3200 ← Total group (people in town)

$$\frac{7}{10} = \frac{x}{3200}$$

Step 3 **Estimate** a reasonable answer. To estimate the answer, notice that 7 out of 10 people is more than half the people, but less than all the people. Half of 3200 people is $3200 \div 2 = 1600$, so our estimate is between 1600 and 3200 people.

Step 4 **Solve** the problem. Solve for the unknown number in the proportion.

$$\frac{7}{10} = \frac{x}{3200}$$

$$10 \cdot x = 7 \cdot 3200 \quad \text{Show that cross products are equal.}$$

$$10 \cdot x = 22{,}400$$

$$\frac{\overset{1}{\cancel{10}} \cdot x}{\underset{1}{\cancel{10}}} = \frac{22{,}400}{10} \quad \text{Divide both sides by 10.}$$

$$x = 2240$$

No rounding is needed here.

Continued on Next Page

1 Set up and solve a proportion for each problem.

GS **(a)** If 2 pounds of fertilizer will cover 50 square feet of garden, how many pounds are needed for 225 square feet?

$$\frac{2 \text{ pounds}}{50 \text{ square feet}} = \frac{\_\_\_ \text{ pounds}}{\_\_\_ \text{ square feet}}$$

(b) A U.S. map has a scale of 1 inch to 75 miles. Lake Superior is 4.75 inches long on the map. What is the lake's actual length to the nearest whole mile ?

(c) A cough syrup is given at the rate of 30 milliliters for each 100 pounds of body weight. How much should be given to a 34-pound child? Round to the nearest whole milliliter.

Answers

1. (a) $\frac{2 \text{ pounds}}{50 \text{ square feet}} = \frac{x \text{ pounds}}{225 \text{ square feet}}$
$x = 9$ pounds

(b) $\frac{1 \text{ inch}}{75 \text{ miles}} = \frac{4.75 \text{ inches}}{x \text{ miles}}$
$x = 356.25$ miles, rounds to 356 miles

(c) $\frac{30 \text{ milliliters}}{100 \text{ pounds}} = \frac{x \text{ milliliters}}{34 \text{ pounds}}$
$x \approx 10$ milliliters (rounded)

2 Solve each problem to find a reasonable answer. Then flip one side of your proportion to see what answer you get with an **incorrect** set-up. Explain why the second answer is **unreasonable.**

GS **(a)** A survey showed that 2 out of 3 people would like to lose weight. At this rate, how many people in a group of 150 want to lose weight?

$$\frac{2}{3} = \frac{x}{\quad}$$

Lose weight → 2 (top left); People surveyed → 3 (bottom left); ← Lose weight (top right); ← People in group (bottom right)

(b) In one state, 3 out of 5 college students receive financial aid. At this rate, how many of the 4500 students at Central Community College receive financial aid?

Answers

2. (a) $\frac{2}{3} = \frac{x}{150}$; 100 people (reasonable); incorrect set-up gives 225 people (only 150 people in the group).
(b) 2700 students (reasonable); incorrect set-up gives 7500 students (only 4500 students at the college).

Step 5 **State the answer.** You would expect **2240 people** in town to watch the news on TV.

Be sure to write **people** in your answer.

Step 6 **Check** your work. The answer, 2240 people, is between 1600 and 3200, as called for in the estimate.

CAUTION

Always check that your answer is reasonable. If it is not, look at the way your proportion is set up. Be sure you have matching units in the numerators and matching units in the denominators.

For example, suppose you had set up the last proportion ***incorrectly,*** as shown here.

$$\frac{7}{10} = \frac{3200}{x} \leftarrow \textbf{Incorrect set-up}$$

$$7 \cdot x = 10 \cdot 3200$$

$$\frac{\overset{1}{\cancel{7}} \cdot x}{\underset{1}{\cancel{7}}} = \frac{32{,}000}{7}$$

$$x \approx 4571 \text{ people} \leftarrow \textbf{Unreasonable answer}$$

This answer is ***unreasonable*** because there are only 3200 people in the town; it is ***not*** possible for 4571 people to watch the news.

◀ Work Problem 2 at the Side.

5.5 Exercises

FOR EXTRA HELP

MyMathLab®

CONCEPT CHECK *Complete the proportion for each situation in Exercises 1 and 2. Do not solve the proportion.*

1. When Linda makes potato salad, she uses 6 potatoes and 4 hard-boiled eggs. How many eggs will she need if she uses 12 potatoes? Write the rest of the units in the proportion below.

$$\frac{6 \text{ potatoes}}{4 \text{ \_\_\_\_\_\_}} = \frac{12 \text{ \_\_\_\_\_\_}}{x \text{ \_\_\_\_\_\_}}$$

2. A potato chip bag says that 1 serving is 15 chips. Joe ate 70 chips. How many servings did he eat? Write the rest of the units in the proportion below.

$$\frac{1 \text{ \_\_\_\_\_\_}}{15 \text{ chips}} = \frac{x \text{ \_\_\_\_\_\_}}{70 \text{ \_\_\_\_\_\_}}$$

Set up and solve a proportion for each application problem. ***See Example 1.***

3. Caroline can sketch four cartoon strips in five hours. How long will it take her to sketch 18 strips?

Compares cartoon strips to **hours** $\frac{4 \text{ strips}}{5 \text{ hours}} = \frac{\text{\_\_\_\_ strips}}{\text{\_\_\_\_ hours}}$

4. The Cosmic Toads recorded eight songs on their first album in 26 hours. At this same rate, how long will it take them to record 14 songs for their second album?

Compares songs to **hours** $\frac{8 \text{ songs}}{26 \text{ hours}} = \frac{\text{\_\_\_\_ songs}}{\text{\_\_\_\_ hours}}$

5. Sixty newspapers cost \$27. Find the cost of 16 newspapers.

6. Twenty-two guitar lessons cost \$528. Find the cost of 12 lessons.

7. If three pounds of fescue grass seed cover about 350 square feet of ground, how many pounds are needed for 4900 square feet?

8. Anna earns \$1242.08 in 14 days. How much does she earn in 260 days?

9. Tom makes \$672.80 in 5 days. How much does he make in 3 days?

10. If 5 ounces of a medicine must be mixed with 8 ounces of water, how many ounces of medicine would be mixed with 20 ounces of water?

11. The bag of rice noodles below makes 7 servings. At that rate, how many ounces of noodles do you need for 12 servings, to the nearest ounce?

12. This can of sweet potatoes is enough for 4 servings. How many ounces are needed for 9 servings, to the nearest ounce?

13. Three quarts of a latex enamel paint will cover about 270 square feet of wall surface. How many quarts will you need to cover 350 square feet of wall surface in your kitchen and 100 square feet of wall surface in your bathroom?

14. One gallon of clear gloss wood finish covers about 550 square feet of surface. If you need to apply three coats of finish to 400 square feet of surface, how many gallons do you need, to the nearest tenth?

Use the floor plan shown to complete Exercises 15–18. On the plan, one inch represents four feet.

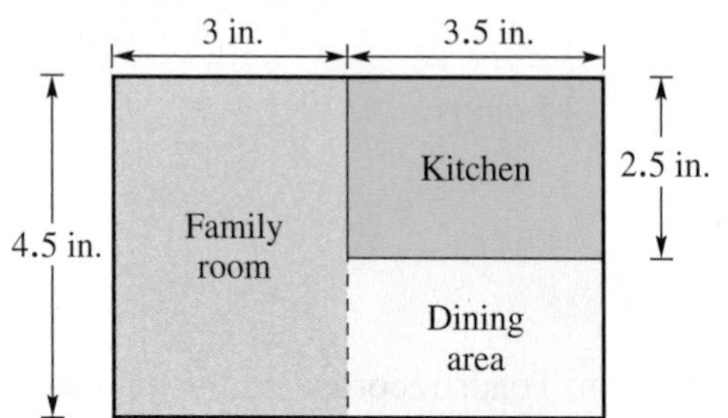

15. What is the actual length and width of the kitchen?

16. What is the actual length and width of the family room?

17. What is the actual length and width of the dining area?

18. What is the actual length and width of the entire floor plan?

The table below lists recommended amounts of food to order for 25 party guests. Use the table to answer Exercises 19 and 20. (Source: Cub Foods.*)*

FOOD FOR 25 GUESTS

Item	Amount
Fried chicken	40 pieces
Lasagna	14 pounds
Deli meats	4.5 pounds
Sliced cheese	$2\frac{1}{3}$ pounds
Bakery buns	3 dozen
Potato salad	6 pounds

19. How much of each food item should Nathan and Amanda order for a graduation party with 60 guests?

20. Taisha is having 20 neighbors over for a Fourth of July picnic. How much food should she buy?

21. CONCEPT CHECK Eight out of 10 students voted to take a fifteen-minute break in the middle of a two-hour class. There are 35 students in the class. How many students voted for the break? Carl set up the proportion below. Explain how to tell if Carl's answer is reasonable or not.

$$\frac{8}{10} = \frac{35}{x} \qquad 8 \cdot x = 350$$

$$\frac{\overset{1}{\cancel{8}} \cdot x}{\underset{1}{\cancel{8}}} = \frac{350}{8}$$

$$x \approx 44 \text{ students (rounded)}$$

22. CONCEPT CHECK Many experts recommend 30 minutes of exercise each day. If Vera follows that advice for 60 days, how many minutes of exercise will she get? Vera set up the proportion below. Explain how to tell if Vera's answer is reasonable or not.

$$\frac{1}{30} = \frac{x}{60} \qquad 30 \cdot x = 60$$

$$\frac{\overset{1}{\cancel{30}} \cdot x}{\underset{1}{\cancel{30}}} = \frac{60}{30}$$

$$x = 2 \text{ minutes}$$

Set up a proportion to solve each problem. Check to see whether your answer is reasonable. Then flip one side of your proportion to see what answer you get with an incorrect set-up. Explain why the second answer is unreasonable. **See Example 2.**

23. About 7 out of 10 people entering a community college need to take a refresher math course. If there are 2950 entering students, how many will probably need refresher math? (*Source:* Minneapolis Community and Technical College.)

24. In a survey, only 3 out of 100 people like their eggs poached. At that rate, how many of the 60 customers who ordered eggs at Soon-Won's restaurant this morning asked to have them poached? Round to the nearest whole person.

25. About 1 out of 3 people choose vanilla as their favorite ice cream flavor. If 250 people attend an ice cream social, how many would you expect to choose vanilla? Round to the nearest whole person.

26. In a test of 200 sewing machines, only one had a defect. At that rate, how many of the 5600 machines shipped from the factory have defects?

Set up and solve a proportion for each problem.

27. The stock market report says that 5 stocks went up for every 6 stocks that went down. If 750 stocks went down yesterday, how many went up?

28. The human body contains 90 pounds of water for every 100 pounds of body weight. How many pounds of water are in a child who weighs 80 pounds?

29. The ratio of the length of an airplane wing to its width is 8 to 1. If the length of a wing is 32.5 meters, how wide must it be? Round to the nearest hundredth.

30. The Rosebud School District wants a student-to-teacher ratio of 19 to 1. How many teachers are needed for 1850 students? Round to the nearest whole number.

31. The number of calories you burn is proportional to your weight. A 150-pound person burns 222 calories during 30 minutes of tennis. How many calories would a 210-pound person burn, to the nearest whole number? (*Source: Wellness Encyclopedia.*)

32. Refer to **Exercise 31.** A 150-pound person burns 189 calories during 45 minutes of grocery shopping. How many calories would a 115-pound person burn, to the nearest whole number? (*Source: Wellness Encyclopedia.*)

33. At 3 P.M., Coretta's shadow is 1.05 meters long. Her height is 1.68 meters. At the same time, a tree's shadow is 6.58 meters long. How tall is the tree? Round to the nearest hundredth.

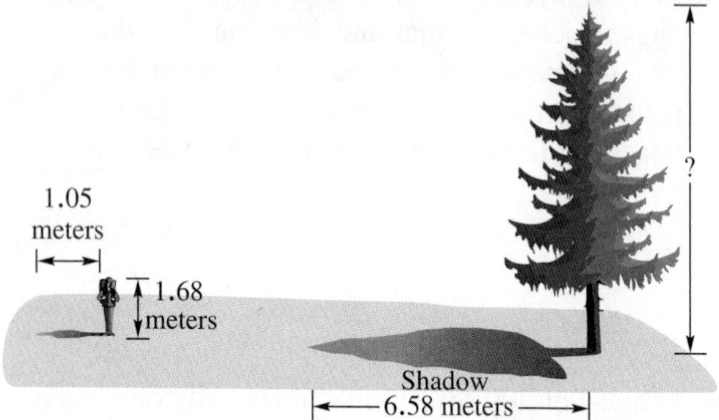

34. Refer to **Exercise 33.** Later in the day, Coretta's shadow was 2.95 meters long. How long a shadow did the tree have at that time? Round to the nearest hundredth.

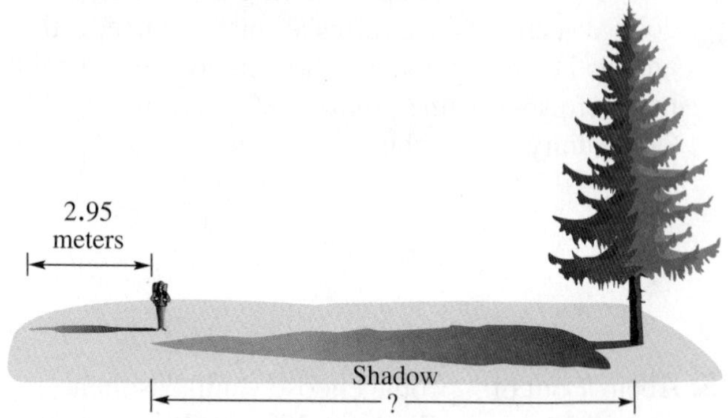

35. Can you set up a proportion to solve this problem? Explain why or why not. Jim is 25 years old and weighs 180 pounds. How much will he weigh when he is 50 years old?

36. Write your own application problem that can be solved by setting up a proportion. Also show the proportion and the steps needed to solve your problem.

37. A survey of college students shows that 4 out of 5 drink coffee. Of the students who drink coffee, 1 out of 8 adds cream to it. How many of the 56,100 students at Ohio State University would be expected to use cream in their coffee?

38. About 9 out of 10 adults think it is a good idea to exercise regularly. But of the ones who think it is a good idea, only 1 in 6 actually exercises at least three times a week. At this rate, how many of the 300 employees in our company exercise regularly?

39. The nutrition information on a bran cereal box says that a $\frac{1}{3}$-cup serving provides 80 calories and 8 grams of dietary fiber. At that rate, how many calories and grams of fiber are in a $\frac{1}{2}$-cup serving? (*Source:* Kraft Foods, Inc.)

40. A $\frac{3}{4}$-cup serving of whole grain penne pasta has 210 calories and 6 grams of dietary fiber. How many calories and grams of fiber would be in a 1-cup serving? (*Source:* Barilla Pasta.)

Relating Concepts (Exercises 41–42) For Individual or Group Work

A box of instant mashed potatoes has the list of ingredients shown in the table. Use this information to ***work Exercises 41–42 in order.***

Ingredient	For 12 Servings
Water	$3\frac{1}{2}$ cups
Margarine	6 Tbsp
Milk	$1\frac{1}{2}$ cups
Potato flakes	4 cups

Source: General Mills.

41. Find the amount of each ingredient needed for 6 servings. Show *two* different methods for finding the amounts. One method should use proportions.

42. Find the amount of each ingredient needed for 18 servings. Show *two* different methods for finding the amounts, one using proportions and one using your answers from **Exercise 41.**

Chapter 5 Summary

Key Terms

5.1

ratio A ratio compares two quantities having the same units. For example, the ratio of 6 apples to 11 apples is written in fraction form as $\frac{6}{11}$. The common units (apples) divide out.

5.2

rate A rate compares two measurements with different units. Examples are 96 dollars for 8 hours, or 450 miles on 18 gallons.

unit rate A unit rate has 1 in the denominator.

cost per unit Cost per unit is a rate that tells how much you pay for one item or one unit. The lowest cost per unit is the best buy.

5.3

proportion A proportion states that two ratios or rates are equal.

cross products Multiply along one diagonal and then multiply along the other diagonal to find the cross products of a proportion. If the cross products are equal, the proportion is true.

Test Your Word Power

See how well you have learned the vocabulary in this chapter.

1. A **ratio**
   A. can be written only as a fraction
   B. compares two quantities that have the same units
   C. compares two quantities that have different units.

2. A **rate**
   A. can be written only as a decimal
   B. compares two quantities that have the same units
   C. compares two quantities that have different units.

3. A **unit rate**
   A. has a numerator of 1
   B. has a denominator of 1
   C. is found by cross multiplying.

4. **Cost per unit** is
   A. the best buy
   B. a ratio written in lowest terms
   C. the price of one item or one unit.

5. A **proportion**
   A. shows that two ratios or rates are equal
   B. contains only whole numbers or decimals
   C. always has one unknown number.

6. **Cross products** are
   A. equal when a proportion is false
   B. used to find the best buy
   C. equal when a proportion is true.

Answers to Test Your Word Power

1. B; *Example:* The ratio of 3 miles to 4 miles is $\frac{3}{4}$; the common units (miles) divide out.

2. C; *Example:* \$4.50 for 3 pounds is a rate comparing dollars to pounds.

3. B; *Example:* $\frac{\$1.79}{1 \text{ pound}}$ is a unit rate. We write it as \$1.79 per pound or \$1.79/pound.

4. C; *Example:* \$3.95 per gallon tells the price of one gallon (one unit).

5. A; *Example:* The proportion $\frac{5}{6} = \frac{25}{30}$ says that $\frac{5}{6}$ is equal to $\frac{25}{30}$.

6. C; *Example:* The cross products for $\frac{5}{6} = \frac{25}{30}$ are $6 \cdot 25 = 150$ and $5 \cdot 30 = 150$.

Quick Review

Concepts	Examples
5.1 Writing a Ratio A ratio compares two quantities that have the same units. A ratio is usually written as a fraction with the number that is mentioned first in the numerator. The common units divide out and are not written in the answer. Check that the fraction is in lowest terms.	Write this ratio as a fraction in lowest terms. 60 ounces of medicine to 160 ounces of medicine $\frac{60 \text{ ounces}}{160 \text{ ounces}} = \frac{60 \div 20}{160 \div 20} = \frac{3}{8}$ ← Ratio in lowest terms ↑ Divide out common units.
5.1 Using Mixed Numbers in a Ratio If a ratio has mixed numbers, change the mixed numbers to improper fractions. Rewrite the problem in horizontal format using the "÷" symbol for division. Finally, multiply by the reciprocal of the divisor.	Write as a ratio of whole numbers in lowest terms. $2\frac{1}{2}$ to $3\frac{3}{4}$ $\frac{2\frac{1}{2}}{3\frac{3}{4}} = \frac{\frac{5}{2}}{\frac{15}{4}}$ Reciprocal $= \frac{5}{2} \div \frac{15}{4} = \frac{5}{2} \cdot \frac{4}{15}$ $= \frac{\overset{1}{\cancel{5}}}{\underset{1}{\cancel{2}}} \cdot \frac{\overset{2}{\cancel{4}}}{\underset{3}{\cancel{15}}} = \frac{2}{3}$ ← Ratio in lowest terms
5.1 Using Measurements in Ratios When a ratio compares measurements, both measurements must be in the *same* units. It is usually easier to compare the measurements using the smaller unit, for example, inches instead of feet.	Write as a ratio in lowest terms. 8 in. to 6 ft Compare using the smaller unit, inches. Because 1 ft has 12 in., 6 ft is $6 \cdot 12 \text{ in.} = 72 \text{ in.}$ The ratio is shown below. $\frac{8 \text{ in.}}{72 \text{ in.}} = \frac{8 \div 8}{72 \div 8} = \frac{1}{9}$ ↑ Divide out common units.
5.2 Writing Rates A rate compares two measurements with different units. The units do *not* divide out, so you must write them as part of the rate.	Write the rate as a fraction in lowest terms. 475 miles in 10 hours $\frac{475 \text{ miles} \div 5}{10 \text{ hours} \div 5} = \frac{95 \text{ miles}}{2 \text{ hours}}$ ← Must write units: miles and hours
5.2 Finding a Unit Rate A unit rate has 1 in the denominator. To find the unit rate, divide the numerator by the denominator. Write unit rates using the word per or a / mark.	Write as a unit rate: \$1278 in 9 days. $\frac{\$1278}{9 \text{ days}}$ ← The fraction bar indicates division. $9\overline{)1278}$ = 142 so $\frac{\$1278 \div 9}{9 \text{ days} \div 9} = \frac{\$142}{1 \text{ day}}$ Write the answer as \$142 per day or \$142/day.

Concepts	Examples
5.2 Finding the Best Buy The best buy is the item with the lowest cost per unit. Divide the price by the number of units. Round to thousandths, if necessary. Then compare to find the lowest cost per unit.	Find the best buy on grapes. You have a coupon for 50¢ off on 2 pounds or 75¢ off on 3 pounds. 2 pounds for \$2.75 3 pounds for \$4.15 Find the cost per unit (cost per pound) after subtracting the coupon. 2 pounds cost \$2.75 − \$0.50 = \$2.25 $\frac{\$2.25}{2} = \$1.125$ per pound 3 pounds cost \$4.15 − \$0.75 = \$3.40 $\frac{\$3.40}{3} \approx \$1.133$ per pound The lower cost per pound is \$1.125, so 2 pounds of grapes is the best buy.
5.3 Writing Proportions A proportion states that two ratios or rates are equal. The proportion "5 is to 6 as 25 is to 30" is written as shown below. $\frac{5}{6} = \frac{25}{30}$ To see whether a proportion is true or false, multiply along one diagonal, then multiply along the other diagonal. If the two cross products are equal, the proportion is true. If the two cross products are unequal, the proportion is false.	Write as a proportion: 8 is to 40 as 32 is to 160 $\frac{8}{40} = \frac{32}{160}$ Is this proportion true or false? $\frac{6}{8\frac{1}{2}} = \frac{24}{34}$ Find the cross products. $\frac{6}{8\frac{1}{2}} = \frac{24}{34}$ $8\frac{1}{2} \cdot 24 = \frac{17}{\underset{1}{\cancel{2}}} \cdot \frac{\overset{12}{\cancel{24}}}{1} = 204$ $6 \cdot 34 = 204$ ← Equal The cross products are equal, so the proportion is true.
5.4 Solving Proportions Solve for an unknown number in a proportion by using the steps shown on the next page.	Find the unknown number. $\frac{12}{x} = \frac{6}{8}$ $\frac{12}{x} = \frac{3}{4}$ Write $\frac{6}{8}$ in lowest terms as $\frac{3}{4}$. *(Continued)*

Concepts	Examples
5.4 Solving Proportions (*Continued*)	
Step 1 Find the cross products. (If desired, you can rewrite the ratios in lowest terms before finding the cross products.)	*Step 1* $\frac{12}{x} = \frac{3}{4}$ $x \cdot 3$, $12 \cdot 4$ Find cross products.
Step 2 Show that the cross products are equal.	*Step 2* $x \cdot 3 = 12 \cdot 4$ Show that cross products are equal. $x \cdot 3 = 48$
Step 3 Divide both products by the number multiplied by x (the number next to x).	*Step 3* $\frac{x \cdot \overset{1}{\cancel{3}}}{\underset{1}{\cancel{3}}} = \frac{48}{3}$ Divide both sides by 3. $x = 16$ The solution is 16.
Step 4 Check by writing the solution in the *original* proportion and finding the cross products.	*Step 4* x is 16. → $\frac{12}{16} = \frac{6}{8}$ $16 \cdot 6 = 96$, $12 \cdot 8 = 96$ Equal The cross products are equal, so **16 is the correct solution** (**not** 96).
5.5 Solving Application Problems with Proportions	
Decide what is being compared. Set up and solve a proportion using the two rates described in the problem. Be sure that *both* rates compare things in the *same order*. Use a letter, like x, to represent the unknown number. Use the six problem-solving steps.	If 3 pounds of grass seed cover 450 square feet of lawn, how much seed is needed for 1500 square feet of lawn?
Step 1 **Read** the problem carefully.	*Step 1* The problem asks for the pounds of grass seed needed for 1500 square feet of lawn.
Step 2 **Work out a plan.**	*Step 2* Pounds of seed is compared to square feet of lawn. Set up and solve a proportion using the two given rates. Be sure that pounds of seed is in both numerators and square feet of lawn is in both denominators.
Step 3 **Estimate** a reasonable answer.	*Step 3* Because 1500 square feet is about three times as much lawn as 450 square feet, about three times as much seed is needed. So, $3 \cdot 3$ pounds $= 9$ pounds as our estimate.
Step 4 **Solve** the problem.	*Step 4* With the proportion set up correctly, solve for the unknown number. Matching units $\frac{\textbf{3 pounds}}{\textbf{450 square feet}} = \frac{\textbf{x pounds}}{\textbf{1500 square feet}}$ Matching units

Concepts	Examples
5.5 Solving Application Problems with Proportions (*Continued*)	Both sides compare pounds to square feet. Ignore the units while finding the cross products and solving for x. $450 \cdot x = \underbrace{3 \cdot 1500}$ Show that cross products are equal. $450 \cdot x = 4500$ $\frac{\overset{1}{\cancel{450}} \cdot x}{\underset{1}{\cancel{450}}} = \frac{4500}{450}$ Divide both sides by 450. $x = 10$
Step 5 **State the answer.**	***Step 5*** **10 pounds** of grass seed are needed.
Step 6 **Check** your work.	***Step 6*** The exact answer, 10 pounds of seed, is close to our estimate of 9 pounds, so it is reasonable.

Chapter 5 Review Exercises

5.1 *Write each ratio as a fraction in lowest terms. Change to the same units when necessary, using the table of measurement comparisons in the first section of this chapter. Use the information in the graph to answer Exercises 1–3.*

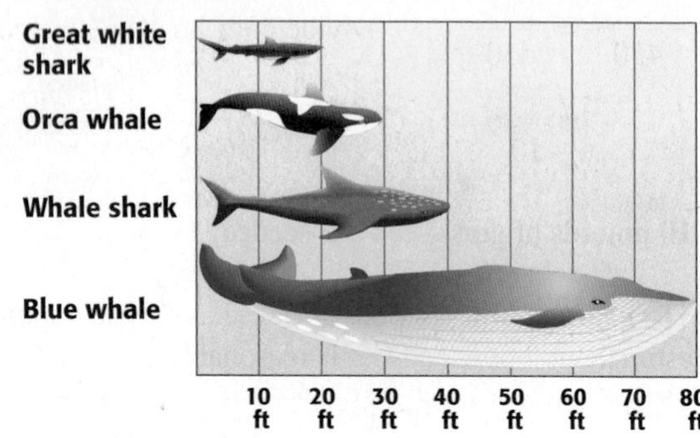

Source: *Grolier Multimedia Encyclopedia.*

1. Ratio of orca whale's length to whale shark's length

2. Ratio of blue whale's length to great white shark's length

3. Which two animals' lengths give a ratio of $\frac{1}{2}$? There are several answers.

4. \$2.50 to \$1.25

5. \$0.30 to \$0.45

6. $1\frac{2}{3}$ cups to $\frac{2}{3}$ cup

7. $2\frac{3}{4}$ miles to $16\frac{1}{2}$ miles

8. 5 hours to 100 minutes

9. 9 in. to 2 ft

10. 1 ton to 1500 pounds

11. 8 hours to 3 days

12. Jake sold \$350 worth of his kachina figures. Ramona sold \$500 worth of her pottery. What is the ratio of Ramona's sales to Jake's sales?

13. Ms. Wei's new car gets 35 miles per gallon. Her old car got 25 miles per gallon. Find the ratio of the new car's mileage to the old car's mileage.

14. This fall, 6000 students are taking math courses and 7200 students are taking English courses. Find the ratio of math students to English students.

5.2 *Write each rate as a fraction in lowest terms.*

15. $88 for 8 dozen

16. 96 children in 40 families

17. When entering data into his computer, Patrick can type four pages in 20 minutes. Give his rate in pages per minute and minutes per page.

18. Elena made $60 in three hours. Give her earnings in dollars per hour and hours per dollar.

Find the best buy.

19. Minced onion

8 ounces for $4.98

3 ounces for $2.49

2 ounces for $1.89

20. Dog food; you have a coupon for $1 off on 8 pounds or more.

35.2 pounds for $36.96

17.6 pounds for $18.69

3.5 pounds for $4.25

5.3 *Use either the method of writing in lowest terms or the method of finding cross products to decide whether each proportion is true or false. Show your work and then write* true *or* false.

21. $\frac{6}{10} = \frac{9}{15}$

22. $\frac{6}{48} = \frac{9}{36}$

23. $\frac{47}{10} = \frac{98}{20}$

24. $\frac{64}{36} = \frac{96}{54}$

25. $\frac{1.5}{2.4} = \frac{2}{3.2}$

26. $\frac{3\frac{1}{2}}{2\frac{1}{3}} = \frac{6}{4}$

5.4 *Find the unknown number in each proportion. Round answers to the nearest hundredth, if necessary.*

27. $\frac{4}{42} = \frac{150}{x}$

28. $\frac{16}{x} = \frac{12}{15}$

29. $\frac{100}{14} = \frac{x}{56}$

30. $\frac{5}{8} = \frac{x}{20}$

31. $\frac{x}{24} = \frac{11}{18}$

32. $\frac{7}{x} = \frac{18}{21}$

33. $\frac{x}{3.6} = \frac{9.8}{0.7}$

34. $\frac{13.5}{1.7} = \frac{4.5}{x}$

35. $\frac{0.82}{1.89} = \frac{x}{5.7}$

5.5 *Set up and solve a proportion for each application problem.*

36. The ratio of cats to dogs at the animal shelter is 3 to 5. If there are 45 dogs, how many cats are there?

37. Danielle had 8 hits in 28 times at bat during last week's games. If she continues to hit at the same rate, how many hits will she gets in 161 times at bat?

38. If 3.5 pounds of ground beef cost \$9.77, what will 5.6 pounds cost? Round to the nearest cent.

39. About 4 out of 10 students are expected to vote in campus elections. There are 8247 students. How many are expected to vote? Round to the nearest whole number.

40. The scale on Brian's model railroad is 1 in. to 16 ft. One of the scale model boxcars is 4.25 in. long. What is the length of a real boxcar in feet?

41. Marvette makes necklaces to sell at a local gift shop. She made 2 dozen necklaces in $16\frac{1}{2}$ hours. How long will it take her to make 40 necklaces?

42. A 180-pound person burns 284 calories playing basketball for 25 minutes. At this rate, how many calories would the person burn in 45 minutes, to the nearest whole number? (*Source: Wellness Encyclopedia.*)

43. In the hospital pharmacy, Michiko sees that a medicine is to be given at the rate of 3.5 milligrams for every 50 pounds of body weight. How much medicine should be given to a patient who weighs 210 pounds?

Mixed Review Exercises

Find the unknown number in each proportion. Round answers to the nearest hundredth, if necessary.

44. $\frac{x}{45} = \frac{70}{30}$

45. $\frac{x}{52} = \frac{0}{20}$

46. $\frac{64}{10} = \frac{x}{20}$

47. $\frac{15}{x} = \frac{65}{100}$

48. $\frac{7.8}{3.9} = \frac{13}{x}$

49. $\frac{34.1}{x} = \frac{0.77}{2.65}$

Find cross products to decide whether each proportion is true or false. Show the cross products and then circle true *or* false.

50. $\frac{55}{18} = \frac{80}{27}$

True False

51. $\frac{5.6}{0.6} = \frac{18}{1.94}$

True False

52. $\frac{\frac{1}{5}}{2} = \frac{1\frac{1}{6}}{11\frac{2}{3}}$

True False

Write each ratio as a fraction in lowest terms. Change to the same units when necessary.

53. 4 dollars to 10 quarters

54. $4\frac{1}{8}$ in. to 10 in.

55. 10 yd to 8 ft

56. \$3.60 to \$0.90

57. 12 eggs to 15 eggs

58. 37 meters to 7 meters

59. 3 pints to 4 quarts

60. 15 minutes to 3 hours

61. $4\frac{1}{2}$ miles to $1\frac{3}{10}$ miles

62. Nearly 7 out of 8 fans buy something to drink at rock concerts. How many of the 28,500 fans at today's concert would be expected to buy a beverage? Round to the nearest hundred fans.

63. Emily spent \$150 on car repairs and \$400 on car insurance. What is the ratio of the amount spent on insurance to the amount spent on repairs?

64. Antonio is choosing among three packages of plastic wrap. Is the best buy 25 ft for \$0.78; 75 ft for \$1.99; or 100 ft for \$2.59? He has a coupon for 50¢ off either of the larger two packages.

65. On this scale drawing of a backyard patio, 0.5 in. represents 6 ft. If the patio measures 1.75 in. long and 1.25 in. wide on the drawing, what will be the actual length and width of the patio when it is built?

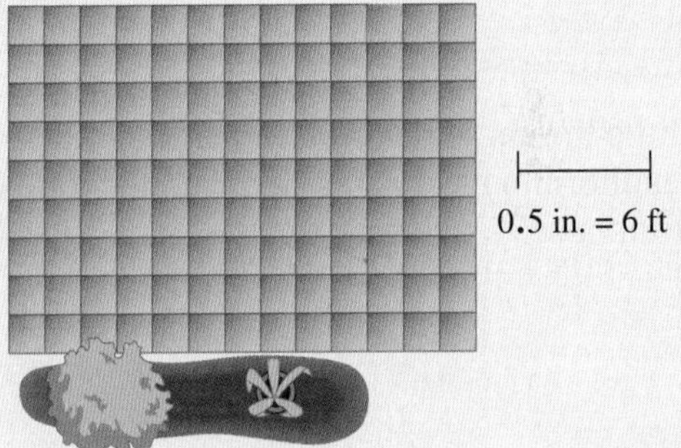

66. A lawn mower uses 0.8 gallon of gas every 3 hours. How long can the mower run on 2 gallons of gas?

67. An antibiotic is given at the rate of $1\frac{1}{2}$ teaspoons for every 24 pounds of body weight. How much should be given to an infant who weighs 8 pounds?

68. Charles made 251 points during 169 minutes of playing time last year. At that same rate, how many points would you expect him to make if he plays 14 minutes in tonight's game? Round to the nearest whole number.

69. Refer to **Exercise 67.** Explain each step you took in solving the problem. Be sure to tell how you decided which way to set up the proportion and how you checked your answer.

70. A vitamin supplement for cats is given at the rate of 1000 milligrams for a 5-pound cat. (*Source:* St. Jon Pet Care Products.)

(a) How much should be given to a 7-pound cat?

(b) How much should be given to an 8-ounce kitten?

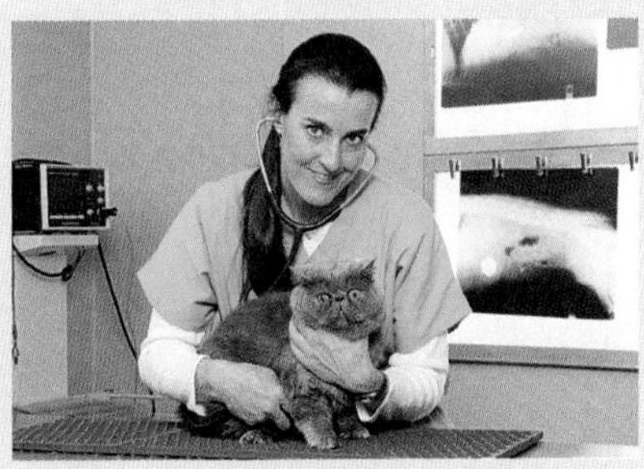

Chapter 5 Test

The Chapter Test Prep Videos with test solutions are available on DVD, in MyMathLab, and on YouTube—search "LialBasicCollegeMath" and click on "Channels."

Write each rate or ratio as a fraction in lowest terms. Change to the same units when necessary.

1. 16 fish to 20 fish

2. 300 miles on 15 gallons

3. \$15 for 75 minutes

4. 3 hours to 40 minutes

5. The little theater at our college has 320 seats. The auditorium has 1200 seats. Find the ratio of auditorium seats to theater seats.

6. Use the information in the table about Quizno's honey mustard chicken sub sandwich to find the best buy.

Size	Length of Sub	Price
Small	5 inches	\$ 5.99
Regular	8 inches	\$ 6.89
Large	11 inches	\$10

Source: Quizno's

7. Find the best buy on spaghetti sauce. You have a coupon for 75¢ off Brand X and a coupon for 50¢ off Brand Y.

26 ounces of Brand X for \$3.89

16 ounces of Brand Y for \$1.89

14 ounces of Brand Z for \$1.29

8. Suppose the ratio of your income last year to your income this year is 3 to 2. Explain what this means. Give an example of the dollars earned last year and this year that fits the 3 to 2 ratio.

Determine whether each proportion is true or false. Show your work and then write true *or* false.

9. $\frac{6}{14} = \frac{18}{45}$

10. $\frac{8.4}{2.8} = \frac{2.1}{0.7}$

Find the unknown number in each proportion. In Problems 11–13, round the answers to the nearest hundredth, if necessary.

11. $\frac{5}{9} = \frac{x}{45}$

12. $\frac{3}{1} = \frac{8}{x}$

13. $\frac{x}{20} = \frac{6.5}{0.4}$

14. $\frac{2\frac{1}{3}}{x} = \frac{\frac{8}{9}}{4}$

Set up and solve a proportion for each application problem.

15. Pedro entered 18 orders into his computer in thirty minutes at his job. At that rate, how many orders could he enter in forty minutes?

16. Just 0.8 ounce of wildflower seeds is enough for 50 square feet of ground. What weight of seeds is needed for a garden with 225 square feet? (*Source:* White Swan Ltd.)

17. About 2 out of every 15 people are left-handed. How many of the 650 students in our school would you expect to be left-handed? Round to the nearest whole number.

18. A student set up the proportion for **Problem 17** this way and arrived at an answer of 4875.

$\frac{2}{15} = \frac{650}{x}$

CHECK $\frac{2}{15} = \frac{650}{4875}$

$15 \cdot 650 = 9750$

$2 \cdot 4875 = 9750$

Because the cross products are equal, the student said the answer is correct. Is the student right? Explain why or why not.

19. A medication is given at the rate of 8.2 grams for every 50 pounds of body weight. How much should be given to a 145-pound person? Round to the nearest tenth of a gram.

20. On a scale model, 1 in. represents 8 ft. If a building in the model is 7.5 in. tall, what is the actual height of the building in feet?

Chapters 1–5 Cumulative Review Exercises

Round each number as indicated.

1. 9903 to the nearest hundred

2. 617.0519 to the nearest tenth

3. \$99.81 to the nearest dollar

4. \$3.0555 to the nearest cent

Simplify.

5. $30 - 0.66$

6. Write the answer using R for the remainder. $33\overline{)20{,}157}$

7. $(1.9)(0.004)$

8. $3020 - 708$

9. $0.401 + 62.98 + 5$

10. $1.39 \div 0.025$

11. $36 + 18 \div 6$

12. $8 \div 4 + (10 - 3^2) \cdot 4^2$

13. $88 \div \sqrt{121} \cdot 2^3$

14. $(16.2 - 5.85) - 2.35(4)$

15. Write 0.0105 in words.

16. Write sixty and seventy-one thousandths in numbers.

In a survey of 1000 adults, 550 drank coffee daily, 250 drank coffee occasionally, and 200 never drank coffee. Use this information for Exercises 17–18. Write each ratio as a fraction in lowest terms.

17. What is the ratio of those who do not drink coffee to the total number of people in the survey?

18. Find the ratio of all the adults who drink coffee to those who never drink it.

Nest boxes for birds are made with different sizes of entry holes. Then, only certain types of birds can use the nest box and it helps prevent entry by predators. Use the information in the table to answer Exercises 19–20. The diameter is the distance across the opening at its widest point.

Eastern bluebird nest box with 1.5-inch opening

Bird	Diameter of Opening
Eastern bluebird	1.5 inches
Western bluebird	$1\frac{9}{16}$ inches
Chickadee	1.25 inches
Swallow	$1\frac{3}{8}$ inches
Wren	$1\frac{1}{8}$ inches

Source: Duncraft.

19. List the nest box openings in order from smallest to largest.

20. What is the difference in diameter between the openings for an eastern bluebird and a western bluebird? Write your answer as a fraction and as a decimal.

Find the unknown number in each proportion. Round your answers to the nearest hundredth, when necessary.

21. $\frac{9}{12} = \frac{x}{28}$

22. $\frac{7}{12} = \frac{10}{x}$

23. $\frac{6.7}{x} = \frac{62.8}{9.15}$

Solve each application problem.

24. The college honor society has a goal of collecting 1500 pounds of food to fill Thanksgiving baskets. So far it has collected $\frac{5}{6}$ of the goal. How many more pounds are needed?

25. Tara has a photo that is 10 centimeters wide by 15 centimeters long. If the photo is enlarged to a length of 40 centimeters, find the new width, to the nearest tenth.

26. The distance around Dunning Pond is $1\frac{1}{10}$ miles. Norma ran around the pond four times in the morning and $2\frac{1}{2}$ times in the afternoon. How far did she run in all?

27. Find the best buy on instant mashed potatoes. You have a coupon for 50¢ off either the 34-serving or 65-serving box.

A box that makes 20 servings for $2.15

A box that makes 34 servings for $3.35

A box that makes 65 servings for $10.41

28. In a survey, 5 out of 8 apartment residents said they are sometimes bothered by noise from their neighbors. How many of the 224 residents at Harris Towers would you expect to be bothered by noise?

29. The directions on a bottle of plant food call for $\frac{1}{2}$ teaspoon in two quarts of water. How much plant food is needed for five quarts? (*Source:* Schultz Company.)

Use the information in the table on international long-distance calling card rates to answer Exercises 30–34. Round answers to the nearest whole minute when necessary.

$20 International Phone Cards
No Connection Fee!

Place a call to	Cost per minute
Mexico	$0.02
Philippines	$0.084
Japan (to a cell phone)	$0.099
Hong Kong	$0.018
Canada	$0.10
India (to a cell phone)	$0.106
Afghanistan	$0.134

Source: noblecom.com.

30. List the per-minute rates from least to greatest.

31. If you buy a $20 calling card, how long a call could you make to a cell phone in Japan?

32. How many $20 cards would you have to buy to make six hours' worth of calls to a cell phone in India?

33. What is the ratio of minutes you can call Mexico for $20 to minutes you can call Canada for $20?

34. You have already made a 95-minute call to the Philippines. With the amount left on one $20 phone card, how long a call could you make to Afghanistan?

Math in the Media

FEEDING HUMMINGBIRDS

After getting a hummingbird feeder, the next step is to fill it! You have two choices at this point: you can either buy one of the commercial mixtures or you can make your own solution. See the recipe at the right.

The concentration of the sugar is important. The 1 to 4 ratio of sugar to water is recommended because it approximates the ratio of sugar to water found in the nectar of many hummingbird flowers.

Boiling the solution helps slow down fermentation. Sugar-and-water solutions are subject to rapid spoiling, especially in hot weather.

Source: *The Hummingbird Book.*

Recipe for Homemade Mixture:
1 part sugar (not honey)
4 parts water
Boil for 1 to 2 minutes. Cool.
Store extra in refrigerator.

A recipe can be used to make as much of a mixture as you need as long as the ingredients are kept proportional. Use the recipe for a homemade mixture of sugar water for hummingbird feeders to answer these problems.

1. What is the ratio of sugar to water in the recipe? What is the ratio of water to sugar?

2. Complete each table.

Sugar	Water
1 cup	4 cups
	5 cups
	6 cups
	7 cups
2 cups	8 cups

Sugar	Water
1 cup	4 cups
	3 cups
	2 cups
	1 cup

3. How much water would you need if you used

(a) 3 cups of sugar?

(b) 4 cups of sugar?

(c) $\frac{1}{3}$ cup of sugar?

4. As you change the amounts of water and sugar, should you change the length of time that you boil the mixture? Explain your answer.

6 Percent

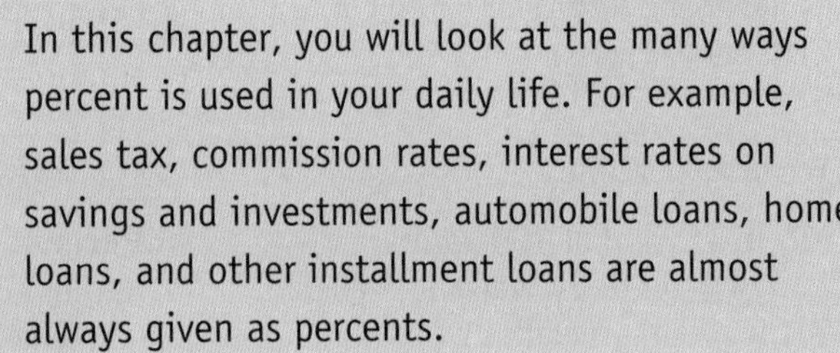

In this chapter, you will look at the many ways percent is used in your daily life. For example, sales tax, commission rates, interest rates on savings and investments, automobile loans, home loans, and other installment loans are almost always given as percents.

6.1 Basics of Percent

OBJECTIVES

1. Learn the meaning of percent.
2. Write percents as decimals.
3. Write decimals as percents.
4. Understand 100%, 200%, and 300%.
5. Use 50%, 10%, and 1%.

VOCABULARY TIP

Percent means "per 100" or "out of 100," so any **percent** can be converted to a decimal by dividing the percent by 100. For example, $30\% = \frac{30}{100} = 0.30$.

1 Write as percents.

(a) In a group of 100 adults, 74 keep fit by walking. What percent are walking?

(b) The sales tax is \$6 per \$100. What percent is this?

(c) Out of 100 Americans, 32 picked football as their favorite sport. What percent picked football?

Answers

1. (a) 74% **(b)** 6% **(c)** 32%

Notice that the figure below has one hundred squares of equal size. Eleven of the squares are shaded. The shaded portion is $\frac{11}{100}$, or 0.11, of the total figure.

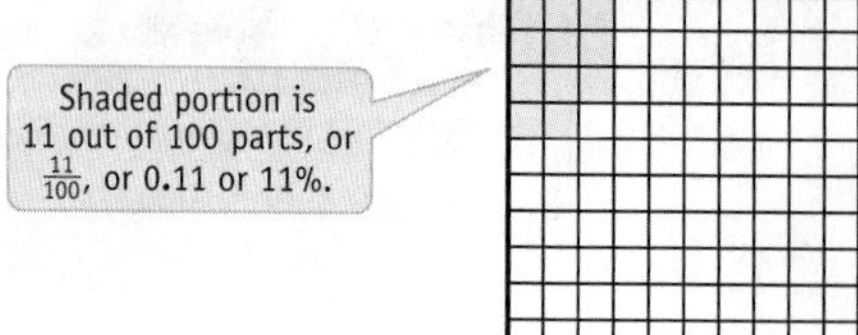
Shaded portion is
11 out of 100 parts, or
$\frac{11}{100}$, or 0.11 or 11%.

The shaded portion is also 11% of the total, or "eleven parts out of 100 parts." Read **11%** as "eleven percent."

OBJECTIVE 1 Learn the meaning of percent. As we just saw, a percent is a ratio with a denominator of 100.

The Meaning of Percent

Percent means *per one hundred.* The "%" symbol is used to show the number of parts out of one hundred parts.

EXAMPLE 1 Understanding Percent

(a) If *43 out of 100* students are men, then *43 per 100* or $\frac{43}{100}$ or **43%** of the students are men.

(b) If a person pays a tax of \$7 on every \$100 of purchases, then the tax rate is \$7 per \$100. The ratio is $\frac{7}{100}$ and the percent of tax is **7%.**

◀ Work Problem **1** at the Side.

OBJECTIVE 2 Write percents as decimals. If 8% means 8 parts out of 100 parts or $\frac{8}{100}$, then $p\%$ means p parts out of 100 parts or $\frac{p}{100}$. Because $\frac{p}{100}$ is another way to write the division $p \div 100$, we have

$$p\% = \frac{p}{100} = p \div 100.$$

Writing a Percent as a Decimal

$$\underbrace{p\% = \frac{p}{100}}_{\text{As a fraction}} \quad \text{or} \quad \underbrace{p\% = p \div 100}_{\text{As a decimal}}$$

EXAMPLE 2 Writing Percents as Decimals

Write each percent as a decimal.

(a) 47%

$$p\% = p \div 100$$

$$47\% = 47 \div 100 = 0.47 \leftarrow \text{Decimal form}$$

0.47 is $\frac{47}{100}$, which is equivalent to 47%.

Continued on Next Page

(b) 76% 76% = 76 ÷ 100 = 0.76 ← Decimal form

(c) 28.2% 28.2% = 28.2 ÷ 100 = 0.282 ← Decimal form

(d) 100% 100% = 100 ÷ 100 = 1.00 ← Decimal form

CAUTION

In **Example 2(d)** above, notice that 100% is 1.00, or 1, which is a whole number. Whenever you have a percent that is *100% or greater*, the equivalent decimal number will be *1 or greater than 1.*

Work Problem 2 at the Side. ▶

The answers in **Example 2** above suggest these steps for writing a percent as a decimal.

Writing a Percent as a Decimal

Step 1 Drop the percent symbol.

Step 2 Divide by 100.

Note

Recall from **Chapter 4** that a quick way to divide a number by 100 is to move the decimal point **two places to the left.**

EXAMPLE 3 Writing Percents as Decimals by Moving the Decimal Point

Write each percent as a decimal by moving the decimal point two places to the left.

(a) 17%

17% = 17.% Decimal point starts at far right side.

0.17 ← Percent symbol is dropped. *(Step 1)*

Decimal point is moved two places to the left. *(Step 2)*

17% = 0.17

(b) 160%

1.60 is equivalent to 1.6 because $1\frac{60}{100}$ simplifies to $1\frac{6}{10}$.

160% = 160.% = 1.60 or 1.6 Decimal point starts at far right side.

(c) 4.9%

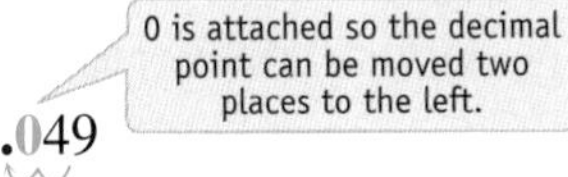
0 is attached so the decimal point can be moved two places to the left.

.049

4.9% = 0.049

Continued on Next Page

2 Write each percent as a decimal.

(a) 68%

68% = 68 ÷ ____ = ____

(b) 34%

(c) 58.5%

(d) 175%

(e) 200%

Answers

2. **(a)** 100; 0.68 **(b)** 0.34 **(c)** 0.585 **(d)** 1.75 **(e)** 2.00 or 2

3 Write each percent as a decimal.

GS **(a)** 96%

96% = 96.% (Decimal point starts here.)

= .96% Move decimal point two places to the left.

= ____

(b) 6%

(c) 24.8%

GS **(d)** 0.9%

0.9% = 0.___ ___9

Answers

3. **(a)** 0.96 **(b)** 0.06 **(c)** 0.248 **(d)** 0; 0; 0.009

(d) 0.6%

0.6% = 0.006 (Two zeros are attached so the decimal point can be moved two places to the left.)

CAUTION

Look at **Example 3(d)** above, where 0.6% is less than 1%. Because 0.6% is $\frac{6}{10}$ of 1%, it is *less than 1%*. Any fraction of a percent is *less than 1%*.

◀ **Work Problem 3 at the Side.**

OBJECTIVE 3 Write decimals as percents. You can write any decimal as a percent. For example, the decimal 0.78 is the same as the fraction

$$\frac{78}{100}.$$

This fraction means 78 out of 100 parts, or 78%. The following steps give the same result.

Writing a Decimal as a Percent

Step 1 Multiply by 100.

Step 2 Attach a percent symbol.

Note

A quick way to divide or multiply a number by 100 is to move the decimal point two places to the left or two places to the right, respectively.

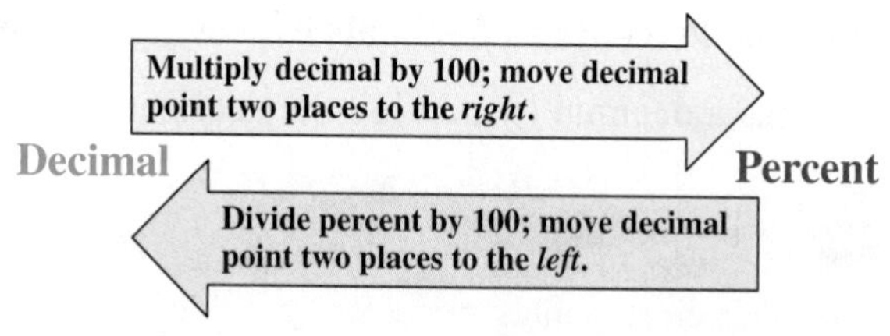

EXAMPLE 4 Writing Decimals as Percents by Moving the Decimal Point

Write each decimal as a percent by moving the decimal point two places to the right.

(a) 0.21

0.21

Decimal point is moved two places to the right. *(Step 1)*

0.21 = 21% ← Percent symbol is attached. *(Step 2)* (Remember to attach the percent (%) symbol.)

Decimal point is not written with whole number percents.

Continued on Next Page

(b) 0.529 = 52.9%

Move the decimal point two places to the right. Then attach a % symbol.

(c) 1.92 = 192%

(d) 2.5

2.50 — 0 is attached so the decimal point can be moved two places to the right.

2.5 = 250% ← Attach % symbol.

When necessary, attach zeros so you can move the decimal point.

(e) 3

3. = 3.00 — Two zeros are attached so the decimal point can be moved two places to the right.

so 3 = 300% ← Attach % symbol.

CAUTION

Look at **Examples 4(c), 4(d), and 4(e)** above, where 1.92, 2.5, and 3 are greater than 1. Because the number 1 is equivalent to 100%, all numbers greater than 1 will be *greater than 100%.*

Work Problem 4 at the Side. ▶

OBJECTIVE 4 Understand 100%, 200%, and 300%. When working with percents, it is helpful to have several reference points. 100%, 200%, and 300% are three such helpful reference points.

100% means 100 parts out of 100 parts. That's ***all*** of the parts. If 100% of the 18 people attending last week's meeting attended this week's meeting, then 18 people (***all*** of them) attended this week.

If attendance at the meeting this week is 200% of last week's attendance of 18 people, then this week's attendance is 36 people, or *two* times as many people (2 • 18 = 36). Likewise, if attendance is 300% of last week's attendance, then *three* times as many people, or 54 people, attended (3 • 18 = 54).

EXAMPLE 5 Finding 100%, 200%, and 300% of a Number

Answer the following.

Notice that 100% of something is all of it (the whole thing).

(a) What is 100% of 82 people?

100% is ***all*** of the people. So, 100% of 82 people is 82 people.

(b) What is 200% of $63?

200% is twice (2 times) as much money.

So, 200% of $63 is 2 • $63 = $126.

(c) What is 300% of 32 employees?

300% is 3 times as many employees.

So, 300% of 32 employees is 3 • 32 = 96 employees.

Work Problem 5 at the Side. ▶

4 Write each decimal as a percent.

(a) 0.74 (b) 0.15

(c) 0.09

GS (d) 0.617

0.__ __ __

So, 0.617 = ____ %.

(e) 0.834 (f) 5.34

(g) 2.8 (h) 4

5 Answer the following.

(a) What is 100% of $7.80?

(b) What is 100% of 1850 workers?

GS (c) What is 200% of 24 photographs?

200% is twice (____ times).

So, 2 • 24 = ______.

(d) What is 300% of 8 miles?

Answers

4. (a) 74% (b) 15% (c) 9% (d) 6; 1; 7; 61.7% (e) 83.4% (f) 534% (g) 280% (h) 400%

5. (a) $7.80 (b) 1850 workers (c) 2; 48 photographs (d) 24 miles

OBJECTIVE 5 Use 50%, 10%, and 1%. 50% means 50 parts out of 100 parts, which is ***half*** of the parts ($\frac{50}{100} = \frac{1}{2}$). So, 50% of \$18 is \$9 (***half*** of the money).

When using 10%, we have 10 parts out of 100 parts, which is $\frac{1}{10}$ of the parts ($\frac{10}{100} = \frac{1}{10}$). To find 10% or $\frac{1}{10}$ of a number, we move the decimal point **one** place to the left. 10% of \$285 is \$28.50 (because \$28.5. = \$28.50).

To find 1% of a number ($\frac{1}{100}$), we move the decimal point **two** places to the left. 1% of \$198 is \$1.98 (because \$1.98. = \$1.98).

EXAMPLE 6 Finding 50%, 10%, and 1% of a Number

Answer the following.

(a) What is 50% of 24 hours?

Think: 50% of something is $\frac{50}{100}$ or $\frac{1}{2}$ of it.

50% is half of the hours.

So, 50% of 24 hours is **12 hours**.

(b) What is 10% of 280 pages?

10% is $\frac{1}{10}$ of the pages. Move the decimal point *one* place to the left.

So, 10% of 280. pages is **28 pages**.

(c) What is 1% of \$540?

1% is $\frac{1}{100}$ of the money. Move the decimal point *two* places to the left.

So, 1% of \$540. is **\$5.40**.

◀ **Work Problem 6 at the Side.**

Note

Two other frequently used percents are 25% and 75%.

25% means 25 parts out of 100 parts, which is $\frac{1}{4}$ of the parts $\left(\frac{25}{100} = \frac{1}{4}\right)$. For example, 25% of \$32 is $\frac{1}{4}$ of the money, or \$8.

75% means 75 parts out of 100 parts, which is $\frac{3}{4}$ of the parts $\left(\frac{75}{100} = \frac{3}{4}\right)$. For example, 75% of \$32 is $\frac{3}{4}$ of the money, or \$24.

---

6 Answer the following.

GS **(a)** What is 50% of 200 patients?

50% is half of the patients.

$\frac{1}{2}$ of 200 patients is ______.

(b) What is 50% of 64 tweets?

GS **(c)** What is 10% of 3850 elm trees?

10% is $\frac{1}{10}$ of the trees. Move the decimal point ____ place to the ____.

10% of 3850. is ______.

(d) What is 10% of 7 pounds?

GS **(e)** What is 1% of 240 ft?

1% is $\frac{1}{100}$ of the length.

Move the decimal point ____ places to the ____.

1% of 240. is ______.

(f) What is 1% of \$3000?

Answers

6. **(a)** 100 patients **(b)** 32 tweets **(c)** one; left; 385 elm trees **(d)** 0.7 pound **(e)** two; left; 2.4 ft **(f)** \$30

6.1 Exercises

FOR EXTRA HELP

MyMathLab®

CONCEPT CHECK *Fill in each blank with the correct response.*

1. To write a percent as a decimal, drop the ________ symbol and then divide by ________.

2. A quick way to divide a number by 100 is to move the ________ two places to the ________.

Write each percent as a decimal. **See Examples 2 and 3.**

3. 12%
   12.% = __.12

4. 57%
   57.% = 0.__ __

5. 70%

6. 40%

7. 25%

8. 35%

9. 140%

10. 250%

11. 5.5%

12. 6.7%

13. 100%

14. 600%

15. 0.5%

16. 0.25%

17. 0.35%

18. 0.75%

CONCEPT CHECK *Fill in each blank with the correct response.*

19. To write a decimal as a percent, multiply by ______ and then attach a ______ symbol.

20. A quick way to multiply a number by 100 is to move the ________ two places to the ________.

Write each decimal as a percent. **See Example 4.**

21. 0.6
    0.6 = 0.60
    So, 0.6 = 60%

22. 0.9
    0.9 = 0.__ __
    So, 0.9 = __ __%

23. 0.01

24. 0.07

25. 0.375

26. 0.625

27. 2

28. 5

29. 3.7

30. 2.2

31. 0.0312

32. 0.0625

33. 4.162

34. 8.715

35. 0.0028

36. 0.0064

37. Fractions, decimals, and percents are all used to describe a part of something. The use of percents is much more common than that of fractions and decimals. Why do you suppose this is true?

38. List five uses of percent that are or will be part of your life. Consider the activities of working, shopping, saving, and planning for the future.

Over the next 10 years, the projected growth in the workforce for several occupations is shown. Write each percent as a decimal and each decimal as a percent. **See Examples 2–4.** *(Source: Bloomberg BusinessWeek.)*

39. Truck drivers 13%

40. Registered nurses 22.2%

41. Physicians and surgeons 0.218

42. Waiters and waitresses 0.064

43. Postsecondary teachers 15.1%

44. Security guards 14.2%

45. Customer service representatives 0.177

46. Carpenters 0.129

47. Cooks 14.6%

48. Management analysts 23.9%

Write each percent as a decimal and each decimal as a percent. **See Examples 2–4.**

49. Only 0.08 of the total population has the professional training.

50. The church building fund has 0.8 of the money needed.

51. The patient's blood pressure was 30% above normal.

52. Success with the diet was 170% greater than anticipated.

CONCEPT CHECK *Fill in the blanks. Remember that 100% is all of something, 200% is two times as many, and 300% is three times as many.* **See Example 5.**

53. The Saturday morning tae kwon do class has 12 children enrolled. If 100% of the children are present, how many children are there? __________

54. When 500 adults were asked, "Do you think your taxes are too high," 100% said yes. How many said yes? __________

55. Last year we had 210 employees. This year we have 200% of that number. How many employees do we have this year? ______________

56. Last season Active Sports sold 380 fishing licenses. This season they sold 200% of that number. How many fishing licenses did they sell this season?

____

57. Last week 90 chairs were used for the meeting. This week we need 300% of that number of chairs. We'll need ______________.

57. One month ago she had 28 friends on Facebook. Now she has 300% of that number. The new number of friends is ______________.

Fill in the blanks. Remember that 50% is half of something, 10% is found by moving the decimal point one place to the left, and 1% is found by moving the decimal point two places to the left. ***See Example 6.***

59. Jacob owes $755 for tuition. Financial aid will pay 50% of the cost. Financial aid will pay ______________.

60. Linda Redding needs $2320 for school and living expenses this semester. She applied for a student loan to cover 50% of this amount. The amount of the loan is ______________.

61. Only 10% of 8200 commuters are carpooling to work. How many commuters carpool? ______________

Move the decimal point one place to the left. 8200**.**

62. Sarah Bryn expects that 10% of the 240 dozen plants in her greenhouse will not be sold. The expected number of unsold plants is ______________.

63. The naturalist said that 1% of the 2600 plants in the park are poisonous. How many plants are poisonous?

____

64. Of the 4800 accidents, only 1% were caused by mechanical failure. How many accidents were caused by mechanical failure? ______________

Move the decimal point two places to the left. 4800**.**

65. **(a)** Describe a shortcut method of finding 100% of a number.

(b) Show an example using your shortcut.

66. **(a)** Describe a shortcut method of finding 50% of a number.

(b) Show an example using your shortcut.

67. **(a)** Describe a shortcut method of finding 200% of a number.

(b) Show an example using your shortcut.

68. **(a)** Describe a shortcut method of finding 300% of a number.

(b) Show an example using your shortcut.

69. **(a)** Describe a shortcut method of finding 10% of a number.

(b) Show an example using your shortcut.

70. **(a)** Describe a shortcut method of finding 1% of a number.

(b) Show an example using your shortcut.

More than 7.4 million households dress up their pets (dogs and cats) in Halloween costumes. The bar graph shows the ranking of the top pet costumes and the percent of pet owners selecting each costume. Use this graph to answer Exercises 71–74. Write each answer as a percent and as a decimal. (Source: BIGresearch survey of 8877 adult pet owners.*)*

71. What portion of the pet owners selected the devil costume for their pet?

72. What portion of the pet owners selected the pirate costume for their pet?

73. **(a)** What was the third-most-popular costume?

(b) Write the portion of the pet owners who selected this costume.

74. **(a)** What was the second-most-popular costume?

(b) Write the portion of the pet owners who selected this costume.

PETS GET DRESSED UP

This year, 7.4 million households plan to put their furry friends into a Halloween costume. Top outfits:

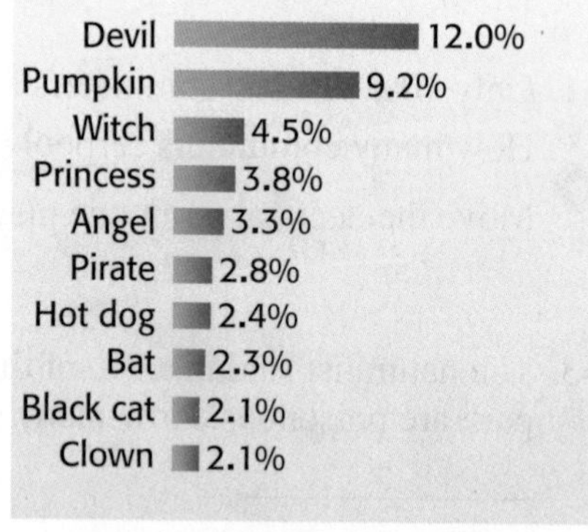

Source: BIGresearch survey of 8877 adult pet owners.

Motorists were asked which traffic violation they felt should be given the biggest fine. The circle graph shows the percent of motorists who chose each traffic violation. Use this graph to answer Exercises 75–78. Write each answer as a percent and a decimal.

75. What portion of the motorists chose "Tailgating" for the biggest fine?

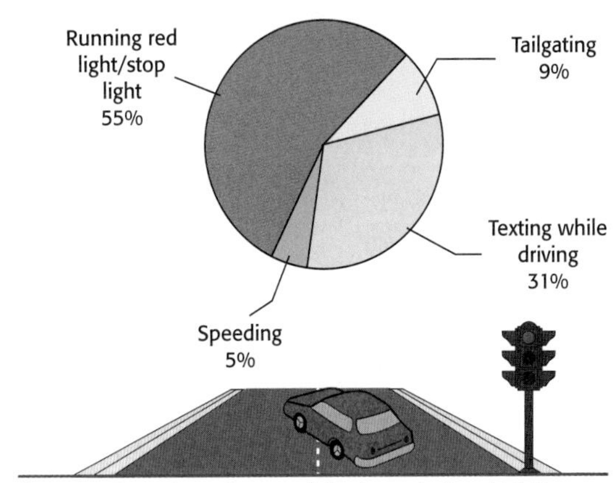

Source: American Automobile Association (AAA).

76. What portion of the motorists chose "Texting while driving" for the biggest fine?

77. (a) Which violation was chosen least often for the biggest fine?

(b) What portion was this?

78. (a) Which violation was chosen most often for the biggest fine?

(b) What portion was this?

Food researchers suspect that childhood obesity starts early. The pictograph shows the percent of children 19 to 24 months old who eat each type of food at least once a day. Use this graph to answer Exercises 79–82. Write each answer as a percent and as a decimal.

79. What portion of the children eat french fries?

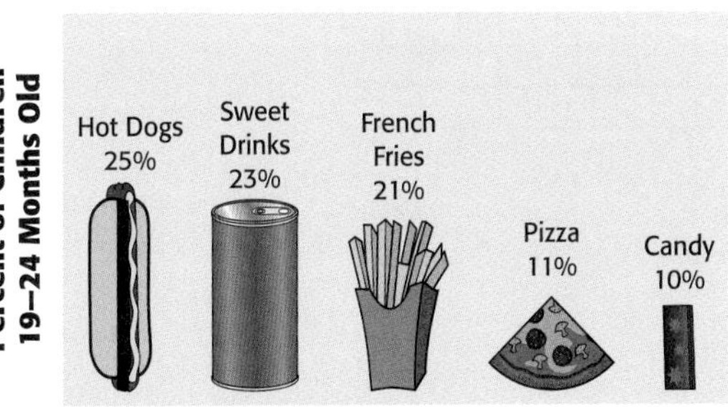

Source: Mathematica Policy Research for Gerber Products Company.

80. What portion of the children eat pizza?

81. (a) Which food is eaten by the lowest portion of children?

(b) What portion is this?

82. (a) Which food is eaten by the highest portion of children?

(b) What portion is this?

Write a percent for both the shaded and unshaded portions of each figure.

83. 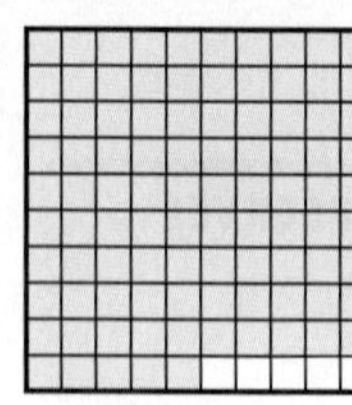

95 out of 100 parts are shaded = ____ %

5 out of 100 parts are not shaded = ____ %

84. 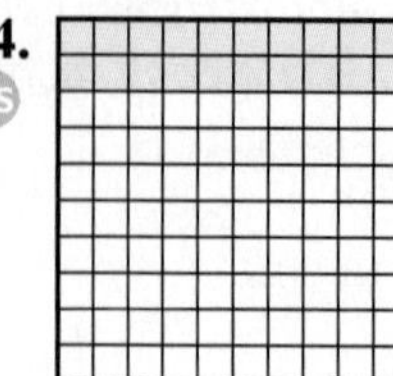

20 out of 100 parts are shaded = ____ %

80 out of 100 parts are not shaded = ____ %

85. 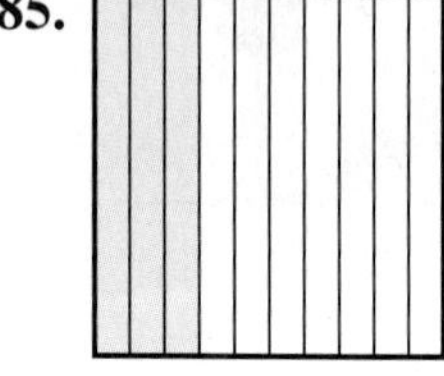

86. 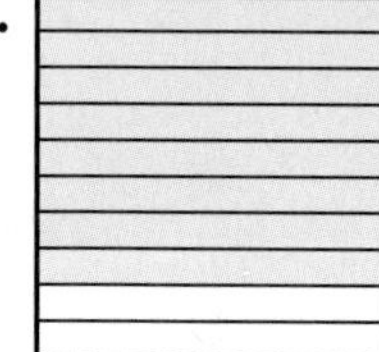

87. 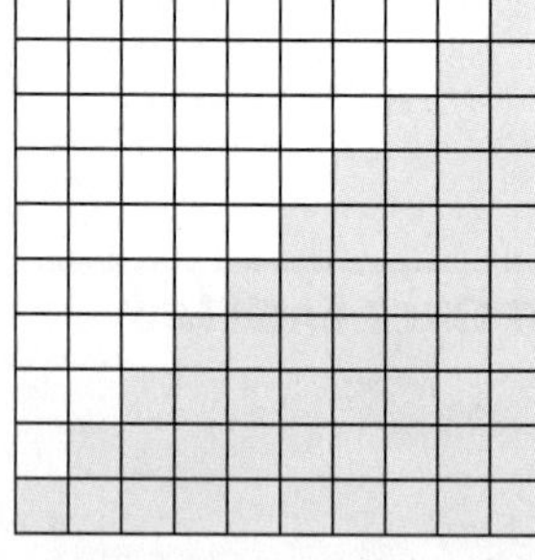

88. 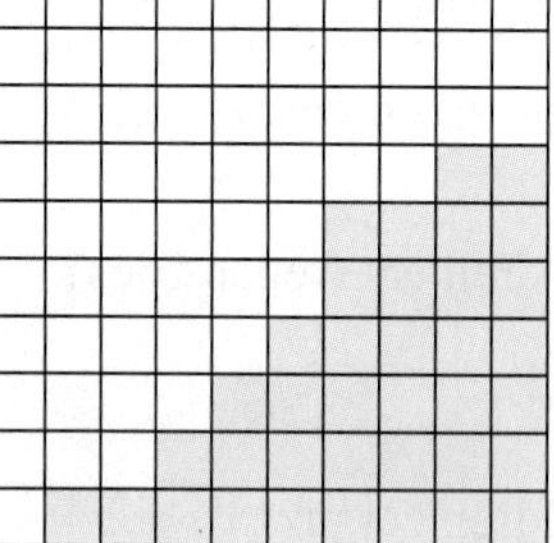

89. 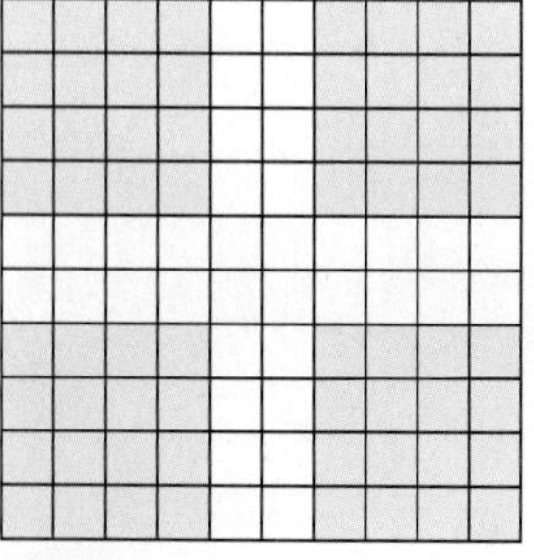

90. 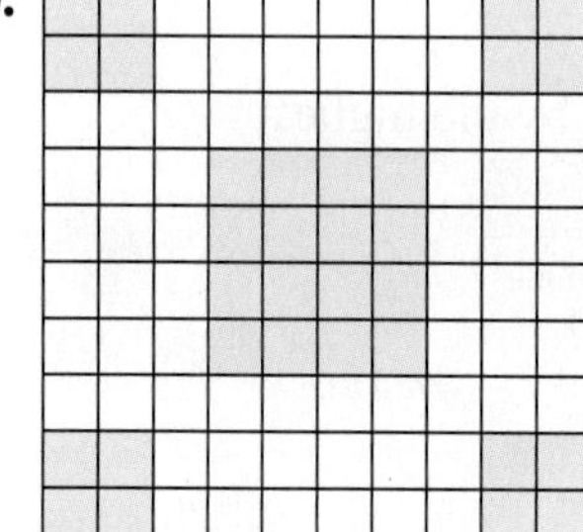

6.2 Percents and Fractions

OBJECTIVES

1. Write percents as fractions.
2. Write fractions as percents.
3. Use the table of percent equivalents.

OBJECTIVE 1 **Write percents as fractions.** Percents can be written as fractions by using what we learned in the previous section.

Writing a Percent as a Fraction

$$p\% = \frac{p}{100}, \quad \text{as a fraction}$$

EXAMPLE 1 **Writing Percents as Fractions**

Write each percent as a fraction or mixed number in lowest terms.

(a) 25%

As we saw in the last section, 25% can be written as a decimal.

$$25\% = 25 \div 100 = 0.25 \quad \text{Percent symbol dropped}$$

Because 0.25 means 25 hundredths,

$$0.25 = \frac{25}{100} = \frac{25 \div 25}{100 \div 25} = \frac{1}{4} \leftarrow \text{Lowest terms}$$

It is not necessary, however, to write 25% as a decimal first.

$$25\% = \frac{25}{100} \quad \text{25 per 100}$$

$$= \frac{1}{4} \leftarrow \text{Lowest terms}$$

(b) 76%

Write 76% as $\frac{76}{100}$.

The percent becomes the numerator.

The *denominator* is always 100 because percent means *parts per 100.*

Write $\frac{76}{100}$ in lowest terms.

To write a fraction in lowest terms, divide numerator and denominator by the *same number*.

$$\frac{76 \div 4}{100 \div 4} = \frac{19}{25} \leftarrow \text{Lowest terms}$$

(c) 150%

$$150\% = \frac{150}{100} = \frac{150 \div 50}{100 \div 50} = \frac{3}{2} = 1\frac{1}{2} \leftarrow \text{Mixed number}$$

($\frac{3}{2}$: Lowest terms)

Note

Remember that percent means ***per 100.***

Work Problem 1 at the Side. ▶

1 Write each percent as a fraction or mixed number in lowest terms.

GS **(a)** 50%

$$50\% = \frac{50}{100} = \frac{50 \div 50}{\_\_\_ \div 50}$$

$$= \_\_\_$$

(b) 75%

(c) 48%

(d) 23%

GS **(e)** 125%

$$125\% = \frac{\_\_\_}{100}$$

$$= \frac{\_\_\_ \div 25}{100 \div 25}$$

$$= \frac{\_\_\_}{4} = \_\_\_$$

(f) 250%

Answers

1. (a) 100; $\frac{1}{2}$ (b) $\frac{3}{4}$ (c) $\frac{12}{25}$ (d) $\frac{23}{100}$
(e) 125; 125; 5; $1\frac{1}{4}$ (f) $\frac{5}{2} = 2\frac{1}{2}$

2 Write each percent as a fraction in lowest terms.

GS **(a)** 37.5%

$$37.5\% = \frac{37.5}{100}$$
$$= \frac{37.5\ (10)}{100\ (\_\_\_)}$$
$$= \frac{375 \div \_\_\_}{1000 \div 125}$$
$$= \_\_\_$$

(b) 62.5%

(c) 4.5%

(d) $66\frac{2}{3}\%$

(e) $10\frac{1}{3}\%$

(f) $87\frac{1}{2}\%$

Answers

2. **(a)** 10; 125; $\frac{3}{8}$ **(b)** $\frac{5}{8}$ **(c)** $\frac{9}{200}$ **(d)** $\frac{2}{3}$ **(e)** $\frac{31}{300}$ **(f)** $\frac{7}{8}$

The next example shows how to write decimal and fraction percents as fractions.

EXAMPLE 2 Writing Decimal or Fraction Percents as Fractions

Write each percent as a fraction in lowest terms.

(a) 15.5%
Write 15.5 over 100.

$$15.5\% = \frac{15.5}{100}$$

The denominator is 100 because percent means **parts per 100.**

To get a whole number in the numerator, multiply the numerator and denominator by 10. (Recall that multiplying by $\frac{10}{10}$ is the same as multiplying by 1.)

$$\frac{15.5}{100} = \frac{15.5\,(10)}{100\,(10)} = \frac{155}{1000}$$

Move the decimal point 1 place to the *right* to multiply by 10.

Now write the fraction in lowest terms.

$$\frac{155 \div 5}{1000 \div 5} = \frac{31}{200}$$

(b) $33\frac{1}{3}\%$
Write $33\frac{1}{3}$ over 100.

$$33\frac{1}{3}\% = \frac{33\frac{1}{3}}{100}$$

When there is a mixed number in the numerator, rewrite the mixed number as an improper fraction.

$$\frac{33\frac{1}{3}}{100} = \frac{\frac{100}{3}}{100}$$

Write $33\frac{1}{3}$ as $\frac{100}{3}$.

Next, rewrite the division problem in a horizontal form. Finally, multiply by the reciprocal of the divisor.

Reciprocals

$$\frac{\frac{100}{3}}{100} = \frac{100}{3} \div 100 = \frac{100}{3} \div \frac{100}{1} = \frac{\overset{1}{\cancel{100}}}{3} \cdot \frac{1}{\underset{1}{\cancel{100}}} = \frac{1}{3} \leftarrow \text{Lowest terms}$$

Note

In **Example 2(a)** at the top of the page, we could have changed 15.5% to $15\frac{1}{2}\%$ and then written it as the improper fraction $\frac{31}{2}$ over 100. But it is usually easier *not* to change decimals to fractions and to work with decimal percents as they are.

◀ **Work Problem 2 at the Side.**

OBJECTIVE 2 Write fractions as percents. We will use the formula from the beginning of this section to write fractions as percents.

$$p\% = \frac{p}{100}$$

EXAMPLE 3 Writing Fractions as Percents

Write each fraction as a percent. Round to the nearest tenth if necessary.

(a) $\frac{3}{5}$

Write $\frac{3}{5}$ as a percent by solving for p in the proportion below.

$$\frac{3}{5} = \frac{p}{100}$$

Find cross products and show that they are equal.

$$5 \cdot p = 3 \cdot 100$$

$$5 \cdot p = 300$$

$$\frac{\overset{1}{\cancel{5}} \cdot p}{\underset{1}{\cancel{5}}} = \frac{300}{5} \quad \text{Divide both sides by 5.}$$

$$p = 60$$

This result means that $\frac{3}{5} = \frac{60}{100}$ or 60%.

Note

Solving proportions can be reviewed in **Chapter 5.**

(b) $\frac{7}{8}$

Write a proportion.

$$\frac{7}{8} = \frac{p}{100}$$

$$8 \cdot p = 7 \cdot 100 \quad \text{Show that cross products are equal.}$$

$$8 \cdot p = 700$$

$$\frac{\overset{1}{\cancel{8}} \cdot p}{\underset{1}{\cancel{8}}} = \frac{700}{8} \quad \text{Divide both sides by 8.}$$

$$p = 87.5$$

So, $\frac{7}{8} = 87.5\%$.

Note

If you think of $\frac{700}{8}$ as an improper fraction, changing it to a mixed number gives an answer of $87\frac{1}{2}$. So $\frac{7}{8} = 87.5\%$ or $87\frac{1}{2}\%$.

Continued on Next Page

3 Write as percents. Round to the nearest tenth if necessary.

GS **(a)** $\frac{1}{4}$

$\frac{1}{4} = \frac{p}{100}$

$4 \bullet p = 1 \bullet$ ____

$\frac{\overset{1}{\cancel{4}} \bullet p}{\underset{1}{\cancel{4}}} = \frac{100}{4}$

$p =$ ____

(b) $\frac{3}{10}$

(c) $\frac{6}{25}$

(d) $\frac{5}{8}$

(e) $\frac{1}{6}$

(f) $\frac{2}{9}$

Answers

3. **(a)** 100; 25% **(b)** 30% **(c)** 24%
(d) 62.5%
(e) 16.7% (rounded); $16\frac{2}{3}$% (exact)
(f) 22.2% (rounded); $22\frac{2}{9}$% (exact)

(c) $\frac{5}{6}$

Start with a proportion.

$$\frac{5}{6} = \frac{p}{100}$$

$6 \bullet p = 5 \bullet 100$ Show that cross products are equal.

$6 \bullet p = 500$

$\frac{\overset{1}{\cancel{6}} \bullet p}{\underset{1}{\cancel{6}}} = \frac{500}{6}$ Divide both sides by 6.

$p = 83.\overline{3}$ (A bar over the 3 indicates that the decimal keeps repeating 83.3333 forever.)

$p \approx 83.3$ Round to the nearest tenth.

So, $\frac{5}{6} = 83.\overline{3}\% \approx 83.3\%$ (rounded) (The "≈" symbol shows that $83.\overline{3}$% is rounded.)

Note

You can change $\frac{500}{6}$ to a mixed number to get an exact answer of $83\frac{1}{3}$%.

◀ **Work Problem 3 at the Side.**

OBJECTIVE 3 **Use the table of percent equivalents.** Knowing how to find the equivalents of fractions, decimals, and percents is important. However, the table on the next page shows common fractions and mixed numbers and their decimal and percent equivalents.

EXAMPLE 4 Using the Table of Percent Equivalents

Find the following in the table.

(a) $\frac{1}{6}$ as a percent

Find $\frac{1}{6}$ in the "fraction" column. The equivalent percent is 16.7% (rounded) or $16\frac{2}{3}$% (exact).

(b) 0.375 as a fraction

Look in the "decimal" column for 0.375. The equivalent fraction is $\frac{3}{8}$.

(c) $\frac{7}{8}$ as a percent

Find $\frac{7}{8}$ in the "fraction" column. The equivalent percent is 87.5% or $87\frac{1}{2}$%.

Calculator Tip

Example 4 (c) above can be solved on a calculator as shown below.

7 (÷) 8 (=) 0.875 (×) 100 (=) 87.5

Multiplying by 100 changes the decimal to a percent.
Or, if your calculator has a percent key, follow these steps.

7 (÷) 8 (%) 87.5

Press % key.

On scientific calculators, you may need to press the (2nd) key to access the % function, and some models require pressing (=) to get the answer.

Note

When a fraction like $\frac{1}{3}$ is changed to a decimal, it is a *repeating decimal* that goes on forever, 0.333333…. In the table below these decimals are rounded to the nearest thousandth. When the decimal is then changed to a percent, it will be to the nearest tenth of a percent. Decimals that do not repeat are usually not rounded.

Work Problem 4 at the Side. ▶

PERCENT, DECIMAL, AND FRACTION EQUIVALENTS

Percent (rounded to tenths when necessary)	Decimal	Fraction
1%	0.01	$\frac{1}{100}$
5%	0.05	$\frac{1}{20}$
10%	0.1	$\frac{1}{10}$
12.5% or $12\frac{1}{2}$%	0.125	$\frac{1}{8}$
16.7% (rounded) or $16\frac{2}{3}$% (exact)	$0.1\overline{6}$ rounds to 0.167	$\frac{1}{6}$
20%	0.2	$\frac{1}{5}$
25%	0.25	$\frac{1}{4}$
33.3% (rounded) or $33\frac{1}{3}$% (exact)	$0.\overline{3}$ rounds to 0.333	$\frac{1}{3}$
37.5% or $37\frac{1}{2}$%	0.375	$\frac{3}{8}$
40%	0.4	$\frac{2}{5}$
50%	0.5	$\frac{1}{2}$
60%	0.6	$\frac{3}{5}$
62.5% or $62\frac{1}{2}$%	0.625	$\frac{5}{8}$
66.7% (rounded) or $66\frac{2}{3}$% (exact)	$0.\overline{6}$ rounds to 0.667	$\frac{2}{3}$
75%	0.75	$\frac{3}{4}$
80%	0.8	$\frac{4}{5}$
87.5% or $87\frac{1}{2}$%	0.875	$\frac{7}{8}$
100%	1.0	1
150%	1.5	$1\frac{1}{2}$
200%	2.0	2

4 Find the following fractions, mixed numbers, decimals, and percents in the table on this page. If you already know the answer or can solve for the answer quickly, don't use the table.

(a) $\frac{3}{4}$ as a percent

(b) 10% as a fraction

(c) $0.\overline{6}$ as a fraction

(d) $37\frac{1}{2}$% as a fraction

(e) $\frac{7}{8}$ as a percent

(f) $\frac{1}{2}$ as a percent

(g) $33\frac{1}{3}$% as a fraction

(h) $1\frac{1}{2}$ as a percent

Answers

4. **(a)** 75% **(b)** $\frac{1}{10}$ **(c)** $\frac{2}{3}$ **(d)** $\frac{3}{8}$ **(e)** 87.5% **(f)** 50% **(g)** $\frac{1}{3}$ **(h)** 150%

6.2 Exercises

FOR EXTRA HELP

Download the MyDashBoard App

MyMathLab®

CONCEPT CHECK *Is each percent equivalent to the given fraction? Write* true *or* false.

1. $25\% = \frac{1}{2}$

2. $30\% = \frac{3}{10}$

3. $75\% = \frac{3}{4}$

4. $80\% = \frac{5}{6}$

Write each percent as a fraction in lowest terms and as a mixed number when possible.
See Examples 1 and 2.

5. 85%
GS
$0.85 = \frac{85}{100} = \frac{17}{\_\_\_\_}$

6. 45%
GS
$0.45 = \frac{45}{100} =$

7. 62.5%

8. 87.5%

9. 6.25%

10. 43.75%

11. $16\frac{2}{3}\%$

12. $66\frac{2}{3}\%$

13. $6\frac{2}{3}\%$

14. $46\frac{2}{3}\%$

15. 0.5%

16. 0.8%

17. 180%

18. 140%

19. 375%

20. 225%

CONCEPT CHECK *Is each fraction equivalent to the given percent? Write* true *or* false.

21. $\frac{1}{2} = 50\%$

22. $\frac{1}{100} = 10\%$

23. $\frac{4}{5} = 75\%$

24. $\frac{3}{10} = 30\%$

Write each fraction as a percent. Round percents to the nearest tenth if necessary.
See Example 3.

25. $\frac{7}{10} = \frac{p}{100}$
GS
Solve for p.
$p =$ ____%

26. $\frac{3}{4} = \frac{p}{100}$
GS
Solve for p.
$p =$ ____%

27. $\frac{37}{100}$

28. $\frac{63}{100}$

29. $\frac{5}{8}$

30. $\frac{1}{8}$

31. $\frac{7}{8}$

32. $\frac{3}{8}$

33. $\frac{12}{25}$

34. $\frac{15}{25}$

35. $\frac{23}{50}$

36. $\frac{18}{50}$

37. $\frac{7}{20}$

38. $\frac{9}{20}$

39. $\frac{5}{6}$

40. $\frac{1}{6}$

41. $\frac{5}{9}$

42. $\frac{7}{9}$

43. $\frac{1}{7}$

44. $\frac{5}{7}$

45. CONCEPT CHECK Which of the following is equal to $\frac{3}{4}$?

(a) 65% **(b)** 0.75 **(c)** 0.5 **(d)** 25%

46. CONCEPT CHECK Which of the following is equal to 80%?

(a) 0.85 **(b)** $\frac{3}{5}$ **(c)** $\frac{7}{10}$ **(d)** $\frac{4}{5}$

Complete the chart. Round decimals to the nearest thousandth and percents to the nearest tenth if necessary. ***See Examples 3 and 4.***

	Fraction	Decimal	Percent
47.	________	0.5	________
48.	________	0.8	________
49.	________	________	87.5%
50.	________	________	60%
51.	$\frac{1}{6}$	________	________
52.	$\frac{1}{3}$	________	________
53.	________	0.7	________

	Fraction	Decimal	Percent
54.			37.5%
55.			12.5%
56.		0.625	
57.	$\frac{2}{3}$		
58.	$\frac{5}{6}$		
59.	$\frac{3}{50}$		
60.	$\frac{3}{10}$		
61.	$\frac{8}{100}$		
62.			100%
63.	$\frac{1}{200}$		
64.	$\frac{1}{400}$		

	Fraction	Decimal	Percent
65.	______	2.5	______
66.	______	1.7	______
67.	$3\frac{1}{4}$	______	______
68.	$2\frac{4}{5}$	______	______

69. Select a decimal percent and write it as a fraction. Select a different fraction and write it as a percent. Write an explanation of each step of your work.

70. Prepare a table showing fraction, decimal, and percent equivalents for five fractions and mixed numbers of your choice.

In the following application problems, write the answer as a fraction in lowest terms, as a decimal, and as a percent.

71. Many pet owners say they have used the Internet to find pet information. Of 500 people who used the Internet for this purpose, 90 said they used it when buying a pet. What portion used the Internet when buying a pet? (*Source:* American Animal Hospital Association.)

72. About $\frac{1}{3}$ of all books purchased last year were for children. Of these children's books, 27 of every 100 purchased included a coloring activity. What portion of the children's books included a coloring activity? (*Source:* Consumer Research Study on Book Publishing, the American Booksellers, and the Book Industry Study Group.)

73. Only 13 out of every 100 adults ages 19–50 consume the recommended 1000 milligrams of calcium daily. What portion consumes the recommended daily amount? (*Source:* Market Facts for Milk Mustache.)

74. In a survey on how people learn to parent, 360 parents out of 800 said they were most influenced by relatives, friends, and spouses. What portion learned to parent this way? (*Source:* Bama Research.)

75. In a recent survey, 750 workers were asked if they would hire their own boss if they were in charge. A total of 150 workers said no, they would not. What portion of the workers said no? (*Source:* Marlin's 13th annual Attitudes in the Workplace Survey.)

76. When 1500 adults were asked what was the most important factor to consider when relocating after retirement, 675 said that climate was most important. What portion consider climate to be most important? (*Source:* Longevity Alliance Retirement and Relocation Survey.)

77. An insurance office has 80 employees. If 64 of the employees have iPhones, find

(a) the portion of the employees who have iPhones.

(b) the portion of the employees who do not have iPhones.

78. For every 640 tons of food purchased, 160 tons are discarded by Americans. (*Source:* U.S. Department of Agriculture.)

(a) What portion of the food is discarded?

(b) What portion of the food is not discarded?

79. An antibiotic is used to treat 380 people. If 342 people do not have side effects from the antibiotic, find the portion that do have side effects.

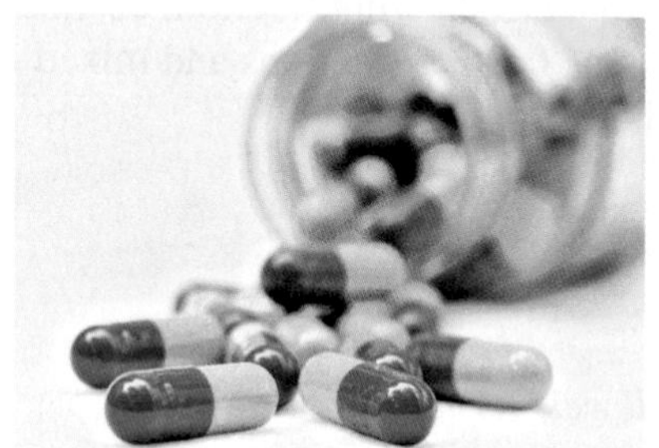

80. A survey of 340 doctors found that 119 of the doctors used a basic system of electronic record keeping.

(a) What portion of the doctors used electronic record keeping?

(b) What portion did not use it?

The circle graph shows the number of people in a survey of 5400 adults who picked each season as their favorite. Use this graph to answer Exercises 81–84, giving each answer as a fraction, as a decimal, and as a percent.

WEATHER OR NOT

Americans Pick Their Favorite Season (5400 adults surveyed)

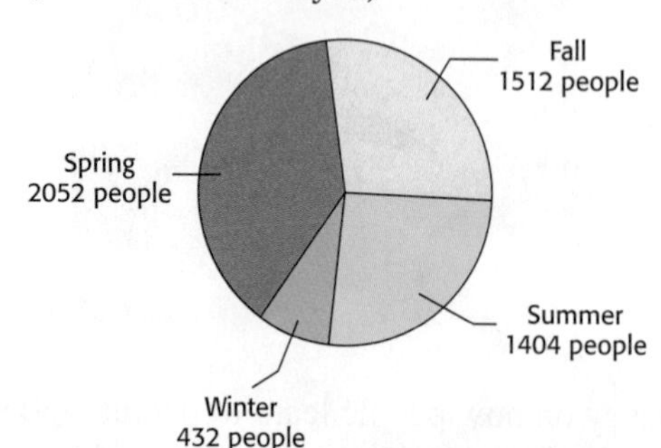

Source: Strategy One Survey for Back to Nature Foods.

81. What portion of the adults picked Winter as their favorite season?

82. What portion of the adults picked Spring as their favorite season?

83. Find the portion of the adults who say Summer is their favorite season.

84. Find the portion of the adults who say Fall is their favorite season.

Relating Concepts (Exercises 85–94) For Individual or Group Work

To review the basics of percent, ***work Exercises 85–94 in order.***

85. 100% of a number means all of the parts or ___________ parts out of ___________ parts. 200% means two times as many parts and 300% means three times as many parts.

86. Fill in the blanks.

(a) 100% of 765 workers is ___________.

(b) 200% of 48 letters is ___________.

(c) 300% of 7 DVDs is ___________.

87. 50% of a number is ___________ parts out of ___________ parts, which is ___________ of the parts.

88. 10% of a number is ___________ parts out of ___________ parts and can be found quickly by moving the decimal point ___________ place to the ___________.

89. 1% of a number is ___________ part out of ___________ parts and can be found quickly by moving the decimal point ___________ places to the ___________.

90. Fill in the blanks.

(a) 50% of 1050 homes is ___________.

(b) 10% of 370 printers is ___________.

(c) 1% of $8 is ___________.

In Exercises 91–94, use the shortcut methods for finding 1%, 10%, 50%, 100%, 200%, *and* 300%.

91. Describe a shortcut method for finding 15% of a number. Use your method to find 15% of $160.

92. Describe a shortcut method for finding 150% of a number. Use your method to find 150% of $160.

93. Explain a shortcut method for finding 90% of a number. Show how to find 90% of $450 using your method.

94. Explain a shortcut method for finding 210% of a number. Show how to find 210% of $800 using your method.

6.3 Using the Percent Proportion and Identifying the Components in a Percent Problem

OBJECTIVES

1. Learn the percent proportion.
2. Solve for an unknown value in a percent proportion.
3. Identify the percent.
4. Identify the whole.
5. Identify the part.

1 As a review of proportions, use the method of comparing cross products to decide whether each proportion is *true* or *false*. Show the cross products.

GS (a) $\frac{1}{2} = \frac{25}{50}$

$\frac{1}{2} = \frac{25}{50}$ $2 \cdot 25 =$ ____ $1 \cdot 50 =$ ____

The proportion is (true/false).

(b) $\frac{3}{4} = \frac{150}{200}$

(c) $\frac{7}{8} = \frac{180}{200}$

(d) $\frac{112}{41} = \frac{332}{123}$

Answers

1. (a) 50; 50; true (b) $600 = 600$; true
(c) $1440 \neq 1400$; false
(d) $13{,}612 \neq 13{,}776$; false

There are two ways to solve percent problems. One method uses proportions and is discussed in this and the next section. The other method uses the percent equation and is explained later in this chapter.

OBJECTIVE 1 Learn the percent proportion. We have seen that a statement of two equivalent ratios is called a proportion.

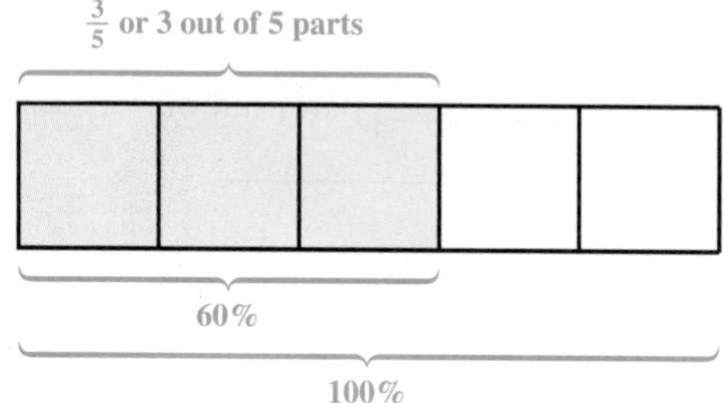

For example, the fraction $\frac{3}{5}$ is the same as the ratio 3 to 5, and 60% is the same as the ratio 60 to 100. As the figure above shows, these two ratios are equivalent and make a proportion.

◀ **Work Problem 1 at the Side.**

The percent proportion can be used to solve percent problems.

Percent Proportion

Part is to *whole* as *percent* is to *100*.

$$\frac{\text{part}}{\text{whole}} = \frac{\text{percent}}{100} \leftarrow \text{Always 100 because percent means } per\ 100$$

In the figure at the top of the page, the **whole** is 5 (the entire quantity), the **part** is 3 (the part of the whole), and the **percent** is 60. Write the percent proportion as follows.

$$\frac{\text{part}}{\text{whole}} \begin{matrix}\rightarrow\\ \rightarrow\end{matrix} \frac{3}{5} = \frac{60}{100} \begin{matrix}\leftarrow\\ \leftarrow\end{matrix} \frac{\text{percent}}{100}$$

Remember: Percent means per 100.

OBJECTIVE 2 Solve for an unknown value in a percent proportion. As shown in **Chapter 5**, if any three of the four values in a proportion are known, the fourth can be found by solving the proportion.

EXAMPLE 1 Using the Percent Proportion

Use the percent proportion to solve for x, representing the unknown value.

(a) part $= 12$, percent $= 25$; find the whole.

$$\frac{\text{part}}{\text{whole}} = \frac{\text{percent}}{100}$$ Percent proportion

Percent

$$\text{Part} \rightarrow \frac{12}{x} = \frac{25}{100} \quad \text{or} \quad \frac{12}{x} = \frac{1}{4}$$
Whole (unknown) $\rightarrow$

$\frac{25}{100}$ in lowest terms is $\frac{1}{4}$.

Continued on Next Page

First find the cross products.

$$\frac{12}{x} = \frac{1}{4} \qquad x \cdot 1 \qquad 12 \cdot 4$$

The unknown in this proportion is the *whole*.

Show that the cross products are equal.

$$x \cdot 1 = 12 \cdot 4$$
$$x = 48$$

The whole is 48.

CAUTION

In a **proportion,** you **cannot** divide out a common factor from the numerator of one ratio and the denominator of the other ratio. Dividing out common factors is done ***only*** when you are multiplying fractions.

(b) part = 30, whole = 50; find the percent.
Use the percent proportion.

Percent (unknown) ↓ — The unknown in this proportion is the percent.

Part → 30, Whole → 50

$$\frac{30}{50} = \frac{x}{100} \quad \text{Percent proportion}$$

$$\frac{3}{5} = \frac{x}{100} \quad \text{Write } \tfrac{30}{50} \text{ as } \tfrac{3}{5} \text{ in lowest terms.}$$

$$5 \cdot x = 3 \cdot 100 \quad \text{Find the cross products.}$$

$$5 \cdot x = 300$$

$$\frac{\overset{1}{\cancel{5}} \cdot x}{\underset{1}{\cancel{5}}} = \frac{300}{5} \quad \text{Divide both sides by 5.}$$

$$x = 60$$

The percent is 60, written as 60%.

Remember to write the % symbol when solving for an unknown percent.

(c) whole = 150, percent = 18; find the part.

Percent ↓

Part (unknown) → x, Whole → 150

$$\frac{x}{150} = \frac{18}{100} \text{ or } \frac{x}{150} = \frac{9}{50} \quad \text{Write } \tfrac{18}{100} \text{ as } \tfrac{9}{50} \text{ in lowest terms.}$$

$$x \cdot 50 = 150 \cdot 9 \quad \text{Find the cross products.}$$

$$x \cdot 50 = 1350$$

$$\frac{x \cdot \overset{1}{\cancel{50}}}{\underset{1}{\cancel{50}}} = \frac{1350}{50} \quad \text{Divide both sides by 50.}$$

$$x = 27$$

The part is 27.

Work Problem 2 at the Side. ▶

2 Use the percent proportion $\left(\frac{\text{part}}{\text{whole}} = \frac{\text{percent}}{100}\right)$ and solve for the unknown value.

GS **(a)** part = 12, percent = 16

$$\frac{12}{x} = \frac{16}{100} \text{ or } \frac{12}{x} = \frac{4}{25}$$

$$x \cdot 4 = 12 \cdot \_\_\_\_$$

$$x \cdot 4 = \_\_\_\_$$

$$\frac{x \cdot \overset{1}{\cancel{4}}}{\underset{1}{\cancel{4}}} = \frac{300}{4}$$

$$x = \_\_\_\_$$

(b) part = 30, whole = 120

(c) whole = 210, percent = 20

(d) whole = 4000, percent = 32

(e) part = 74, whole = 185

Answers

2. (a) 25; 300; 75; whole = 75
(b) percent = 25 (so, the percent is 25%)
(c) part = 42 **(d)** part = 1280
(e) percent = 40 (so, the percent is 40%)

3 Identify the percent.

GS **(a)** Of the 900 blood glucose tests, 25% will be completed by Adrian.

The number preceding the ____ symbol is the percent.

So, ____ is the percent.

(b) Of the 620 preschool students, 65% will be served breakfast and lunch.

(c) Find the amount of sales tax by multiplying \$590 and $6\frac{1}{2}$ percent.

(d) 8500 tons of recyclables is 42% of what number of tons?

(e) What percent of the 1450 parents use child care?

Answers

3. (a) % or percent; 25 **(b)** 65 **(c)** $6\frac{1}{2}$ **(d)** 42 **(e)** The percent is unknown.

As a help in solving percent problems, keep in mind this basic idea.

Percent Problems

All percent problems involve a comparison between a part of something and the whole.

Solving these problems requires identifying the three components of a percent proportion: part, whole, and percent.

OBJECTIVE 3 Identify the percent. Look for the percent first. It is the easiest to identify.

Percent

The **percent** is the ratio of a part to a whole, with 100 as the denominator. In a problem, the percent appears with the word **percent** or with the symbol "%" after it.

EXAMPLE 2 Finding the Percent in Percent Problems

Find the percent in the following.

(a) 32% of the 900 men were too large for the imported car.

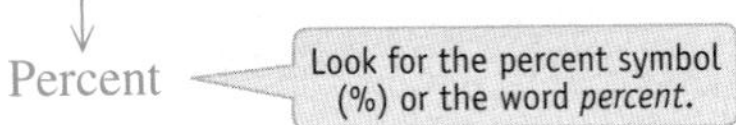
Percent
Look for the percent symbol (%) or the word *percent*.

The percent is 32. The number 32 appears with the percent symbol (%).

(b) \$150 is 25 percent of what number?

Percent

The percent is 25 because 25 appears with the word *percent.*

(c) What percent of 7000 pounds is 3500 pounds?

Percent (unknown)

The word *percent* has no number with it, so the percent is the unknown part of the problem.

◀ **Work Problem 3 at the Side.**

OBJECTIVE 4 Identify the whole. Next, look for the whole.

Whole

The **whole** is the entire quantity. In a percent problem, the whole often appears after the word **of**.

EXAMPLE 3 Finding the Whole in Percent Problems

Identify the whole in the following.

(a) 32% of the 900 men were too large for the imported car.

Whole

The whole is 900. The number 900 appears after the word *of*.

(b) \$150 is 25 percent of what number?

Whole — The whole is the unknown part of the problem.

(c) What percent of 7000 pounds is 3500 pounds?

Whole

Work Problem 4 at the Side. ▶

OBJECTIVE 5 Identify the part. Finally, look for the part.

Part

The **part** is the portion being compared with the whole.

Note

If you have trouble identifying the part, find the percent and whole first. The remaining number is the part.

EXAMPLE 4 Finding the Part in Percent Problems

Identify the part. Then set up the percent proportion. (Do **not** solve the proportions.)

(a) 54% of 700 students is 378 students.
First find the percent and the whole.

54% of 700 students is 378 students.
(54% → Percent; with % symbol) (700 → Whole; follows "of")

Find the percent and then the whole. The remaining number, 378, is the part.

The remaining number, 378, is the part.

54% of 700 students is 378 students.
(54% → Percent) (700 → Whole) (378 → Part)

$$\frac{\text{Part} \rightarrow 378}{\text{Whole} \rightarrow 700} = \frac{54 \leftarrow \text{Percent}}{100 \leftarrow \text{Always } 100}$$

(b) \$150 is 25% of what number?
(25% → Percent) (what number → Whole (unknown))

$$\frac{\text{Part} \rightarrow 150}{\text{Whole} \rightarrow \text{unknown}} = \frac{25 \leftarrow \text{Percent}}{100 \leftarrow \text{Always } 100}$$

\$150 is the remaining number, so the part is \$150.

(c) 85% of 7000 is what number?
(85% → Percent) (7000 → Whole) (what number → Part (unknown))

$$\frac{\text{Part} \rightarrow \text{unknown}}{\text{Whole} \rightarrow 7000} = \frac{85 \leftarrow \text{Percent}}{100 \leftarrow \text{Always } 100}$$

Work Problem 5 at the Side. ▶

4 Identify the whole.

(a) Of the 900 blood glucose tests, 25% will be completed by Adrian.

(b) Of the 620 preschool students, 65% will be served breakfast and lunch.

(c) 8500 tons of recyclables is 42% of what number of tons?

5 Identify the part, then set up the percent proportion.

(a) Of the 900 blood glucose tests, 25% or 225 will be completed by Adrian.

(b) Of the 620 preschool students, 65% or 403 will be served breakfast and lunch.

(c) 8500 tons of recyclables is 42% of what number of tons?

Answers

4. (a) 900 **(b)** 620
(c) what number (an unknown)

5. (a) 225; $\frac{\text{Part} \rightarrow 225}{\text{Whole} \rightarrow 900} = \frac{25 \leftarrow \text{Percent}}{100 \leftarrow \text{Always } 100}$

(b) 403; $\frac{\text{Part} \rightarrow 403}{\text{Whole} \rightarrow 620} = \frac{65 \leftarrow \text{Percent}}{100 \leftarrow \text{Always } 100}$

(c) 8500; $\frac{\text{Part} \rightarrow 8500}{\text{Whole} \rightarrow \text{unknown}} = \frac{42 \leftarrow \text{Percent}}{100 \leftarrow \text{Always } 100}$

6.3 Exercises

FOR EXTRA HELP

MyMathLab®

CONCEPT CHECK *Identify the component that is unknown. Circle the correct answer. Then, set up the percent proportion. Do **not** solve the proportion.*

1. part = 5, percent = 10
part whole percent

2. part 1.5, whole = 4.5
part whole percent

3. part = 36, whole = 24
part whole percent

4. whole = 72, percent = 30
part whole percent

5. whole = 160, percent = 35
part whole percent

6. part = 20, percent = 25
part whole percent

Find the unknown value in the percent proportion $\frac{part}{whole} = \frac{percent}{100}$. *Round to the nearest tenth if necessary. If the answer is a percent, be sure to include a percent symbol (%).* ***See Example 1.***

7. part = 30, percent = 20

8. part = 25, percent = 25

9. part = 28, percent = 40

10. part = 11, percent = 5

11. part = 15, whole = 60

12. part = 105, whole = 35

13. part = 9.25, whole = 27.75

14. part = 12.8, whole = 9.6

15. whole = 52, percent = 50

16. whole = 115, percent = 38

17. whole = 94.4, part = 25

18. whole = 89.6, part = 50

Solve each problem. If the answer is a percent, be sure to include a percent sign (%). ***See Examples 2–4.***

19. Find the whole if the part is 46 and the percent is 40.

$$\frac{\text{part}}{\text{whole}} = \frac{\text{percent}}{100}$$

$$\frac{46}{x} = \frac{40}{100}$$

$40 \cdot x$

4600

$40 \cdot x = 4600$

$x =$ ______

20. The percent is 45 and the whole is 160. Find the part.

$$\frac{\text{part}}{\text{whole}} = \frac{\text{percent}}{100}$$

$$\frac{x}{160} = \frac{45}{100}$$

7200

$100 \cdot x$

$100 \cdot x = 7200$

$x =$ ______

21. The whole is 5000 and the part is 20. Find the percent.

22. Suppose the part is 15 and the whole is 2500. Find the percent.

23. Find the percent if the whole is 4300 and the part is $107\frac{1}{2}$.

24. What is the part, if the percent is $12\frac{3}{4}$ and the whole is 5600?

25. The whole is 6480 and the part is 19.44. Find the percent.

26. Suppose the part is 281.25 and the percent is $1\frac{1}{4}$. Find the whole.

CONCEPT CHECK *Fill in each blank with the correct response.*

27. In a percent problem, the percent can be identified because it appears with the word ____________ or with the ____________ symbol after it.

28. In a percent problem, the whole is the ____________ quantity and often appears after the word ____________.

29. In a percent problem, the part is the portion being compared with the ____________.

30. To solve a percent proportion, you must cross ____________.

In Exercises 31–42, set up the percent proportion, and write "unknown" for any value that is not given. Recall that the percent proportion is $\frac{part}{whole} = \frac{percent}{100}$. *Do* ***not*** *try to solve for the unknowns.* ***See Examples 2–4.***

31. 10% of how many bicycles is 60 bicycles?

Part → $\frac{60}{\text{unknown}} = \frac{\quad}{\quad}$ ← Percent
Whole → ← Always 100

32. 58% of how many preschoolers is 203 preschoolers?

Part → $\frac{\quad}{\quad} = \frac{58}{100}$ ← Percent
Whole → ← Always 100

33. What % of \$800 is \$600?

34. What % of \$1500 is \$1395?

35. What is 25% of \$970?

36. What is 61% of 830 homes?

37. 12 injections is 20% of what number of injections?

38. 410 pallets is $33\frac{1}{3}$% of how many pallets.

39. 54.34 is 3.25% of what number?

40. 16.74 is 11.9% of what number?

41. 0.68% of \$487 is what amount?

42. What amount is 6.21% of \$704.35?

43. Identify the three components in a percent problem. In your own words, write a sentence telling how you will identify each of these three components.

44. Write one short sentence using numbers and words. The sentence should include a percent, a whole, and a part. Identify each of these three components.

*Set up the percent proportion for each application problem. Do **not** try to solve for any unknowns.*

45. Fry's Electronics sold 1262 computers in a recent promotion. If 730 of these computers were laptop computers, what percent of the computers were laptops?

Part → $\dfrac{730}{}$ = $\dfrac{}{100}$ ← Percent
Whole → ← Always 100

46. Ivory Soap is $99\frac{44}{100}\%$ pure. If a bar of Ivory Soap weighs 4 ounces, how many ounces are pure? (*Source:* Procter & Gamble.)

Part → $\dfrac{\text{unknown}}{}$ = $\dfrac{}{100}$ ← Percent
Whole → ← Always 100

47. Of the 142 people attending a movie theater, 86 bought buttered popcorn. What percent bought buttered popcorn?

48. On her first check from the Pizza Hut Restaurant, 15% was withheld from Maria's total earnings of \$225. What amount was withheld?

49. Of the customers buying a salad at McDonald's, 23% prefer Newman's Own Light Salad Dressing. If the total number of salad customers is 610, find the number who prefer Newman's Own Light Salad Dressing.

50. At a community college campus, it was found that 2322 of the students work full-time. If this was 27% of the students, find the total number of students on campus. (*Source*: American Association of Community Colleges.)

51. A survey of 8600 full-time community college students found that 4300 of them work part-time. What percent of the full-time students work part-time? (*Source*: American Association of Community Colleges.)

52. There have been 36 cups of coffee served from a banquet-sized coffee pot. If this is 30% of the capacity of the pot, find the capacity of the pot.

53. In a recent survey of 480 adults, 55% said that they would prefer to have their wedding at a religious site. How many said they would prefer the religious site? (*Source:* National Family Opinion Research.)

54. Sue Ann needs 64 credits to graduate. If she has completed 48 of the credits needed, what percent of the credits has she already completed?

55. In a poll of 822 people, 49.5% said that they get their news from television. Find the number of people who said they get their news from television. (*Source: Brill's Content.*)

56. The sales tax on a new car is $1575. If the sales tax rate is 7%, find the price of the car before the sales tax is added.

57. When asked "What co-worker behaviors annoy you the most," 12% of those surveyed said "Being perpetually late." Find the number of people in the survey if 168 people gave this response. (*Source*: Accountemps.)

58. The state troopers tested 924 cars for safety. There were 231 cars that failed the safety test for one or more reasons. Find the percent of cars that failed the test.

59. Jerry Azzaro has listed 680 antique toys on eBay. If 45% of these were antique toy trains, find the number that were toy trains.

60. In a survey of 27,400 American sports fans, ages 12 and up, 3562 picked Major League baseball as their favorite spectator sport. What percent picked Major League baseball as their favorite sport? (*Source*: ESPN Sports Poll.)

6.4 Using Proportions to Solve Percent Problems

OBJECTIVES

1. Use the percent proportion to find the part.
2. Find the whole using the percent proportion.
3. Find the percent using the percent proportion.

This is the percent proportion that you learned about in the previous section.

$$\frac{\text{part}}{\text{whole}} = \frac{\text{percent}}{100}$$

Recall that if one of the values is unknown, you can find it by solving the percent proportion.

OBJECTIVE 1 **Use the percent proportion to find the part.** The first example shows how to use the percent proportion to find the part.

EXAMPLE 1 Finding the Part with the Percent Proportion

Find 15% of \$160.

Here the percent is 15 and the whole is 160. (Recall that the whole often comes after the word ***of***.) Now find the part. Let x represent the unknown part.

$$\frac{\text{part}}{\text{whole}} = \frac{\text{percent}}{100} \quad \text{so} \quad \frac{x}{160} = \frac{15}{100} \quad \text{or} \quad \frac{x}{160} = \frac{3}{20}$$

Write $\frac{15}{100}$ as $\frac{3}{20}$ in lowest terms.

Find the cross products in the proportion and show that they are equal.

$x \bullet 20 = 160 \bullet 3$ Cross products

$x \bullet 20 = 480$

$\frac{x \bullet \overset{1}{\cancel{20}}}{\underset{1}{\cancel{20}}} = \frac{480}{20}$ Divide both sides by 20.

$x = 24$ The unknown part is 24.

15% of \$160 is \$24.

◀ Work Problem 1 at the Side.

Just as with some of the fraction application problems in **Chapter 2,** the word ***of*** may be an indicator word meaning ***multiply.*** Here is an example.

15% of 160

↓

15% • 160

In this type of example, there is another way to find the part.

Finding the Part Using Multiplication

To find the part:

Step 1 Identify the percent. Write the percent as a decimal.

Step 2 Multiply this decimal by the whole.

---

1 Use the percent proportion to find the part.

GS **(a)** 8% of 400 patients

$$\frac{x}{400} = \frac{8}{100} \quad \text{or} \quad \frac{x}{400} = \frac{2}{25}$$

$x \bullet 25 = 400 \bullet$ ____

$x \bullet 25 =$ ____

$\frac{x \bullet \overset{1}{\cancel{25}}}{\underset{1}{\cancel{25}}} = \frac{800}{25}$

$x =$ ________

(b) 15% of \$3220

(c) 7% of 2700 miles

(d) 48% of 1580 kilowatts

Answers

1. **(a)** 2; 800; 32 patients **(b)** \$483 **(c)** 189 miles **(d)** 758.4 kilowatts

EXAMPLE 2 **Finding the Part Using Multiplication**

Use multiplication to find the part.

(a) Find 42% of 830 yards.

Step 1 Here, the percent is 42. Write 42% as the decimal 0.42.

Step 2 Multiply 0.42 and the whole, which is 830.

$$\text{part} = (0.42)(830) = 348.6 \text{ yd}$$

When the percent and whole are given, the part must be found.

It is a good idea to estimate the answer, to make sure no mistakes were made with decimal points. Round 42% to 40% or 0.4, and round 830 to 800. Next, 40% of 800 is

$$(0.4)(800) = 320 \leftarrow \text{Estimate}$$

so the exact answer of 348.6 is reasonable.

(b) Find 25% of 1680 cars.

Identify the percent as 25. Write 25% in decimal form as 0.25. Now, multiply 0.25 and 1680.

$$\text{part} = (0.25)(1680) = 420 \text{ cars} \quad \text{Multiply.}$$

You can also use a shortcut to find the answer. Since 25% means 25 parts out of 100 parts, this is the same as $\frac{1}{4}$ of the whole $\left(\frac{25}{100} = \frac{1}{4}\right)$. Do you see a shortcut here? You can find $\frac{1}{4}$ of a number by dividing the number by 4. So, this shortcut gives us the exact answer, $1680 \div 4 = 420$.

(c) Find 140% of 60 miles.

In this problem, the percent is 140. Write 140% as the decimal 1.40. Next, multiply 1.40 and 60.

$$\text{part} = (1.40)(60) = 84 \text{ miles} \quad \text{Multiply.}$$

You can estimate the answer by realizing that 140% is close to 150% (which is $1\frac{1}{2}$) and $1\frac{1}{2}$ times 60 is 90. So, 84 miles is a reasonable answer.

(d) Find 0.4% of 50 kilometers.

$$\text{part} = (0.004)(50) = 0.2 \text{ kilometer} \quad \text{Multiply.}$$

↑ Write 0.4% as a decimal.

Estimate the answer by realizing that 0.4% is less than 1%.

$$1\% \text{ of } 50 \text{ kilometers} = 50. = 0.5 \text{ kilometer}$$

So our exact answer should be *less than* 0.5 kilometer, and 0.2 kilometer fits this requirement.

Work Problem 2 at the Side. ▶

EXAMPLE 3 **Solving for the Part in an Application Problem**

Raley's Markets has 850 employees. Of these employees, 28% are students. How many of the employees are students? Use the six problem-solving steps.

Continued on Next Page

2 Use multiplication to find the part.

GS **(a)** 55% of 10,000 x-rays

Write 55% in decimal form as 0.55.

part = (0.55) (________)

= ________

(b) 16% of 120 miles

(c) 135% of 60 dosages

(d) 0.5% of \$238

Answers

2. **(a)** 10,000; 5500 x-rays **(b)** 19.2 miles **(c)** 81 dosages **(d)** \$1.19

3 Use the six problem-solving steps to solve each problem.

GS **(a)** One day on Jacob's mail route there were 2920 pieces of mail. If 45% of those were advertising pieces, find the number of advertising pieces.

Notice the blue word of as an indicator for multiplication. Write 45% in decimal form as _____.

part = (_____) • (2920)

= _________

(b) There are 9750 students at the college. If 12% of them wear glasses or contact lenses, how many students wear glasses or contact lenses?

Answers

3. **(a)** 0.45; 0.45; 1314 advertising pieces
(b) 1170 wear glasses or contact lenses

Step 1 **Read** the problem. The problem asks us to find the number of employees who are students.

Step 2 **Work out a plan.** Look for the word *of* as an indicator word for multiplication.

28% of the employees are students.
↑ Indicator word

The total number of employees is 850, so the whole is 850. The percent is 28. To find the number of students, find the part.

Step 3 **Estimate** a reasonable answer. You can estimate the answer by rounding 28% to 25% and 850 to 900. Remember that 25% is 25 parts out of 100, which is equivalent to $\frac{1}{4}$. So divide 900 by 4.

$$900 \div 4 = 225 \text{ students} \leftarrow \text{Estimate}$$

Step 4 **Solve** the problem.

$$\text{part} = (0.28)(850) = 238 \quad \text{Multiply.}$$

↑ Write 28% as a decimal.

Notice that the decimal point was moved two places to the left.

Step 5 **State the answer.** Raley's Markets has 238 student employees.

Step 6 **Check.** The exact answer, 238 students, is close to our estimate of 225 students.

◀ **Work Problem 3 at the Side.**

Calculator Tip

If you are using a calculator, you could solve **Example 3** above like this.

0.28 ⊗ 850 ⊜ 238

Or, if your calculator has a % key, check the instructions for an alternate method using the % key.

OBJECTIVE 2 Find the whole using the percent proportion. The next example shows how to use the percent proportion to find the whole.

Note

Remember, the *whole* is the entire quantity.

EXAMPLE 4 Finding the Whole with the Percent Proportion

(a) 8 iPods is 4% of what number of iPods?

Here the percent is 4, the whole is unknown, and the part is 8. Use the percent proportion to find the whole. Let x represent the unknown whole.

$$\frac{8}{x} = \frac{4}{100} \quad \text{or} \quad \frac{8}{x} = \frac{1}{25}$$

Write $\frac{4}{100}$ as $\frac{1}{25}$ in lowest terms.

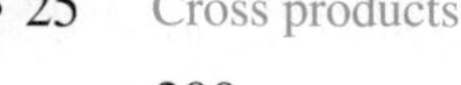

$$x \cdot 1 = 8 \cdot 25 \quad \text{Cross products}$$

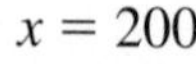

$$x = 200$$

8 iPods is 4% of 200 iPods.

Continued on Next Page

(b) 135 tourists is 15% of what number of tourists?
The percent is 15 and the part is 135.

$$\text{Part} \rightarrow \frac{135}{x} = \frac{15}{100} \begin{array}{l} \leftarrow \text{Percent} \\ \leftarrow \text{Always } 100 \end{array}$$
Whole (unknown) $\rightarrow x$

If the part and percent are given, the *whole* must be found.

$$\frac{135}{x} = \frac{3}{20}$$ Write $\frac{15}{100}$ as $\frac{3}{20}$ in lowest terms.

$$x \cdot 3 = 135 \cdot 20 \quad \text{Cross products}$$

$$x \cdot 3 = 2700$$

$$\frac{x \cdot \overset{1}{\cancel{3}}}{\underset{1}{\cancel{3}}} = \frac{2700}{3} \quad \text{Divide both sides by 3.}$$

$$x = 900$$

135 tourists is 15% of **900 tourists**.

Work Problem 4 at the Side. ▶

EXAMPLE 5 Applying the Percent Proportion

At Newark Salt Works, 78 employees are absent because of illness. If this is 5% of the total number of employees, how many employees does the company have? Use the six problem-solving steps.

Step 1 **Read** the problem. The problem asks for the total number of employees.

Step 2 **Work out a plan.** From the information in the problem, the percent is 5 and the part of the total number of employees is 78. The total number of employees or entire quantity, which is the whole, is the unknown.

Step 3 **Estimate** a reasonable answer. Round the number of employees from 78 to 80. Then, 5% is equivalent to the fraction $\frac{1}{20}$, and 80 is $\frac{1}{20}$ of the total number of employees.

$$80 \cdot 20 = 1600 \text{ employees} \leftarrow \text{Estimate}$$

Step 4 **Solve** the problem. Use the percent proportion to find the whole (the total number of employees).

$$\text{Part} \rightarrow \frac{78}{x} = \frac{5}{100} \begin{array}{l} \leftarrow \text{Percent} \\ \leftarrow \text{Always } 100 \end{array}$$
Whole (unknown) $\rightarrow x$

$$\frac{78}{x} = \frac{1}{20} \quad \text{Write } \frac{5}{100} \text{ as } \frac{1}{20} \text{ in lowest terms.}$$

$$x \cdot 1 = 78 \cdot 20 \quad \text{Cross products}$$

$$x = 1560$$

Step 5 **State the answer.** The company has **1560 employees**.

Step 6 **Check.** The exact answer, 1560 employees, is close to our estimate of 1600 employees.

Continued on Next Page

4 Use the percent proportion to find the unknown whole.

GS **(a)** 75 *American Idol* contestants are only 5% of what number who auditioned?

$$\frac{75}{x} = \frac{5}{100}$$ Write $\frac{5}{100}$ in lowest terms.

$$\frac{75}{x} = \frac{1}{\_\_\_}$$

$$x \cdot 1 = 75 \cdot 20$$

$$x = \_\_\_\_\_$$

(b) 28 antiques is 35% of what number of antiques?

(c) 387 customers is 36% of what number of customers?

(d) 292.5 miles is 37.5% of what number of miles?

Answers

4. (a) 20; 1500 auditioned **(b)** 80 antiques **(c)** 1075 customers **(d)** 780 miles

5 Use the six problem-solving steps and the percent proportion to solve each problem.

GS **(a)** A freeze resulted in a loss of 52% of an avocado crop. If the loss was 182 tons, find the total number of tons in the crop.

$$\frac{182}{x} = \frac{52}{100}$$ Write $\frac{52}{100}$ in lowest terms.

$$\frac{182}{x} = \frac{13}{\_\_\_\_}$$

$$\frac{x \cdot \cancel{13}^{1}}{\cancel{13}_{1}} = \frac{4550}{13}$$

$$x = \_\_\_\_\_\_\_\_$$

(b) A factory batch of cake mix contains 900 pounds of sugar, which is 18%, by weight, of the entire batch. What is the total weight of the batch?

Answers

5. (a) 25; 350 tons (b) 5000 pounds

Note

To estimate the answer to **Example 5** on the previous page, 5% was changed to its fraction equivalent, $\frac{1}{20}$. Because 80 (rounded) is $\frac{1}{20}$ of the total employees, 80 was multiplied by 20 to get 1600, the estimated answer.

◀ Work Problem **5** at the Side.

OBJECTIVE 3 Find the percent using the percent proportion. If the part and the whole are known, the percent proportion can be used to find the percent.

EXAMPLE 6 Using the Percent Proportion to Find the Percent

(a) 13 coupons is what percent of 52 coupons?
The whole is 52 (follows *of*) and the part is 13. Next, find the percent.

The whole often follows the word "of".

$$\frac{\text{part}}{\text{whole}} = \frac{\text{percent}}{100}$$

Part → 13, Whole → 52; x ← Percent (unknown), 100 ← Always 100

$$\frac{13}{52} = \frac{x}{100}$$

Write $\frac{13}{52}$ as $\frac{1}{4}$ in lowest terms.

$$\frac{1}{4} = \frac{x}{100}$$

Find the cross products.

$$4 \cdot x = 1 \cdot 100 \quad \text{Cross products}$$

$$\frac{\cancel{4}^{1} \cdot x}{\cancel{4}_{1}} = \frac{100}{4} \quad \text{Divide both sides by 4.}$$

$$x = 25$$

13 coupons is 25% of 52 coupons.

(b) What percent of \$500 is \$100?
The whole is 500 (follows *of*) and the part is 100.

$$\frac{100}{500} = \frac{x}{100} \quad \leftarrow \text{Percent (unknown)}$$

Write $\frac{100}{500}$ as $\frac{1}{5}$ in lowest terms.

$$\frac{1}{5} = \frac{x}{100}$$

$$5 \cdot x = 1 \cdot 100 \quad \text{Cross products}$$

$$5 \cdot x = 100$$

$$\frac{\cancel{5}^{1} \cdot x}{\cancel{5}_{1}} = \frac{100}{5} \quad \text{Divide both sides by 5.}$$

$$x = 20$$

Remember to write the % symbol in the answer.

20% of \$500 is \$100.

Continued on Next Page

CAUTION

When finding the percent, be sure to label your answer with the percent symbol (%).

Work Problem 6 at the Side. ▶

EXAMPLE 7 Applying the Percent Proportion

A roof is expected to last 20 years before needing replacement. If the roof is now 15 years old, what percent of the roof's life has been used?

Step 1 **Read** the problem. The problem asks for the percent of the roof's life that is already used.

Step 2 **Work out a plan.** The expected life of the roof is the entire quantity or *whole,* which is 20. The *part* of the roof's life that is already used is 15. Use the percent proportion to find the percent of the roof's life used.

Step 3 **Estimate** a reasonable answer. Since the roof is 15 years old, it is $\frac{15}{20}$ or $\frac{3}{4}$ used. Remember that $\frac{3}{4}$ is equivalent to 75%, so our estimate is 75%.

Step 4 **Solve** the problem. Let x represent the unknown percent.

$$\begin{matrix}\text{Part} \rightarrow \\ \text{Whole} \rightarrow\end{matrix} \frac{15}{20} = \frac{x}{100} \quad \text{or} \quad \frac{3}{4} = \frac{x}{100} \qquad \text{Write } \tfrac{15}{20} \text{ as } \tfrac{3}{4} \text{ in lowest terms.}$$

$$4 \cdot x = 3 \cdot 100 \qquad \text{Cross products}$$

$$4 \cdot x = 300$$

$$\frac{\overset{1}{\cancel{4}} \cdot x}{\underset{1}{\cancel{4}}} = \frac{300}{4} \qquad \text{Divide both sides by 4.}$$

$$x = 75$$

Step 5 **State the answer.** 75% of the roof's life has been used.

Step 6 **Check.** The exact answer, 75%, matches our estimate of 75%.

Work Problem 7 at the Side. ▶

EXAMPLE 8 Applying the Percent Proportion

Rainfall this year was 33 inches, while normal rainfall is only 30 inches. What percent of normal rainfall is this year's rainfall?

Step 1 **Read** the problem. The problem asks us to find what percent this year's rainfall is of normal rainfall.

Step 2 **Work out a plan.** The normal rainfall is the *whole,* which is 30. This year's rainfall is *all of normal rainfall and more,* or 33 (part = 33). You need to find the percent that this year's rainfall is of normal rainfall.

Continued on Next Page

6 Use the percent proportion to solve each problem.

GS **(a)** \$21 is what percent of \$105?

Write $\frac{21}{105}$ in lowest terms.

$$\frac{21}{105} = \frac{x}{100}$$

$$\frac{1}{\_\_\_} = \frac{x}{100}$$

$$5 \cdot x = 100$$

$$\frac{\overset{1}{\cancel{5}} \cdot x}{\underset{1}{\cancel{5}}} = \frac{100}{5}$$

$$x = \_\_\_\%$$

(b) What percent of 320 Internet companies is 48 Internet companies?

(c) What percent of 2280 court trials is 1026 trials?

7 Solve each problem.

(a) The bid price on an auction item is \$289 while the minimum acceptable price is \$425. The bid price is what percent of the minimum?

(b) A laboratory technician completes 80 tests in one day. If 52 of these tests were completed in the morning, what percent of the tests were completed in the morning?

Answers

6. (a) 5; 20% **(b)** 15% **(c)** 45%

7. (a) 68% **(b)** 65%

8 Solve each problem.

GS **(a)** A new Toyota Prius Hybrid gets 32 miles per gallon on the highway and 48 miles per gallon around town. What percent of the highway mileage does the car get around town?

Write $\frac{48}{32}$ in lowest terms.

$$\frac{48}{32} = \frac{x}{100}$$

$$\frac{3}{2} = \frac{x}{100}$$

$$2 \cdot x = 3 \cdot 100$$

$$\frac{\overset{1}{\cancel{2}} \cdot x}{\underset{1}{\cancel{2}}} = \frac{\_\_\_\_}{2}$$

$$x = \_\_\_\_\%$$

(b) The service department set a goal of 360 service calls this week. If they made 432 service calls, find the percent of their goal that they achieved.

Answers

8. **(a)** 300; 150% **(b)** 120%

Step 3 **Estimate** a reasonable answer. The increase in rainfall is 3 inches and the whole is 30 inches. The increase is $\frac{3}{30}$ or $\frac{1}{10}$ which is 10%. The whole is 100%, so 100% + 10% = 110%, our estimate.

Step 4 **Solve** the problem. Let x represent the unknown percent.

$$\frac{33}{30} = \frac{x}{100} \quad \text{or} \quad \frac{11}{10} = \frac{x}{100} \qquad \text{Write } \tfrac{33}{30} \text{ as } \tfrac{11}{10} \text{ in lowest terms.}$$

$$10 \cdot x = 11 \cdot 100 \qquad \text{Cross products}$$

$$10 \cdot x = 1100$$

$$\frac{\overset{1}{\cancel{10}} \cdot x}{\underset{1}{\cancel{10}}} = \frac{1100}{10} \qquad \text{Divide both sides by 10.}$$

$$x = 110$$

If the part is *greater than* the whole, the percent is greater than 100.

Step 5 **State the answer.** This year's rainfall is **110%** of normal rainfall.

Step 6 **Check.** The exact answer, 110%, matches our estimate of 110%.

◀ **Work Problem 8 at the Side.**

Note

In **Example 8**, the part (33 inches) is *greater* than the whole (30 inches). This can occur when there is an increase or a gain and we are comparing this *greater* amount (part) to the original amount (whole).

6.4 Exercises

FOR EXTRA HELP

Download the MyDashBoard App

MyMathLab®

1. CONCEPT CHECK To find the part using the multiplication shortcut, use the formula

part = ________ • ________.

2. CONCEPT CHECK Solve for the part when the percent is 38 and the whole is 200. Circle the correct answer.

7.6 76 760 7600

Find the part using the multiplication shortcut. ***See Example 2.***

3. 35% of 120 test tubes
GS part = (0.35)(120)
part = ____________

4. 20% of 1800 rentals
GS part = (0.20)(1800)
part = ____________

5. 45% of 4080 military personnel

6. 12% of 3650 websites

7. 4% of 120 ft

8. 9% of $150

9. 150% of 210 files

10. 130% of 60 trees

11. 52.5% of 1560 trucks

12. 38.2% of 4250 loads

13. 2% of $164

14. 6% of $434

15. 225% of 680 tables

16. 110% of 150 apartments

17. 17.5% of 1040 cell phones

18. 46.1% of 843 kilograms

19. 0.9% of $2400

20. 0.3% of $1400

CONCEPT CHECK *Set up a percent proportion for each application problem. Do* ***not*** *try to solve for any unknowns.*

21. 80 e-mails is 25% of what number of e-mails?

22. 32 medical exams is 5% of what number of medical exams?

Find the whole using the percent proportion. ***See Example 4.***

23. 30% of what number of hay bales is 48 hay bales?
GS part is 48; percent is 30; whole is unknown

$$\frac{48}{x} = \frac{30}{100} \quad \text{or} \quad \frac{48}{x} = \frac{3}{10}$$

$3 \cdot x = 480$ Find cross products.

$$\frac{\overset{1}{\cancel{3}} \cdot x}{\underset{1}{\cancel{3}}} = \frac{480}{3}$$ Divide both sides by 3.

$x =$ ____

24. 55% of what number of experiments is 209 experiments?
GS

$$\frac{209}{x} = \frac{55}{100} \quad \text{or} \quad \frac{209}{x} = \frac{11}{20}$$

$11 \cdot x = 4180$ Find cross products.

$$\frac{\overset{1}{\cancel{11}} \cdot x}{\underset{1}{\cancel{11}}} = \frac{4180}{11}$$ Divide both sides by 11.

$x =$ ____

25. 495 successful students is 90% of what number of students?

26. 168 text messages is 28% of what number of text messages?

27. 462 mountain bikes is 140% of what number of mountain bikes?

28. 1496 graduates is 110% of what number of graduates?

29. $12\frac{1}{2}\%$ of what number is 350?
$\left(\textit{Hint: } \text{Write } 12\frac{1}{2}\% \text{ as } 12.5\%.\right)$

30. $5\frac{1}{2}\%$ of what number is 176?
$\left(\textit{Hint: } \text{Write } 5\frac{1}{2}\% \text{ as } 5.5\%.\right)$

CONCEPT CHECK *Set up a percent proportion for each application problem. Do* **not** *try to solve for any unknowns.*

31. 18 bean burritos is what percent of 36 bean burritos?

32. 62 hospital rooms is what percent of 248 hospital rooms?

Find the percent using the percent proportion. Round answers to the nearest tenth if necessary. ***See Example 6.***

33. 780 hybrid cars is what percent of 1500 hybrid cars?

34. 14 tweets is what percent of 700 tweets?

35. GS 27 downloaded songs is what percent of 1800 downloaded songs?

part = 27; whole = 1800

$$\frac{27}{1800} = \frac{x}{100} \quad \text{or} \quad \frac{3}{200} = \frac{x}{100}$$

$$200 \cdot x = 300$$

$$\frac{\overset{1}{\cancel{200}} \cdot x}{\underset{1}{\cancel{200}}} = \frac{300}{200}$$

$x =$ ____

The percent is ____.

36. GS 60 cartons is what percent of 2400 cartons?

$$\frac{60}{2400} = \frac{x}{100} \quad \text{or} \quad \frac{1}{40} = \frac{x}{100}$$

$$40 \cdot x = 100$$

$$\frac{\overset{1}{\cancel{40}} \cdot x}{\underset{1}{\cancel{40}}} = \frac{100}{40}$$

$x =$ ____

The percent is ____.

37. What percent of \$344 is \$64?

38. What percent of \$398 is \$14?

39. What percent of 250 tires is 23 tires?

40. What percent of 105 employees is 54 employees?

41. A student turned in the following answers on a test. You can tell that two of the answers are incorrect without even working the problems. Find the incorrect answers and explain how you identified them (without actually solving the problems).

50% of $84 is <u>$42</u>.

150% of $30 is <u>$20</u>.

25% of $16 is <u>$32</u>.

100% of $217 is <u>$217</u>.

42. Write a percent problem on any topic you choose. Be sure to include only two of the three components so that you can solve for the third component. Identify each component of the problem and then solve it.

Solve each application problem. Round percent answers to the nearest tenth if necessary. ***See Examples 3, 5, 7, and 8.***

43. Bonnie Boehme, who works part-time, earns $240 per week and has 22% of this amount withheld for taxes, Social Security, and Medicare. Find

(a) the amount withheld.

(b) the amount remaining after the withholdings.

44. An estimated 29.5% of automobile crashes are caused by driver distractions such as using a cell phone or texting. (*Source:* National Conference of State Legislatures.) If there are 16,450 automobile crashes in a study, find

(a) the number of crashes caused by driver distractions.

(b) the number caused by other factors. Round to the nearest whole number.

The graph below shows when people check their e-mail. This data was gathered from 3020 adults and those surveyed could give more than one response. Use this graph to answer Exercises 45–48. Round to the nearest number.

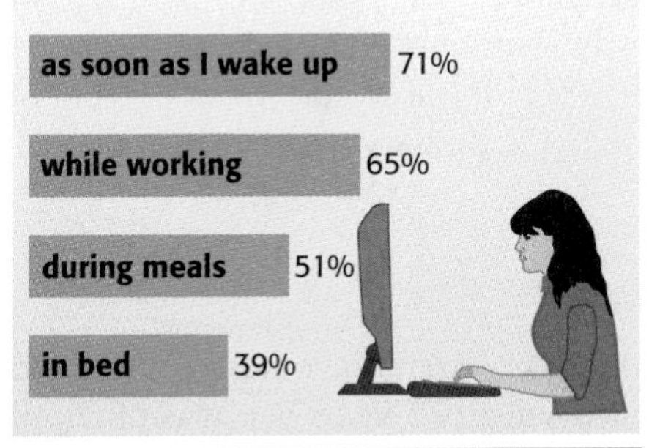

45. Find the number of people who check their e-mail while working.

46. How many people check their e-mail while in bed?

47. Find the number of people who check their e-mail as soon as they wake up.

48. How many people check their e-mail during meals?

49. Each day in Los Angeles there are 48 million commuter trips. (*Source:* CNN.) If only 2% of these trips use public transportation, find

(a) the number of trips using public transportation.

(b) the number of trips not using public transportation.

50. At work, 68% of those surveyed said that they recycle bottles and cans while 56% recycle paper products and newspapers. (*Source*: Microsystems by Harris Interactive.) If 1750 workers were surveyed,

(a) how many recycle bottles and cans?

(b) how many recycle paper products and newspapers?

The bar graph below shows the percent of children 6–11 years of age who are neglecting dental hygiene. Use the information to answer Exercises 51–54.

DOWN IN THE MOUTH

The percent* of children ages 6–11 who are neglecting dental hygiene by:

Going to bed without brushing	Brushing less than one minute	Not flossing	Brushing only once a day
67%	61%	49%	41%

*Respondents allowed to choose multiple answers.
Source: Services for Crest.

51. What percent of the children brush their teeth before going to bed?

52. What percent of the children floss their teeth?

53. If 3400 children answered the questions for this survey, how many of the children brush less than one minute?

54. How many of the 3400 children in the survey brush only once a day?

55. A recent study examined 48,000 military jobs, such as Army attack helicopter pilot or Navy gunner's mate. It was found that only 960 of these jobs are filled by women. What percent of these jobs are filled by women? (*Source:* Rand's National Defense Research Institute.)

56. There are more than 55,000 words in *Webster's Dictionary,* but most educated people can identify only 20,000 of these words. What percent of the words in the dictionary can these people identify?

57. Ebony Durrant has 7.5% of her earnings deposited into her retirement plan. If $240 per month is deposited in the plan, find her monthly and yearly earnings.

58. The number of federal income tax returns that were filed electronically 5 years ago was 68.3 million, or 51% of all returns. Find the total number of income tax returns that were filed 5 years ago. Round to the nearest tenth of a million. (*Source:* Internal Revenue Service.)

The circle graph shows the percent of various ice cream brands purchased by the 1582 Americans in a recent survey. Use this information to answer Exercises 59–62.

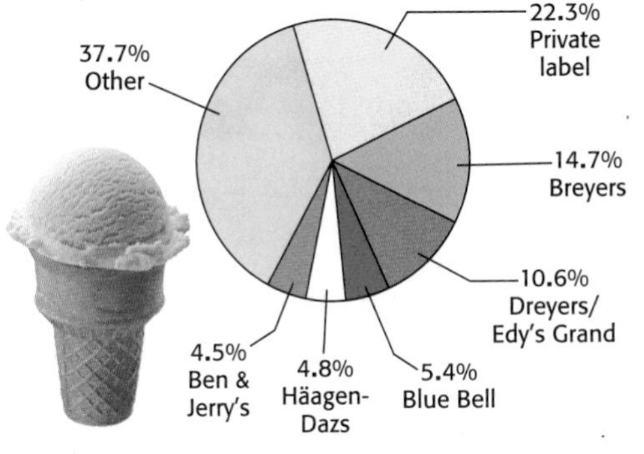

Source: Information Resources Inc.; NPD Group.

59. Of the specific brands purchased (not "Other" or "Private Label"), which brand was purchased most often?

60. What percent of ice cream purchases were "Private Label" or "Other" brands?

61. Find the number of people in the survey who said they purchase Häagen-Dazs. Round to the nearest whole number.

62. How many more people said they purchase Blue Bell brand than Ben & Jerry's brand? Round to the nearest whole number.

The bar graph below shows the percent of drivers in each age group who were stopped by police last year. Use the information to answer Exercises 63–66.

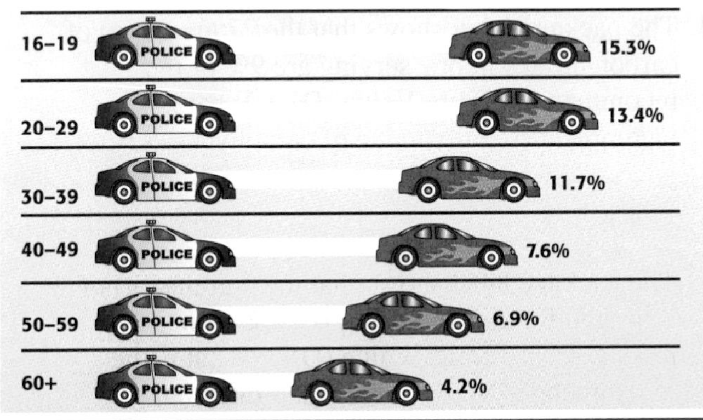

Source: Bureau of Justice Statistics.

63. What percent of drivers in the 16–19 age group were not pulled over by police?

64. What percent of drivers in the 60+ age group were not pulled over by police?

65. If 8000 drivers in the study were in the 20–29 age group, how many were pulled over by police?

66. How many of the 6000 drivers in the 30–39 age group were pulled over by police?

67. Marketing Intelligence Service says that there were 15,401 new products introduced last year. If 86% of the products introduced last year failed to reach their business objectives, find the number of products that were successful. (Round to the nearest whole number.)

68. The income earned on an investment is 8.5% of the amount invested. If the income is $12,750, find the amount of the investment.

69. Kathy West owns Banjo Shirt Company and sells handmade clothing online. If 85% of her 1540 customers paid for their orders using Pay Pal, how many of her customers used some other method of payment?

70. A family of four with a monthly income of \$2900 spends 90% of its earnings and saves the rest. Find

(a) the monthly savings.

(b) the annual savings of this family.

Relating Concepts (Exercises 71–78) For Individual or Group Work

Knowing and using the percent proportion is useful when solving percent problems.
Work Exercises 71–78 in order.

71. In the percent proportion, part is to ______ as percent is to ______.

72. All percent problems involve a comparison between a part of something and the ______.

Use this Ramen Noodles label of nutrition facts to answer Exercises 73–78. Read the label very carefully. Round to the nearest tenth of a percent.

Ramen Noodles

Nutrition Facts	Amount/serving	%DV*	Amount/serving	%DV*
Serving Size 1/2 Pkg. (15 oz/42.5 g)	**Total Fat** 8g	12%	**Total Carbohydrates** 27g	9%
Servings Per Package 2	Saturated Fat 4g	20%	Dietary Fiber Less Than 1g	3%
Calories 190	**Cholesterol** 0mg	0%	Sugars Less Than 1g	
Calories from Fat 70	**Sodium** 670mg	28%	**Protein** 4g	
	Vitamin A 0% • Vitamin C 0% • Calcium 2% • Iron 4%			
*Percent Daily Values (DV) are based on a 2,000 calorie diet.	**Calories Per Gram** Fat 9 • Carbohydrates 4 • Protein 4			

Source: Ramen Noodles package.

73. How many calories per serving are from total carbohydrates?

74. The package label shows that the 27 grams (g) of carbohydrates in one serving are 9% of the recommended Daily Value (DV). What is the recommended daily value of carbohydrates?

75. Find the number of grams of total fat that are needed to meet the recommended Daily Value (DV) of fat. Round to the nearest gram.

76. The package label shows that the 4 grams (g) of saturated fat in one serving is 20% of the recommended Daily Value (DV). What is the recommended daily value of saturated fat?

77. Will a person eating two packages of Ramen Noodles in one day exceed their recommended daily value of sodium? Explain your answer.

78. How many packages of Ramen Noodles must be eaten in a day to meet the recommended daily value of fiber? Would this be possible? Would this result in good nutrition? (Round to the nearest whole package.)

6.5 Using the Percent Equation

OBJECTIVES

1. Use the percent equation to find the part.
2. Find the whole using the percent equation.
3. Find the percent using the percent equation.

In the last section you used a proportion to solve percent problems. In this section we show another way to solve these problems by using the **percent equation.** The percent equation is a rearrangement of the percent proportion.

Percent Equation

part = percent • whole

Be sure to write the percent as a decimal before using the equation.

When using the percent proportion, we did *not* have to write the percent as a decimal because 100 was used in the denominator of the proportion. However, because there is no 100 in the percent *equation, we must* first write the percent as a decimal by dividing by 100.

Some of the examples solved earlier will be reworked using the percent equation. For comparison, you can look back in the previous section to see how these same problems were solved using proportions.

OBJECTIVE 1 Use the percent equation to find the part. The first example shows how to find the part.

EXAMPLE 1 Finding the Part

(a) Find 15% of $160.

Write 15% as the decimal 0.15. The whole, which comes after the word *of,* is 160. Next, use the percent equation. Let x represent the unknown part.

part = percent • whole

$$x = (0.15)(160)$$

Multiply 0.15 and 160.

15% *must* be written as the decimal 0.15

$$x = 24$$

15% of $160 is $24.

(b) Find 110% of 80 cases.

Write 110% as the decimal 1.10. The whole is 80. Let x represent the unknown part.

part = percent • whole

$$x = (1.10)(80)$$

$$x = 88$$

Write 110% as a decimal. 110% = 1.10

110% of 80 cases is 88 cases.

(c) Find 0.4% of 250 patients.

Write 0.4% as the decimal 0.004. The whole is 250. Let x represent the unknown part.

part = percent • whole

$$x = (0.004)(250)$$

$$x = 1$$

Write 0.4% as a decimal. 0.4% = 0.004

0.4% of 250 patients is 1 patient.

Continued on Next Page

1 Use the percent equation to find the part.

GS **(a)** 15% of 880 policyholders

part = percent • whole

part = (____) (____)

↑

Write 15% in decimal form.

part = ____

(b) 23% of 840 gallons

(c) 120% of $220

GS **(d)** 135% of $1080

part = percent • whole

part = (____) (____)

(e) 0.5% of 1200 fruit cups

(f) 0.25% of 1600 lab tests

Answers

1. **(a)** (0.15)(880); 132 policyholders
(b) 193.2 gallons **(c)** $264
(d) (1.35)(1080); $1458 **(e)** 6 fruit cups
(f) 4 lab tests

To estimate the answer, think of 0.4% as approximately 0.5% or $\frac{1}{2}$ of 1%. Because 1% is $\frac{1}{100}$, you can use the shortcut from earlier in this chapter: Move the decimal point two places to the left.

1% of 250. = 2.5

Since 1% of 250 is 2.5, then 0.5% of 250 is 1.25 (because 2.5 ÷ 2 = 1.25). So, the exact answer of 1 patient is reasonable.

◀ **Work Problem 1 at the Side.**

CAUTION

When using the percent equation, the percent must always be *changed to a decimal* before multiplying.

OBJECTIVE 2 **Find the whole using the percent equation.** The next example shows how to use the percent equation to find the whole.

Note

When the word *of* follows a percent, it is an indicator word for *multiply.*

EXAMPLE 2 **Solving for the Whole**

(a) 8 tables is 4% of what number of tables?

The part is 8 and the percent is 4% or the decimal 0.04. The whole is unknown.

8 is 4% of what number?

Now, use the percent equation. (The word "of" can be used to identity the whole.)

part = percent • whole

$8 = (0.04)(x)$ Let x represent the unknown whole. Write 4% in decimal form as 0.04.

$\frac{8}{0.04} = \frac{(0.04)(x)}{0.04}$ Divide both sides by 0.04.

$200 = x$ ← Whole

8 tables is 4% of **200 tables**.

(b) 135 tourists is 15% of what number of tourists?

Write 15% as 0.15. The part is 135. Use the percent equation to find the whole.

part = percent • whole

$135 = (0.15)(x)$ Let x represent the unknown whole. Write 15% in decimal form as 0.15.

$\frac{135}{0.15} = \frac{(0.15)(x)}{0.15}$ Divide both sides by 0.15.

$900 = x$ ← Whole

135 tourists is 15% of **900 tourists**.

Continued on Next Page

(c) $8\frac{1}{2}\%$ of what number is 102?

Write $8\frac{1}{2}\%$ as 8.5%, or the decimal 0.085. The part is 102. Use the percent equation.

Write 8.5% as a decimal. 8.5% = 0.085%

$$\text{part} = \text{percent} \cdot \text{whole}$$

$$102 = (0.085)(x) \quad \text{Let } x \text{ represent the unknown whole.}$$

$$\frac{102}{0.085} = \frac{\overset{1}{\cancel{(0.085)}}(x)}{\underset{1}{\cancel{0.085}}} \quad \text{Divide both sides by 0.085.}$$

$$1200 = x \longleftarrow \text{Whole}$$

102 is $8\frac{1}{2}\%$ of **1200**.

Estimate the answer. Notice that $8\frac{1}{2}\%$ is close to 10%. If 102 is 10% of a number, then the number is 10 times 102, or 1020. So the exact answer, 1200, is reasonable.

CAUTION

In **Example 2(c)** above, $8\frac{1}{2}\%$ was first written as 8.5%, which is the decimal form of $8\frac{1}{2}\%$ ($8\frac{1}{2} = 8.5$). **The percent sign still remained** in 8.5%. Then 8.5% was changed to the decimal 0.085 so the equation could be solved.

Work Problem 2 at the Side. ▶

OBJECTIVE 3 Find the percent using the percent equation. The final example shows how to use the percent equation to find the percent.

EXAMPLE 3 Finding the Percent

(a) 13 auto mechanics is what percent of 52 auto mechanics?

Because 52 follows *of*, the whole is 52. The part is 13, and the percent is unknown. Use the percent equation.

$$\text{part} = \text{percent} \cdot \text{whole}$$

$$13 = x \cdot 52 \quad \text{Let } x \text{ represent the unknown percent.}$$

$$\frac{13}{52} = \frac{x \cdot \overset{1}{\cancel{52}}}{\underset{1}{\cancel{52}}} \quad \text{Divide both sides by 52.}$$

$$0.25 = x$$

$$0.25 \text{ is } 25\%$$

You *must* write the decimal answer as a percent. 0.25 = 25%

13 auto mechanics is **25%** of 52 auto mechanics.

The equation can also be set up using *of* as an indicator word for multiplication and *is* as an indicator word for "is equal to."

13 is what percent of 52?

$$13 = \quad x \quad \cdot 52$$

$$13 = x \cdot 52 \quad \text{Same equation as above}$$

Continued on Next Page

2 Find the whole using the percent equation.

GS **(a)** 18 dancers is 45% of what number of dancers?

$$\text{part} = \text{percent} \cdot \text{whole}$$

$$18 = (\_\_\_)(x)$$

$$\frac{18}{0.45} = \frac{\overset{1}{\cancel{0.45}}(x)}{\underset{1}{\cancel{0.45}}}$$

$$\_\_\_ = x$$

(b) 67.5 containers is 27% of what number of containers?

(c) 666 inoculations is 45% of what number of inoculations?

(d) $5\frac{1}{2}\%$ of what number of policies is 66 policies?

Answers

2. (a) 0.45; 40; 40 dancers **(b)** 250 containers **(c)** 1480 inoculations **(d)** 1200 policies

3 Find the percent using the percent equation.

GS **(a)** What percent of 35 Facebook friends is 7 Facebook friends?

part = percent • whole

$7 = x \cdot$ ____

$\frac{7}{35} = \frac{x \cdot \cancel{35}^{1}}{\cancel{35}_{1}}$

$0.20 = x$

0.20 is ____%

(b) 34 post office boxes is what percent of 85 post office boxes?

(c) What percent of 920 invitations is 1288 invitations?

(d) 9 world-class runners is what percent of 1125 runners?

Answers

3. **(a)** 35; 20% **(b)** 40% **(c)** 140% **(d)** 0.8%

(b) What percent of \$500 is \$100?
The whole is 500 and the part is 100. Let x represent the unknown percent.

part = percent • whole

$100 = x \cdot 500$ Let x represent the unknown percent.

$\frac{100}{500} = \frac{x \cdot \cancel{500}^{1}}{\cancel{500}_{1}}$ Divide both sides by 500.

$0.20 = x$

0.20 is 20% (Write the decimal as a percent. 0.20 = 20%)

20% of \$500 is \$100.

(c) What percent of \$300 is \$390?
The whole is 300 and the part is 390. Let x represent the unknown percent.

part = percent • whole

$390 = x \cdot 300$ Let x represent the unknown percent.

$\frac{390}{300} = \frac{x \cdot \cancel{300}^{1}}{\cancel{300}_{1}}$ Divide both sides by 300.

$1.3 = x$

1.3 is 130% (Write the decimal as a percent. 1.3 = 1.30 = 130%)

130% of \$300 is \$390.

(d) 6 ladders is what percent of 1200 ladders?
Since 1200 follows *of,* the whole is 1200. The part is 6.

part = percent • whole

$6 = x \cdot 1200$ Let x represent the unknown percent.

$\frac{6}{1200} = \frac{x \cdot \cancel{1200}^{1}}{\cancel{1200}_{1}}$ Divide both sides by 1200.

$0.005 = x$

0.005 is 0.5% (Write the decimal as a percent. 0.005 = 0.5%)

6 ladders is 0.5% of 1200 ladders.

You can estimate the answer because 1% of 1200 ladders is found by moving the decimal point two places to the left in 1200, resulting in 12. Since 6 ladders is half of 12 ladders, our answer should be $\frac{1}{2}$ of 1% or 0.5%. Our exact answer matches the estimate.

CAUTION

When you use the percent equation to solve for an unknown percent, the answer will always be in decimal form. Notice that in **Example 3 (a), (b), (c), and (d)** above, **the decimal answer had to be changed to a percent** by multiplying by 100 and attaching the percent symbol. The answers became: (a) 0.25 = 25%; (b) 0.20 = 20%; (c) 1.3 = 130%; and (d) 0.005 = 0.5%.

◀ **Work Problem 3 at the Side.**

6.5 Exercises

FOR EXTRA HELP

Download the MyDashBoard App

MyMathLab®

CONCEPT CHECK *When solving for the part in a percent problem, the percent and the whole must be identified. Write either* percent *or* whole *next to the values taken from each problem.*

1. 46% of 780 text messages

46 ______ 780 ______

2. 14% of 1800 forums

1800 ______ 14 ______

Find the part using the percent equation. **See Example 1.**

3. 25% of 1080 blood donors

part = percent • whole

$x = (0.25)(1080)$

$x =$ ____

4. 19% of 700 MP3 players

part = percent • whole

$x = (0.19)(700)$

$x =$ ____

5. 45% of 3000 bath towels

6. 75% of 360 dosages

7. 32% of 260 quarts

8. 44% of 430 liters

9. 140% of 2500 air bags

10. 145% of 580 hamburgers

11. 12.4% of 8300 meters

12. 26.4% of 4700 miles

13. 0.8% of \$520

14. 0.3% of \$480

CONCEPT CHECK *When solving for the whole in a percent problem, the part and the percent must be identified. Write either* part *or* percent *next to the values taken from each problem.*

15. 70% of what number of backpackers is 476 backpackers?

70 ______ 476 ______

16. 270 lab tests is 45% of what number of lab tests?

270 ______ 45 ______

Find the whole using the percent equation. **See Example 2.**

17. 24 patients is 15% of what number of patients?

part = percent • whole

$24 = (0.15)(x)$

$\frac{24}{0.15} = \frac{\overset{1}{\cancel{(0.15)}}(x)}{\underset{1}{\cancel{(0.15)}}}$

$160 = x$

18. 32 classrooms is 20% of what number of classrooms?

part = percent • whole

$32 = (0.2)(x)$

$\frac{32}{0.2} = \frac{\overset{1}{\cancel{(0.2)}}(x)}{\underset{1}{\cancel{(0.2)}}}$

$160 = x$

19. 40% of what number of salads is 130 salads?

20. 75% of what number of wrenches is 675 wrenches?

21. $12\frac{1}{2}\%$ of what number of people is 135 people?

22. $18\frac{1}{2}\%$ of what number of batteries is 370 batteries?

23. $1\frac{1}{4}\%$ of what number of gallons is 3.75 gallons?

24. $2\frac{1}{4}\%$ of what number of files is 9 files?

CONCEPT CHECK *Fill in each blank with the correct response.*

25. When using the percent equation to solve for percent, you must always change the decimal answer to _______ by moving the decimal point _______ places to the _______ and attaching a _______ symbol.

26. When using the percent equation to solve for part or whole, you must always change the percent to a decimal by moving the _______ point _______ places to the _______ and dropping the _______ symbol.

Find the percent using the percent equation. ***See Example 3.***

27. 114 tuxedos is what percent of 150 tuxedos?

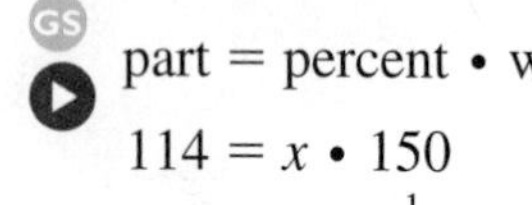

part = percent • whole

$114 = x \cdot 150$

$\frac{114}{150} = \frac{x \cdot \overset{1}{\cancel{150}}}{\underset{1}{\cancel{150}}}$

$0.76 = x$

change 0.76 to a percent = _______

28. 75 iPods is what percent of 125 iPods?

part = percent • whole

$75 = x \cdot 125$

$\frac{75}{125} = \frac{x \cdot \overset{1}{\cancel{125}}}{\underset{1}{\cancel{125}}}$

$0.6 = x$

0.60 is _______

29. What percent of 160 liters is 2.4 liters?

30. What percent of 600 meters is 7.5 meters?

31. 170 cartons is what percent of 68 cartons?

32. 612 orders is what percent of 425 orders?

33. When using the percent equation, the percent must always be changed to a decimal before doing any calculations. Show and explain how to change a fraction percent to a decimal. Use $2\frac{1}{2}\%$ in your explanation.

34. Suppose a problem on your homework assignment was, "Find $\frac{1}{2}\%$ of \$1300." Your classmates got answers of \$0.65, \$6.50, \$65, and \$650. Which answer is correct? How and why are they getting all of these answers? Explain.

Solve each application problem.

35. A study of office workers found that 27% would like more storage space. If there are 14 million office workers, how many want more storage space? (*Source:* Steelcase Workplace Index.)

36. Most shampoos contain 75% to 90% water. If a 16-ounce bottle of shampoo contains 78% water, find the number of ounces of water in the bottle. Round to the nearest tenth of an ounce.

37. The results of a recent poll are shown below. If 12,500 people were included in the poll, find

(a) the number who would rather admit their age.

(b) the number who would rather admit their weight.

(c) the number who would rather admit their salary.

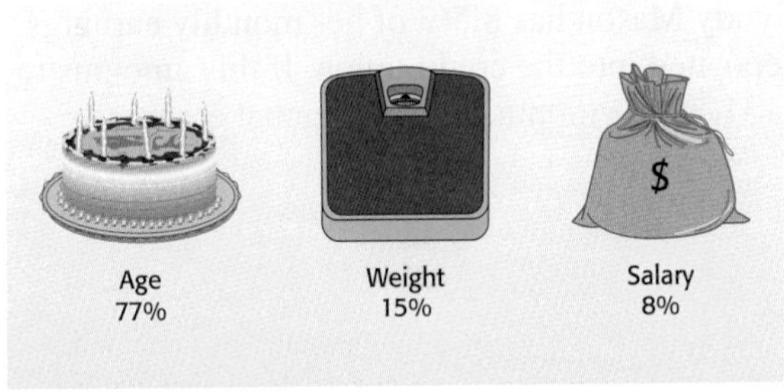

Source: Parade magazine.

38. Three Spam Mobiles travel throughout the country promoting Spam and Spam Lite. By using these kitchens on wheels, the annual goal is to give 1.5 million taste samples to the public. If 58.6% of the goal has been met, find the number of samples that have been given. (*Source:* Hormel Foods Corporation.)

39. In the United States, 98% of all households have a refrigerator. (*Source:* American Housing Authority.) Out of 18,000 households,

(a) how many are expected to have a refrigerator?

(b) how many are expected to not have a refrigerator?

40. In the United States, 56% of the households have a dishwasher. (*Source:* American Housing Authority.) Out of 214,500 households,

(a) how many are expected to have a dishwasher?

(b) how many are expected to not have a dishwasher?

41. In the United States, 2 of the 50 states (Alaska and Louisiana) do not have any drive-in movies. The remaining states do have drive-in movies. (*Source: USA Today.*)

(a) What percent of the states do not have drive-in movies?

(b) What percent have drive-in movies?

42. In a recent survey of 3450 attorneys, 897 said that they take work home with them during the week and 1311 said that they take work home with them on the weekend. Find

(a) the percent of the attorneys who take work home during the week.

(b) the percent of attorneys who take work home on the weekend.

43. In a survey of 1250 Americans, 461 rated their health as excellent. What percent of these Americans rate their health as excellent? Round to the nearest tenth of a percent. (*Source:* National Health Interview Survey.)

44. General Nutrition Center now has 3200 stores and plans to add 450 more stores. Find the percent of additional stores that they have planned. Round to the nearest tenth of a percent.

The graph shows the type of day care used by families for their preschool children. Assume that 7800 families were surveyed to gather the data. Use this graph to answer Exercises 45–48.

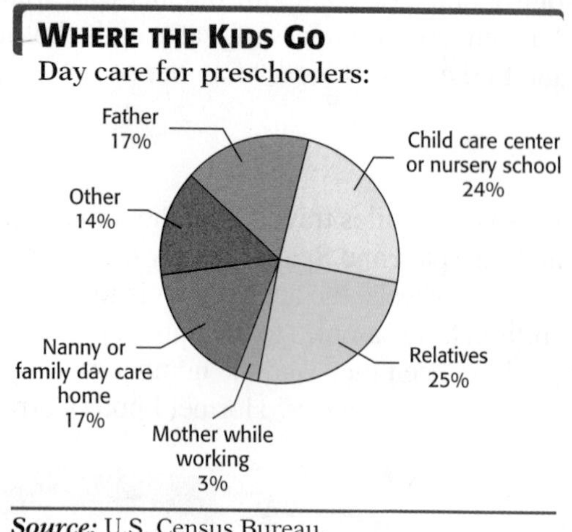

Source: U.S. Census Bureau.

45. **(a)** Which type of day care was used most often?

(b) Find the number of families who used this type of day care.

46. **(a)** Which type of day care was used least often?

(b) What number of families used this type of day care?

47. How many families used a nanny or family day care home?

48. Find the number of families who left their preschoolers with a child care center or nursery school.

49. Last year, college seniors who graduated carried an average of \$22,650 in student loan debt. This year the average amount of loan debt has increased 6%. Find the average amount of student loan debt this year. (*Source:* U.S. Department of Education.)

50. Cyndy Mason has 8.5% of her monthly earnings deposited into the credit union. If this amounts to \$131.75 per month, find her annual earnings.

51. The Chevy Camaro was introduced in 1967. Sales that year were 220,917 Camaros, which was 46.2% of the number of Ford Mustangs sold in the same year. Find the number of Mustangs sold in 1967. Round to the nearest whole number.

52. Chris Goodwin is a waiter and has sales of \$822.25 on Saturday. If this is 28.6% of his sales for the week, find his weekly sales.

53. J & K Mustang has increased the sale of auto parts by $32\frac{1}{2}\%$ over last year. If the sale of parts last year amounted to \$385,200, find the amount of sales this year.

54. An ad for steel-belted radial tires promises 15% better mileage. If mileage had been 25.6 miles per gallon in the past, what mileage could be expected after these tires are installed? Round to the nearest tenth of a mile.

55. A Polaris Vac-Sweep is priced at \$524 with an allowed trade-in of \$125 on an old unit. If sales tax of $7\frac{3}{4}\%$ is charged on the price of the new Polaris unit before the trade-in, find the total cost to the customer after receiving the trade-in. (*Hint:* Trade-in is subtracted last.)

56. General Motors car sales in China were 36.7% greater than last year's sales of 629,778 cars. Find this year's sales. Round to the nearest whole number. (*Source:* General Motors Corporation.)

Summary Exercises *Using the Percent Proportion and Percent Equation*

CONCEPT CHECK *Write each percent as a decimal and each decimal as a percent.*

1. 6.25%

2. 380%

3. 0.375

4. 0.006

CONCEPT CHECK *Write each percent as a fraction or mixed number in lowest terms and each fraction as a percent.*

5. 87.5%

6. 160%

7. $\frac{5}{8}$

8. $\frac{1}{125}$

The circle graph shows what 1100 adults dread the most about the Thanksgiving holiday. Use this graph to answer Exercises 9–12. Write each answer as a fraction in lowest terms, as a decimal, and as a percent.

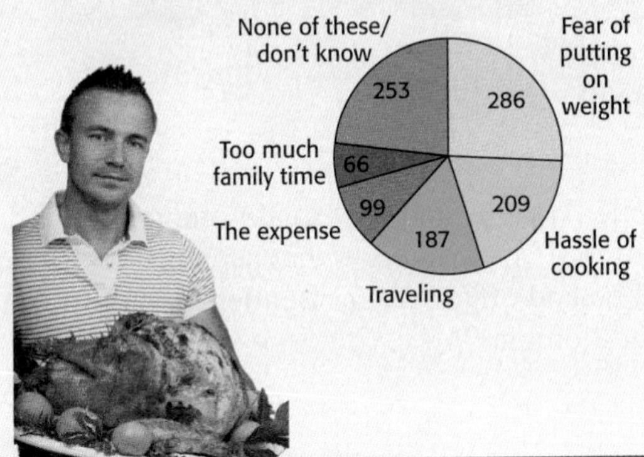

Source: Opinion Research Corporation.

9. What portion of the adults have a fear of putting on weight?

10. What portion of the adults dread traveling?

11. Find the portion who dread too much family time.

12. Find the portion who dread the expense.

Find the part, whole, or percent as indicated. Round percent answers to the nearest tenth if necessary.

13. 6% of \$780 is what amount?

14. 343 cable customers is 35% of how many cable customers?

15. What percent of 320 policies is 176 policies?

16. 0.8% of 3500 screening exams is how many exams?

17. 1016.4 acres is 280% of what number of acres?

18. What percent of 658 circuits is 18 circuits?

Solve each application problem. Round to the nearest cent or nearest tenth of a percent if necessary.

19. In a poll of 1582 people, 10.6% said that they prefer to purchase Dreyer's/Edy's brand ice cream. Find the number of people who said they prefer Dreyer's/Edy's brand ice cream. Round to the nearest whole number. (*Source:* Information Resources Inc; NPD Group.)

20. A real estate broker wants to purchase a smart phone priced at $399. If the sales tax rate is 7.75%, find the total price including the sales tax. (*Source:* Real Estate Technology.)

21. The number of outstanding student loans last year was 3,370,000. If 7.2% of these loans were in default, find the number of student loans that were in default. Round to the nearest thousand. (*Source:* U.S. Department of Education.)

22. In a recent survey of dog owners, it was found that 901, or 34%, of the owners take their dogs on vacation with them. Find the number of dog owners in the survey who do not take their dogs on vacation. (*Source:* American Animal Hospital Association.)

23. The distance around Lake Tahoe is 65 miles. If Lino Delgadillo has completed 39 miles of the Lake Tahoe run, what percent of the run remains?

24. Beutler Heating and Air Conditioning plans to lay off 45 of its 1215 workers. What percent of the workers will be laid off? (*Source:* Beutler Heating and Air Conditioning.)

25. Last year there were 98.3 million tax returns filed electronically. If this was 69% of all returns filed, how many total returns were filed? Round to the nearest tenth of a million. (*Source*: Internal Revenue Service.)

26. The number of ski-lift tickets sold last week was 3820. This week's sales are down 10.5%. Find the number of ski-lift tickets sold this week. Round to the nearest whole number.

6.6 Solving Application Problems with Percent

Percent has many applications in our daily lives. This section discusses percent as it applies to sales tax, commissions, discounts, and the percent of change (increase and decrease).

OBJECTIVES

1. Find sales tax.
2. Find commissions.
3. Find the discount and sale price.
4. Find the percent of change.

OBJECTIVE 1 **Find sales tax.** States, countries, and cities often collect taxes on sales to customers. The **sales tax** is a percent of the cost of an item. The following formula for finding sales tax is based on the percent equation.

Sales Tax Formula

part = percent • whole

amount of sales tax = rate of tax • cost of item

VOCABULARY TIP

Sales tax The sales tax is always part of the cost of an item. The cost of the item is the **whole,** the sales tax rate is the **percent,** and the amount of the sales tax is the **part.**

EXAMPLE 1 Solving for Sales Tax

Office Max sells a laptop computer for $499. If the sales tax rate is 5%, how much tax is paid? What is the total cost of the laptop computer? Use the six problem-solving steps.

Step 1 **Read** the problem. The problem asks for the total cost of the laptop computer, including the sales tax.

Step 2 **Work out a plan.** Use the sales tax formula to find the amount of sales tax. Write the tax rate (5%) as a decimal (0.05). The cost of the item is $499. Use the letter *a* to represent the unknown *amount* of tax. Add the sales tax to the cost of the item.

Step 3 **Estimate** a reasonable answer. Round $499 to $500. Recall that 5% is equivalent to $\frac{1}{20}$, so divide $500 by 20 to estimate the tax.

$$\$500 \div 20 = \$25 \text{ tax}$$

The total estimated cost is $500 + $25 = $525. ← Estimate

Step 4 **Solve** the problem.

part = percent • whole

amount of sales tax = rate of tax • cost of item

$$a = (5\%)(\$499)$$

$$a = (0.05)(\$499) \quad \text{Write 5\% as a decimal 0.05.}$$

$$a = \$24.95 \quad \text{Sales tax}$$

The tax paid on the laptop computer is $24.95. The customer would pay a total cost of $499 + $24.95 = $523.95.

Step 5 **State the answer.** The total cost of the laptop computer is $523.95.

Step 6 **Check.** The exact answer, $523.95, is close to our estimate of $525.

Work Problem 1 at the Side. ▶

1 Suppose the sales tax rate in your state is 6%. Find the amount of the tax and the total you would pay for each item.

GS **(a)** $29 Little League bat

Sales tax = (0.06)($____)

= $____

Total cost = $29 + $____

= $____

(b) $349 home theater system

(c) $1287 leather chair and ottoman

(d) $24,500 pickup truck

Answers

1. (a) $29; $1.74; $1.74; $30.74
(b) $20.94; $369.94 (c) $77.22; $1364.22
(d) $1470; $25,970

2 Find the rate of sales tax.

GS **(a)** The tax on a $320 patio set is $25.60.

$25.60 = r \cdot \_\_\_\_$

$$\frac{25.6}{320} = \frac{r \cdot \overset{1}{\cancel{320}}}{\underset{1}{\cancel{320}}}$$

$r = 0.08 = \_\_\_\_$

(b) The tax on a $96 park bench is $6.24.

(c) The tax on a $22,995 Dodge Charger is $919.80.

Answers

2. **(a)** $320; 8% **(b)** 6.5% or $6\frac{1}{2}$% **(c)** 4%

EXAMPLE 2 Finding the Sales Tax Rate

The sales tax on a $24,200 Ford Edge is $1573. Find the rate of the sales tax.

Step 1 **Read** the problem. This problem asks us to find the sales tax rate.

Step 2 **Work out a plan.** Use the sales tax formula.

sales tax = rate of tax • cost of item

Solve for the rate of tax, which is the percent. The cost of the Ford Edge (the whole) is $24,200, and the amount of sales tax (the part) is $1573. Use the letter r to represent the unknown *rate* of tax (the percent).

Step 3 **Estimate** a reasonable answer. Round $24,200 to $24,000 and round $1573 to $1600. The sales tax is $\frac{1600}{24,000}$ or $\frac{1}{15}$ of the cost of the car. So divide 1 by 15 to estimate the percent (rate) of sales tax.

$$\frac{1}{15} = 0.06\overline{6} \approx 7\% \leftarrow \text{Rounded estimate}$$

Step 4 **Solve** the problem.

sales tax = rate of tax • cost of item

$$\$1573 = (r)\ (\$24,200)$$

$$\frac{1573}{24,200} = \frac{(r)(\overset{1}{\cancel{24,200}})}{\underset{1}{\cancel{24,200}}} \quad \text{Divide both sides by 24,200.}$$

$$0.065 = r$$

0.065 is 6.5%

Write the decimal as a percent. 0.065 = 6.5%.

Step 5 **State the answer.** The sales tax rate is 6.5% or $6\frac{1}{2}$%.

Step 6 **Check.** The exact answer, $6\frac{1}{2}$%, is close to our estimate of 7%.

◀ Work Problem **2** at the Side.

Note

You can use the sales tax formula to find the amount of sales tax, the cost of an item, or the rate of sales tax (the percent).

OBJECTIVE 2 Find commissions. Many salespeople are paid by *commission* rather than an hourly wage. If you are paid by **commission,** you are paid a certain percent of your total sales dollars. The formula below for finding the commission is based on the percent equation.

Commission Formula

part = percent • whole

amount of commission = rate of commission • amount of sales

EXAMPLE 3 **Determining the Amount of Commission**

Lynn Cochran had automotive tool sales of \$42,500 last month. If Cochran's commission rate is 9%, find the amount of his commission.

Step 1 **Read** the problem. The problem asks for the amount of commission that Cochran earned.

Step 2 **Work out a plan.** Use the commission formula. Write the rate of commission (9%) as a decimal (0.09). The amount of Cochran's sales (\$42,500) is the whole. Use the letter c to represent the unknown *amount* of commission.

Step 3 **Estimate** a reasonable answer. Round the commission rate of 9% to 10%. Round the amount of sales from \$42,500 to \$40,000. Since 10% is equivalent to $\frac{1}{10}$, divide \$40,000 by 10 to estimate the amount of commission: \$40,000 ÷ 10 gives \$4000 as our estimate.

Step 4 **Solve** the problem.

amount of commission = rate of commission • amount of sales

$$c = (9\%)(\$42{,}500)$$

$$c = (0.09)(\$42{,}500)$$

$$c = \$3825 \quad \text{Amount of commission}$$

Step 5 **State the answer.** Cochran earned a commission of \$3825.

Step 6 **Check.** The exact answer, \$3825, is close to our estimate of \$4000.

Work Problem 3 at the Side. ▶

3 Find the amount of commission.

(a) Jill Owens sells dental equipment at a commission rate of 11% and has sales for the month of \$38,700.

Use the letter c to represent the amount of commission.

$c = (11\%)(\$\_\_\_\_\_\_)$

$c = (0.11)(38{,}700)$

$c = \$\_\_\_\_\_\_$

(b) Last month Alyssa Paige sold a home for \$189,500 and earned a commission of 6%.

Answers

3. **(a)** \$38,700; \$4257 **(b)** \$11,370

EXAMPLE 4 **Finding the Rate of Commission**

Carol Merrigan earned a commission of \$510 for selling \$17,000 worth of shipping supplies. Find the rate of commission.

Step 1 **Read** the problem. We must find the rate (percent) of commission.

Step 2 **Work out a plan.** You could use the commission formula. Another approach is to use the percent proportion. The *whole* is \$17,000, the *part* is \$510, and the *percent* (rate of commission) is the unknown.

Step 3 **Estimate** a reasonable answer. Round the commission, \$510, to \$500, and round \$17,000 to \$20,000. The commission in fraction form is $\frac{\$500}{\$20{,}000}$, which simplifies to $\frac{1}{40}$. Changing $\frac{1}{40}$ to a percent gives $2\frac{1}{2}\%$ (rounded), as our estimate.

Step 4 **Solve** the problem.

$$\frac{\text{part}}{\text{whole}} = \frac{x}{100} \quad \leftarrow \text{Percent (unknown)}$$

Think: The rate of the commission is the percent.

$$\frac{510}{17{,}000} = \frac{x}{100}$$

$$17{,}000 \cdot x = 510 \cdot 100 \quad \text{Cross products}$$

$$\frac{\overset{1}{\cancel{17{,}000}} \cdot x}{\underset{1}{\cancel{17{,}000}}} = \frac{51{,}000}{17{,}000} \quad \text{Divide both sides by 17,000.}$$

$$x = 3$$

Continued on Next Page

4 Find the rate of commission.

GS **(a)** A commission of \$450 is earned on the sale of computer products worth \$22,500.

$$\frac{450}{22{,}500} = \frac{x}{100}$$

$$22{,}500 \cdot x = 450 \cdot 100$$

$$\frac{\overset{1}{\cancel{22{,}500}} \cdot x}{\underset{1}{\cancel{22{,}500}}} = \frac{45{,}000}{22{,}500}$$

$$x = \_\_\_\_$$

(b) Jamal Story earned \$2898 for selling office furniture worth \$32,200.

5 Find the amount of the discount and the sale price.

GS **(a)** An Easy-Boy leather recliner originally priced at \$950 is offered at a 42% discount.

$$\text{Discount} = (0.42)(\$950)$$
$$= \$399$$
$$\text{Sale price} = \$950 - \$\_\_\_\_$$
$$= \$\_\_\_\_$$

(b) Walmart has women's sweater sets on sale at 35% off. One sweater set was originally priced at \$30.

Answers

4. **(a)** 2% **(b)** 9%
5. **(a)** \$399; \$551 **(b)** \$10.50; \$19.50

Step 5 **State the answer.** The rate of commission is 3%.

Step 6 **Check.** The exact answer of 3% is close to our estimate of $2\frac{1}{2}$%.

◄ **Work Problem 4 at the Side.**

OBJECTIVE 3 Find the discount and sale price. Most of us prefer buying things when they are on sale. A store will reduce prices, or **discount,** to attract additional customers. Use the following formula to find the discount and the sale price.

Discount Formula and Sale Price Formula

amount of discount = rate (or percent) of discount • original price

sale price = original price – amount of discount

EXAMPLE 5 Finding a Sale Price

Whitings Oak Furniture Store has a home theater cabinet with an original price of \$840 on sale at 15% off. Find the sale price of the cabinet.

Step 1 **Read** the problem. This problem asks for the price of a home theater cabinet after a discount of 15%.

Step 2 **Work out a plan.** The problem is solved in two steps. First, find the amount of the discount, that is, the amount that will be "taken off" (subtracted), by multiplying the original price (\$840) by the rate of the discount (15%). The second step is to subtract the amount of discount from the original price. This gives you the sale price, which is what you will actually pay for the home theater cabinet.

Step 3 **Estimate** a reasonable answer. Round the original price from \$840 to \$800, and the rate of discount from 15% to 20%. Since 20% is equivalent to $\frac{1}{5}$, the estimated discount is $\$800 \div 5 = \$160$, so the estimated sale price is $\$800 - \$160 = \$640$.

Step 4 **Solve** the problem. First find the exact amount of the discount.

amount of discount = rate of discount • original price

$a = (0.15)(\$840)$ Write 15% as a decimal.

$a = \$126$ Amount of discount

Now find the sale price of the home theater cabinet by subtracting the amount of the discount (\$126) from the original price.

sale price = original price – amount of discount

$= \$840 - \$126$ Remember to subtract the discount from the original price to get the sale price.

$= \$714$ Sale price

Step 5 **State the answer.** The sale price of the home theater cabinet is \$714.

Step 6 **Check.** The exact answer, \$714, is close to our estimate of \$640.

◄ **Work Problem 5 at the Side.**

Calculator Tip

In **Example 5** on the previous page, you can use a scientific calculator to find the amount of discount and subtract the discount from the original price, all in one step.

840 (−) .15 (×) 840 (=) 714

Original price ↑ ; Amount of discount (.15 × 840) ; Sale price ↑

A scientific calculator observes the order of operations, so it will automatically do the multiplication before the subtraction.

OBJECTIVE ④ Find the percent of change. We are often interested in looking at increases or decreases in sales, production, population, and many other items. This type of problem involves finding the *percent of change.* Use the following steps to find the **percent of increase.**

Finding the Percent of Increase

Step 1 Use subtraction to find the amount of increase.

Step 2 Use the percent proportion to find the percent of increase.

$$\frac{\textbf{amount of increase (part)}}{\textbf{original value (whole)}} = \frac{\textbf{percent}}{\textbf{100}}$$

EXAMPLE 6 Finding the Percent of Increase

Attendance at county parks climbed from 18,300 last month to 56,730 this month. Find the percent of increase.

Step 1 **Read** the problem. The problem asks for the percent of increase.

Step 2 **Work out a plan.** Subtract the attendance last month (18,300) from the attendance this month (56,730) to find the amount of increase in attendance. Next, use the percent proportion. The whole is 18,300 (last month's original attendance), the part is 38,430 (amount of increase in attendance), and the percent is unknown.

Step 3 **Estimate** a reasonable answer. Round 18,300 to 20,000 and 56,730 to 60,000. The amount of increase is 60,000 − 20,000 = 40,000. Since 40,000 (the increase) is *twice* as large as the original amount, the estimated percent of increase is 200%.

Step 4 **Solve** the problem.

$$56{,}730 - 18{,}300 = 38{,}430$$

Subtract to find the **amount of increase** in attendance.

Amount of increase →
$$\frac{38{,}430}{18{,}300} = \frac{x}{100}$$
Percent proportion

Use the *original* value of 18,300 (**not** 56,730).

Solve this proportion to find that $x = 210$.

Step 5 **State the answer.** The percent of increase is 210%.

Step 6 **Check.** The exact answer, 210%, is close to our estimate of 200%.

Work Problem ⑥ at the Side. ▶

VOCABULARY TIP

Percent of increase or decrease The change from the original value is the amount of increase or decrease. The amount of increase or decrease is the **part** and the original value is **always the whole.**

⑥ Find the percent of increase.

(a) A manufacturer of snowboards increased production from 14,100 units last year to 19,035 this year.

19,035 − 14,100 = ______ increase

Increase → ; Original value
$$\frac{\_\_\_\_}{14{,}100} = \frac{x}{100}$$

$$\frac{\overset{1}{\cancel{14{,}100}} \cdot x}{\underset{1}{\cancel{14{,}100}}} = \frac{493{,}500}{14{,}100}$$

$$x = 35$$

Percent of increase is ____.

(b) The number of flu cases rose from 496 cases last week to 620 this week.

Answers

6. **(a)** 4935; 4935; 35% **(b)** 25%

7 Find the percent of decrease.

GS **(a)** The number of service calls fell from 380 last month to 285 this month.

$380 - 285 =$ ____ decrease

Decrease → $\frac{\_\_\_\_}{380} = \frac{x}{100}$ ← Original value

$$\frac{\overset{1}{\cancel{380}} \cdot x}{\underset{1}{\cancel{380}}} = \frac{9500}{380}$$

$$x = 25$$

Percent of decrease is ____.

(b) The number of workers applying for unemployment fell from 4850 last month to 3977 this month.

Answers

7. **(a)** 95; 95; 25% **(b)** 18%

Use the following steps to find the **percent of decrease.**

Finding the Percent of Decrease

Step 1 Use subtraction to find the amount of decrease.

Step 2 Use the percent proportion to find the percent of decrease.

$$\frac{\textbf{amount of decrease (part)}}{\textbf{original value (whole)}} = \frac{\textbf{percent}}{\textbf{100}}$$

EXAMPLE 7 Finding the Percent of Decrease

The number of production employees this week fell to 1406 people from 1480 people last week. Find the percent of decrease.

Step 1 **Read** the problem. The problem asks for the percent of decrease.

Step 2 **Work out a plan.** Subtract the number of employees this week (1406) from the number of employees last week (1480) to find the amount of decrease. Then, use the percent proportion. The whole is 1480 (last week's *original* number of employees), the part is 74 (amount of decrease in employees), and the percent is unknown.

Step 3 **Estimate** a reasonable answer. Estimate the answer by rounding 1406 to 1400 and 1480 to 1500. The decrease is $1500 - 1400 = 100$. Since 100 is $\frac{1}{15}$ of 1500, our estimate is $1 \div 15 \approx 0.07$ or 7%.

Step 4 **Solve** the problem.

$$1480 - 1406 = 74$$ ← Subtract to find the **amount of decrease** in number of employees.

Amount of decrease → $\frac{74}{1480} = \frac{x}{100}$ Percent proportion

Use the original value of 1480 (**not** 1406).

Solve this proportion to find that $x = 5$.

Step 5 **State the answer.** The percent of decrease is 5%.

Step 6 **Check.** The exact answer, 5%, is close to our estimate of 7%.

CAUTION

When solving for percent of increase or decrease, the ***whole is always the original value*** or ***value before the change occurred.*** The part is the change in values, that is, how much something went up or went down.

◀ **Work Problem 7 at the Side.**

6.6 Exercises

FOR EXTRA HELP

Download the MyDashBoard App

MyMathLab®

CONCEPT CHECK *Underline the correct answer.*

1. When solving a sales tax problem, the cost of an item is the (*tax rate/whole*), the sales tax rate is the (*percent/whole*), and the amount of sales tax is the (*part/whole*).

CONCEPT CHECK *Fill in each blank with the correct response.*

2. To find the amount of sales tax, multiply the sales tax __________ by the __________. The total cost is the cost of the item plus the __________ tax.

Find the amount of sales tax or the tax rate and the total cost (amount of sale + amount of tax = total cost). Round money answers to the nearest cent if necessary. ***See Examples 1 and 2.***

	Cost of Item	Tax Rate	Amount of Tax	Total Cost
3.	\$6	4%	__________	__________
4.	\$45	5%	__________	__________
5.	\$425	__________	\$12.75	__________
6.	\$84	__________	\$5.88	__________
7.	\$12,229	$5\frac{1}{2}\%$	__________	__________
8.	\$11,789	$7\frac{1}{2}\%$	__________	__________

CONCEPT CHECK *Underline the correct answer.*

9. When solving commission problems, the whole is the (*commission/sales amount*), the percent is the (*commission rate/sales amount*), and the part is the (*commission/sales amount*).

CONCEPT CHECK *Fill in each blank with the correct response.*

10. To find the commission, multiply the rate of __________ by the __________.

Find the commission earned or the rate of commission. Round money answers to the nearest cent if necessary. ***See Examples 3 and 4.***

	Sales	Rate of Commission	Commission
11.	\$280	8%	__________
12.	\$660	10%	__________

GS Commission = rate of commission • sales

$= (10\%)(\$\_\_\_\_\_\_) = (0.\_\_\_\_)(\$660)$

	Sales	Rate of Commission	Commission
13.	\$3000	__________	\$600

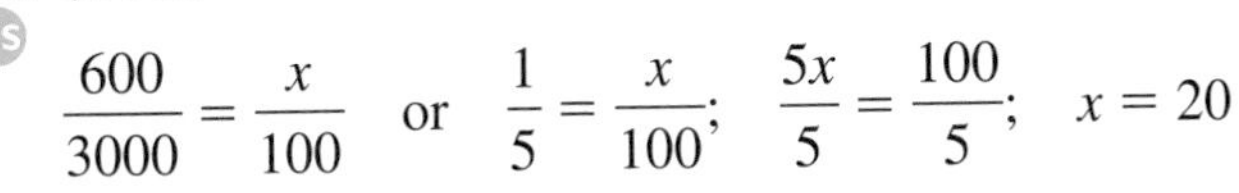

GS $\frac{600}{3000} = \frac{x}{100}$ or $\frac{1}{5} = \frac{x}{100}$; $\frac{5x}{5} = \frac{100}{5}$; $x = 20$

	Sales	Rate of Commission	Commission
14.	\$7800	__________	\$1170
15.	\$6183.50	3%	__________
16.	\$4416.70	7%	__________

CONCEPT CHECK *Underline the correct answer.*

17. When solving retail sales discount problems, the *original* price is always the (*part/whole*), the rate of discount is the (*percent/whole*), and the amount of discount is the (*percent/part*).

CONCEPT CHECK *Fill in each blank with the correct response.*

18. To find the sale price in a retail discount problem, subtract the amount of __________ from the __________ price.

Find the amount or rate of discount and the sale price after the discount. Round money answers to the nearest cent if necessary. ***See Example 5.***

	Original Price	Rate of Discount	Amount of Discount	Sale Price
19.	$199.99	10%	________	________
20.	$29.95	15%	________	________
21.	$180	________	$54	________
22.	$38	________	$9.50	________
23.	$58.40	15%	________	________
24.	$99.80	30%	________	________

25. You are trying to decide between Company A paying a 10% commission and Company B paying an 8% commission. For which company would you prefer to work? What considerations other than commission rate would be important to you?

26. Give four examples of where you might use the percent of increase or the percent of decrease in your own personal activities. Think in terms of work, school, home, hobbies, and sports.

Solve each application problem. Round money answers to the nearest cent and rates to the nearest tenth of a percent if necessary. ***See Examples 1–7.***

Country Store has a unique selection of merchandise that it sells by mail and over the Internet. Use the shipping and insurance delivery chart at the right and a sales tax rate of 5% to solve Exercises 27–30. There is no sales tax on shipping and insurance. (*Source:* Country Store Catalog.)

SHIPPING AND INSURANCE DELIVERY CHART

Order amount	Shipping
Up to $15.00	add $4.99
$15.01 to $25.00	add $6.99
$25.01 to $35.00	add $7.99
$35.01 to $50.00	add $8.99
$50.01 to $70.00	add $10.99
$70.01 to $99.99	add $12.99
$100.00 or more	add $14.99

27. Find the total cost of six Small Fry Handi-Pan electric skillets priced at $29.99 each.

28. A customer ordered five sets of flour-sack towels priced at $12.99 per set. What is the total cost?

29. Find the total cost of three pop-up hampers at $9.99 each and four nonstick mini doughnut pans at $10.99 each.

30. What is the total cost of five coach lamp bird feeders at $19.99 each and six garden weather centers at $14.99 each?

31. An Anderson wood-frame French door is priced at $1980 with a sales tax of $99. Find the rate of sales tax.

32. Six gallons of Dutch Boy Dirt Fighter exterior paint cost $155.94 plus sales tax of $7.80. Find the sales tax rate. (*Source*: Orchard Supply Hardware.)

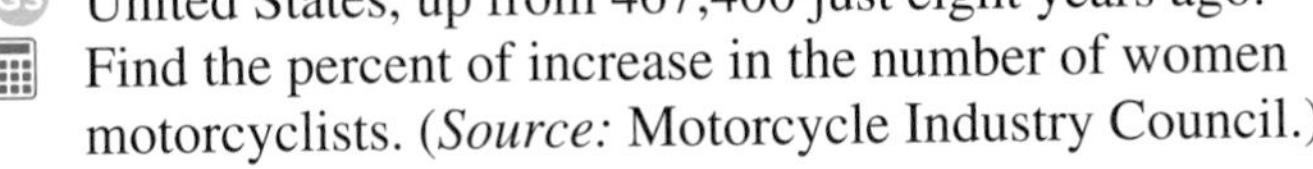

33. Today there are 635,000 women motorcyclists in the United States, up from 467,400 just eight years ago. Find the percent of increase in the number of women motorcyclists. (*Source:* Motorcycle Industry Council.)

Increase = 635,000 − 467,400 = 167,600

$$\frac{167{,}600 \text{ (increase)}}{467{,}400 \text{ (original)}} = \frac{x}{100}$$

34. This year Jo O'Neill has earned \$46,750, up from \$32,500 last year. Find the percent of increase in her annual earnings.

Increase = \$46,750 − \$32,500 = \$14,250

$$\frac{14{,}250 \text{ (increase)}}{32{,}500 \text{ (original)}} = \frac{x}{100}$$

35. Misty Downes sold her Toyota Prius, which got 48 miles per gallon on the highway, and purchased a Chevrolet Cruze, which gets 36 miles per gallon. What is the percent of decrease in mileage? (*Source:* edmonds.com)

36. Rich Williams is considering selling his Corvette, which gets 26 miles per gallon on the highway, and purchasing a new Corvette ZR1, which gets 20 miles per gallon. If he buys the new Corvette, find the percent of decrease in mileage. (*Source:* edmonds.com)

37. A preholiday sale at Macy's offers 60% off on all women's wear. What is the sale price of a wool coat normally priced at \$335?

38. What is the sale price of a \$1098 Kenmore washer/dryer set with a discount of 25%?

A weekly sales report for the top four sales people at Active Sports is shown below. Use this information to answer Exercises 39–42.

Employee	Sales	Rate of Commission	Commission
Strong, A.	\$18,960	3%	______
Ferns, K.	\$21,460	3%	______
Keyes, B.	\$17,680	______	\$707.20
Vargas, K.	\$23,104	______	\$1152.20

39. Find the commission for Strong.

40. Find the commission for Ferns.

41. What is the rate of commission for Keyes?

42. What is the rate of commission for Vargas?

43. A Sony Micro Hi Fi Component System was priced at \$390 and is on sale at 22% off. Find the discount and the sale price.

44. A Honda Pilot is offered at 12% off the manufacturer's suggested retail price. Find the discount and the sale price of this SUV, originally priced at \$32,500.

45. The price of a video game was marked down from \$45.50 to \$35.49. Find the percent of discount.

46. In the past five years, the cost of generating electricity from the sun has been brought down from 24 cents per kilowatt hour to 8 cents (less than the newest nuclear power plants). Find the percent of decrease.

47. College students are offered a 6% discount on a dictionary that sells for \$18.50. If the sales tax is 6%, find the cost of the dictionary, including the sales tax.

48. A fax machine priced at \$398 is marked down 7% to promote the new model. If the sales tax is also 7%, find the cost of the fax machine, including sales tax.

49. A plumbing supply sales representative has sales of \$380,450 and is paid a commission of 2% of sales. If 20% of his commission is deducted for office space and other expenses, how much does the sales representative get?

50. The local real estate agents' association collects a fee of 2% on all money received by its members. The members charge 6% of the selling price of a property as their fee. How much does the association get, if its members sell property worth a total of \$8,680,000?

51. What is the total price of a ski boat with an original price of \$15,321, if it is sold at a 15% discount? The sales tax rate is $7\frac{3}{4}\%$.

52. A commercial security alarm system originally priced at \$10,800 is discounted 22%. Find the total price of the system if the sales tax rate is $7\frac{1}{4}\%$.

Relating Concepts (Exercises 53–58) For Individual or Group Work

Knowing how to use the percent equation is important when solving application problems involving sales tax. ***Work Exercises 53–58 in order.***

53. The percent equation is

part = ______ • ______.

54. The formula used to find sales tax is an application of the percent equation. The sales tax formula is

sales tax = ____________ • ____________.

In the United States there are certain items on which an excise tax is charged in addition to a sales tax. A table of federal excise taxes is shown here. Use this table to answer Exercises 55–58. (Excise tax is calculated on the amount of the sale ***before*** *sales tax is added.) Round answers to the nearest cent.*

FEDERAL EXCISE TAXES*

Product or Service	Rate	Product or Service	Rate
Local telephone service	3%	Bows and arrows	11%
Cigarettes	36¢ per pack	Gasoline	18.4¢/gal
Tires	\$0.0945 for each 10 pounds of the maximum rated load capacity over 3500 pounds	Diesel fuel and kerosene	24.4¢/gal
		Aviation fuel	19.4¢/gal
		Truck and trailer, chassis and bodies	12%
Air transportation	7.5%		
International air travel	\$15.40 per person	Inland waterways fuel	20¢/gal
Air freight	6.25%	Ship passenger tax	\$3/passenger
Fishing rods	10%	Vaccines	78¢/dose.

*In addition to the federal excise taxes shown here, there are a number of additional excise taxes that apply to alcoholic beverages, tobacco products, and firearms.

Source: Publication 510, I.R.S., Excise Taxes.

55. Julia Lauren purchased a fly fishing rod for \$59 and a spinning rod for \$36. Use the federal excise tax table and a sales tax rate of $6\frac{1}{2}\%$ to find the cost of the equipment, including both taxes. Sales tax is not charged on the \$9.50 federal excise tax. (Round to the nearest cent.)

56. Refer to **Exercise 55.** Calculate the two taxes separately and then add them together. Now, add the two tax rates together and then find the tax. Are your answers the same? Why or why not? (*Hint:* Recall the commutative and associative properties of multiplication.)

57. The price of an international airline ticket is \$1248. Use the federal excise tax table and a sales tax rate of $7\frac{3}{4}\%$ to find the total cost of one ticket. Sales tax is not charged on the \$15.40 federal excise tax. (Round to the nearest cent.)

58. Refer to **Exercise 57.** Can the federal excise tax be added to the sales tax rate to find the total tax? Why or why not?

6.7 Simple Interest

When we open a savings account, we are actually lending money to the bank or credit union. The bank or credit union will in turn lend this money to individuals and businesses. These people then become borrowers. The bank or credit union pays a fee to the savings account holders and charges a higher fee to its borrowers. These fees are called *interest.*

Interest is a fee paid or a charge made for lending or borrowing money. The amount of money borrowed is called the **principal.** The charge for interest is often given as a percent, called the interest rate or **rate of interest.** The rate of interest is assumed to be *per year,* unless stated otherwise. Time is always expressed in years or fractions of a year.

OBJECTIVES

1. Find the simple interest on a loan.
2. Find the total amount due on a loan.

OBJECTIVE 1 **Find the simple interest on a loan.** In most cases, interest on a loan is computed on the *original principal* and is called **simple interest.** We use the following **interest formula** to find simple interest.

Formula for Simple Interest

Interest = principal • rate • time

The formula is usually written using letters.

$$I = p \cdot r \cdot t$$

Note

Simple interest is used for most short-term business loans, most real estate loans, and many automobile and consumer loans.

EXAMPLE 1 **Finding Simple Interest for a Year**

Find the interest on $5000 at 3% for 1 year.

The amount borrowed, or principal (p), is $5000. The interest rate (r) is 3%, which is 0.03 as a decimal, and the time of the loan (t) is 1 year. Use the formula.

$$I = p \cdot r \cdot t$$
$$I = (5000)(0.03)(1)$$
$$I = \$150$$

Notice that "1" is used as the time for 1 year.

The interest rate is always written in decimal form.

The interest is $150.

Work Problem 1 at the Side. ▶

EXAMPLE 2 **Finding Simple Interest for More Than a Year**

Find the interest on $4200 at 4% for three and a half years.

The principal (p) is $4200. The rate ($r$) is 4%, or 0.04 as a decimal, and the time (t) is $3\frac{1}{2}$ or 3.5 years. Use the formula.

$$I = p \cdot r \cdot t$$
$$I = (4200)(0.04)(3.5)$$
$$I = 588$$

3.5 years is equivalent to $3\frac{1}{2}$ years because 3.5 is $3\frac{5}{10}$ which simplifies to $3\frac{1}{2}$.

4% is changed to the decimal 0.04

The interest is $588.

Work Problem 2 at the Side. ▶

VOCABULARY TIP

Interest is what the borrower pays for the use of the lender's money.

1 Find the interest.

GS **(a)** $1000 at 3% for 1 year

$= p \cdot r \cdot t$

$= (1000)(\_\_\_\_)(1)$

$= \_\_\_\_$

(b) $3650 at 2% for 1 year

VOCABULARY TIP

Rate of interest The number with the % symbol or **"percent"** after it is the rate of interest.

2 Find the interest.

GS **(a)** $820 at 1% for $3\frac{1}{2}$ years

$I = p \cdot r \cdot t$

$= (\_\_\_\_)(\_\_\_\_)(3.5)$

$= \_\_\_\_$

(b) $4850 at 4% for $2\frac{1}{2}$ years

(c) $16,800 at 3% for $2\frac{3}{4}$ years

Answers

1. (a) 0.03; $30 (b) $73
2. (a) $820; 0.01; $28.70 (b) $485 (c) $1386

3 Find the interest.

GS **(a)** \$1800 at 3% for 4 months

$I = p \cdot r \cdot t$

$= (1800)(0.03)(\_\_\_\_)$

$= \_\_\_\_$

(b) \$28,000 at $7\frac{1}{2}\%$ for 3 months

4 Find the total amount due on each loan.

GS **(a)** \$3800 at $6\frac{1}{2}\%$ for 6 months

$I = p \cdot r \cdot t$

$= (3800)(0.065)(\_\_\_\_)$

$= \_\_\_\_$

Total amount due = principle + interest
= \$3800 + \$123.50
= ____

(b) \$12,400 at 5% for 5 years

(c) \$2400 at $4\frac{1}{2}\%$ for $2\frac{3}{4}$ years

Answers

3. (a) $\frac{4}{12}$ or $\frac{1}{3}$; \$18 **(b)** \$525

4. (a) $\frac{6}{12}$ or $\frac{1}{2}$; \$123.50; \$3923.50 **(b)** \$15,500 **(c)** \$2697

CAUTION

It is best to rewrite fractions of percents or fractions of years in their decimal form. In **Example 2**, $3\frac{1}{2}$ years is rewritten as 3.5 years.

Interest rates are given ***per year.*** For loan periods of less than one year, be careful to express time as a fraction of a year.

If time is given in months, for example, use a denominator of 12, because there are 12 months in a year. A loan for 9 months would be for $\frac{9}{12}$ of a year.

EXAMPLE 3 Finding Simple Interest for Less Than 1 Year

Find the interest on \$840 at $4\frac{1}{2}\%$ for 9 months.

The principal is \$840. The rate is $4\frac{1}{2}\%$ or 0.045 as a decimal, and the time is $\frac{9}{12}$ of a year. Use the formula $I = p \cdot r \cdot t$.

$$I = (840)(0.045)\left(\frac{9}{12}\right)$$

9 months = $\frac{9}{12}$ of a year

Rewrite $4\frac{1}{2}\%$ as 0.045

$$= (37.8)\left(\frac{3}{4}\right)$$

Write $\frac{9}{12}$ in lowest terms as $\frac{3}{4}$.

$$= \frac{(37.8)(3)}{4}$$

$$= \frac{113.4}{4} = 28.35$$

The interest is \$28.35.

Calculator Tip

The calculator solution to **Example 3** above uses chain calculations.

840 (×) .045 (×) 9 (÷) 12 (=) 28.35

◀ **Work Problem 3 at the Side.**

OBJECTIVE 2 Find the total amount due on a loan. When a loan is repaid, the interest is added to the original principal to find the total amount due.

Formula for Total Amount Due

Total amount due = principal + interest

EXAMPLE 4 Calculating the Total Amount Due

A loan of \$3240 was made at 6% for 3 months. Find the total amount due.

First find the interest. Then add the principal and the interest to find the total amount due.

$$I = (3240)(0.06)\left(\frac{3}{12}\right)$$

3 months = $\frac{3}{12}$ of a year.

$$I = \$48.60$$

The interest is \$48.60.

Remember that the *total amount due* is the amount of the loan plus the interest.

Total amount due = principal + interest

= \$3240 + \$48.60 = \$3288.60

The total amount due is \$3288.60.

◀ **Work Problem 4 at the Side.**

6.7 Exercises

FOR EXTRA HELP

MyMathLab®

CONCEPT CHECK *Rewrite all interest rates and times in decimal form.*

1. $3\frac{1}{2}\%$ **2.** $7\frac{1}{2}\%$ **3.** $2\frac{1}{4}$ years **4.** $5\frac{3}{4}$ years

Find the interest. ***See Examples 1 and 2.***

	Principal	Rate	Time in Years	Interest
5.	\$100	6%	1	______
	$I = p \cdot r \cdot t$ $= (100)(0.06)(1)$			
6.	\$200	3%	1	______
	$I = p \cdot r \cdot t$ $= (200)(0.03)(1)$			
7.	\$700	5%	3	______
8.	\$900	2%	4	______
9.	\$2300	$4\frac{1}{2}\%$	$2\frac{1}{2}$	______
	$I = p \cdot r \cdot t$ $= (2300)(0.045)(2.5)$			
10.	\$4700	$5\frac{1}{2}\%$	$1\frac{1}{2}$	______
	$I = p \cdot r \cdot t$ $= (4700)(0.055)(1.5)$			
11.	\$10,800	$7\frac{1}{2}\%$	$2\frac{3}{4}$	______
12.	\$12,400	$1\frac{1}{2}\%$	$3\frac{3}{4}$	______

CONCEPT CHECK *Change all times in months to fractions of a year and to decimal form.*

13. 3 months **14.** 6 months **15.** 9 months **16.** 15 months

Find the interest. Round to the nearest cent if necessary. ***See Example 3.***

	Principal	Rate	Time in Months	Interest
17.	\$400	3%	6	______
	$I = p \cdot r \cdot t$ $= (400)(0.03)\left(\frac{1}{2}\right)$ 6 months $= \frac{1}{2}$ year			

	Principal	Rate	Time in Months	Interest
18.	\$780	5%	24	________
19.	\$940	3%	18	________
20.	\$178	4%	12	________
21.	\$1225	$5\frac{1}{2}\%$	3	________
22.	\$2660	$7\frac{1}{2}\%$	3	________
23.	\$15,300	$7\frac{1}{4}\%$	7	________
24.	\$13,700	$3\frac{3}{4}\%$	11	________

Work for 18 (GS):

$I = p \bullet r \bullet t$

$= (780)(0.05)(2)$ 24 months = 2 years

Find the total amount due on the following loans. Round to the nearest cent if necessary. ***See Example 4.***

	Principal	Rate	Time	Total Amount Due
25.	\$200	5%	1 year	________
26.	\$400	2%	6 months	________
27.	\$740	6%	9 months	________
28.	\$1180	3%	2 years	________

Work for 25 (GS):

$I = p \bullet r \bullet t$

$= (200)(0.05)(1)$

Total amount due = principal + interest

= \$200 + interest

Work for 26 (GS):

$I = p \bullet r \bullet t$

$= (400)(0.02)\left(\frac{1}{2}\right)$ 6 months $= \frac{1}{2}$ year

Total amount due = principal + interest

= \$400 + interest

	Principal	Rate	Time	Total Amount Due
29.	$1800	9%	18 months	________
30.	$9000	6%	7 months	________
31.	$3250	$3\frac{1}{2}\%$	6 months	________
32.	$7600	$4\frac{1}{2}\%$	1 year	________
33.	$16,850	$7\frac{1}{2}\%$	9 months	________
34.	$19,450	$5\frac{1}{2}\%$	6 months	________

35. The amount of interest paid on savings accounts and charged on loans can vary from one institution to another. However, when the amount of interest is calculated, three factors are used in the calculation. Name these three factors and describe them in your own words.

36. Interest rates are usually given as a rate per year (annual rate). Explain what must be done when time is given in months. Write your own problem where time is given in months and then show how to solve it.

Solve each application problem. Round to the nearest cent if necessary.

37. Paul Plescia has a savings account of $8642 at his credit union. If the account pays interest of 2%, how much interest will he earn in 2 years?

38. Esther Albert, a professional dancer, deposits $68,000 of her earnings at 4% for 5 years. How much interest will she earn?

39. To increase distribution of her office supply products, Stella Glitter borrowed $150,000 at 7% for 30 months. Find the amount of interest on this loan.

40. The Jidobu family invests $18,000 at 3% for 9 months. What amount of interest will the family earn?

41. To complete her fourth semester of college, Yohko Kitagawa borrowed \$4650 at 5%. If she repaid the loan in 5 months, find the total amount due.

42. Sarah Brynski borrows \$5500 from her dad for a used car. The loan will be paid back with 4% interest at the end of 9 months. Find the total amount due.

43. Nicholas Thomas deposits \$14,800 in his school credit union account for 10 months. If the credit union pays $2\frac{1}{4}\%$ interest, find the amount of interest he will earn.

44. Sid and Shirley Kordell, owners of the Nut House, borrow \$54,000 to update their store. If the loan is for 42 months at $7\frac{1}{4}\%$, find the amount of interest they will owe.

45. After the sale of her home, the buyer still owes Tiina Luig a balance of \$8800. If the interest rate on this amount is $7\frac{1}{4}\%$, find the amount of interest she will earn every 3 months.

46. Pat Carper owes \$1900 in taxes. She is charged a penalty of $9\frac{1}{4}\%$ annual interest and pays the taxes and penalty after 6 months. Find the amount of the penalty.

47. Business has been good at Rocky's Grill, and the owner deposits \$5400 into an account paying $2\frac{3}{4}\%$ interest. Find

(a) the amount of interest.

(b) the total amount in the account at the end of $1\frac{1}{2}$ years.

48. Moises Guardado, the owner of MG Painting, borrowed \$29,400 to purchase another truck and other equipment for his business. The loan is for $3\frac{1}{2}$ years at $5\frac{1}{4}\%$ interest. Find

(a) the amount of interest.

(b) the total amount he must repay.

49. Jake's Exotic Pets bought four new aviary environments (bird cages) at a cost of \$980 each. The owner borrowed 70% of the cost of the cages at an interest rate of $7\frac{1}{2}\%$. Find the total amount owed at the end of $2\frac{1}{2}$ years.

50. The owners of Baily and Daughters Excavating purchased four earth movers at a cost of \$485,000 each. If they borrowed 80% of the total purchase price for $2\frac{1}{2}$ years at $10\frac{1}{2}\%$ interest, find the total amount due.

6.8 Compound Interest

The interest we studied in the last section was *simple interest* (interest only on the original principal). A common type of interest used with savings accounts and most investments is **compound interest** or interest paid on past interest as well as on the principal.

OBJECTIVES

1. Understand compound interest.
2. Understand compound amount.
3. Find the compound amount.
4. Use a compound interest table.
5. Find the compound amount and the amount of interest.

OBJECTIVE 1 **Understand compound interest.** Suppose that you make a single deposit of $1000 in a savings account that earns 5% per year. What will happen to your savings over 3 years? At the end of the first year, 1 year's interest on the original deposit is calculated. Use the simple interest formula.

Interest = principal • rate • time

Year 1 ($1000)(0.05)(1) = $50

Add the interest to the $1000 to find the amount in your account at the end of the first year. $1000 + $50 = **$1050** ← In year 1, interest is calculated on principal.

The interest for the second year is found on $1050, that is, the interest is **compounded.**

Year 2 (**$1050**)(0.05)(1) = $52.50

Add this interest to the $1050 to find the amount in your account at the end of the second year. **$1050** + $52.50 = $1102.50.

The interest for the third year is found on $1102.50.

Year 3 ($1102.50)(0.05)(1) ≈ $55.13 ← In year 2 and thereafter, interest is calculated on the principal and all past interest.

Add this interest to the $1102.50. So, $1102.50 + $55.13 = **$1157.63.**

At the end of 3 years, you will have **$1157.63** in your savings account. The $1157.63 that you have in your account is called the **compound amount.**

If you had earned only *simple* interest for 3 years, your interest would be as follows.

$$I = (\$1000)(0.05)(3)$$
$$= \$150 \leftarrow \text{Simple interest}$$

At the end of 3 years, you would have $1000 + $150 = $1150 in your account. Compounding the interest increased your earnings by $7.63 because $1157.63 − $1150 = $7.63.

With *compound* interest, the interest earned during the second year is greater than that earned during the first year, and the interest earned during the third year is greater than that earned during the second year. This happens because the interest earned each year is *added* to the principal, and the new total is used to calculate the amount of interest in the next year.

Compound Interest

Interest paid on principal plus past interest is called **compound interest.**

VOCABULARY TIP

Simple interest With simple interest, there is usually only one interest calculation for a specific period of time at a specific rate of interest.

VOCABULARY TIP

Compound interest/Compounded *Compound* means to mix, combine, or put together. When interest has been compounded, it means that it has been combined with the original principal and all previous interest on the principal.

VOCABULARY TIP

Compound amount The compound amount at the end of any compound interest period is the total of the original principal and all the interest combined.

1 Find the compound amount given the following deposits. Round to the nearest cent if necessary.

GS **(a)** $400 at 4% for 2 years

yr 1: ($400)(0.04)(1) = $16

$400 + $16 = $416

yr 2: ($416)(0.04)(1) = ______

$416 + $16.64 = ______

(b) $2000 at 2% for 3 years

2 Find the compound amount by multiplying the original deposit by 100% plus the compound interest rate. Round to the nearest cent if necessary.

GS **(a)** $1800 at 2% for 3 years

(1800)(1.02)(1.02)(1.02) =

______

(b) $900 at 3% for 2 years

(c) $2500 at 5% for 4 years

Answers

1. **(a)** $16.64; $432.64 **(b)** $2122.42 (rounded)
2. **(a)** $1910.17 (rounded) **(b)** $954.81 **(c)** $3038.77 (rounded)

OBJECTIVE 2 **Understand compound amount.** Find the compound amount as shown below.

EXAMPLE 1 Finding the Compound Amount

Mavis Chamski deposits $3400 in an account that pays 3% interest compounded annually for 4 years. Find the compound amount. Round to the nearest cent when necessary.

Year	Interest	Compound Amount
1	($3400)(0.03)(1) = $102 $3400 + $102 =	$3502
2	($3502)(0.03)(1) = $105.06 $3502 + $105.06 =	$3607.06
3	($3607.06)(0.03)(1) ≈ $108.21 $3607.06 + $108.21 =	$3715.27
4	($3715.27)(0.03)(1) ≈ $111.46 $3715.27 + $111.46 =	$3826.73

The compound amount is $3826.73.

◀ Work Problem **1** at the Side.

OBJECTIVE 3 **Find the compound amount.** A more efficient way of finding the compound amount is to add the interest rate to 100% and then multiply by the original deposit. Notice that in **Example 1** above, at the end of the first year, you will have $3400 (100% of the original deposit) plus 3% (of the original deposit) or 103% (because 100% + 3% = 103%).

EXAMPLE 2 Finding the Compound Amount

Find the compound amount in **Example 1** using multiplication.

Year 1 Year 2 Year 3 Year 4

($3400)(1.03)(1.03)(1.03)(1.03) ≈ $3826.73

↑ Original deposit; 100% + 3% = 103% = 1.03; ↑ Compound amount

This method works well with a calculator.

Our answer, $3826.73, is the same as in **Example 1** above.

◀ Work Problem **2** at the Side.

Note

By adding the compound interest rate to 100%, we can then multiply by the original deposit. This will give us the compound amount at the end of each compound interest period.

Calculator Tip

If you use a calculator for **Example 2** above, you can use the y^x key (exponent key). (On some calculators, you use the ^ key instead.)

3400 × 1.03 y^x 4 = 3826.73 (rounded)

The 4 following the y^x key represents the number of compound interest periods.

OBJECTIVE 4 Use a compound interest table. The calculation of compound interest can be quite tedious. For this reason, compound interest tables have been developed.

Suppose you deposit \$1 in a savings account today that earns 4% compounded annually and you allow the deposit to remain for 3 years. The diagram below shows the compound amount at the end of each of the 3 years.

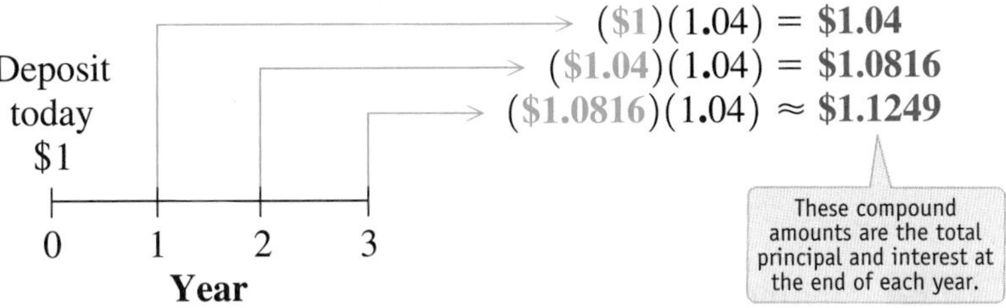

Using the compound amounts for \$1, a table can be formed. Look at the table below and find the column headed 4%. The first three numbers for years 1, 2, and 3 are the same as those we have calculated for \$1 at 4% for 3 years. This table, giving the compound amounts on a \$1 deposit for given lengths of time and interest rates, can be used for finding the compound amount on any amount of deposit.

COMPOUND INTEREST TABLE

Time Periods	2.00%	2.50%	3.00%	3.50%	4.00%	4.50%	5.00%	5.50%	6.00%	Time Periods
1	1.0200	1.0250	1.0300	1.0350	1.0400	1.0450	1.0500	1.0550	1.0600	1
2	1.0404	1.0506	1.0609	1.0712	1.0816	1.0920	1.1025	1.1130	1.1236	2
3	1.0612	1.0769	1.0927	1.1087	**1.1249**	1.1412	1.1576	1.1742	1.1910	3
4	1.0824	1.1038	1.1255	1.1475	1.1699	1.1925	1.2155	1.2388	1.2625	4
5	1.1040	1.1314	1.1593	1.1877	1.2167	1.2462	1.2763	1.3070	1.3382	5
6	1.1261	1.1597	1.1941	**1.2293**	1.2653	1.3023	1.3401	1.3788	1.4185	6
7	1.1486	1.1887	1.2299	1.2723	1.3159	1.3609	1.4071	1.4547	1.5036	7
8	1.1717	1.2184	1.2668	1.3168	1.3686	1.4221	1.4775	1.5347	1.5938	8
9	1.1951	1.2489	1.3048	1.3629	1.4233	1.4861	1.5513	1.6191	1.6895	9
10	1.2190	1.2801	1.3439	1.4106	1.4802	1.5530	1.6289	1.7081	1.7908	10
11	1.2434	1.3121	1.3842	1.4600	1.5395	1.6229	1.7103	1.8021	1.8983	11
12	1.2682	1.3449	1.4258	1.5111	1.6010	1.6959	1.7959	1.9012	2.0122	12

EXAMPLE 3 Using a Compound Interest Table

Find the compound amount using the compound interest table.

A ruler or straightedge helps align the column and row.

(a) \$1 is deposited at a 5% interest rate for 10 years.

Look down the column headed 5%, and across to row 10 (because 10 years = 10 time periods). At the intersection of the column and row, read the compound amount, 1.6289.

(b) \$1 is deposited at $3\frac{1}{2}$% for 6 years.

The intersection of the $3\frac{1}{2}$% (3.50%) column and row 6 shows **1.2293** as the compound amount.

Work Problem 3 at the Side. ▶

3 Find the compound amount using the compound interest table.

GS **(a)** \$1 at 3% for 6 years

(1)(1.1941) = ________

Look down the 3% column in the table to row 6.

(b) \$1 at 2% for 8 years

(c) \$1 at $4\frac{1}{2}$% for 12 years

Answers

3. (a) \$1.19 (rounded) **(b)** \$1.17 (rounded) **(c)** \$1.70 (rounded)

4 Use the compound interest table to find the compound amount and the interest.

(a) $4000 at 3% for 10 years

Table shows 1.3439.

(4000)(1.3439) = ________

________ − 4000

= ________ interest

(b) $12,600 at $3\frac{1}{2}$% for 8 years

(c) $32,700 at $4\frac{1}{2}$% for 12 years

Answers

4. **(a)** $5375.60; $5375.60; $1375.60
(b) $16,591.68; $3991.68
(c) $55,455.93; $22,755.93

OBJECTIVE 5 Find the compound amount and the amount of interest. Find the compound amount and interest as follows.

Finding the Compound Amount and the Interest

Compound Amount
Multiply the principal by the compound amount for $1 (from the table on the previous page).

Interest
Find the interest earned on a deposit by subtracting the original deposit from the compound amount.

EXAMPLE 4 Finding Compound Amount and Interest

Use the compound interest table to find the compound amount and the interest.

(a) $1000 at $5\frac{1}{2}$ % interest for 12 years

Look in the table on the previous page for $5\frac{1}{2}$% (5.50%) and 12 periods to find the number 1.9012 but do *not* round it. Multiply this number and the principal of $1000.

Never round the numbers found in the table.

($1000) (1.9012) = **$1901.20**

The account will contain $1901.20 after 12 years.

Find the interest by subtracting the original deposit from the compound amount.

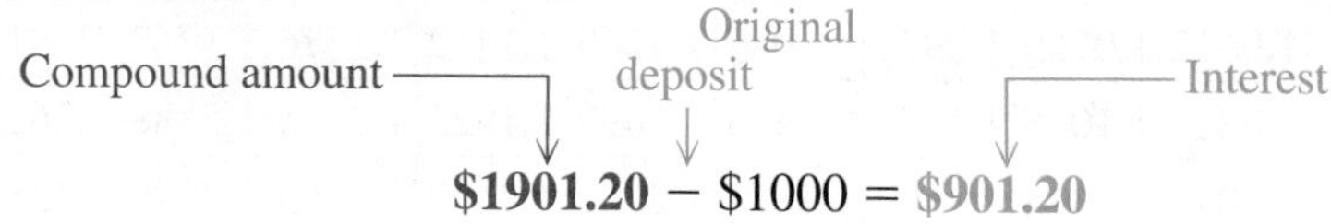

(b) $6400 at 2% for 7 years

Look in the table for 2% and 7 periods to find 1.1486. Multiply.

($6400) (1.1486) = **$7351.04** Compound amount

Subtract the original deposit from the compound amount.

$7351.04 − $6400 = **$951.04** Interest

Remember: Compound amount − Principal = Interest.

A total of $951.04 in interest was earned.

◀ **Work Problem 4 at the Side.**

6.8 Exercises

FOR EXTRA HELP

MyMathLab®

1. CONCEPT CHECK *Write* true *or* false.
For short periods of time, five years or less, the amount of interest earned will be the same whether you use simple or compound interest.

2. CONCEPT CHECK *Underline the correct answer.*
When using compound interest, the amount of (*interest/principal*) earned at the end of a compound interest period will be added to the amount of (*interest/principal*) at the beginning of that period.

Find the compound amount given the following deposits. Calculate the interest each year, then add it to the previous year's amount. ***See Example 1.***

3. \$500 at 4% for 2 years

$I = p \cdot r \cdot t$

Year 1 $\begin{cases} (\$500)(0.04)(1) = \$20 \\ \$500 + \$20 = \$520 \end{cases}$

Year 2 $\begin{cases} (\$520)(0.04)(1) = \$20.80 \\ \$520 + \$20.80 = \_\_\_\_\_\_\_\_ \end{cases}$

4. \$1500 at 5% for 3 years

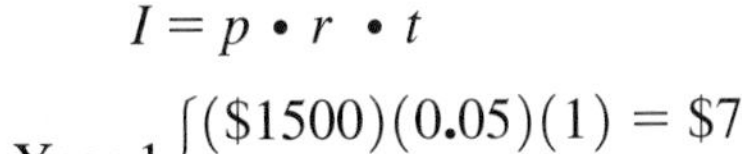

$I = p \cdot r \cdot t$

Year 1 $\begin{cases} (\$1500)(0.05)(1) = \$75 \\ \$1500 + \$75 = \$1575 \end{cases}$

Year 2 $\begin{cases} (\$1575)(0.05)(1) = \$78.75 \\ \$1575 + \$78.75 = \$1653.75 \end{cases}$

Year 3 $\begin{cases} (\$1653.75)(0.05)(1) = \$82.69 \text{ (rounded)} \\ \$1653.75 + \$82.69 = \_\_\_\_\_\_\_\_ \end{cases}$

5. \$1800 at 3% for 3 years

6. \$2000 at 2% for 3 years

7. \$3500 at 7% for 4 years

8. \$5500 at 6% for 4 years

9. CONCEPT CHECK *Fill in the blanks.*
To find the compound amount for a deposit of \$2000 at 2% for 3 years, you can multiply

(2000)____ ____ ____ = \$2122.42 (rounded)

10. CONCEPT CHECK *Fill in the blanks.*
To find the compound amount for a deposit of \$3000 at 5% for 2 years, you can multiply

(\$3000)____ ____ = \$3307.50

Find each compound amount by multiplying the original deposit by 100% plus the compound rate given in the following. ***See Example 2.*** *Round answers to the nearest cent if necessary.*

11. \$1000 at 5% for 2 years

Year 1 Year 2
(\$1000)(1.05)(1.05) = ________

12. \$500 at 4% for 3 years

(\$500)(1.04)(1.04)(1.04) = ________

13. \$1400 at 6% for 5 years

14. \$2500 at 3% for 4 years

15. \$1180 at 7% for 8 years

16. \$12,800 at 6% for 7 years

17. \$10,940 at 4% for 6 years

18. \$15,710 at 2% for 8 years

CONCEPT CHECK *In the compound interest table, locate the compound amount for \$1 for each of the following situations.*

19. 3% for 6 years

20. 4.50% for 10 years

21. 2% for 8 years

22. 5.50% for 3 years

Use the table to find the compound amount and the interest. Interest is compounded annually. Round answers to the nearest cent if necessary. ***See Examples 3 and 4.***

23. \$1000 at 4% for 5 years

GS Table shows 1.2167.

(\$1000)(1.2167) = \$1216.70 compound amount

\$1216.70 − \$1000 = ______

↑ Compound amount ↑ Original deposit ↑ Interest

24. \$10,000 at 3% for 4 years

GS Table shows 1.1255.

(\$10,000)(1.1255) = \$11,255 compound amount

\$11,255 − \$10,000 = ______

↑ Compound amount ↑ Original deposit ↑ Interest

25. \$8000 at 2% for 10 years

26. \$7800 at 5% for 8 years

27. \$8428.17 at $4\frac{1}{2}$% for 6 years

28. \$10,472.88 at $5\frac{1}{2}$% for 12 years

29. Write a definition for compound interest. Describe in your own words what compound interest means to you.

30. What is the difference between the compound amount and the interest?

Use the compound interest table in this section to solve each application problem. Round answers to the nearest cent if necessary. ***See Examples 3 and 4.***

31. After winning \$50,000 in the Veterans of Foreign Wars (VFW) lottery, Jerry Murray opened a money market account. He deposited all of his winnings and earned 4% interest compounded annually. Find the amount he will have (compound amount) in the account at the end of 5 years.

32. Yen Lee borrowed \$32,800 from her family to open a small restaurant named Shanghai Winds. Her plan is to repay the loan at the end of 4 years at $3\frac{1}{2}\%$ interest compounded annually. Find the total amount that she must repay.

33. Al Granard lends \$76,000 to the owner of Rick's Limousine Service. He will be repaid at the end of 9 years at 6% interest compounded annually. Find

(a) the total amount that he should be repaid.

(b) the amount of interest earned.

34. Recent sales of their music gave Bob and Cindy Kilpatrick \$28,500 in additional income. If they invest all of it at 4% interest compounded annually for 10 years, find

(a) the total amount they will have at the end of 10 years.

(b) the amount of interest earned.

35. Jennifer Barrister deposits \$30,000 at 6% interest compounded annually. Two years after she makes the first deposit, she deposits another \$40,000, also at 6% compounded annually.

(a) What total amount will she have five years after her first deposit?

(b) What amount of interest will she have earned?

36. Christine Campbell invested \$25,000 at 4% interest compounded annually. Three years after she made the first deposit, she deposited another \$25,000, also at 4% compounded annually.

(a) What total amount will she have five years after her first deposit?

(b) What amount of interest will she earn?

Relating Concepts (Exercises 37–42) For Individual or Group Work

Knowing how to solve interest problems is important to businesspeople and consumers alike. ***Work Exercises 37–40 in order.***

37. Simple interest calculation is used for most short-term business loans, most real estate loans, and many automobile and consumer loans. The formula for simple interest is

Interest = ______ • ______ • ______

or $I =$ ______ • ______ • ______.

38. When a loan is repaid, the interest is added to the original principal. The formula for the total amount due is

Amount due = ______ + ______.

39. Compound interest is paid on most savings accounts and many other types of investments. Compound interest is interest calculated on ______ plus past ______.

40. The compound amount is the total amount in an account at the end of a period of time. Compound amount is the original ______ + compound ______.

Use the compound interest table to answer Exercises 41 and 42.

41. Julie Maxey has two choices. She can invest \$4350 at 6% simple interest for 6 years, or she can invest the same amount at 6% interest compounded annually for 6 years.

(a) Find the difference in the amount of interest earned in these two accounts.

(b) If the length of time is doubled from 6 to 12 years, will the difference in the interest earned also double?

(c) Use your own example to determine that what you found in part (b) is true with a different interest rate and length of time.

42. One account is opened with \$10,000 at 5% simple interest for 10 years. Another account is opened with \$9500 at 5% interest compounded annually for 10 years.

(a) At the end of the 10 years, which account has the higher balance and by how much?

(b) What does this tell you about compound interest? Why?

Study Skills

PREPARING FOR YOUR FINAL EXAM

OBJECTIVES

1. Create a final exam week plan.
2. Break studying into chunks and study over several days.
3. Practice all types of problems.

Your math final exam is likely to be a **comprehensive exam.** This means that it will cover material from the **entire term.** The end of the term will be less stressful if you **make a plan** for how you will prepare for each of your exams.

First, figure out the **score you need to earn on the final exam** to get the course grade you want. Check your course syllabus for grading policies, or ask your instructor if you are not sure of them. This allows you to set a goal for yourself.

How many points do you need to earn on your mathematics final exam to get the grade you want? ______________________________

____

Create a Plan

Second, create a **final exam week plan for your work and personal life.** If you need to make an adjustment in your work schedule, do it in advance, so you aren't scrambling at the last minute. If you have family members to care for, you might want to enlist some help from others so you can spend extra time studying. Try to plan in advance so you don't create additional stress for yourself. You will have to set some priorities, and studying has to be at the top of the list! Although life doesn't stop for finals, some things can be ignored for a short time. You don't want to "burn out" during final exam week; **get enough sleep and healthy food so you can perform your best.**

What adjustments in your personal life do you need to make for final exam week? ______________________________

____

Study and Review

Third, use the following suggestions to guide your studying and reviewing.

- **Know exactly which chapters and sections will be on the final exam.**
- **Divide up the chapters,** and decide how much you will review each day.
- Begin your reviewing **several days** before the exam.
- **Use returned quizzes and tests** to review earlier material (if you have them).
- **Practice all types of problems,** but emphasize the types that are most difficult for you. Use the **Cumulative Reviews** that are at the end of each chapter in your textbook.
- **Rewrite your notes or make mind maps** to create summaries.
- **Make study cards for all types of problems.** Be sure to use the same **direction words** (such as *simplify, solve, estimate*) that your exam will use. Carry the cards with you and review them whenever you have a few spare minutes.

Study Skills

Continued from page 459

Managing Stress

Of course, a week of final exams produces stress. **Students who develop skills for reducing and managing stress do better on their final exams and are less likely to "bomb" an exam.** You already know the damaging effect of adrenaline on your ability to think clearly. But several days (or weeks) of elevated stress is also harmful to your brain and your body. You will feel better if you make a conscious effort to reduce your stress level. Even if it takes you away from studying for a little while each day, the time will be well spent.

Reducing Physical Stress

Examples of ways to reduce **physical stress** are listed below. Can you add any of your own ideas to the list?

Which techniques will you try?

____

____

____

- *Laugh until your eyes water.* Watching your favorite funny movie, exchanging a joke with a friend, or viewing a comedy bit on the Internet are all ways to generate a healthy laugh. Laughing raises the level of *calming* chemicals (endorphins) in your brain.
- *Exercise for 20 to 30 minutes.* If you normally exercise regularly, do NOT stop during final exam week! Exercising helps relax muscles, diffuses adrenaline, and raises the level of endorphins in your body. If you don't exercise much, get some gentle exercise, such as a daily walk, to help you relax.
- *Practice deep breathing.* Several minutes of deep, smooth breathing will calm you. Close your eyes too.
- *Visualize a relaxing scene.* Choose something that you find peaceful and picture it. Imagine what it feels like and sounds like. Try to put yourself in the picture.
- If you feel stress in your muscles, such as in your shoulders or back, *slowly squeeze the muscles as much as you can, and then release them.* Sometimes we don't realize we are clenching our teeth or holding tension in our shoulders until we consciously work with them. Try to notice what it feels like when they are relaxed and loose. Squeezing and then releasing muscles is also something you can do during an exam if you feel yourself tightening up.

Reducing Mental Stress

Mental stress reduction is also a powerful tool both before and during an exam. In addition to these suggestions, do you have any of your own techniques?

Which techniques will you try?

____

____

____

- *Talk positively to yourself.* Tell yourself you will get through it.
- *Reward yourself.* Give yourself small breaks, a little treat—something that makes you happy—every day of final exam week.
- *Make a list of things to do* and feel the sense of accomplishment when you cross each item off.
- When you take time to relax or exercise, *make sure you are relaxing your mind too.* Use your mind for something *completely* different from the kind of thinking you do when you study. Plan your garden, play your favorite music, walk your dog, read a good book.
- *Visualize.* Picture yourself completing exams and projects successfully. Picture yourself taking the test calmly and confidently.

Chapter 6 Summary

Key Terms

6.1

percent Percent means "per one hundred." A percent is a ratio with a denominator of 100.

6.3

percent proportion The proportion $\frac{\text{part}}{\text{whole}} = \frac{\text{percent}}{100}$ is used to solve percent problems.

whole The whole in a percent problem is the entire quantity, the total, or the base.

part The part in a percent problem is the portion being compared with the whole.

6.5

percent equation The percent equation is: part = percent • whole. It is another way to solve percent problems.

6.6

sales tax Sales tax is a percent of the cost of an item charged as a tax.

commission Commission is a percent of the dollar value of total sales paid to a salesperson.

discount Discount is often expressed as a percent of the original price; it is then deducted from the original price, resulting in the sale price.

percent of increase or decrease Percent of increase or decrease is the amount of change (increase or decrease) expressed as a percent of the original amount.

6.7

interest Interest is a fee paid or a charge made for lending or borrowing money.

principal Principal is the amount of money on which interest is earned.

rate of interest Often referred to as "rate," it is the charge for interest and is given as a percent.

simple interest Interest that is computed only on the original principal is simple interest.

interest formula The interest formula is used to calculate interest. It is: interest = principal • rate • time, or $I = p \cdot r \cdot t$.

6.8

compound interest Compound interest is interest paid both on the past interest and on the principal.

compounding Interest that is compounded once each year is compounded annually.

compound amount Compound amount is the total amount in an account, including compound interest and the original principal.

New Symbols

% percent (per one hundred)

New Formulas

To write percents as decimals: $p\% = p \div 100$ **To write percents as fractions:** $p\% = \frac{p}{100}$

Percent proportion: $\frac{\text{part}}{\text{whole}} = \frac{\text{percent}}{100}$ **Percent equation:** part = percent • whole

Amount of sales tax: amount of sales tax = rate of tax • cost of item

Amount of commission: amount of commission = rate of commission • amount of sales

Amount of discount: amount of discount = rate of discount • original price

Sale price: sale price = original price − amount of discount

Percent of increase: $\frac{\text{amount of increase}}{\text{original value}} = \frac{\text{percent}}{100}$

Percent of decrease: $\frac{\text{amount of decrease}}{\text{original value}} = \frac{\text{percent}}{100}$

Interest: Interest = principal • rate • time or $I = p \cdot r \cdot t$

Total amount due: total amount due = principal + interest

Test Your Word Power

See how well you have learned the vocabulary in this chapter.

1 To write a **percent as a decimal,** you drop the percent symbol
- **A.** after finding the decimal point
- **B.** and move the decimal point two places to the right
- **C.** and move the decimal point two places to the left.

2 To write a **decimal as a percent,** you attach the percent symbol
- **A.** after removing the decimal point
- **B.** after moving the decimal point two places to the right
- **C.** after moving the decimal point two places to the left.

3 **Percent** means
- **A.** the same as interest
- **B.** per one thousand
- **C.** per one hundred.

4 When you use $\frac{\textbf{part}}{\textbf{whole}} = \frac{\textbf{percent}}{\textbf{100}}$ to solve percent problems, you are using the
- **A.** simple interest formula
- **B.** percent proportion
- **C.** percent equation.

5 The **percent equation** is
- **A.** part = percent • whole
- **B.** $I = p \cdot r \cdot t$
- **C.** $p\% = \frac{p}{100}$

6 In a **percent of increase** problem, the increase is a percent of
- **A.** the largest amount
- **B.** the original amount
- **C.** the new or most recent amount.

7 In the formula $\boldsymbol{I = p \cdot r \cdot t}$**,** the ***p*** stands for
- **A.** proportion
- **B.** principal
- **C.** percent.

8 The term **rate** in an interest problem represents the
- **A.** whole
- **B.** percent
- **C.** part.

Answers to Test Your Word Power

1. C; *Example:* 50% written as a decimal is 0̮50. or 0.5.
2. B; *Example:* 0.25 written as a percent is 0.25 or 25%.
3. C; *Example:* 8% means 8 per 100.
4. B; *Example:* Part = 4, and whole = 25. To find the percent, $\frac{4}{25} = \frac{x}{100}$; $x = 16$ or 16%.
5. A; *Example:* Percent = 25, and whole = 300. To find the part, $(0.25)(300) = 75$.
6. B; *Example:* Original value = \$200, and amount of increase = \$40. To find the percent of increase, $\frac{40}{200} = \frac{x}{100}$; $x = 0.20$ or 20% increase.
7. B; *Example:* Principal (p) = \$800, rate (r) = 5%, and time (t) = 1 year. Then, $I = p \cdot r \cdot t$ so $I = (\$800)(0.05)(1) = \$40$.
8. B; *Example:* Principal (p) = \$1650, rate (r) = 4%, and time $(t) = \frac{1}{2}$ year. To find the interest, $I = (\$1650)(0.04)\left(\frac{1}{2}\right) = \$33$.

Quick Review

Concepts	Examples
6.1 Basics of Percent **Writing a Percent as a Decimal** To write a percent as a decimal, drop the percent symbol and move the decimal point two places to the left. **Writing a Decimal as a Percent** To write a decimal as a percent, move the decimal point two places to the right and attach a % symbol.	50% (.50%) = 0.50 or just 0.5 3%(.03%) = 0.03 12.5% (12.5%) = 0.125 0.75 (0.75) = 75% 0.875(0.875) = 87.5% 3.6 (3.60) = 360%
6.2 Writing a Fraction as a Percent Use a proportion and solve for p to change a fraction to percent.	$\frac{2}{5} = \frac{p}{100}$ Proportion $5 \cdot p = 2 \cdot 100$ Cross products $5 \cdot p = 200$ $\frac{\overset{1}{\cancel{5}} \cdot p}{\underset{1}{\cancel{5}}} = \frac{200}{5}$ Divide both sides by 5. $p = 40$ $\frac{2}{5} = 40\%$ Attach % symbol.

Concepts	Examples
6.3 Learning the Percent Proportion Part is to whole as percent is to 100. $\frac{\text{part}}{\text{whole}} = \frac{\text{percent}}{100}$ ← Always 100 because percent means "per 100."	Use the percent proportion to solve for the unknown value. part = 30, whole = 50; find the percent. Part → 30, Whole → 50; x ← Percent (unknown); 100 ← Always 100 $\frac{30}{50} = \frac{x}{100}$ $\frac{3}{5} = \frac{x}{100}$ Write $\frac{30}{50}$ as $\frac{3}{5}$ in lowest terms. $5 \cdot x = 3 \cdot 100$ Cross products $5 \cdot x = 300$ $\frac{\overset{1}{\cancel{5}} \cdot x}{\underset{1}{\cancel{5}}} = \frac{300}{5}$ Divide both sides by 5. $x = 60$ The percent is 60, which is written as 60%.
6.3 Identifying Percent, Whole, and Part in a Percent Problem The percent appears with the word **percent** or with the symbol **%**. The whole often appears after the word **of**. The whole is the entire quantity or total. The part is the portion of the total. If the percent and the whole are found first, the remaining number is the part.	Find the percent, whole, and part in the following. 10% (Percent) **of** the 500 (Whole) pies is how many pies? (Part (unknown)) 20 cats (Part) is 5% (Percent) **of** what number of cats? (Whole (unknown)) What percent (Percent (unknown)) **of** $220 (Whole) is $33? (Part)
6.4 Applying the Percent Proportion Read the problem and identify the percent, whole, and part. Use the percent proportion to solve for the unknown quantity.	A liquid mixture in a tank contains 35% distilled water. If 28 gallons of distilled water are in the tank when it is full, find the capacity of the tank. percent = 35 and part = 28 Use the percent proportion to find the whole. $\frac{\text{part}}{x} = \frac{\text{percent}}{100}$ ← Whole (unknown) → x $\frac{28}{x} = \frac{35}{100}$ $\frac{28}{x} = \frac{7}{20}$ Write $\frac{35}{100}$ as $\frac{7}{20}$ in lowest terms. $x \cdot 7 = 560$ Cross products $\frac{x \cdot \overset{1}{\cancel{7}}}{\underset{1}{\cancel{7}}} = \frac{560}{7}$ Divide both sides by 7. $x = 80$ The capacity of the tank is 80 gallons.

Concepts	Examples

6.5 Using the Percent Equation

The percent equation is part = percent • whole. Identify the percent, whole, and part and solve for the unknown quantity. Always write the percent as a decimal before using the equation.

Solve each problem.

(a) Find 20% of 220 applicants.

$$\text{part (unknown)} = \text{percent} \cdot \text{whole}$$
$$x = (0.2)(220)$$
$$x = 44$$

20% of 220 applicants is 44 applicants.

(b) 8 balls is 4% of what number of balls?

$$\text{part} = \text{percent} \cdot \text{whole (unknown)}$$
$$8 = (0.04)(x)$$
$$\frac{8}{0.04} = \frac{\overset{1}{\cancel{(0.04)}}(x)}{\underset{1}{\cancel{0.04}}}$$
$$x = 200$$

8 balls is 4% of 200 balls.

(c) \$13 is what percent of \$52?

$$\text{part} = \text{percent (unknown)} \cdot \text{whole}$$
$$13 = x \cdot 52$$
$$\frac{13}{52} = \frac{x \cdot \overset{1}{\cancel{52}}}{\underset{1}{\cancel{52}}}$$
$$x = 0.25 = 25\%$$

\$13 is 25% of \$52.

6.6 Solving Application Problems with Proportions

To solve for **sales tax,** use this formula.

amount of sales tax = rate of tax • cost of item

The price of a 46-inch plasma HD television is \$699, and the sales tax is 5%. Find the sales tax.

$$\text{amount of sales tax} = (5\%)(\$699)$$
$$= (0.05)(\$699) = \$34.95$$

To find **commissions,** use this formula.

amount of commission =
rate of commission • amount of sales

The sales are \$92,000 with a commission rate of 3%. Find the commission.

$$\text{amount of commission} = (3\%)(\$92{,}000)$$
$$= (0.03)(\$92{,}000)$$
$$= \$2760$$

To find the **discount** and the **sale price,** use these formulas.

amount of discount = rate of discount • original price

sale price = original price − amount of discount

A gas oven originally priced at \$480 is offered at a 25% discount. Find the amount of the discount and the sale price.

$$\text{discount} = (0.25)(\$480) = \$120$$
$$\text{sale price} = \$480 - \$120 = \$360$$

Concepts	Examples
6.6 Solving Application Problems with Proportions *(Continued)*	
To find the **percent of change,** subtract to find the amount of change (increase or decrease), which is the part. The whole is the *original* value or value *before* the change.	The number of parking violations rose from 1980 violations to 2277. Find the percent of increase. $2277 - 1980 = 297$ Increase $\frac{297 \text{ (Increase)}}{1980 \text{ (Original value)}} = \frac{\text{percent}}{100}$ Solve the proportion to find that the percent = 15, so the percent of increase is 15%.
6.7 Finding Simple Interest	
Use the formula $I = p \cdot r \cdot t$ Interest = principal • rate • time Time (t) is in years. When the time is given in months, use a fraction with 12 in the denominator because there are 12 months in a year.	\$2800 is deposited at 2% for 3 months. Find the amount of interest. $I = p \cdot r \cdot t$ $= (2800)(0.02)\left(\frac{3}{12}\right)$ $= (56)\left(\frac{1}{4}\right) = \frac{(56)(1)}{4} = \$14$
6.8 Finding Compound Amount and Compound Interest	
There are three methods for finding the compound amount.	Find the compound amount and interest if \$1500 is deposited at 5% interest for 3 years.
1. Calculate the interest for each compound interest period, then add it back to the principal.	**1.** Year 1: Interest (\$1500)(0.05)(1) = \$75; Compound Amount \$1500 + \$75 = **\$1575** Year 2: Interest (\$1575)(0.05)(1) = \$78.75; Compound Amount \$1575 + \$78.75 = **\$1653.75** Year 3: Interest (\$1653.75)(0.05) ≈ \$82.69; Compound Amount \$1653.75 + \$82.69 = **\$1736.44**
2. Multiply the original deposit by 100% plus the compound interest rate.	**2.** (\$1500)(1.05)(1.05)(1.05) ≈ \$1736.44 \$1500: Original deposit; (1.05)(1.05)(1.05): 100% + 5% = 105% = 1.05; \$1736.44: Compound amount
3. Use the compound interest table to find the interest on \$1. Then, multiply the table value by the principal. The interest is found with this formula. Interest = compound amount − original deposit	**3.** Locate 5% across the top of the table and 3 periods at the left. The table value is 1.1576. compound amount = (\$1500)(1.1576) = **\$1736.40*** interest = \$1736.40 − \$1500 = \$236.40 * The difference in the compound amount results from rounding in the table.

Chapter 6 Review Exercises

6.1 *Write each percent as a decimal and each decimal as a percent.*

1. 35% **2.** 150% **3.** 99.44% **4.** 0.085%

5. 3.15 **6.** 0.02 **7.** 0.875 **8.** 0.002

6.2 *Write each percent as a fraction or mixed number in lowest terms and each fraction as a percent.*

9. 15% **10.** 37.5% **11.** 175% **12.** 0.25%

13. $\frac{3}{4}$ **14.** $\frac{5}{8}$ **15.** $3\frac{1}{4}$ **16.** $\frac{1}{200}$

Complete this chart.

Fraction	Decimal	Percent
$\frac{1}{8}$	**17.** _____	**18.** _____
19. _____	0.25	**20.** _____
21. _____	**22.** _____	180%

6.3 *Find the unknown value in the percent proportion* $\frac{part}{whole} = \frac{percent}{100}$.

23. part = 25, percent = 10

24. whole = 480, percent = 5

Identify each component and then set up each problem using the percent proportion $\frac{part}{whole} = \frac{percent}{100}$. *Do **not** try to solve for the unknown value.*

25. 35% of 820 mailboxes is 287 mailboxes.

26. 73 DVDs is what percent of 90 DVDs?

27. Find 14% of 160 mountain bikes.

28. 418 curtains is 16% of what number of curtains?

29. A golfer lost three of his eight golf balls. What percent were lost?

30. Only 88% of the door keys cut will operate properly. If there are 1280 keys cut, find the number of keys that will operate properly.

6.4 *Find the part using the percent proportion or the multiplication shortcut.*

31. 18% of 950 programs

32. 60% of 1450 reference books

33. 0.6% of 5200 acres

34. 0.2% of 1400 kilograms

Find the whole using the percent proportion.

35. 105 crates is 14% of what number of crates?

36. 348 test tubes is 15% of what number of test tubes?

37. 677.6 miles is 140% of what number of miles?

38. 2.5% of what number of cases is 425 cases?

Find the percent using the percent proportion. Round percent answers to the nearest tenth if necessary.

39. 649 tulip bulbs is what percent of 1180 tulip bulbs?

40. What percent of 1620 dinner rolls is 85 dinner rolls?

41. What percent of 380 pairs of socks is 36 pairs?

42. What percent of 650 soup cans is 200 soup cans?

6.1–6.4 *Solve each application problem. Round percent answers to the nearest tenth if necessary.*

43. Last year there was a total of 63 shark attacks on humans in the world (6 were fatal). This year there was an increase of 25.4% in the number of these attacks. Find the number of shark attacks on humans this year. (Round to the nearest whole number.) (*Source*: MSNBC.com)

44. Each week 3200 people shop at the three Crescent City Farmers Markets in New Orleans, Louisiana. If 896 of these shoppers are seniors, what percent of the shoppers are seniors? (*Source: USA Today.*)

6.5 *Use the percent equation to answer each question.*

45. 32% of $454 is what amount?

46. 155% of 120 trucks is how many trucks?

47. 0.128 ounce is what percent of 32 ounces?

48. 304.5 meters is what percent of 174 meters?

49. 33.6 miles is 28% of what number of miles?

50. $92 is 16% of what amount?

6.6 *Find the amount of sales tax or the tax rate and the total cost. Round to the nearest cent if necessary.*

	Cost of Item	Tax Rate	Amount of Tax	Total Cost
51.	$630	5%	____________	____________
52.	$780	____________	$58.50	____________

Find the commission earned or the rate of commission.

	Sales	Rate of Commission	Commission
53.	$3450	8%	____________
54.	$65,300	____________	$3265

Find the amount or rate of discount and the sale price. Round to the nearest cent if necessary.

	Original Price	Rate of Discount	Amount of Discount	Sale Price
55.	$112.50	30%	____________	____________
56.	$252	____________	$63	____________

6.7 *Find the simple interest due on each loan.*

	Principal	Rate	Time in Years	Interest
57.	$200	4%	1	____________
58.	$1080	5%	$1\frac{1}{4}$	____________

Find the simple interest paid on each investment.

	Principal	Rate	Time in Months	Interest
59.	$400	$3\frac{1}{2}\%$	3	____________
60.	$1560	$6\frac{1}{2}\%$	18	____________

Find the total amount due on each simple interest loan.

	Principal	Rate	Time	Total Amount Due
61.	$750	$5\frac{1}{2}\%$	2 years	____________
62.	$1560	3%	9 months	____________

6.8 *Find the compound amount and the interest. Interest is compounded annually. You may use the compound interest table. Round answers to the nearest cent if necessary.*

	Principal	Rate	Time in Years	Compound Amount	Interest
63.	$4000	3%	10	____________	____________
64.	$1870	4%	4	____________	____________
65.	$3600	$4\frac{1}{2}\%$	3	____________	____________
66.	$12,500	$5\frac{1}{2}\%$	5	____________	____________

Mixed Review Exercises

Find the unknown value in the percent proportion $\frac{\textit{part}}{\textit{whole}} = \frac{\textit{percent}}{100}$.

67. whole = 80, percent = 15

68. part = 738, percent = 45

Use the percent proportion or percent equation to answer each question.

69. 12% of 194 meters is how many meters?

70. 327 cars is what percent of 218 cars?

71. 0.6% of $85 is what amount?

72. 99 employees is 5% of what number of employees?

73. 76 chickens is what percent of 190 chickens?

74. 214.484 liters is 43% of what number of liters?

Write each percent as a decimal and each decimal as a percent.

75. 55%

76. 300%

77. 5

78. 4.71

79. 8.6%

80. 0.621

81. 0.375%

82. 0.0006

Write each percent as a fraction in lowest terms and each fraction or mixed number as a percent.

83. $\frac{3}{4}$

84. 42%

85. 87.5%

86. $\frac{3}{8}$

87. $32\frac{1}{2}\%$

88. $\frac{3}{5}$

89. 0.25%

90. $3\frac{3}{4}$

Solve each application problem. Round percent answers to the nearest tenth and money answers to the nearest cent if necessary.

91. Eva Jacob deposits $20,500 in her credit union savings account. If she earns $6\frac{1}{2}\%$ simple interest for 30 months, how much interest will be earned?

92. Hap Pishke, owner of Mardi Gras Barbers, borrows $14,750 to remodel his shop. He agrees to an 8% simple interest rate and will repay the loan in 18 months. Find the total amount due.

93. A study of 1005 adults found that 965 of them have a smoke alarm in their home and 461 have a carbon monoxide detector. (*Source*: Liberty Mutual International Association of Firefighters.)

(a) Find the percent of adults who have a smoke alarm in their home. Round to the nearest tenth of a percent.

(b) Find the percent of adults who have a carbon monoxide detector in their home. Round to the nearest tenth of a percent.

94. Alan Zagorin borrows \$148,000 to expand his business, located at the end of historic Route 66. The loan has an interest rate of 5% compounded annually and will be repaid in 4 years.

(a) Find the compound amount of this loan at the end of 4 years. Do not use the table.

(b) Find the amount of interest that he owes.

95. Tom Dugally, a real estate agent, sold two properties, one for \$125,000 and the other for \$290,000. He receives a commission of $1\frac{1}{2}\%$ of total sales. Find the commission that he earned.

96. Vending machines on campus must include healthy food choices such as fruits, fruit juices, and healthy snacks. Sales of healthy foods in the vending machines increased from 4320 items last month to 5107 items this month. Find the percent of increase.

97. A Sears Kenmore washer/dryer set priced at \$958 is marked down 18%. If the sales tax is 8%, find the cost of the washer/dryer set, including the sales tax.

98. In a recent insurance company study of boaters who had lost items overboard, 88 boaters or 8% said that they lost their cell phones. Find the total number of boaters in the survey. (*Source:* Progressive Groups of Insurance Companies.)

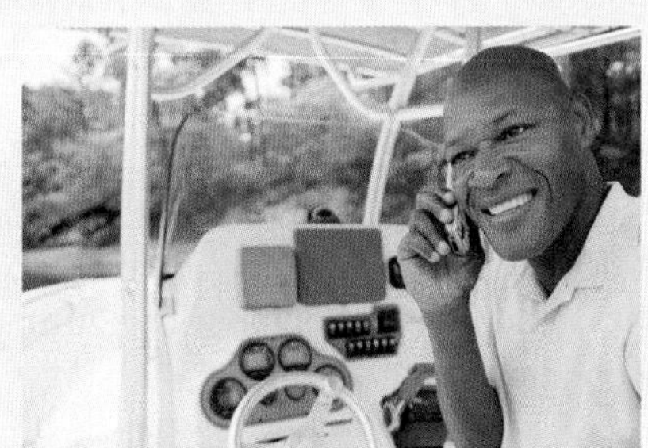

99. Jack and Jill Ahearn begin to budget 25% for rent, 20% for food, 7% for education, 5% for clothing, 10% for transportation, 12% for travel and recreation, 4% for miscellaneous, and the remainder for savings. Jack takes home \$2850 per month, and Jill takes home \$42,300 per year. How much money will the couple save in a year?

100. The mileage on a hybrid car dropped from 42.8 miles per gallon in the city to 28.5 miles per gallon on the highway. Find the percent of decrease.

Chapter 6 Test

The Chapter Test Prep Videos with test solutions are available on DVD, in MyMathLab, and on YouTube —search "LialBasicCollegeMath" and click on "Channels."

Write each percent as a decimal and each decimal as a percent.

1. 65%

2. 0.8

3. 1.75

4. 0.875

5. 300%

6. 2%

Write each percent as a fraction in lowest terms.

7. 12.5%

8. 0.25%

Write each fraction or mixed number as a percent.

9. $\frac{3}{5}$

10. $\frac{5}{8}$

11. $2\frac{1}{2}$

Solve each problem.

12. 32 sacks is 4% of what number of sacks?

13. \$680 is what percent of \$3400?

14. There are still 100,000 households in the United States that do not have electricity. If this is 0.08% of the homes, find the total number of households. (*Source: Time* magazine.)

15. The price of a diamond engagement ring is \$3240 plus sales tax of $6\frac{1}{2}$%. Find the total cost of the engagement ring including sales tax.

16. An insurance company pays its salespeople on commission. If a commission of \$628 is earned on insurance sales of \$7850, find the rate of commission.

17. Attendance at the homecoming game decreased from 5760 fans last year to 4320 fans this year. Find the percent of decrease.

18. A problem includes last year's salary, this year's salary, and asks for the percent of increase. Explain how you would identify the part, the whole, and the percent in the problem. Show the percent proportion that you would use.

19. Write the formula used to find interest. Explain the difference in what to do if the time is expressed in months or in years. Write a problem that involves finding interest for 9 months and another problem that involves finding interest for $2\frac{1}{2}$ years. Use your own numbers for the principal and the rate. Show how to solve your problems.

Find the amount of discount and the sale price.

	Original Price	Rate of Discount	Amount of Discount	Sale Price
20.	$96	12%	________	________
21.	$280	32.5%	________	________

Find the simple interest on each loan.

	Principal	Rate	Time	Simple Interest
22.	$4200	6%	$1\frac{1}{2}$ years	________
23.	$6400	9%	4 months	________

24. Sita Lalchandani takes out a short-term loan of $19,200 to pay for her daughter's medical school expenses. The interest rate on the loan is 7%, and the loan is for 15 months. Find the total amount needed to repay the loan.

25. The River City School PTA Emergency Fund deposited $4000 at 6% interest compounded annually. Two years after the first deposit, they deposit another $5000, also at 6% interest compounded annually. Use the compound interest table.

(a) What total amount will they have 4 years after their first deposit? Round to the nearest dollar.

(b) What amount of interest will they have earned?

Chapters 1–6 Cumulative Review Exercises

First use front end rounding to round each number and estimate the answer. Then find the exact answer.

1. *Estimate:* $-$ ______ *Exact:* $75{,}078 - 46{,}090$

2. *Estimate:* $-$ ______ *Exact:* $7.8 - 3.5029$

3. *Estimate:* $\times$ ______ *Exact:* 6538×708

4. *Estimate:* $\times$ ______ *Exact:* 65.3×8.7

5. *Estimate:* ______ *Exact:* $43\overline{)38{,}786}$

6. *Estimate:* ______ *Exact:* $0.8\overline{)6.76}$

Use the order of operations to simplify each expression.

7. $6^2 - 3(6)$

8. $\sqrt{49} + 5 \cdot 4 - 8$

9. $9 + 6 \div 3 + 7(4)$

Round each number.

10. 7,583,281 to the nearest hundred-thousand

11. \$513.499 to the nearest dollar

12. \$362.735 to the nearest cent

Add, subtract, multiply, or divide as indicated. Write answers in lowest terms and as whole or mixed numbers when possible.

13. $5\frac{3}{4} + 7\frac{5}{8}$

14. $8\frac{3}{8} - 4\frac{1}{2}$

15. $36 \cdot \frac{4}{5}$

16. $12 \div \frac{3}{4}$

17. The size of a prison cell at Alcatraz Prison in the San Francisco Bay is 5 feet by 9 feet. The average size of a shark cage is 5 feet by $6\frac{1}{2}$ feet. How many more square feet are there in the floor of the prison cell than the shark cage? (*Source:* Discovery Channel, Monster Garage Factoid.)

18. To prepare for the state real estate exam, Mia Dawson studied $5\frac{1}{2}$ hours on the first day, $6\frac{1}{4}$ hours on the second day, $3\frac{3}{4}$ hours on the third day, and 7 hours on the fourth day. How many hours did she study altogether?

Write $<$ or $>$ to make each statement true.

19. $\frac{5}{8}$ ___ $\frac{2}{3}$

20. $\frac{8}{15}$ ___ $\frac{11}{20}$

21. $\frac{2}{3}$ ___ $\frac{7}{12}$

Simplify each expression. Use the order of operations as needed.

22. $\frac{2}{3}\left(\frac{7}{8} - \frac{1}{2}\right)$

23. $\frac{7}{8} \div \left(\frac{3}{4} + \frac{1}{8}\right)$

24. $\left(\frac{5}{6} - \frac{5}{12}\right) - \left(\frac{1}{2}\right)^2 \cdot \frac{2}{3}$

Write each fraction as a decimal. Round to the nearest thousandth if necessary.

25. $\frac{3}{4}$

26. $\frac{3}{8}$

27. $\frac{7}{12}$

28. $\frac{11}{20}$

Find the unknown value in each proportion.

29. $\frac{1}{5} = \frac{x}{30}$ **30.** $\frac{224}{32} = \frac{28}{x}$ **31.** $\frac{8}{x} = \frac{72}{144}$ **32.** $\frac{x}{120} = \frac{7.5}{30}$

Write each percent as a decimal. Write each decimal as a percent.

33. 3% **34.** 200% **35.** 0.87 **36.** 3.8

Write each percent as a fraction or mixed number in lowest terms. Write each fraction or mixed number as a percent.

37. 8% **38.** 62.5% **39.** 175% **40.** $\frac{7}{8}$ **41.** $4\frac{1}{5}$

Solve.

42. 65% of 3280 DVDs is how many DVDs?

43. $4\frac{1}{2}$% of what number of miles is 76.5 miles?

44. 252 hours is what percent of 180 hours?

45. Diane McKinney paid $29.90 in sales tax on a $460 purchase. What was the tax rate?

46. While enrolled in college full-time, Theresa Goldsmith is a part-time waitress. Her tips average 14%, and on a recent shift she had total sales of $837. Find the amount of her tips.

47. A box spring and mattress originally priced at $456 is discounted 45%. Find the amount of discount and the sale price.

48. A loan of $46,300 is made at 3% simple interest for 9 months. Find the total amount to be repaid.

Set up and solve a proportion for each problem.

49. Gerri Junso gives school readiness tests to 9 children in 4 hours. Find the number of children she can test in 20 hours.

50. If 12.5 ounces of Roundup weed and grass killer is needed to make 5 gallons of spray, how much Roundup is needed for 102 gallons of spray?

The number of existing single-family homes sold in four regions of the country in the same month of two separate years is shown in the figure below. Use this information to answer Exercises 51–54. Round answers to the nearest tenth of a percent if necessary.

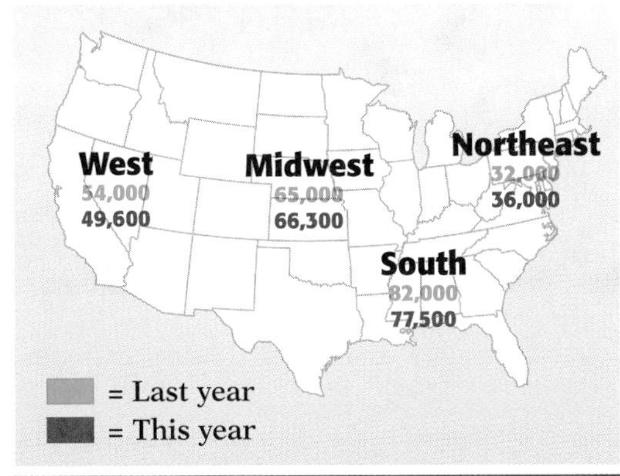

51. Find the percent of increase in sales in the northeastern region.

52. Find the percent of increase in sales in the midwestern region.

53. What is the percent of decrease in sales in the southern region?

54. What is the percent of decrease in sales in the western region?

Math in the Media

EDUCATIONAL TAX INCENTIVES

The government sponsors tax incentive programs to make education more affordable. To qualify for the programs, you must have an adjusted gross income below a certain level (most recently $90,000). You can find specific information at the Internal Revenue Service website (irs.gov).

- The Hope Scholarship offers 100% of the first $2000 spent for certain expenses, such as tuition and books, during the first year of college, plus 25% of the next $2000. The scholarship money is payable as a tax refund. The student cannot have completed the first two years of post-secondary education and must meet certain educational goals and workload criteria.

Suppose you are paying your own educational costs, and your adjusted gross income meets the guidelines to qualify for the Hope Scholarship. Your goals are to earn an Associate of Arts degree from a community college and then transfer to a state university to complete a Bachelor's degree. Tuition costs for resident students at American River Community College in California are used as an example of educational expenses.

Residents of the college district pay an enrollment fee of $46 per semester hour plus a parking permit fee of $30 each semester. Assume that you must study a total of 15 semester hours in developmental work in mathematics, reading, and writing, and an additional 60 semester hours to complete an Associate of Arts degree. You decide to limit your course load to 15 credit hours each semester. Assume that one course is 3 semester hours, and you will have to purchase books at an approximate cost of $95 per course.

1. How many semesters and how many courses will it take you to finish the requirements for an Associate of Arts degree?

2. What is the total cost to complete the Associate of Arts degree for (**a**) books and (**b**) tuition and fees?

3. (**a**) Calculate the total cost for enrollment fees, parking fees, and books to complete the Associate of Arts degree (5 semesters).
(**b**) What is the maximum tax incentive payable to you under the Hope Scholarship?

Appendix: Inductive and Deductive Reasoning

Inductive and Deductive Reasoning

OBJECTIVES

1. Use inductive reasoning to analyze patterns.
2. Use deductive reasoning to analyze arguments.
3. Use deductive reasoning to solve problems.

OBJECTIVE 1 Use inductive reasoning to analyze patterns. In many scientific experiments, conclusions are drawn from specific outcomes. After many repetitions and similar outcomes, the findings are generalized into statements that appear to be true. When general conclusions are drawn from specific observations, we are using a type of reasoning called **inductive reasoning.** The next examples illustrate this type of reasoning.

EXAMPLE 1 Using Inductive Reasoning

Find the next number in the sequence 3, 7, 11, 15, ...

To discover a pattern, calculate the difference between each pair of successive numbers.

$$7 - 3 = 4$$
$$11 - 7 = 4$$
$$15 - 11 = 4$$

Notice that the difference is always 4. Each number is 4 greater than the previous one. Thus, the next number in the pattern is $15 + 4$, or 19.

Work Problem 1 at the Side. ▶

EXAMPLE 2 Using Inductive Reasoning

Find the next number in this sequence.

$$7, 11, 8, 12, 9, 13, \ldots$$

The pattern in this example involves addition and subtraction.

$$7 + 4 = 11$$
$$11 - 3 = 8$$
$$8 + 4 = 12$$
$$12 - 3 = 9$$
$$9 + 4 = 13$$

To get the second number, we add 4 to the first number. To get the third number, we subtract 3 from the second number. To obtain subsequent numbers, we continue the pattern. The next number is $13 - 3 = 10$.

Work Problem 2 at the Side. ▶

1. Find the next number in the sequence 2, 8, 14, 20, ... Describe the pattern.

2. Find the next number in the sequence 6, 11, 7, 12, 8, 13, ... Describe the pattern.

Answers

1. 26; add 6 each time.
2. 9; add 5, subtract 4.

3 Find the next number in the sequence 2, 6, 18, 54, ... Describe the pattern.

EXAMPLE 3 Using Inductive Reasoning

Find the next number in the sequence 1, 2, 4, 8, 16, ...

Each number after the first is obtained by multiplying the previous number by 2. So the next number would be $16 \cdot 2 = 32$.

◀ **Work Problem 3 at the Side.**

EXAMPLE 4 Using Inductive Reasoning

(a) Find the next geometric shape in this sequence.

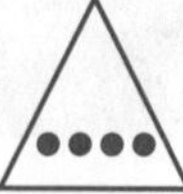

The figures alternate between a blue circle and a red triangle. Also, the number of dots increases by 1 in each subsequent figure. Thus, the next figure should be a blue circle with five dots inside it.

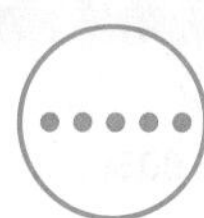

(b) Find the next geometric shape in this sequence.

4 Find the next shape in this sequence.

The first two shapes consist of vertical lines with horizontal lines at the bottom extending first *left* and then *right.* The third shape is a vertical line with a horizontal line at the top extending to the *left.* Therefore, the next shape should be a vertical line with a horizontal line at the top extending to the *right.*

◀ **Work Problem 4 at the Side.**

OBJECTIVE 2 Use deductive reasoning to analyze arguments. In the previous discussion, specific cases were used to find patterns and predict the next event. There is another type of reasoning called **deductive reasoning,** which moves from general cases to specific conclusions.

EXAMPLE 5 Using Deductive Reasoning

Does the conclusion follow from the premises in this argument?

All Hondas are automobiles. ← Premise

All automobiles have horns. ← Premise

$\therefore$ All Hondas have horns. ← Conclusion

In this example, the first two statements are called *premises* and the third statement (below the line) is called a *conclusion.* The symbol $\therefore$ is a mathematical symbol meaning "**therefore.**" The entire set of statements is called an *argument.*

Continued on Next Page

Answers

3. 162; multiply by 3.

4.

The focus of deductive reasoning is to determine whether the conclusion follows (is valid) from the premises. A set of circles called **Euler circles** is used to analyze the argument.

In **Example 5,** the statement "All Hondas are automobiles" can be represented by two circles, one for Hondas and one for automobiles. Note that the circle representing Hondas is totally inside the circle representing automobiles because the first premise states that *all* Hondas are automobiles.

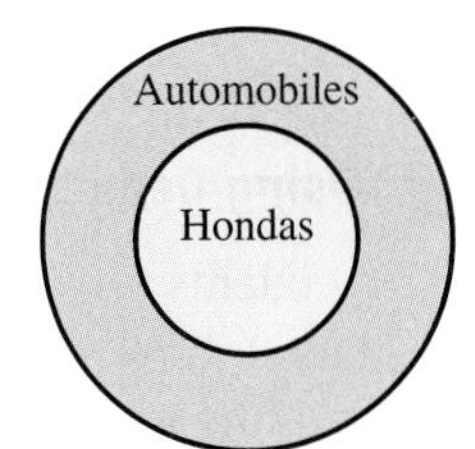

Now, a circle is added to represent the second statement, vehicles with horns. This circle must completely surround the circle representing automobiles because the second premise states that *all* automobiles have horns.

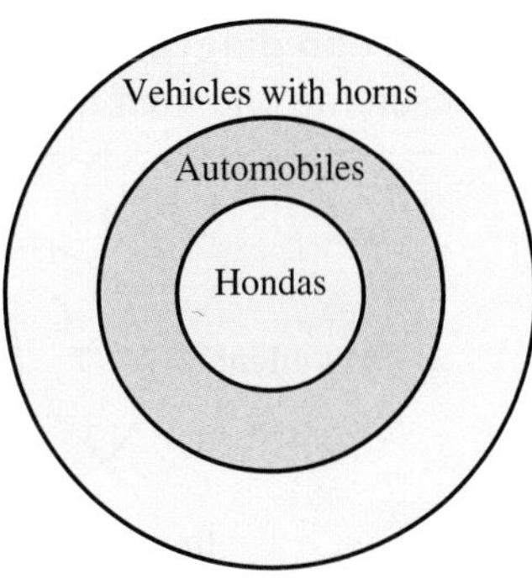

To analyze the conclusion, notice that the circle representing Hondas is *completely* inside the circle representing vehicles with horns. Therefore, it must follow that all Hondas have horns. ***The conclusion is valid.***

Work Problem **5** at the Side. ▶

5 Does the conclusion follow from the premises in the following argument?

All cars have four wheels.
All Fords are cars.

∴ All Fords have four wheels.

6 Does each conclusion follow from the premises?

(a) All animals are wild.
All cats are animals.

∴ All cats are wild.

(b) All students use math.
All adults use math.

∴ All adults are students.

EXAMPLE 6 Using Deductive Reasoning

Does the conclusion follow from the premises in this argument?

All tables are round.
All glasses are round.

∴ All glasses are tables.

Use Euler circles. Draw a circle representing tables *inside* a circle representing round objects, because the first premise states that *all* tables are round.

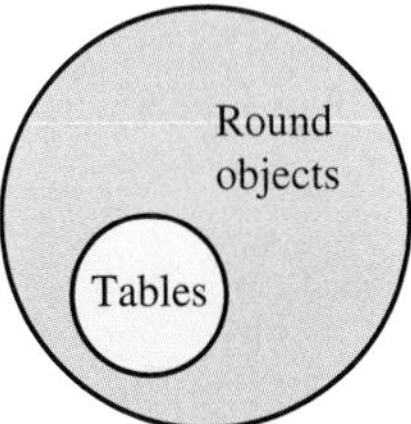

The second statement requires that a circle representing glasses must now be drawn inside the circle representing round objects, but not necessarily inside the circle representing tables. Therefore, the conclusion does ***not*** follow from the premises. This means that ***the conclusion is invalid.***

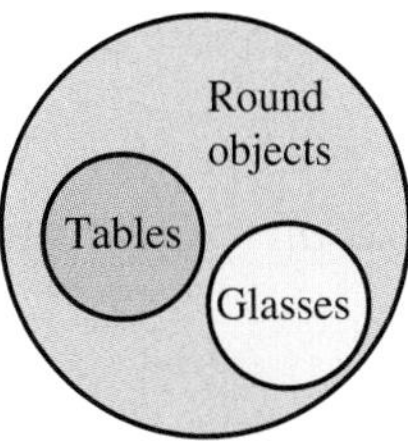

Work Problem **6** at the Side. ▶

Answers

5. The conclusion follows from the premises; it is valid.

6. (a) The conclusion follows from the premises; it is valid.

(b) The conclusion does *not* follow from the premises; it is invalid.

7 In a college class of 100 students, 35 take both math and history, 50 take history, and 40 take math. How many take neither math nor history? Draw a Venn diagram.

8 A Chevy, BMW, Cadillac, and Ford are parked side by side. The known facts are:

(a) The Ford is on the right end.

(b) The BMW is next to the Cadillac.

(c) The Chevy is between the Ford and the Cadillac.

Which car is parked on the left end?

Answers

7.

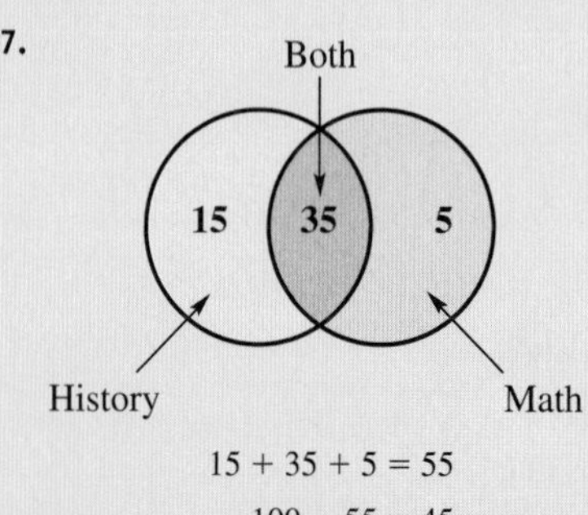

$15 + 35 + 5 = 55$

$100 - 55 = 45$

45 students take neither math nor history.

8. BMW

OBJECTIVE 3 **Use deductive reasoning to solve problems.** Another type of deductive reasoning problem occurs when a set of facts is given in a problem and a conclusion must be drawn using these facts.

EXAMPLE 7 Using Deductive Reasoning

There were 25 students enrolled in a ceramics class. During the class, 10 of the students made a bowl and 8 students made a birdbath. Three students made both a bowl and a birdbath. How many students did not make either a bowl or a birdbath?

This type of problem is best solved by organizing the data using a drawing called a **Venn diagram.** Two overlapping circles are drawn, with each circle representing one item made by the students, as shown below.

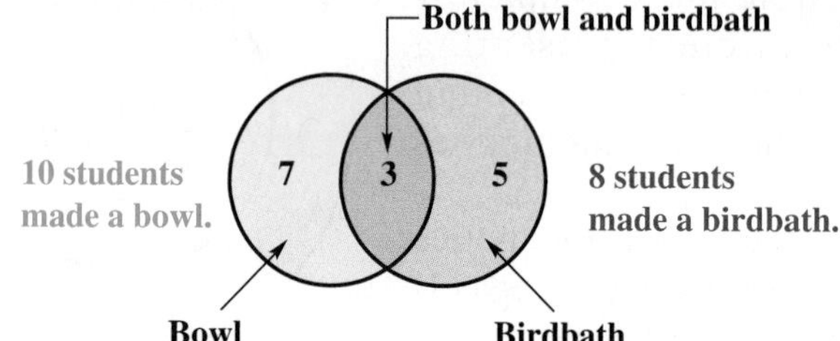

In the region where the circles overlap, write the number of students who made *both* items, namely, 3. In the remaining portion of the birdbath circle, write the number 5, which when added to 3 will give the total number of students who made a birdbath, namely, 8. In a similar manner, write 7 in the remaining portion of the bowl circle, since $7 + 3 = 10$, the total number of students who made a bowl. The total of all three numbers written in the circles is 15. Since there were 25 students in the class, this means $25 - 15$ or 10 students did not make either a birdbath or a bowl.

◀ **Work Problem 7 at the Side.**

EXAMPLE 8 Using Deductive Reasoning

Four cars in a race finish first, second, third, and fourth. The following facts are known.

(a) Car A beat Car C.

(b) Car D finished between Cars C and B.

(c) Car C beat Car B.

In which order did the cars finish?

To solve this type of problem, it is helpful to use a line diagram.

1. *Write A before C,* because Car A beat Car C (fact **a**).

A C

2. *Write B after C,* because Car C beat Car B (fact **c**).

A C B

3. *Write D between C and B,* because Car D finished between Car C and Car B (fact **b**).

The correct order of finish is shown below.

A C D B

◀ **Work Problem 8 at the Side.**

Appendix Exercises

FOR EXTRA HELP

Download the MyDashBoard App

MyMathLab®

Find the next number in each sequence. Describe the pattern in each sequence. **See Examples 1–3.**

1. 2, 9, 16, 23, 30, . . .

2. 5, 8, 11, 14, 17, . . .

3. 0, 10, 8, 18, 16, . . .

4. 3, 9, 7, 13, 11, . . .

5. 1, 2, 4, 8, . . .

6. 1, 4, 16, 64, . . .

7. 1, 3, 9, 27, 81, . . .

8. 3, 6, 12, 24, 48, . . .

9. 1, 4, 9, 16, 25, . . .

10. 6, 7, 9, 12, 16, . . .

Find the next shape in each sequence. **See Example 4.**

11. 

12.  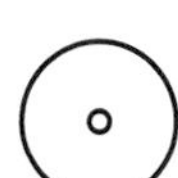 

13. 

14. 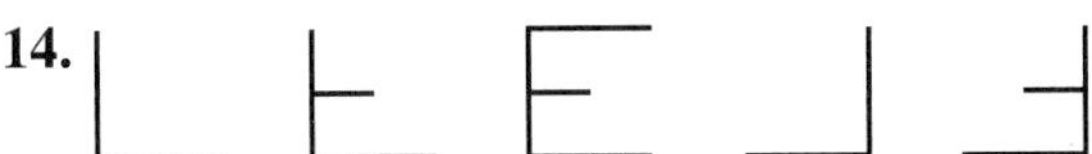

For each argument, draw Euler circles and then state whether or not the conclusion follows from the premises. **See Examples 5 and 6.**

15. All animals are wild.
All lions are animals.
∴ All lions are wild.

16. All students are hard workers.
All business majors are students.
∴ All business majors are hard workers.

17. All teachers are serious.
All mathematicians are serious.
∴ All mathematicians are teachers.

18. All boys ride bikes.
All Americans ride bikes.
∴ All Americans are boys.

Solve each application problem. **See Examples 7 and 8.**

19. In a given 30-day period, a husband watched television 20 days and his wife watched television 25 days. If they watched television together 18 days, how many days did neither watch television? Draw a Venn diagram.

20. In a class of 40 students, 21 students take both calculus and physics. If 30 students take calculus and 25 students take physics, how many do not take either calculus or physics? Draw a Venn diagram.

21. Tom, Dick, Mary, and Joan all work for the same company. One is a secretary, one is a computer operator, one is a receptionist, and one is a mail clerk.

(a) Tom and Joan eat dinner with the computer operator.

(b) Dick and Mary carpool with the secretary.

(c) Mary works on the same floor as the computer operator and the mail clerk.

Who is the computer operator?

22. Four cars—a Ford, a Buick, a Mercedes, and an Audi—are parked in a garage in four spaces.

(a) The Ford is in the last space.

(b) The Buick and Mercedes are next to each other.

(c) The Audi is next to the Ford but not next to the Buick.

Which car is in the first space?

Answers to Selected Exercises

In this section we provide the answers that we think most students will obtain when they work the exercises using the methods explained in the text. If your answer does not look exactly like the one given here, it is not necessarily wrong. In many cases there are equivalent forms of the answer that are correct. For example, if the answer section shows $\frac{3}{4}$ and your answer is 0.75 you have obtained the right answer but written it in a different (yet equivalent) form. Unless the directions specify otherwise, 0.75 is just as valid an answer as $\frac{3}{4}$.

In general, if your answer does not agree with the one given in the text, see whether it can be transformed into the other form. If it can, then it is the correct answer. If you still have doubts, talk with your instructor.

CHAPTER 1 Whole Numbers

SECTION 1.1 (pages 8–9)

1. c **2.** b **3.** 1; 0 **5.** 8; 2 **7.** 687 **8.** 321 **9.** 3; 561; 435 **11.** Evidence suggests that this is true. It is common to count using fingers. **12.** No doubt there is a relationship here. One answer might be that people could count using their fingers and toes and, therefore, thought of them as numbers or digits. **13.** false (no "and") **14.** true **15.** three hundred forty-six thousand, nine **17.** twenty-five million, seven hundred fifty-six thousand, six hundred sixty-five **19.** 63,163 **21.** 10,000,223 **23.** 3,200,000 parachute jumps **25.** 50,051,507 cans **27.** 54,750,000 Hot Wheels **29.** 800,000,621,020,215 **31.** public transportation; six million, sixty-nine thousand, five hundred eighty-nine workers **33.** seven million, eight hundred ninety-four thousand, nine hundred eleven workers

SECTION 1.2 (pages 18–21)

1. 97 **3.** 89 **5.** 889 **7.** 889 **9.** 7785 **11.** correct **12.** incorrect **13.** incorrect **14.** correct **15.** 78,446 **17.** 8928 **19.** 59,224 **21.** correct **22.** incorrect **23.** correct **24.** incorrect **25.** incorrect **27.** 145 **29.** 102 **31.** 1651 **33.** 1154 **35.** 413 **37.** 1771 **39.** 1410 **41.** 6391 **43.** 11,624 **45.** 17,611 **47.** 15,954 **49.** 10,648 **51.** 15,594 **53.** 11,557 **55.** 12,078 **57.** 4250 **59.** 12,268 **61.** correct **63.** incorrect; should be 769 **65.** correct **67.** incorrect; should be 11,577 **69.** correct **71.** Changing the order in which numbers are added does not change the sum. You can add from bottom to top when checking addition. **72.** Grouping the addition of numbers in any order does not change the sum. You can add numbers in any order. For example, you can add pairs of numbers that add to 10. **73.** 33 miles **75.** 38 miles **77.** $16,342 **79.** 699 people **81.** 20,157 students **83.** 970 ft **85.** 72 ft **87.** 9421 **88.** 1249 **89.** 77,762 **90.** 22,267 **91.** 9,994,433 **92.** 3,334,499 **93.** Write the largest digits on the left, using the smaller digits as you move right. **94.** Write the smallest digits on the left, using the larger digits as you move right.

SECTION 1.3 (pages 28–31)

1. 32 **2.** 4 **3.** 86 **4.** 35 **5.** 17 **7.** 213 **9.** 101 **11.** 7111 **13.** 3412 **15.** 2111 **17.** 13,160 **19.** 41,110 **21.** correct **23.** incorrect; should be 62 **25.** incorrect; should be 121 **27.** correct **29.** incorrect; should be 7222 **31.** minuend; subtrahend; below **32.** **(b) and (d)** **33.** 38 **35.** 45 **37.** 19 **39.** 281 **41.** 519 **43.** 7059 **45.** 7589 **47.** 7 **49.** 19 **51.** 2833 **53.** 7775 **55.** 503 **57.** 156 **59.** 2184 **61.** 5687 **63.** 19,038 **65.** 31,556 **67.** 6584 **68.** check **69.** subtraction **70.** addition **71.** correct **73.** correct **75.** correct **77.** correct **79.** Possible answers are 1. $3 + 2 = 5$ could be changed to $5 - 2 = 3$ or $5 - 3 = 2$ 2. $6 - 4 = 2$ could be changed to $2 + 4 = 6$ or $4 + 2 = 6$. **80.** No, you cannot. Numbers must be subtracted in the order given. The difference found in subtraction is the result of subtracting the subtrahend from the minuend. Changing the order of the minuend and subtrahend does change the answer. **81.** 47 calories **83.** 69 more tornadoes **85.** 246 feet **87.** 3270 jobs eliminated **89.** 9539 flags **91.** $263 **93.** 141 miles **95.** 138,888 patients **97.** 310 fewer calories; 26 fewer fat grams **99.** 370 calories; 10 fat grams

SECTION 1.4 (pages 38–41)

1. the same **2.** commutative **3.** zero **4.** zeros; right **5.** 24 **7.** 0 **9.** 24 **11.** 40 **13.** Factors may be multiplied in any order to get the same answer. They are the same; you may add or multiply numbers in any order. **14.** You may shift the parentheses in a multiplication problem. Just as in addition, the different grouping results in the same answer. **15.** 210 **17.** 238 **19.** 3210 **21.** 1872 **23.** 8612 **25.** 10,084 **27.** 20,488 **29.** 258,447 **31.** 86; 172; two; 17,200 **32.** 45; 3510; three; 3,510,000 **33.** 480 **35.** 2220 **37.** 3600 **39.** 3750 **41.** 65,400 **43.** 270,000 **45.** 86,000,000 **47.** 48,500 **49.** 720,000 **51.** 1,940,000 **53.** 476 **55.** 2400 **57.** 3735 **59.** 2378 **61.** 6164 **63.** 15,792 **65.** 21,665 **67.** 15,730 **69.** 82,320 **71.** 183,996 **73.** 2,468,928 **75.** 66,005 **77.** 86,028 **79.** 19,422,180 **81.** 2,278,410 **83.** To multiply by 10, 100, or 1000, just attach one, two, or three zeros, respectively, to the number you are multiplying and that's your answer.

84. $\begin{array}{r} 291 \\ \times\ 307 \\ \hline 2037 \\ 000 \\ 873 \\ \hline 89{,}337 \end{array}$ $\quad\begin{array}{r} 291 \\ \times\ 307 \\ \hline 2037 \\ 8730 \\ \hline 89{,}337 \end{array}$ **85.** 3000 balls **87.** $2328 **89.** 24,090 gallons **91.** $600 **93.** $1560 **95.** $112,888 **97.** 50,568 **99.** 38,250 trees **101.** 7,746,712 people **103.** $14,160 **105.** **(a)** 452 **(b)** 452 **106.** commutative **107.** **(a)** 281 **(b)** 281 **108.** associative **109.** **(a)** 15,840 **(b)** 15,840 **110.** commutative **111.** **(a)** 6552 **(b)** 6552 **112.** associative **113.** No. Some examples are: $7 - 5 = 2$, but $5 - 7$ does not equal 2; $12 - 6 = 6$, but $6 - 12$ does not equal 6; $(8 - 2) - 5 = 1$, but $8 - (2 - 5)$ does not equal 1. **114.** No. Some examples are: $10 \div 2 = 5$, but $2 \div 10$ does not equal 5; $(16 \div 8) \div 2 = 1$, but $16 \div (8 \div 2)$ does not equal 1.

SECTION 1.5 (pages 51–54)

1. • ; × ; ( ) ( ) **2.** ÷ ; $\overline{)}$; —— (fraction bar) **3.** $\begin{array}{r} 6 \\ 4\overline{)24} \end{array}$ $\frac{24}{4} = 6$ **5.** $\begin{array}{r} 5 \\ 9\overline{)45} \end{array}$ $45 \div 9 = 5$ **7.** $16 \div 2 = 8$ $\frac{16}{2} = 8$ **9.** number **10.** 0

11. 1 **13.** 7 **15.** undefined **17.** 24 **19.** 0 **21.** undefined **23.** 2; 3; 5; 10 **24.** 3; 5 **25.** 2; 3 **26.** 2; 3; 5; 10 **27.** 25 **29.** 18 **31.** 304 **33.** 627 R1 **35.** 1522 R5 **37.** 309 **39.** 3005 **41.** 5006 **43.** 811 R1 **45.** 2589 R2 **47.** 2630 **49.** 12,458 R3 **51.** 10,253 R5 **53.** 18,377 R6 **55.** correct **57.** incorrect; should be 1908 R1 **59.** incorrect; should be 670 R2 **61.** incorrect; should be 3568 R1 **63.** correct **65.** correct **67.** incorrect; should be 9628 R3 **69.** correct **71.** Multiply the quotient by the divisor and add any remainder. The result should be the dividend. **72.** Three choices might be: A number is divisible by 2 if it ends in a 0, 2, 4, 6, or 8; A number is divisible by 5 if it ends in 0 or 5; A number is divisible by 10 if it ends in 0. **73.** 328 tables **75.** 9600 each hour **77.** $48,500 **79.** 165 locations **81.** $1,137,500 **83.** 862,500 berries **85.** ✓✓✓✓ **87.** ✓XXX **89.** XX✓X **91.** X✓XX **93.** ✓✓XX **95.** XXXX

SECTION 1.6 (pages 60–62)

1. 53 **3.** 250 **5.** 120 R7 **7.** 1105 R5 **9.** 7134 R12 **11.** 900 R100 **13.** 73 R5 **15.** 476 R15 **17.** 2407 R1 **19.** 1146 R15 **21.** 3331 R82 **23.** 850 **25.** incorrect; should be 101 R14 **27.** incorrect; should be 658 **29.** incorrect; should be 62 **31.** 117 episodes **33.** 56 floor clocks **35.** $355 **37.** 43,200 rings **39.** **(a)** 648 each year **(b)** less than 2 each day **41.** $0 **42.** 0 **43.** undefined **44.** impossible; if you have 6 cookies, it is not possible to divide them among 0 people. **45.** **(a)** 14 **(b)** 17 **(c)** 38 **46.** Yes. Some examples are 18 • 1 = 18; 26 • 1 = 26; 43 • 1 = 43. **47.** **(a)** 3200 **(b)** 320 **(c)** 32 **48.** Drop the same number of zeros that appear in the divisor. The result is the quotient. With the divisor 10, drop one 0; with 100, drop two zeros; with 1000, drop three zeros.

SUMMARY EXERCISES Whole Numbers Computation (pages 63–65)

1. 3; 4 **2.** 6; 0 **3.** 1; 6 **4.** eighty-six thousand, two **5.** four hundred twenty-five million, two hundred eight thousand, seven hundred thirty-three **6.** 97 **7.** 905 **8.** 21 **9.** 409 **10.** 17,573 **11.** 82,164 **12.** 677 **13.** 37,674 **14.** 35,889 **15.** 560 **16.** 5600 **17.** 350,000 **18.** 252,000 **19.** 1,238,201 **20.** 5,549,375 **21.** 1 **22.** 0 **23.** undefined **24.** 15 **25.** 56 **26.** 0 **27.** 96 **28.** 304 **29.** 2750 R2 **30.** 761 R3 **31.** 3380 **32.** 220,545 **33.** 2016 **34.** 1476 **35.** 78 **36.** 210 **37.** 18,038,816 **38.** 506 R28 **39.** 52 **40.** 1208 R3 **41.** 573 R3 **42.** 41 **43.** 208,530 **44.** 1,101,744 **45.** 2,154,000,000 **46.** 40,000,000 **47.** 1520 hours **48.** 7500 tons **49.** Four hundred twenty-two million, four hundred thousand feet **50.** **(a)** More than a mile; 3701 ft more **(b)** Less than a mile; 1080 ft less

SECTION 1.7 (pages 72–75)

1. ten **2.** ten **3.** hundred **4.** hundred **5.** thousand **6.** thousand **7.** ten-thousand **8.** ten-thousand **9.** 860 **11.** 6800 **13.** 28,500 **15.** 6000 **17.** 16,000 **19.** 8,000,000,000 **21.** 600,000 **23.** 5,000,000 **25.** 4480; 4500; 4000 **27.** 3370; 3400; 3000 **29.** 6050; 6000; 6000 **31.** 5340; 5300; 5000 **33.** 19,540; 19,500; 20,000 **35.** 26,290; 26,300; 26,000 **37.** 93,710; 93,700; 94,000 **39.** 1. Locate the place to be rounded and underline it. 2. Look only at the next digit to the right. If this digit is 5 or more, increase the underlined digit by 1. 3. Change all digits to the right of the underlined place to zeros. **40.** 1. Locate the place to be rounded and underline it. 2. Look only at the next digit to the right. If this digit is 4 or less do not change the underlined digit. 3. Change all digits to the right of the underlined place to zeros **41.** 30 60 50 80 220; 219 **43.** 80 40 40; 35 **45.** 70 30 2100; 2278 **47.** 900 700 400 800 2800; 2828 **49.** 900 400 500; 435 **51.** 800 400 320,000; 282,000 **53.** 8000 60 700 4000 12,760; 12,605 **55.** 700 500 200; 158 **57.** 900 30 27,000; 27,231 **59.** Perhaps the best explanation is that 3492 is closer to 3500 than 3400, but 3492 is closer to 3000 than 4000. **60.** Rounding numbers usually allows for faster calculation and results in an estimated answer prior to getting an exact answer. One example is

	400		432
	− 200		− 209
Estimate:	200	*Exact:*	223

61. 80 million people; 310 million people **63.** 349,000 streets; 350,000 streets **65.** 39,840,000 tickets; 39,800,000 tickets; 40,000,000 tickets **67.** 19,266,000 players; 19,270,000 players; 19,300,000 players **69.** 71,500 **70.** 72,499 **71.** 7500 **72.** 8499 **73.** 3930; 11,240; 15,970; 17,920; 534,880; 2,788,000 **74.** 4000; 10,000; 20,000; 20,000; 500,000; 3,000,000 **75.** **(a)** When using front end rounding, all digits are 0 except the first digit. These numbers are easier to work with when estimating answers. **(b)** Sometimes when using front end rounding; the estimated answer can vary greatly from the exact answer.

SECTION 1.8 (pages 78–81)

1. exponent 2; base 3 **2.** exponent 3; base 2 **3.** exponent 2; base 5 **4.** exponent 2; base 4 **5.** 2; 8; 64 **7.** 2; 15; 225 **9.** 4 **11.** 8 **13.** 10 **15.** 12 **17.** false: 5^2 means that 5 is used as a factor 2 times, $5^2 = 5 \cdot 5 = 25$ **18.** false: $4^2 = 4 \cdot 4 = 16$ **19.** false: 1 raised to any power is 1. In this example, $1 \cdot 1 \cdot 1 = 1$ **20.** false: a number raisecd to the first power is the number itself, so $6^1 = 6$ **21.** 36; 36 **23.** 625; 625 **25.** 10,000; 10,000 **27.** A perfect square is the square of a whole number. The number 25 is the square of 5 because $5 \cdot 5 = 25$. The number 50 is not a perfect square. There is no whole number that can be squared to get 50. **28.** 1. Do all operations inside parentheses or other grouping symbols. 2. Simplify any expressions with exponents and find any square roots. 3. Multiply or divide proceeding from left to right. 4. Add or subtract proceeding from left to right. **29.** true **30.** true **31.** false: $6 + 8 \div 2 = 10$. Multiplications and divisions and performed from left to right, then additions and subtractions. **32.** false: $4 + 5(6 - 4) = 4 + 5 \cdot 2 = 4 + 10 = 14$. A common error is adding 4 to 5 first and then multiplying by 2. Follow the order of operations. **33.** 12 **35.** 20 **37.** 45 **39.** 63 **41.** 118 **43.** 22 **45.** 30 **47.** 102 **49.** 9 **51.** 63 **53.** 33 **55.** 70 **57.** 7 **59.** 17 **61.** 55 **63.** 108 **65.** 26 **67.** 26 **69.** 27 **71.** 16 **73.** 16 **75.** 21 **77.** 7 **79.** 20 **81.** 14 **83.** 25 **85.** 16 **87.** 23 **89.** 233

SECTION 1.9 (pages 86–89)

1. 8; 500; $8 \cdot 500 = 4000$ **2.** $8\frac{1}{2}$; 500; $8\frac{1}{2} \cdot 500 = 4250$ **3.** 4500 stores **5.** Dollar General; about 5750 stores **7.** 1250 fewer stores **9.** 100 adults **10.** 7 carrer paths **11.** 9 people **13.** **(a)** Saw ad **(b)** 25 people **15.** 9 people **17.** 2014; 7000 installations **19.** 4500 installations **21.** Possible answers are 1. shortage of units to install

2. lack of qualified workers 3. poor economy 4. less demand for solar products. **22.** Possible answers are 1. greater demand 2. more units available 3. many qualified workers 4. tax incentives for solar projects. **23.** $(7 - 2) \cdot 3 - 6$ **24.** $(4 + 2) \cdot (5 + 1)$ **25.** $36 \div (3 \cdot 3) \cdot 4$ **26.** $56 \div (2 \cdot 2 \cdot 2) + \frac{0}{6}$ **27. (a)** $7920 + 1320 + 2640 + (5280 - 1320 - 1320) + 2640 + 1320 + 7920 + 5280$ **(b)** $31,680 \times 3 = 95,040$ ft **(c)** 18 miles

SECTION 1.10 (pages 94–97)

1. (c) There are only 24 hours in a day. **2. (b)** $100 per hour is not reasonable **3. (a)** $5 is reasonable **4. (b)** 25 mpg is reasonable **5.** *Estimate:* 600 + 900 + 1000 + 800 + 2000 = 5300 sandwiches; *Exact:* 5208 sandwiches **7.** *Estimate:* $300 − $200 = $100 saved; *Exact:* $104 saved **9.** *Estimate:* 200 × 20 = 4000 *kits*; *Exact:* 5664 kits **11.** *Estimate:* 3000 ÷ 700 ≈ 4 toys; *Exact:* 4 toys **13.** *Estimate:* $30 × 5 = $150; *Exact:* $170 **15.** *Estimate:* 100,000 + 300,000 = 400,000 deaths: *Exact:* 360,222 deaths **17. (a)** *Estimate:* 300,000 − 100,000 = 200,000 deaths; *Exact:* 140,082 deaths **19.** 100,000 + 300,000 + 90,000 + 200,000 = 690,000 deaths *Exact:* 618,642 deaths **21.** *Estimate:* $2000 − $700 − $300 − $400 − $200 − $200 = $200; *Exact:* $350 **23.** *Estimate:* 40,000 × 100 = 4,000,000 square feet; *Exact:* 6,011,280 square feet **25.** *Estimate:* $300 + $200 + $100 + $70 + $400 + $200 = $1270; *Exact:* $1254 **27.** *Estimate:* $300 + $70 + $400 + $200 = $970; $970 − $800 = $170; *Exact:* $185 **29.** *Estimate:* ($1000 × 6) + ($900 × 20) = $24,000; *Exact:* $20,961 **31.** Possible answers are Addition: more; total; gain of. Subtraction: less; loss of; decreased by. Multiplication: twice; of; product. Division: divided by; goes into; per. Equals: is; are. **32.** 1. Read the problem carefully. 2. Work out a plan. 3. Estimate a reasonable answer. 4. Solve the problem. 5. State the answer. 6. Check your work. **33.** $20,009 **35.** 2477 pounds **37.** $378 **39.** $375 **41.** 20 seats

Chapter 1 REVIEW EXERCISES (pages 103–110)

1. 6; 573 **2.** 36; 215 **3.** 105; 724 **4.** 1; 768; 710; 618 **5.** seven hundred twenty-eight **6.** fifteen thousand, three hundred ten **7.** three hundred nineteen thousand, two hundred fifteen **8.** sixty-two million, five hundred thousand, five **9.** 10,008 **10.** 200,000,455 **11.** 110 **12.** 121 **13.** 5464 **14.** 15,657 **15.** 10,986 **16.** 9845 **17.** 40,602 **18.** 49,855 **19.** 36 **20.** 27 **21.** 189 **22.** 184 **23.** 6849 **24.** 4327 **25.** 224 **26.** 25,866 **27.** 49 **28.** 0 **29.** 32 **30.** 64 **31.** 45 **32.** 42 **33.** 56 **34.** 81 **35.** 40 **36.** 45 **37.** 48 **38.** 8 **39.** 0 **40.** 42 **41.** 48 **42.** 0 **43.** 84 **44.** 368 **45.** 522 **46.** 98 **47.** 5000 **48.** 2992 **49.** 5396 **50.** 45,815 **51.** 14,912 **52.** 20,160 **53.** 465,525 **54.** 174,984 **55.** 875 **56.** 2368 **57.** 1176 **58.** 5100 **59.** 15,576 **60.** 30,184 **61.** 887,169 **62.** 500,856 **63.** $360 **64.** $1064 **65.** $20,352 **66.** $684 **67.** 14,000 **68.** 23,800 **69.** 206,800 **70.** 318,500 **71.** 128,000,000 **72.** 90,300,000 **73.** 5 **74.** 7 **75.** 6 **76.** 2 **77.** 6 **78.** 4 **79.** 7 **80.** 0 **81.** undefined **82.** 0 **83.** 8 **84.** 9 **85.** 82 **86.** 98 **87.** 4422 **88.** 352 **89.** 150 R4 **90.** 124 R25 **91.** 820 **92.** 15,200 **93.** 21,000 **94.** 70,000 **95.** 3490; 3500; 3000 **96.** 20,070; 20,100; 20,000 **97.** 98,200; 98,200; 98,000 **98.** 352,120; 352,100; 352,000 **99.** 4 **100.** 7 **101.** 12 **102.** 14 **103.** 3; 7; 343 **104.** 6; 3; 729 **105.** 3; 5; 125 **106.** 5; 4; 1024 **107.** 34 **108.** 26 **109.** 9 **110.** 4 **111.** 9 **112.** 6 **113.** 8 parents **114.** 5 parents **115.** Keeping bedroom clean; 25 parents **116.** Hanging up wet bath towels; 3 parents **117.** *Estimate:* 40 million × 400 = 16,000 million or 16,000,000,000 checks; *Exact:* 14,600 million or 14,600,000,000 checks **118.** *Estimate:* 1000 × 60 = 60,000 revolutions; *Exact:* 84,000 revolutions **119.** *Estimate:* 40,000,000 − 30,000,000 = 10,000,000; *Exact:* 12,073,57 people **120.** *Estimate:* 700,000 − 600,000 = 100,000; *Exact:* 153,223 people **121.** *Estimate:* $3000 + (8 × $90) + 200 = $3920; *Exact:* $3583 **122.** *Estimate:* $60 + ($90 × $2) = $240; *Exact:* $233 **123.** *Estimate:* (30 × $2000) + (30 × $900) = $87,000; *Exact:* $74,052 **124.** *Estimate:* (60 × $20) + (20 × $7) = $1340; *Exact:* $1139 **125.** *Estimate:* $600 − $100 = $500; *Exact:* = $513 **126.** *Estimate:* $2000 − $500 − $400 = $1100; *Exact:* $1019 **127.** *Estimate:* 9000 ÷ 200 = 45 pounds; *Exact:* 50 pounds **128.** *Estimate:* 30,000 ÷ 1000 = 30 hr; *Exact:* 33 hr **129.** *Estimate:* 30,000 ÷ 600 = 50 acres; *Exact:* 52 acres **130.** *Estimate:* 6000 ÷ 200 = 30 homes; *Exact:* 32 homes **131.** 332 **132.** 448 **133.** 253 **134.** 588 **135.** 1041 **136.** 1661 **137.** 32,062 **138.** 24,947 **139.** 3 **140.** 7 **141.** 93,635 **142.** 83,178 **143.** undefined **144.** 7 **145.** 6900 **146.** 2310 **147.** 1,079,040 **148.** 130,212 **149.** 108 **150.** 207 **151.** three hundred seventy-six thousand, eight hundred fifty-three **152.** four hundred eight thousand, six hundred ten **153.** 8700 **154.** 401,000 **155.** 8 **156.** 9 **157.** $5544 **158.** $31,080 **159.** $2288 **160.** $15,782 **161.** 468 cards **162.** 52,700 pounds **163.** $280 **164.** $114,635 **165.** $1905 **166.** $12,420 **167.** 1467 ft **168.** 293 ft **169. (a)** 11,958 ft **(b)** greater than two miles by 1398 ft **170.** more than 4 football fields (a little less than 5)

Chapter 1 TEST (pages 111–112)

1. nine thousand, two hundred five **2.** twenty-five thousand, sixty-five **3.** 426,005 **4.** 8530 **5.** 112,630 **6.** 1045 **7.** 6206 **8.** 168 **9.** 171,000 **10.** 1615 **11.** 4,450,743 **12.** 7047 **13.** undefined **14.** 458 R5 **15.** 160 **16.** 6350 **17.** 76,000 **18.** 41 **19.** 28 **20.** *Estimate:* $500 + $500 + $500 + $400 − $800 = $1100; *Exact:* $1140 **21.** *Estimate:* 90,000 ÷ 400 = 225 acres; *Exact:* 231 acres **22.** *Estimate:* $2000 − $500 − $200 − $200 = $1100; *Exact:* $948 **23.** *Estimate:* (50 × 60 × 4) + (40 × 60 × 3) = 19,200 chicks; *Exact:* 18,000 chicks **24.** 1. Locate the place to which you are rounding and underline it. 2. Look only at the next digit to the right. If this digit is a 4 or less, do not change the underlined digit. If the digit is 5 or more, increase the underlined digit by 1. 3. Change all digits to the right of the underlined place to zeros. Each person's rounding example will vary. **25.** 1. Read the problem carefully. 2. Work out a plan. 3. Estimate a reasonable answer. 4. Solve the problem. 5. State the answer. 6. Check your work.

CHAPTER 2 Multiplying and Dividing Fractions

SECTION 2.1 (pages 116–118)

1. Numerator 4; Denominator 5 **2.** Numerator 5; Denominator 6 **3.** Numerator 9; Denominator 8 **4.** Numerator 7; Denominator 5

5. 3; 8 **6.** 7; 16 **7.** 5; 24 **8.** 24; 32 **9.** $\frac{3}{4}; \frac{1}{4}$ **11.** $\frac{1}{3}; \frac{2}{3}$ **13.** $\frac{7}{5}; \frac{3}{5}$ **15.** $\frac{5}{6}; \frac{4}{6}; \frac{2}{6}$ **17.** $\frac{8}{25}$ **19.** $\frac{303}{520}$

21. Proper $\frac{1}{3}, \frac{5}{8}, \frac{7}{16}$ Improper $\frac{8}{5}, \frac{6}{6}, \frac{12}{2}$

23. Proper $\frac{3}{4}, \frac{9}{11}, \frac{7}{15}$ Improper $\frac{3}{2}, \frac{5}{5}, \frac{19}{18}$

25. One possibility is
3 ← Numerator
4 ← Denominator

The denominator shows the number of equal parts in the whole and the numerator shows how many of the parts are being considered.
26. An example is $\frac{1}{2}$ as a proper fraction and $\frac{3}{2}$ as an improper fraction. A proper fraction has a numerator smaller than the denominator. An improper fraction has a numerator that is equal to or greater than the denominator. Drawings will vary.

SECTION 2.2 (pages 123–125)

1. true **2.** true **3.** false: Multiply 7 by 5 and add the numerator. $(7 \cdot 5) + 2 = 37; \frac{37}{5}$ **4.** false: Any mixed number can be charged to an improper fraction. The reverse is also true. **5.** false: You must add the numerator to the product of the whole number and the denominator. $6\frac{1}{2} = \frac{(6 \cdot 2) + 1}{2} = \frac{13}{2}$ **6.** true **7.** $\frac{5}{4}$ **9.** $\frac{23}{5}$ **11.** $\frac{17}{2}$ **13.** $\frac{81}{8}$ **15.** $\frac{43}{4}$ **17.** $\frac{29}{5}$ **19.** $\frac{43}{5}$ **21.** $\frac{54}{11}$ **23.** $\frac{131}{4}$ **25.** $\frac{221}{12}$ **27.** $\frac{269}{15}$ **29.** $\frac{187}{24}$

31. false: The denominator remains the same. So, $\frac{4}{3} = 1\frac{1}{3}$
32. false: A mixed number always has a value equal to or greater than a whole number. **33.** true **34.** true **35.** $1\frac{1}{3}$ **37.** $2\frac{1}{4}$ **39.** 9

41. $7\frac{3}{5}$ **43.** $15\frac{3}{4}$ **45.** $5\frac{2}{9}$ **47.** $8\frac{1}{8}$ **49.** $16\frac{4}{5}$ **51.** 28 **53.** $26\frac{1}{7}$
55. Multiply the denominator by the whole number and add the numerator. The result becomes the new numerator, which is placed over the original denominator.

$$2\frac{1}{2} \quad (2 \cdot 2) + 1 = 5 \quad \frac{5}{2}$$

56. Divide the numerator by the denominator. The quotient is the whole number of the mixed number and the remainder is the numerator of the fraction part. The denominator is unchanged.

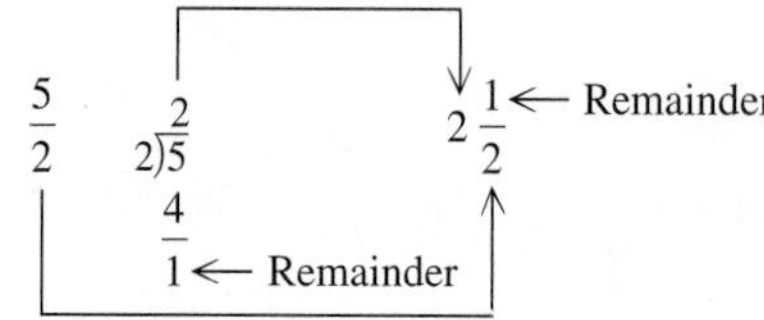

57. $\frac{501}{2}$ **59.** $\frac{1000}{3}$ **61.** $\frac{4179}{8}$ **63.** $154\frac{1}{4}$ **65.** 171 **67.** $122\frac{13}{32}$

69. $\frac{2}{3}, \frac{4}{5}, \frac{3}{4}, \frac{7}{10}$ **70.** **(a)** numerator; denominator

(b) 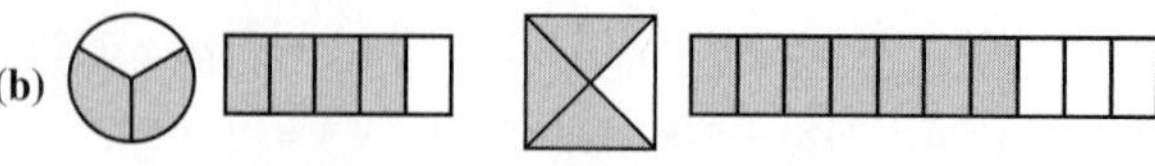

(c) less **71.** $\frac{5}{5}, \frac{10}{3}, \frac{6}{5}$ **72.** **(a)** numerator; denominator

(b) 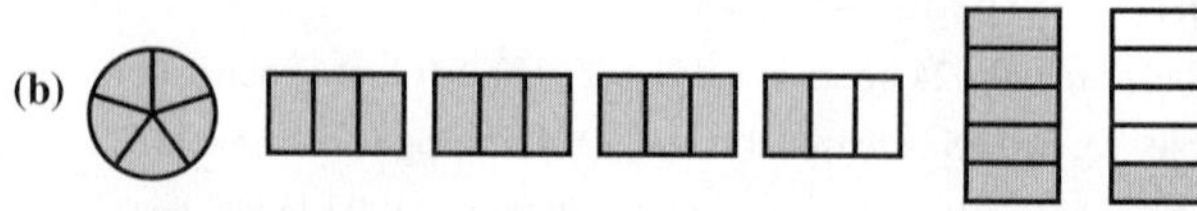

(c) greater **73.** $\frac{5}{3} = 1\frac{2}{3}; \frac{7}{7} = 1; \frac{11}{6} = 1\frac{5}{6}$ **74.** **(a)** improper; greater than or equal to

(b) 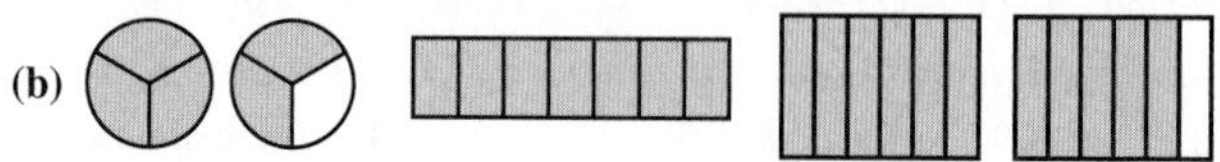

SECTION 2.3 (pages 130–131)

1. false **2.** true **3.** true **4.** false **5.** 1, 2, 3, 4, 6, 8, 12, 16, 24, 48 **7.** 1, 2, 4, 7, 8, 14, 28, 56 **9.** 1, 2, 3, 4, 6, 9, 12, 18, 36 **11.** 1, 2, 4, 5, 8, 10, 20, 40 **13.** 1, 2, 4, 8, 16, 32, 64 **15.** 1, 2, 41, 80 **17.** composite **19.** prime **21.** composite **23.** prime **25.** composite **27.** prime **29.** (b) **30.** (c) **31.** (a) **33.** $2 \cdot 3$ **35.** 5^2 **37.** $2^2 \cdot 17$ **39.** $2^3 \cdot 3^2$ **41.** $2^2 \cdot 11$ **43.** $2^2 \cdot 5^2$ **45.** 5^3 **47.** $2^2 \cdot 3^2 \cdot 5$ **49.** $2^6 \cdot 5$ **51.** $2^3 \cdot 3^2 \cdot 5$ **53.** A prime number is a whole number that has exactly two *different* factors, itself and 1. Examples include 2, 3, 5, 7, 11. A composite number has a factor(s) other than itself or 1. Examples include 4, 6, 8, 9, 10. The numbers 0 and 1 are neither prime nor composite. **54.** No even number other than 2 is prime because all even numbers have 2 as a factor. Many odd numbers are multiples of prime numbers and are not prime. For example, 9, 21, 33, and 45 are all multiples of 3. **55.** All the possible factors of 24 are 1, 2, 3, 4, 6, 8, 12, and 24. This list includes both prime numbers and composite numbers. The prime factors of 24 include only prime numbers. The prime factorization of 24 is $2 \cdot 2 \cdot 2 \cdot 3 = 2^3 \cdot 3$. **56.** No. The order of division does not matter. As long as you use only prime numbers, your answers will be correct. However, it does seem easier to always start with 2 and then use progressively greater prime numbers. The prime factorization of 36 is $36 = 2 \cdot 2 \cdot 3 \cdot 3 = 2^2 \cdot 3^2$. **57.** $2 \cdot 5^2 \cdot 7$ **59.** $2^6 \cdot 3 \cdot 5$ **61.** $2^3 \cdot 3 \cdot 5 \cdot 13$ **63.** $2^2 \cdot 3^2 \cdot 5 \cdot 7$ **65.** 2, 3, 5, 7, 11, 13, 17, 19, 23, 29, 31, 37, 41, 43, 47 **66.** A prime number is a whole number that is evenly divisible by itself and 1 only. **67.** No. Every other even number is divisible by 2 in addition to being divisible by itself and 1. **68.** No. A multiple of a prime number can never be prime because it will always be divisible by the prime number. **69.** $2 \cdot 2 \cdot 3 \cdot 5 \cdot 5 \cdot 7$ **70.** $2^2 \cdot 3 \cdot 5^2 \cdot 7$

SECTION 2.4 (pages 136–137)

1. even **2.** 5; 0 **3.** 0 **4.** 3 **5.** ✓✓✓✓ **7.** ✓✓XX **9.** ✓X✓✓ **11.** ✓✓XX **13.** true **14.** false **15.** false **16.** true **17.** $\frac{3}{5}$ **19.** $\frac{6}{7}$ **21.** $\frac{7}{8}$ **23.** $\frac{6}{7}$ **25.** $\frac{4}{7}$ **27.** $\frac{1}{50}$ **29.** $\frac{8}{11}$

31. $\frac{5}{9}$ **33.** $\frac{\overset{1}{\cancel{2}} \cdot \overset{1}{\cancel{3}} \cdot 3}{\underset{1}{\cancel{2}} \cdot 2 \cdot 2 \cdot \underset{1}{\cancel{3}}} = \frac{3}{4}$ **35.** $\frac{\overset{1}{\cancel{5}} \cdot 7}{2 \cdot 2 \cdot 2 \cdot \underset{1}{\cancel{5}}} = \frac{7}{8}$

37. $\frac{\overset{1}{\cancel{2}} \cdot \overset{1}{\cancel{3}} \cdot \overset{1}{\cancel{5}} \cdot \overset{1}{\cancel{5}}}{\underset{1}{\cancel{2}} \cdot 2 \cdot \underset{1}{\cancel{3}} \cdot \underset{1}{\cancel{5}} \cdot \underset{1}{\cancel{5}}} = \frac{1}{2}$ **39.** $\frac{\overset{1}{\cancel{2}} \cdot \overset{1}{\cancel{2}} \cdot \overset{1}{\cancel{3}} \cdot 3}{\underset{1}{\cancel{2}} \cdot \underset{1}{\cancel{2}} \cdot \underset{1}{\cancel{3}}} = 3$

41. $\frac{2 \cdot 2 \cdot 2 \cdot \overset{1}{\cancel{3}} \cdot \overset{1}{\cancel{3}}}{\underset{1}{\cancel{3}} \cdot \underset{1}{\cancel{3}} \cdot 5 \cdot 5} = \frac{8}{25}$ **43.** $\frac{1}{2} = \frac{1}{2}$; equivalent

45. $\frac{5}{12} \neq \frac{2}{5}$; not equivalent **47.** $\frac{5}{8} \neq \frac{35}{52}$; not equivalent

49. $\frac{7}{8} = \frac{7}{8}$; equivalent **51.** $8 \neq 9$; not equivalent **53.** $\frac{5}{6} = \frac{5}{6}$; equivalent

55. A fraction is in lowest terms when the numerator and the denominator have no common factors other than 1. Some examples are $\frac{1}{2}$, $\frac{3}{8}$, and $\frac{2}{3}$.

56. Two fractions are equivalent when they represent the same portion of a whole. For example, the fractions $\frac{10}{15}$ and $\frac{8}{12}$ are equivalent.

$$\frac{10}{15} = \frac{2 \cdot \overset{1}{\cancel{5}}}{3 \cdot \underset{1}{\cancel{5}}} = \frac{2 \cdot 1}{3 \cdot 1} = \frac{2}{3}$$

Equivalent $\left(\frac{2}{3} = \frac{2}{3}\right)$

$$\frac{8}{12} = \frac{\overset{1}{\cancel{2}} \cdot \overset{1}{\cancel{2}} \cdot 2}{\underset{1}{\cancel{2}} \cdot \underset{1}{\cancel{2}} \cdot 3} = \frac{1 \cdot 1 \cdot 2}{1 \cdot 1 \cdot 3} = \frac{2}{3}$$

57. $\frac{5}{8}$ **59.** $\frac{2}{1} = 2$

SUMMARY EXERCISES Fraction Basics (pages 140–141)

1. $\frac{5}{6}$; $\frac{1}{6}$ **2.** $\frac{1}{3}$; $\frac{2}{3}$ **3.** $\frac{5}{8}$; $\frac{3}{8}$ **4.** 3; 4 **5.** 8; 5 **6.** Proper $\frac{3}{5}$, $\frac{4}{25}$, $\frac{1}{32}$ Improper $\frac{8}{2}$, $\frac{16}{7}$, $\frac{8}{8}$ **7.** $\frac{7}{36}$ **8.** $\frac{24}{36}$ **9.** $\frac{7}{18}$ **10.** $\frac{33}{36}$ **11.** $2\frac{1}{2}$ **12.** $1\frac{3}{8}$ **13.** $1\frac{2}{7}$ **14.** $2\frac{2}{3}$ **15.** 8 **16.** 5 **17.** $7\frac{1}{5}$ **18.** $4\frac{7}{10}$ **19.** $\frac{10}{3}$ **20.** $\frac{43}{8}$ **21.** $\frac{34}{5}$ **22.** $\frac{53}{5}$ **23.** $\frac{51}{4}$ **24.** $\frac{62}{13}$ **25.** $\frac{71}{6}$ **26.** $\frac{189}{8}$ **27.** $2 \cdot 5$ **28.** $5 \cdot 11$ **29.** $2^2 \cdot 3^2$ **30.** 3^4 **31.** $2^3 \cdot 5 \cdot 7$ **32.** $2^3 \cdot 3^2 \cdot 5$ **33.** $\frac{1}{3}$ **34.** $\frac{1}{2}$ **35.** $\frac{1}{4}$ **36.** $\frac{3}{5}$ **37.** $\frac{3}{4}$ **38.** $\frac{5}{6}$ **39.** $\frac{7}{8}$ **40.** $\frac{1}{50}$ **41.** $\frac{1}{8}$ **42.** $\frac{5}{9}$ **43.** $\frac{4}{7}$ **44.** $\frac{5}{9}$ **45.** $\frac{\overset{1}{\cancel{2}} \cdot \overset{1}{\cancel{2}} \cdot 2 \cdot \overset{1}{\cancel{3}}}{\underset{1}{\cancel{2}} \cdot \underset{1}{\cancel{2}} \cdot 3 \cdot \underset{1}{\cancel{3}}} = \frac{2}{3}$

46. $\frac{\overset{1}{\cancel{2}} \cdot \overset{1}{\cancel{2}} \cdot \overset{1}{\cancel{2}} \cdot \overset{1}{\cancel{2}} \cdot \overset{1}{\cancel{5}}}{\underset{1}{\cancel{2}} \cdot \underset{1}{\cancel{2}} \cdot \underset{1}{\cancel{2}} \cdot \underset{1}{\cancel{2}} \cdot 2 \cdot \underset{1}{\cancel{5}}} = \frac{1}{2}$ **47.** $\frac{\overset{1}{\cancel{2}} \cdot \overset{1}{\cancel{3}} \cdot 3 \cdot \overset{1}{\cancel{7}}}{\underset{1}{\cancel{2}} \cdot \underset{1}{\cancel{3}} \cdot \underset{1}{\cancel{7}}} = 3$

48. $\frac{\overset{1}{\cancel{2}} \cdot \overset{1}{\cancel{2}} \cdot \overset{1}{\cancel{2}} \cdot \overset{1}{\cancel{2}} \cdot 2 \cdot 3}{\underset{1}{\cancel{2}} \cdot \underset{1}{\cancel{2}} \cdot \underset{1}{\cancel{2}} \cdot \underset{1}{\cancel{2}} \cdot 7} = \frac{6}{7}$

SECTION 2.5 (pages 148–151)

1. multiply; denominator **2.** numerator; denominator **3.** divide; denominator **4.** lowest terms **5.** $\frac{1}{4}$ **7.** $\frac{2}{35}$ **9.** $\frac{3}{4}$ **11.** $\frac{1}{4}$ **13.** $\frac{5}{12}$ **15.** $\frac{4}{5}$ **17.** $\frac{7}{16}$ **19.** $\frac{7}{48}$ **20.** true **21.** false: $\frac{4}{5} \cdot 8 = \frac{4}{5} \cdot \frac{8}{1} = \frac{32}{5} = 6\frac{2}{5}$ **23.** 15 **25.** $13\frac{1}{2}$ **27.** $31\frac{1}{2}$ **29.** 80 **31.** $94\frac{2}{3}$ **33.** 400 **35.** 810

37. $\frac{1}{4}$ square mile **39.** 9 square meters **41.** $\frac{1}{4}$ square mile

43. Multiply the numerators and multiply the denominators. An example is $\frac{3}{4} \cdot \frac{1}{2} = \frac{3 \cdot 1}{4 \cdot 2} = \frac{3}{8}$ **44.** You must divide a numerator and a denominator by the same number. If you do all possible divisions, your answer will be in lowest terms. One example is $\frac{3}{4} \cdot \frac{2}{3} = \frac{3 \cdot 2}{4 \cdot 3} = \frac{\overset{1}{\cancel{3}} \cdot \overset{1}{\cancel{2}}}{\underset{2}{\cancel{4}} \cdot \underset{1}{\cancel{3}}} = \frac{1}{2}$.

45. $1\frac{1}{2}$ square yards **47.** $3\frac{1}{2}$ square miles

49. They are both the same size: $\frac{3}{64}$ square mile

51. $4000 + 2000 + 1000 + 800 + 400 + 200 + 100 + 80 =$ 8580 supermarkets **52.** 8912 supermarkets

53. *Estimate:* $\frac{4}{5} \cdot 2000 = 1600$; *Exact:* 1762 supermarkets (rounded)

54. *Estimate:* $\frac{3}{8} \cdot 200 = 75$; *Exact:* 61 supermarkets (rounded)

55. $\frac{4}{5} \cdot 2200 = 1760$ supermarkets **56.** $\frac{3}{8} \cdot 160 = 60$ supermarkets

SECTION 2.6 (pages 155–158)

1. times; triple; of; twice; product; twice as much **2.** Check your work. **3.** area **4.** square **5.** $\frac{1}{2}$ square foot **7.** $\frac{8}{9}$ square foot **9.** $\frac{3}{10}$ square yard **11.** \$2568 **13.** \$35 **15. (a)** 910 women **(b)** 650 men **17.** 4 hours; 153 people **19.** $\frac{3}{4}$; 765 people **21.** Because everyone is included and fractions are given for *all* groups, the sum of the fractions must be *1* or *all* of the people. **22.** Answers will vary. Some possibilities are 1. You made an addition error. 2. The fractions on the circle graph are incorrect. 3. The fraction errors were caused by rounding.

23. \$76,000 **25.** \$15,200 **27.** \$4750

29. The correct solution is $\frac{9}{10} \times \frac{20}{21} = \frac{\overset{3}{\cancel{9}}}{\underset{1}{\cancel{10}}} \times \frac{\overset{2}{\cancel{20}}}{\underset{7}{\cancel{21}}} = \frac{6}{7}$

30. Yes, the statements are true. Since whole numbers are 1 or greater, when you multiply, the product will always be greater than either of the numbers multiplied. But, when you multiply two proper fractions, you are finding a fraction of a fraction, and the product will be smaller than either of the two proper fractions. **31.** \$750 **33.** 2 ft **35.** 9000 votes

37. $\frac{1}{32}$ of the estate

SECTION 2.7 (pages 164–167)

1. reciprocal **2.** 1 **3.** invert; multiplication **4.** lowest **5.** $\frac{8}{3}$ **7.** $\frac{6}{5}$ **9.** $\frac{5}{8}$ **11.** $\frac{1}{4}$ **13.** $\frac{2}{3}$ **15.** $2\frac{5}{8}$ **17.** $\frac{9}{20}$ **19.** 4 **21.** 6 **23.** $\frac{13}{16}$

17. *Estimate:* 34 + 19 = 53; *Exact:* $52\frac{1}{10}$

19. *Estimate:* 23 + 15 = 38; *Exact:* $38\frac{5}{28}$

21. *Estimate:* 13 + 19 + 15 = 47; *Exact:* $45\frac{5}{6}$

23. *Estimate:* 15 − 12 = 3; *Exact:* $2\frac{5}{8}$

25. *Estimate:* 13 − 1 = 12; *Exact:* $11\frac{7}{15}$

27. *Estimate:* 28 − 6 = 22; *Exact:* $22\frac{7}{30}$

29. *Estimate:* 17 − 7 = 10; *Exact:* $10\frac{3}{8}$

31. *Estimate:* 19 − 6 = 13; *Exact:* $12\frac{19}{20}$

33. *Estimate:* 20 − 12 = 8; *Exact:* $7\frac{11}{12}$

35. $\frac{15}{4}$ **36.** $\frac{63}{8}$ **37.** $2\frac{2}{5}$ **38.** $3\frac{3}{7}$ **39.** $\frac{43}{8}$ **40.** $\frac{92}{15}$ **41.** $18\frac{2}{3}$

42. $3\frac{5}{8}$ **43.** $9\frac{1}{8}$ **45.** $11\frac{1}{2}$ **47.** $3\frac{5}{6}$ **49.** $6\frac{11}{12}$ **51.** $8\frac{1}{8}$ **53.** $\frac{5}{6}$

55. $2\frac{7}{8}$ **57.** $2\frac{7}{12}$ **59.** $5\frac{9}{20}$ **61.** $3\frac{16}{21}$ **63.** Find the least common denominator. Change the fraction parts so that they have the same denominator. Add the fraction parts. Add the whole number parts. Write the answer as a mixed number. Simplify the answer.

64. You need to regroup when the minuend (top number) is a whole number or when the fraction in the minuend is smaller than the fraction in the subtrahend (bottom number). Examples are shown below.

$$\begin{array}{rl} 10 = & 9\frac{3}{3} \\ -6\frac{2}{3} = & 6\frac{2}{3} \\ \hline & 3\frac{1}{3} \end{array} \qquad \begin{array}{rl} 5\frac{1}{4} = 5\frac{1}{4} = & 4\frac{5}{4} \\ -3\frac{1}{2} = 3\frac{2}{4} = & 3\frac{2}{4} \\ \hline & 1\frac{3}{4} \end{array}$$

65. *Estimate:* 26 − 15 = 11 ft; *Exact:* $11\frac{1}{4}$ ft

67. *Estimate:* 3 − 3 = 0 in.; *Exact:* $\frac{7}{16}$ in.

69. *Estimate:* 3 − 1 = 2 in.; *Exact:* $2\frac{1}{2}$ in.

71. *Estimate:* 3 − 1 = 2 in.; *Exact:* $2\frac{3}{16}$ in.

73. *Estimate:* 16 + 19 + 24 + 31 = 90 ft; *Exact:* $87\frac{8}{4} = 89$ ft

75. *Estimate:* 24 + 35 + 24 + 35 = 118 in.; *Exact:* $116\frac{1}{2}$ in.

77. *Estimate:* 100 − 10 − 14 − 9 − 19 − 12 − 10 − 14 = 12 gallons; *Exact:* $12\frac{5}{8}$ gallons

79. *Estimate:* 527 − 108 − 151 − 139 = 129 ft; *Exact:* 130 ft

81. *Estimate:* 59 + 24 + 17 + 29 + 58 = 187 tons; *Exact:* $186\frac{13}{24}$ tons

83. $4\frac{11}{16}$ in. **85.** $21\frac{3}{8}$ in. **87. (a)** 30 **(b)** 28 **(c)** 25 **(d)** 264

88. least common denominator **89. (a)** $\frac{23}{24}$ **(b)** $\frac{8}{15}$ **(c)** $\frac{43}{48}$ **(d)** $\frac{4}{21}$

90. fraction parts **91.** improper; large

92. (a) $4\frac{5}{8} + 3\frac{6}{8} = 7\frac{11}{8} = 8\frac{3}{8}$; $\frac{37}{8} + \frac{30}{8} = \frac{67}{8} = 8\frac{3}{8}$

(b) $11\frac{56}{40} - 8\frac{35}{40} = 3\frac{21}{40}$; $\frac{496}{40} - \frac{355}{40} = \frac{141}{40} = 3\frac{21}{40}$

SUMMARY EXERCISES Adding and Subtracting Fractions (pages 235–236)

1. proper **2.** improper **3.** improper **4.** proper **5.** $\frac{5}{6}$ **6.** $\frac{7}{8}$ **7.** $\frac{3}{7}$ **8.** $\frac{23}{47}$ **9.** $\frac{1}{2}$ **10.** $\frac{3}{8}$ **11.** 35 **12.** $\frac{5}{6}$ **13.** $1\frac{1}{6}$ **14.** 56 **15.** $1\frac{13}{24}$ **16.** $1\frac{13}{16}$ **17.** $2\frac{1}{12}$ **18.** $\frac{1}{12}$ **19.** $\frac{11}{24}$ **20.** $\frac{2}{15}$

21. *Estimate:* 4 • 2 = 8; *Exact:* $7\frac{7}{8}$

22. *Estimate:* 5 • 3 = 15; *Exact:* $17\frac{15}{32}$

23. *Estimate:* 8 • 6 • 2 = 96; *Exact:* $107\frac{2}{3}$

24. *Estimate:* 4 ÷ 4 = 1; *Exact:* $1\frac{1}{6}$

25. *Estimate:* 7 ÷ 2 = $3\frac{1}{2}$; *Exact:* $3\frac{7}{16}$

26. *Estimate:* 5 ÷ 1 = 5; *Exact:* $6\frac{1}{6}$

27. *Estimate:* 6 + 4 = 10; *Exact:* $9\frac{11}{12}$

28. *Estimate:* 18 + 10 = 28; *Exact:* $28\frac{1}{6}$

29. *Estimate:* 15 + 11 = 26; *Exact:* $25\frac{4}{15}$

30. *Estimate:* 9 − 4 = 5; *Exact:* $4\frac{19}{20}$

31. *Estimate:* 14 − 7 = 7; *Exact:* $6\frac{5}{8}$

32. *Estimate:* 32 − 23 = 9; *Exact:* $9\frac{1}{4}$

33. 40 **34.** 60 **35.** 30 **36.** 48 **37.** 72 **38.** 84 **39.** 35 **40.** 12 **41.** 12 **42.** 25 **43.** 15 **44.** 55

SECTION 3.5 (pages 243–546)

1.–12.

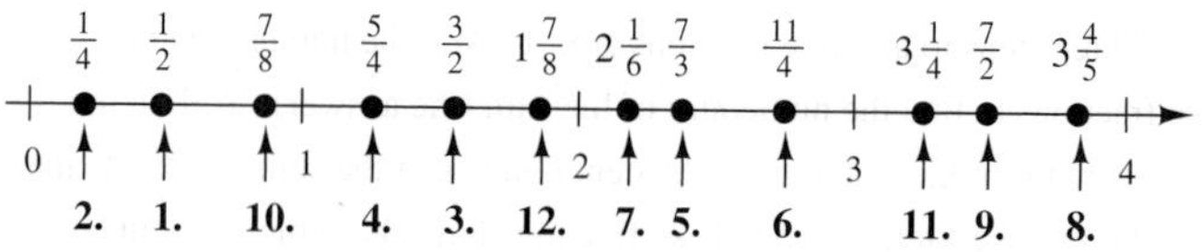

13. > **15.** < **17.** > **19.** > **21.** true; $\left(\frac{1}{2}\right)^2 = \frac{1}{2} \cdot \frac{1}{2} = \frac{1}{4}$

22. false; $\left(\frac{3}{8}\right)^2 = \frac{3}{8} \cdot \frac{3}{8} = \frac{9}{64}$ **23.** false; $\left(\frac{2}{5}\right)^3 = \frac{2}{5} \cdot \frac{2}{5} \cdot \frac{2}{5} = \frac{8}{125}$

24. true; $\left(\frac{5}{6}\right)^3 = \frac{5}{6} \cdot \frac{5}{6} \cdot \frac{5}{6} = \frac{125}{216}$ **25.** $\frac{1}{9}$ **27.** $\frac{25}{64}$ **29.** $\frac{9}{16}$ **31.** $\frac{64}{125}$

33. $\frac{81}{16} = 5\frac{1}{16}$ **35.** $\frac{81}{256}$

37. A number line is a horizontal line with a range of equally spaced whole numbers placed on it. The lowest number is on the left and the greatest number is on the right. It can be used to compare the size or value of numbers.

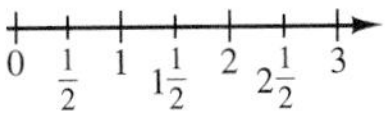

0 $\frac{1}{2}$ 1 $1\frac{1}{2}$ 2 $2\frac{1}{2}$ 3

38. 1. Do all operations inside parentheses or other grouping symbols. 2. Simplify any expressions with exponents or square roots. 3. Multiply or divide, proceeding from left to right. 4. Add or subtract, proceeding from left to right.

39. 4 **41.** 10 **43.** 1 **45.** $\frac{3}{16}$ **47.** $\frac{4}{9}$ **49.** $\frac{1}{3}$ **51.** $\frac{1}{2}$ **53.** $\frac{3}{8}$ **55.** $\frac{1}{4}$ **57.** $1\frac{1}{2}$ **59.** $\frac{1}{12}$ **61.** 3 **63.** $\frac{5}{16}$ **65.** $\frac{1}{4}$ **67.** $\frac{1}{32}$ **69.** $\frac{11}{50}$ in Las Vegas is greater. **71.** <; > **72.** **(a)** like; numerators; numerator **(b)** Answers will vary. **73.** $\frac{2}{45}$ **74.** $2\frac{1}{32}$

75. – 80. $\frac{1}{8}$ $\frac{4}{9}$ $1\frac{1}{8}$ 2 $2\frac{1}{4}$ $2\frac{7}{9}$

0 1 2 3

77. 75. 80. 79. 76. 78.

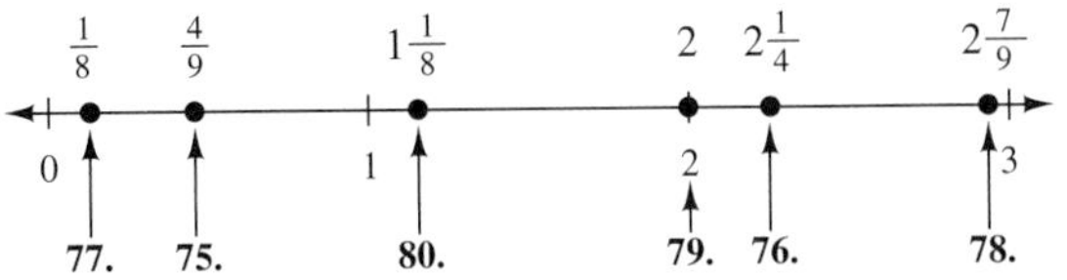

Chapter 3 REVIEW EXERCISES (pages 253–258)

1. $\frac{6}{7}$ **2.** $\frac{7}{9}$ **3.** $\frac{3}{4}$ **4.** $\frac{1}{8}$ **5.** $\frac{4}{5}$ **6.** $\frac{1}{6}$ **7.** $\frac{13}{31}$ **8.** $\frac{1}{3}$

9. $\frac{11}{12}$ of his total income **10.** $\frac{1}{4}$ more of the events **11.** 10 **12.** 12 **13.** 60 **14.** 24 **15.** 120 **16.** 180 **17.** 8 **18.** 21 **19.** 10 **20.** 45 **21.** 32 **22.** 20 **23.** $\frac{5}{6}$ **24.** $\frac{7}{8}$ **25.** $\frac{5}{8}$ **26.** $\frac{5}{12}$ **27.** $\frac{13}{24}$ **28.** $\frac{17}{36}$ **29.** $\frac{9}{10}$ of the students **30.** $\frac{23}{24}$ of her budget

31. *Estimate:* 19 + 14 = 33; *Exact:* $32\frac{3}{8}$

32. *Estimate:* 23 + 15 = 38; *Exact:* $38\frac{1}{9}$

33. *Estimate:* 13 + 9 + 10 = 32; *Exact:* $31\frac{43}{80}$

34. *Estimate:* 32 − 15 = 17; *Exact:* $17\frac{1}{12}$

35. *Estimate:* 34 − 16 = 18; *Exact:* $18\frac{1}{3}$

36. *Estimate:* 215 − 136 = 79; *Exact:* $79\frac{7}{16}$

37. $9\frac{1}{10}$ **38.** $10\frac{5}{12}$ **39.** $3\frac{1}{4}$ **40.** $1\frac{2}{3}$ **41.** $5\frac{1}{2}$ **42.** $2\frac{19}{24}$

43. *Estimate:* 19 − 6 − 7 = 6 miles; *Exact:* $5\frac{19}{24}$ miles

44. *Estimate:* 29 + 25 = 54 tons; *Exact:* $53\frac{5}{12}$ tons

45. *Estimate:* 8 + 3 + 5 + 3 = 19 pounds; *Exact:* $18\frac{1}{2}$ pounds

46. *Estimate:* 1535 − 1476 = 59 pounds; *Exact:* $59\frac{11}{16}$ pounds

47. – 50. $\frac{3}{8}$ $\frac{7}{4}$ $\frac{8}{3}$ $3\frac{1}{5}$

0 1 2 3 4

47. 48. 49. 50.

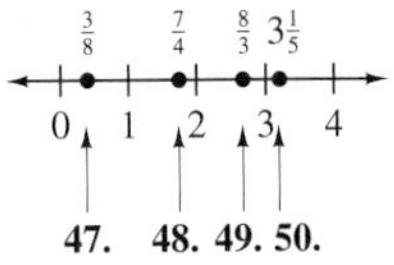

51. < **52.** < **53.** > **54.** > **55.** < **56.** > **57.** < **58.** > **59.** $\frac{1}{4}$ **60.** $\frac{4}{9}$ **61.** $\frac{27}{1000}$ **62.** $\frac{81}{4096}$ **63.** $\frac{1}{2}$ **64.** $6\frac{3}{4}$ **65.** $\frac{1}{16}$ **66.** 1 **67.** $\frac{3}{16}$ **68.** $1\frac{25}{64}$ **69.** $\frac{3}{4}$ **70.** $\frac{2}{5}$ **71.** $\frac{19}{32}$ **72.** $\frac{11}{16}$ **73.** $2\frac{1}{6}$ **74.** $26\frac{1}{4}$ **75.** $5\frac{3}{8}$ **76.** $11\frac{43}{80}$ **77.** $15\frac{5}{12}$ **78.** $\frac{8}{11}$ **79.** $\frac{1}{250}$ **80.** $\frac{1}{2}$ **81.** $\frac{2}{9}$ **82.** $\frac{11}{27}$ **83.** > **84.** < **85.** < **86.** > **87.** 36 **88.** 120 **89.** 126 **90.** 18 **91.** 108 **92.** 60

93. *Estimate:* 93 − 14 − 22 = 57 ft; *Exact:* $56\frac{7}{8}$ ft

94. *Estimate:* 4 • 50 = 200 pounds of sugar; 200 − 69 − 77 − 33 = 21 pounds; *Exact:* $21\frac{5}{8}$ pounds

Chapter 3 TEST (pages 259–260)

1. $\frac{3}{4}$ **2.** $\frac{1}{2}$ **3.** $\frac{2}{5}$ **4.** $\frac{1}{6}$ **5.** 12 **6.** 30 **7.** 108 **8.** $\frac{5}{8}$ **9.** $\frac{23}{36}$ **10.** $\frac{5}{24}$ **11.** $\frac{1}{40}$ **12.** *Estimate:* 8 + 5 = 13; *Exact:* $12\frac{1}{2}$

13. *Estimate:* 16 − 12 = 4; *Exact:* $4\frac{11}{15}$

14. *Estimate:* 19 + 9 + 12 = 40; *Exact:* $40\frac{29}{60}$

15. *Estimate:* 24 − 18 = 6; *Exact:* $5\frac{5}{8}$

16. Answers will vary. One possibility is: Probably addition and subtraction of fractions are more difficult because you have to find the least common denominator and then change the fractions to the same denominator. **17.** Answers will vary. One possibility is: Round mixed numbers to the nearest whole number. Then add or subtract to estimate the answer. The estimate may vary from the exact answer but it lets you know if your answer is reasonable.

18. *Estimate:* 10 + 85 + 37 + 8 = 140 pounds; *Exact:* $140\frac{1}{24}$ pounds

19. *Estimate:* 148 − 69 − 37 = 42 gallons; *Exact:* $41\frac{5}{8}$ gallons

20. > **21.** > **22.** 2 **23.** $\frac{13}{48}$ **24.** $1\frac{3}{4}$ **25.** $1\frac{1}{3}$

Chapters 1–3 CUMULATIVE REVIEW EXERCISES (pages 261–262)

1. 5, 3, 9, 2 **2.** 59,800; 59,800; 60,000

3. *Estimate:* 20,000 − 10,000 = 10,000; *Exact:* 14,389

4. *Estimate:* 100,000 ÷ 40 = 2500; *Exact:* 3211

5. 1,255,609 **6.** 2,801,695 **7.** 160 **8.** 369,408 **9.** 135 **10.** 2693 R2

11. *Estimate:* 20 + 9 + 5 + 20 + 9 + 5 = 68 ft; *Exact:* 64 ft

12. *Estimate:* 20 • 10 = 200 ft^2; *Exact:* 252 ft^2

13. *Estimate:* 47 • 9 = 423 miles; *Exact:* $413\frac{7}{8}$ miles

14. *Estimate:* 70 • 130 = 9100 in.; *Exact:* $9132\frac{1}{2}$ in.

15. 144 **16.** 9 **17.** 44 **18.** $\frac{4}{45}$ **19.** $\frac{9}{10}$ **20.** $1\frac{1}{16}$ **21.** $\frac{1}{2}$

22. $36\frac{3}{4}$ **23.** $2\frac{3}{16}$ **24.** $13\frac{1}{2}$ **25.** *Estimate:* 3 + 5 = 8; *Exact:* $7\frac{7}{8}$

26. *Estimate:* 22 + 4 = 26; *Exact:* $26\frac{7}{24}$

27. *Estimate:* 5 − 2 = 3; *Exact:* $2\frac{5}{8}$

28. – 31. $\frac{1}{9}$ $\frac{5}{3}$ $2\frac{3}{4}$ $\frac{10}{3}$ 0 1 2 3 4 **29.** **30.** **28.** **31.** **32.** < **33.** > **34.** <

CHAPTER 4 Decimals

SECTION 4.1 (pages 269–273)

1. 3 **2.** 4 **3.** 8 tenths **4.** 3 thousandths **5.** 7; 0; 4 **7.** 4; 7; 0 **9.** 1; 8; 9 **11.** 6; 2; 1 **13.** 410.25 **15.** 6.5432 **17.** 5406.045 **19.** $\frac{7}{10}$ **21.** $13\frac{2}{5}$ **23.** $\frac{1}{4}$ **25.** $\frac{33}{50}$ **27.** $10\frac{17}{100}$ **29.** $\frac{3}{50}$ **31.** $\frac{41}{200}$ **33.** $5\frac{1}{500}$ **35.** $\frac{343}{500}$ **37.** five tenths **39.** seventy-eight hundredths **41.** one hundred five thousandths **43.** twelve and four hundredths **45.** one and seventy-five thousandths **47.** 6.7 **49.** 0.32 **51.** 420.008 **53.** 0.0703 **55.** 75.030 **57.** Anne should not say "and" because that denotes a decimal point. **58.** Jerry used "and" twice; only the first "and" is correct. **59.** ten thousandths inch; $\frac{10}{1000} = \frac{1}{100}$ inch **61.** 12 pounds **63.** 3-C **64.** 4-C **65.** 4-A **66.** 3-B **67.** One and six hundred two thousandths centimeters **68.** One and twenty-six thousandths centimeters. **69.** millionths, ten-millionths, hundred-millionths, billionths; these match the words on the left side of the place value chart in Chapter 1 with "ths" attached. **70.** The first place to the left of the decimal point is ones, so the first place to the right could be one*ths,* like tens and ten*ths.* But anything that is 1 or more is to the *left* of the decimal point. **71.** Seventy-two million four hundred thirty-six thousand nine hundred fifty-five hundred-millionths **72.** six hundred seventy-eight thousand five hundred fifty-four billionths **73.** eight thousand six and five hundred thousand one millionths **74.** twenty thousand sixty and five hundred five millionths

75. 0.0302040 **76.** 9,876,543,210.100200300

SECTION 4.2 (pages 279–280)

1. 0 **2.** 9 **3.** Look only at the 6, which is *5 or more.* So round up by adding one thousandth to 5.709 to get 5.710. **4.** Look only at the 2, which is *4 or less.* So the part you are keeping, 10.0, stays the same. **5.** 16.9 **7.** 0.956 **9.** 0.80 **11.** 3.661 **13.** 794.0 **15.** 0.0980 **17.** 9.09 **19.** 82.0002 **21.** $0.82 **23.** $1.22 **25.** $0.70 **27.** $48,650 **29.** $840 **31.** $500 **33.** $1.00 **35.** $1000 **37.** **(a)** 253 miles per hour **(b)** 135 miles per hour **39.** **(a)** 186.0 miles per hour **(b)** 763.0 miles per hour **41.** Rounds to $0 (zero dollars) because $0.499 is closer to $0 than to $1. **42.** Round amounts less than $1.00 to the nearest cent instead of the nearest dollar. **43.** Rounds to $0.00 (zero cents) because $0.0015 is closer to $0.00 than to $0.01. **44.** Both round to $0.60. Rounding to nearest thousandth (tenth of a cent) would allow you to identify $0.597 as less than $0.601.

SECTION 4.3 (pages 285–288)

1. $\begin{array}{r} 6.420 \\ +\ 10.163 \end{array}$ **2.** $\begin{array}{r} 7.000 \\ +\ 9.204 \end{array}$ **3.** $\begin{array}{r} 20.0000 \\ -\ 9.1263 \end{array}$ **4.** $\begin{array}{r} 137.06 \\ -\ 12.00 \end{array}$ **5.** 17.48

7. 7.763 **9.** 77.006 **11.** 20.104 **13.** 0.109 **15.** 330.86895 **17.** **(a)** 24.75 in. **(b)** 3.95 in. **19.** **(a)** 62.27 in. **(b)** 0.39 in. **21.** 6 should be written 6.00; sum is 46.22. **22.** The two problems are done in a different order: 8 − 2.9 is not the same as 2.9 − 8 because subtraction is not commutative. **23.** *Estimate:* $20 − 7 = $13; *Exact:* $13.16 **25.** *Estimate:* 400 + 1 + 20 = 421; *Exact:* 414.645 **27.** *Estimate:* 9 − 4 = 5; *Exact:* 4.849 **29.** *Estimate:* 60 + 500 + 6 = 566; *Exact:* 608.4363 **31.** 0.275 **33.** 6.507 **35.** 1.81 **37.** 6056.7202 **39.** *Estimate:* 80 − 30 = 50 million people; *Exact:* 50.4 million people **41.** *Estimate:* 400 + 200 + 100 + 80 + 40 + 30 + 30 = 880 million people; *Exact:* 940.82 million people **43.** *Estimate:* 2 + 2 + 2 = 6 meters; *Exact:* 6.19 meters, which is less than the rhino by 0.21 meter **45.** *Estimate:* 11 − 10 = 1 ounce; *Exact:* 0.65 ounce **47.** *Estimate:* 20 + 6 + 20 + 6 = 52 in.; *Exact:* 52.1 in. **49.** *Estimate:* $5 − $5 = $0; *Exact:* $0.30 **51.** *Estimate:* $19 + 2 + 2 + 10 + 2 = $35; *Exact:* $35.25 **53.** $2059.36 **55.** $103.97 **57.** $498.22 **59.** b = 1.39 centimeters **61.** q = 7.943 ft

SECTION 4.4 (pages 291–294)

1. 3 **2.** 5 **3.** 2 **4.** 4 **5.** 0.1344 **7.** 159.10 **9.** 15.5844 **11.** $34,500.20 **13.** 43.2 **14.** 43.2 **15.** 0.432 **16.** 0.432 **17.** 0.0432 **18.** 0.0432 **19.** 0.0000312 **21.** 0.000025 **23.** *Estimate:* 40 × 5 = 200; *Exact:* 190.08 **25.** *Estimate:* 40 × 40 = 1600; *Exact:* 1558.2 **27.** *Estimate:* 7 × 5 = 35; *Exact:* 30.038 **29.** *Estimate:* 3 × 7 = 21; *Exact:* 19.24165 **31.** unreasonable; $289.00 **33.** reasonable **35.** unreasonable; $4.19 **37.** unreasonable; 9.5 pounds **39.** 945.87 (rounded) **41.** $2.45 (rounded) **43.** $77.10 (rounded) **45.** $20,265 **47.** **(a)** Area before 1929 ≈ 23.2 in.^{2}; Area today ≈ 16.0 in.^{2} **(b)** 7.2 in.^{2} **49.** **(a)** 0.43 in. **(b)** 4.3 in. **51.** $1264.04; $2723.72 **53.** $4.09 (rounded) **55.** **(a)** 72.05 **(b)** $27.80 **57.** 59.6; 32; 4.76; 803.5; 7226; 9. Multiplying by 10, decimal point moves one place to the right; by 100, two places to the right; by 1000, three places to the right. **58.** 5.96; 0.32; 0.0476; 8.035; 6.5; 52.3. Multiplying by 0.1, decimal point moves one place to the left; by 0.01, two places to the left; by 0.001, three places to the left.

SUMMARY EXERCISES Adding, Subtracting, and Multiplying Decimal Numbers (pages 295–296)

1. $\frac{4}{5}$ **2.** $6\frac{1}{250}$ **3.** $\frac{7}{20}$ **4.** ninety-four and five tenths **5.** two and three ten-thousandths **6.** seven hundred six thousandths **7.** 0.05 **8.** 0.0309 **9.** 10.7 **10.** 6.19 **11.** 1.0 **12.** 0.420 **13.** $0.89 **14.** $3.00 **15.** $100 **16.** 0.945 **17.** 49.6199 **18.** 1.845 **19.** $93.50 **20.** 0.00488 **21.** 2.15 **22.** 10.955 **23.** 18.4009 **24.** 4.3043

25. 0.87 **26.** \$110.84 **27.** 82.84 (rounded) **28.** $P = 3.1$ in.; $A \approx 0.5$ in.^{2} **29.** $P = 3.25$ in.; $A \approx 0.66$ in.^{2} **30.** **(a)** Pumpkin is heaviest; strawberry is lightest. **(b)** 1809.5 pounds **31.** 92.8 pounds (rounded) **32.** **(a)** 40.5 pounds; **(b)** 6.0 or 6 pounds **33.** **(a)** 3.6875 pounds **(b)** 2.4625 pounds **34.** **(a)** 4; 7; 17; 2; 69; 1810; 1; 8 (all pounds) **(b)** *Estimate:* 1918 pounds; *Exact:* 1915.7455 pounds **35.** \$91.28

SECTION 4.5 (pages 303–306)

1. c; $5\overline{)25.5}$ **2.** b; $7\overline{)42.3}$ **3.** $2.2\overline{)8.24}$ **4.** $5.1\overline{)10.5}$ **5.** 3.9 **7.** 0.47 **9.** 400.2 **11.** 36 **13.** 0.06 **14.** 0.6 **15.** 6000 **16.** 6 **17.** 25.3 **19.** 516.67 (rounded) **21.** 26.756 (rounded) **23.** 10,082.647 (rounded) **25.** unreasonable; $40 \div 8 = 5$; Correct answer is 4.725. **27.** reasonable; $50 \div 50 = 1$ **29.** unreasonable; $300 \div 5 = 60$; Correct answer is 60.2. **31.** \$4.00 (rounded) **33.** \$67.08 **35.** \$11.92 per hour **37.** 28.0 miles per gallon (rounded) **39.** \$0.03 per can (rounded) **41.** 7.37 meters (rounded) **43.** 0.08 meter **45.** 22.49 meters **47.** **(a)** Work inside parentheses; subtract $9.5 - 3.1$ to get 6.4. **(b)** Apply the exponent; multiply (2.2)(2.2) to get 4.84. **(c)** Add 4.84 plus 6.4 to get 11.24. **48.** **(a)** Work inside parentheses; add $0.4 + 5.07$ to get 5.47 **(b)** Multiply 5.47 times 3 to get 16.41 **(c)** Subtract $60.41 - 16.41$ to get 44 **49.** 14.25 **51.** 73.4 **53.** 1.205 **55.** 0.334 **57.** **(a)** 1,083,333 pieces (rounded) **(b)** 18,056 pieces (rounded) **(c)** 301 pieces (rounded). **59.** 100,000 box tops **61.** 2632 box tops (rounded) **63.** 0.377; 0.91; 0.0886; 3.019; 40.65; 662.57. **(a)** Dividing by 10, decimal point moves one place to the left; by 100, two places to the left; by 1000, three places to the left. **(b)** The decimal point moved to the *right* when *multiplying* by 10 or 100 or 1000. Here it moves to the *left* when *dividing* by those numbers. **64.** 402; 71; 3.39; 157.7; 460; 8730 **(a)** Dividing by 0.1, decimal point moves one place to the right; by 0.01, two places to the right; by 0.001, three places to the right **(b)** The decimal point moved to the *left* when *multiplying* by 0.1 or 0.01 or 0.001. Here the decimal point moves to the *right* when *dividing* by those numbers.

SECTION 4.6 (pages 311–314)

1. **(a)** not correct, should be $5\overline{)2}$; **(b)** correct **(c)** correct **2.** **(a)** correct **(b)** not correct, should be $3\overline{)1}$; **(c)** correct **3.** $\overset{0.}{4\overline{)3.}}$ To continue dividing, write zeros in the dividend. **4.** $\overset{0.}{8\overline{)1.}}$ To continue dividing, write zeros in the dividend. **5.** **(a)** less than **(b)** greater than **(c)** less than **6.** **(a)** less than **(b)** greater than **(c)** greater than **7.** 0.5 **9.** 0.75 **11.** 0.3 **13.** 0.9 **15.** 0.6 **17.** 0.875 **19.** 2.25 **21.** 14.7 **23.** 3.625 **25.** 6.333 (rounded) **27.** 0.833 (rounded) **29.** 1.889 (rounded) **31.** $\frac{2}{5}$ **33.** $\frac{5}{8}$ **35.** $\frac{7}{20}$ **37.** 0.35 **39.** $\frac{1}{25}$ **41.** 0.2 **43.** $\frac{9}{100}$ **45.** shorter; 0.72 inch **47.** more; 0.05 inch **49.** too much; 0.005 gram **51.** 0.9991 cm, 1.0007 cm **52.** 3.0 ounces, 2.995 ounces, 3.005 ounces **53.** 0.5399, 0.54, 0.5455 **55.** 5.0079, 5.79, 5.8, 5.804 **57.** 0.6009, 0.609, 0.628, 0.62812 **59.** 2.8902, 3.88, 4.876, 5.8751 **61.** 0.006, 0.043, $\frac{1}{20}$, 0.051 **63.** 0.37, $\frac{3}{8}$, $\frac{2}{5}$, 0.4001 **65.** red box **67.** green box **69.** 1.4 in. (rounded) **71.** 0.3 in. (rounded) **73.** 0.4 in. (rounded) **75.** **(a)** A proper fraction like $\frac{5}{9}$ is less than 1, so it cannot be equivalent to a decimal number that is greater than 1. **(b)** $\frac{5}{9}$ means $5 \div 9$ or $9\overline{)5}$ so correct answer is 0.556 (rounded). This makes sense because both the fraction and decimal are less than 1. **76.** **(a)** $2.035 = 2\frac{35}{1000} = 2\frac{7}{200}$, not $2\frac{7}{20}$. **(b)** Adding the whole number part gives $2 + 0.35$, which is 2.35, not 2.035. To check, $2.35 = 2\frac{35}{100} = 2\frac{7}{20}$. **77.** Just add the whole number part to 0.375. So $1\frac{3}{8} = 1.375$; $3\frac{3}{8} = 3.375$; $295\frac{3}{8} = 295.375$. **78.** It works only when the fraction part has a one-digit numerator and a denominator of 10, a two-digit numerator and a denominator of 100, and so on.

Chapter 4 REVIEW EXERCISES (pages 320–323)

1. 0; 5 **2.** 0; 6 **3.** 8; 9 **4.** 5; 9 **5.** 7; 6 **6.** $\frac{1}{2}$ **7.** $\frac{3}{4}$ **8.** $4\frac{1}{20}$ **9.** $\frac{7}{8}$ **10.** $\frac{27}{1000}$ **11.** $27\frac{4}{5}$ **12.** eight tenths **13.** four hundred and twenty-nine hundredths **14.** twelve and seven thousandths **15.** three hundred six ten-thousandths **16.** 8.3 **17.** 0.205 **18.** 70.0066 **19.** 0.30 **20.** 275.6 **21.** 72.79 **22.** 0.160 **23.** 0.091 **24.** 1.0 **25.** \$15.83 **26.** \$0.70 **27.** \$17,625.79 **28.** \$350 **29.** \$130 **30.** \$100 **31.** \$29 **32.** *Estimate:* $6 + 400 + 20 = 426$; *Exact:* 444.86 **33.** *Estimate:* $80 + 1 + 100 + 1 + 30 = 212$; *Exact:* 233.515 **34.** *Estimate:* $300 - 20 = 280$; *Exact:* 290.7 **35.** *Estimate:* $9 - 8 = 1$; *Exact:* 1.2684 **36.** *Estimate:* 90 million − 20 million = 70 million; *Exact:* 77.0 million more cats **37.** *Estimate:* \$400 − \$300 − \$70 = \$30; *Exact:* \$15.80 **38.** *Estimate:* \$2 + \$5 + \$20 = \$27; \$30 − \$27 = \$3; *Exact:* \$4.14 **39.** *Estimate:* $2 + 4 + 5 = 11$ kilometers; *Exact:* 11.55 kilometers **40.** *Estimate:* $6 \times 4 = 24$; *Exact:* 22.7106 **41.** *Estimate:* $40 \times 3 = 120$; *Exact:* 141.57 **42.** 0.0112 **43.** 0.000355 **44.** reasonable; $700 \div 10 = 70$ **45.** unreasonable; $30 \div 3 = 10$; Correct answer is 9.5. **46.** 14.467 (rounded) **47.** 1200 **48.** 0.4 **49.** \$708 (rounded) **50.** \$2.99 (rounded) **51.** 133 shares (rounded) **52.** \$3.47 (rounded) **53.** 29.215 **54.** 10.15 **55.** 3.8 **56.** 0.64 **57.** 1.875 **58.** 0.111 (rounded) **59.** 3.6008, 3.68, 3.806 **60.** 0.209, 0.2102, 0.215, 0.22 **61.** $\frac{1}{8}$, $\frac{3}{20}$, 0.159, 0.17 **62.** 404.865 **63.** 254.8 **64.** 3583.261 (rounded) **65.** 29.0898 **66.** 0.03066 **67.** 9.4 **68.** 175.675 **69.** 9,04 **70.** 19.50 **71.** 8.19 **72.** 0.928 **73.** 35 **74.** 0.259 **75.** 0.3 **76.** \$3.00 (rounded) **77.** \$2.17 (rounded) **78.** \$35.96 **79.** \$199.71 **80.** \$78.50 **81.** **(a)** baked potato with skin **(b)** $\frac{1}{2}$ cup green peas **(c)** 0.59 milligram **82.** **(a)** 2.08 milligrams **(b)** more, by 0.08 milligram

Chapter 4 TEST (pages 324–325)

1. $18\frac{2}{5}$ **2.** $\frac{3}{40}$ **3.** sixty and seven thousandths **4.** two hundred eight ten-thousandths **5.** 725.6 **6.** 0.630 **7.** \$1.49 **8.** \$7860 **9.** *Estimate:* $8 + 80 + 40 = 128$; *Exact:* 129.2028 **10.** *Estimate:* $80 - 4 = 76$; *Exact:* 75.498 **11.** *Estimate:* $6(1) = 6$; *Exact:* 6.948 **12.** *Estimate:* $20 \div 5 = 4$; *Exact:* 4.175 **13.** 839.762

14. 669.004 **15.** 0.0000483 **16.** 480 **17.** 2.625 **18.** 0.44, $\frac{9}{20}$, 0.4506, 0.451 **19.** 35.49 **20.** \$1294.47 **21.** pintails, wigeons, gadwalls **22.** \$5.35 (rounded) **23.** 2.8 degrees **24.** \$3.79 per foot (rounded) **25.** Answer varies.

Chapters 1–4 CUMULATIVE REVIEW EXERCISES (pages 326–327)

1. 500,000 **2.** 602.49 **3.** \$710 **4.** \$0.05 **5.** 9.671 **6.** $1\frac{4}{9}$ **7.** $1\frac{2}{5}$ **8.** 4914 **9.** 93.603 **10.** 404 R3 **11.** $1\frac{17}{24}$ **12.** 233,728 **13.** 0.03264 **14.** 8 **15.** 45 **16.** $\frac{4}{31}$ **17.** $\frac{2}{3}$ **18.** 0.51 (rounded) **19.** 4 **20.** 14 **21.** forty and thirty-five thousandths **22.** 0.0306 **23.** 7.005, 7.5, 7.5005, 7.505 **24.** 0.8, 0.8015, $\frac{21}{25}$, $\frac{7}{8}$ **25.** *Estimate:* \$60 − \$50 − \$1 = \$9; *Exact:* \$11.17 **26.** *Estimate:* 50 − 47 = 3 in.; *Exact:* $3\frac{3}{8}$ in. **27.** *Estimate:* \$10 × 20 = \$200; *Exact:* \$191.90 (rounded) **28.** *Estimate:* (8 × 20) + (10 × 30) = 160 + 300 = 460 students; *Exact:* 488 students **29.** *Estimate:* 2 + 4 = 6 yards; *Exact:* $6\frac{5}{24}$ yards **30.** *Estimate:* \$30 + \$200 − \$40 − \$40 = \$150; *Exact:* \$174.50 **31.** *Estimate:* \$6 ÷ 3 = \$2 per pound; *Exact:* \$2.29 per pound (rounded) **32.** *Estimate:* \$80,000 ÷ 100 = \$800; *Exact:* \$729 (rounded) **33.** size M (medium) **34.** XXS $1\frac{1}{2}$ in.; XS $\frac{3}{4}$ in.; S $\frac{3}{4}$ in.; M $\frac{7}{8}$ in.; L $\frac{3}{4}$ in. **35.** 21.125 − 20.25 = 0.875 in.; 7 ÷ 8 = 0.875 in. or $\frac{875}{1000} = \frac{875 \div 125}{1000 \div 125} = \frac{7}{8}$ in. **36.** Answers will vary. Many people prefer using decimals because you do not need to find a common denominator or rewrite answers in lowest terms.

CHAPTER 5 Ratio and Proportion

SECTION 5.1 (pages 335–338)

1. Answers will vary. One possibility is: A ratio compares two quantities with the same units. Examples will vary. **2.** You can divide out the common (same) units. **3.** 20; $\frac{4}{1}$ **4.** 5; $\frac{4}{15}$ **5.** $\frac{8}{9}$ **7.** $\frac{2}{1}$ **9.** $\frac{1}{3}$ **11.** $\frac{8}{5}$ **13.** $\frac{3}{8}$ **15.** $\frac{9}{7}$ **17.** $\frac{6}{1}$ **19.** $\frac{5}{6}$ **21.** ounces; takes fewer steps to solve and you do not have to work with a mixed number **22.** hours; takes fewer steps to solve and you do not have to work with a mixed number **23.** $\frac{8}{5}$ **25.** $\frac{1}{12}$ **27.** $\frac{5}{16}$ **29.** $\frac{4}{1}$ **31.** $\frac{1}{6}$ **33.** $\frac{15}{1}$ **35.** Answers will vary. One possibility is stocking cards of various types in the same ratios as those in the table. **36.** Answers will vary. Examples: a person may send Valentine's Day cards every year but graduation cards only once every few years; Valentine's Day may have more advertising. **37.** $\frac{6}{5}$; $\frac{36}{17}$ **39.** *White Christmas* to *It's Now or Never; White Christmas* to *I Will Always Love You; Candle in the Wind* to *I Want to Hold Your Hand.* **41.** $\frac{7}{5}$ **43.** $\frac{6}{1}$ **45.** $\frac{38}{17}$ **47.** $\frac{1}{4}$ **49.** $\frac{34}{35}$ **51.** $\frac{1}{1}$; as long as the sides all have the same length, any measurement you choose will maintain the ratio.

52. Answers will vary. Some possibilities are: $\frac{4}{5} = \frac{8}{10} = \frac{12}{15} = \frac{16}{20} = \frac{20}{25} = \frac{24}{30} = \frac{28}{35}$. **53.** It is not possible. Amelia would have to be older than her mother to have a ratio of 5 to 3. **54.** Answers will vary, but a ratio of 3 to 1 means your income is 3 times your friend's income.

SECTION 5.2 (pages 343–346)

1. $\frac{5 \text{ cups}}{3 \text{ people}}$ **3.** $\frac{3 \text{ feet}}{7 \text{ seconds}}$ **5.** $\frac{18 \text{ miles}}{1 \text{ gallon}}$ **7.** Divide; correct set-up is $\frac{\$5.85}{3 \text{ boxes}}$ or $3\overline{)5.85}$ **8.** 5 rooms for 1 nurse. **9.** \$12 per hour or \$12/hour **11.** 1.25 pounds/person **13.** 325.9; 21.0 (rounded) **15.** 338.6; 20.9 (rounded) **17.** \$1.125/oz; \$0.913/oz (rounded); \$0.913/oz; best buy is 4 oz for \$3.65. **19.** 14 ounces for \$2.89, about \$0.206/ounce **21.** 18 ounces for \$1.79, about \$0.099/ ounce **23.** Answers will vary. For example, you might choose Brand B because you like more chicken, so the cost per chicken chunk may actually be the same as or less than Brand A. **24.** Answers will vary. For example, if you use only half of the larger bag, you really pay \$0.30 per pound, so the smaller bag is the better buy. **25.** 1.75 pounds/week **27.** \$12.26/hour **29. (a)** Penny Saver, \$0.415; Most Minutes, \$0.503 (rounded); USA Card, \$0.30 **(b)** Penny Saver, \$0.083/min; Most Minutes, \$0.101/min (rounded); USA Card, \$0.06/min; USA Card is the best buy. **31.** Penny Saver total \$0.54, \$0.018/min; Most Minutes total \$0.565, \$0.019/min (rounded); USA card total \$0.55, \$0.018/min (rounded); all three unit rates are very similar. **33.** one battery for \$1.79; like getting 3 batteries so \$1.79 ÷ 3 ≈ \$0.597 per battery **35.** Brand P with the 50¢ coupon is the best buy. (\$3.39 − \$0.50 = \$2.89; \$2.89 ÷ 16.5 ounces ≈ \$0.175 per ounce) **37.** Plan A: \$0.09/min; Plan B: \$0.07/min; Plan B is the better buy. **38. (a)** 15 min/day **(b)** 30 min/day

39. (60 extra min) (\$0.45) = \$27 overage

\$39.95 + \$27 overage = \$66.95

$$\frac{\$66.95}{450 \text{ min} + 60 \text{ min}} = \frac{\$66.95}{510 \text{ min}} \approx \$0.13 \text{ min}$$

40. (90 extra min) (\$0.40) = \$36 overage

\$59.95 + \$36 overage = \$95.95

$$\frac{\$95.95}{900 \text{ min} + 90 \text{ min}} = \frac{\$95.95}{990 \text{ min}} \approx \$0.10/\text{min}$$

41. From Exercise 39: $\frac{\$69.95}{510 \text{ min}} \approx \$0.14$ min

From Exercise 40: $\frac{\$69.95}{990 \text{ min}} \approx \$0.07$ min

SECTION 5.3 (pages 350–351)

1. $\frac{\$9}{12 \text{ cans}} = \frac{\$18}{24 \text{ cans}}$ **3.** $\frac{200 \text{ adults}}{450 \text{ children}} = \frac{4 \text{ adults}}{9 \text{ children}}$ **5.** $\frac{120}{150} = \frac{8}{10}$ **7.** $\frac{3}{5} = \frac{3}{5}$; true **9.** $\frac{5}{8} = \frac{5}{8}$; true **11.** $\frac{3}{4} \neq \frac{2}{3}$; false **13.** $\frac{14}{5} = \frac{14}{5}$; true **15.** $\frac{16}{9} = \frac{16}{9}$; true **17.** $\frac{7}{6} \neq \frac{9}{8}$; false **19.** Answers may vary. One example: A proportion shows that two ratios are equal. The multiplications on the proportion should show 6 • 45 = 270 and 30 • 9 = 270. **20.** Answers will vary. One possibility: If the cross products are equal, the proportion is true. If the cross products are not equal, the proportion is false. **21.** 54 = 54; True **23.** 336 ≠ 320; False **25.** 2880 ≠ 2970; False

27. 28 = 28; True **29.** 44.8 ≠ 45; False **31.** 66 = 66; True

33. $68\frac{1}{4} = 68\frac{1}{4}$; True **35.** $5\frac{2}{5} \neq 5\frac{1}{3}$; False

37. $2\frac{1}{80} \neq 2\frac{7}{100}$ or 2.0125 ≠ 2.07; False

39. $\frac{68 \text{ hits}}{200 \text{ at bats}} = \frac{153 \text{ hits}}{450 \text{ at bats}}$ $200 \cdot 153 = 30{,}600$; $68 \cdot 450 = 30{,}600$

Cross products are *equal* so the proportion is *true;* they hit equally well.

40. Left-hand ratio compares hours to cartons, but right-hand ratio compares cartons to hours. Correct proportion is shown.

$\frac{3.5 \text{ hours}}{91 \text{ cartons}} = \frac{5.25 \text{ hours}}{126 \text{ cartons}}$

91 • 5.25 = 477.75 Cross products are *not* equal so the proportion is *false*;
3.5 • 126 = 441 the men do not work equally fast.

SECTION 5.4 (pages 356–357)

1. Finding the cross products; 10 • 5 = 50 and 7 • x

2. Divide both sides by 7; $\frac{\overset{1}{\cancel{7}} \cdot x}{\underset{1}{\cancel{7}}} = \frac{50}{7}$ **3.** 12; 12; x = 4 **5.** x = 2

7. x = 88 **9.** x = 91 **11.** x = 5 **13.** x = 10 **15.** x ≈ 24.44 (rounded)

17. x = 50.4 **19.** x ≈ 17.64 (rounded) **21.** x = 1 **23.** $x = 3\frac{1}{2}$

25. x = 0.2 or $x = \frac{1}{5}$ **27.** x = 0.005 or $x = \frac{1}{200}$

29. Find cross products: 20 ≠ 30, so the proportion is false.

$\frac{6\frac{2}{3}}{4} = \frac{5}{3}$ or $\frac{10}{6} = \frac{5}{3}$ or $\frac{10}{4} = \frac{7.5}{3}$ or $\frac{10}{4} = \frac{5}{2}$

30. Find cross products: 192 ≠ 180, so the proportion is false.

$\frac{6.4}{8} = \frac{24}{30}$ or $\frac{6}{7.5} = \frac{24}{30}$ or $\frac{6}{8} = \frac{22.5}{30}$ or $\frac{6}{8} = \frac{24}{32}$

SUMMARY EXERCISES Ratios, Rates, and Proportions (pages 358–359)

1. $\frac{50 \text{ million}}{25 \text{ million}} = \frac{2}{1}$ **2.** $\frac{15 \text{ million}}{20 \text{ million}} = \frac{3}{4}$ **3.** $\frac{20 \text{ million} + 25 \text{ million}}{50 \text{ million}} = \frac{9}{10}$

4. $\frac{380 \text{ million}}{110 \text{ million}} = \frac{38}{11}$ **5.** Comparing the violin to piano, guitar, organ, clarinet, and drums gives ratios of $\frac{1}{11}, \frac{1}{10}, \frac{1}{3}, \frac{1}{2}$, and $\frac{2}{3}$ respectively.

6. (a) guitar to clarinet **(b)** organ to drums, or clarinet to violin **7.** 2.1 points/min; 0.5 min/point (both rounded) **8.** 1.7 points/min; 0.6 min/point (both rounded) **9.** \$16.32/hour; \$24.48/hour for overtime **10.** \$0.20/channel; \$0.17/channel; \$0.18/channel (all answers rounded) **11.** 12 ounces at \$0.60/ounce (rounded) **12.** Brand P with the \$2 coupon is the best buy at \$0.57 per pound. **13.** $\frac{4}{3} = \frac{4}{3}$ or 924 = 924; true **14.** 2.0125 ≠ 2.07;false **15.** $68\frac{1}{4} = 68\frac{1}{4}$; true **16.** x = 28 **17.** x = 3.2 **18.** x = 182 **19.** x ≈ 3.64 (rounded) **20.** x ≈ 0.93 (rounded) **21.** x = 1.56 **22.** x ≈ 0.05 (rounded) **23.** x = 1 **24.** $x = \frac{3}{4}$

SECTION 5.5 (pages 363–366)

1. $\frac{6 \text{ potatoes}}{4 \text{ eggs}} = \frac{12 \text{ potatoes}}{x \text{ eggs}}$ **2.** $\frac{1 \text{ serving}}{15 \text{ chips}} = \frac{x \text{ serving}}{70 \text{ chips}}$ **3.** 18; x; 22.5 hours **5.** \$7.20 **7.** 42 pounds **9.** \$403.68 **11.** 10 ounces (rounded) **13.** 5 quarts **15.** 14 ft, 10 ft **17.** 14 ft, 8 ft **19.** 96 pieces chicken; 33.6 pounds lasagna; 10.8 pounds deli meats; $5\frac{3}{5}$ pounds cheese; 7.2 dozen (about 86) buns; 14.4 pounds of salad **21.** 44 students is an *unreasonable* answer because there are only 35 students in the class. **22.** 2 minutes is an *unreasonable* answer because it is less than the 30 minutes she gets in just one day. **23.** 2065 students (reasonable); about 4214 students with incorrect set-up (only 2950 students in the group) **25.** about 83 people (reasonable); about 750 people with incorrect set-up (only 250 people attended) **27.** 625 stocks **29.** 4.06 meters (rounded) **31.** 311 calories (rounded) **33.** 10.53 meters (rounded) **35.** You cannot solve this problem using a proportion because the ratio of age to weight is not constant. As Jim's age increases, his weight may decrease, stay the same, or increase. **36.** Answers will vary; Exercises 3–34 are all examples of application problems. **37.** 5610 students use cream **39.** 120 calories and 12 grams of fiber **41.** $1\frac{3}{4}$ cups water, 3 Tbsp margarine, $\frac{3}{4}$ cup milk, 2 cups flakes **42.** $5\frac{1}{4}$ cups water, 9 Tbsp margarine, $2\frac{1}{4}$ cups milk, 6 cups flakes

Chapter 5 REVIEW EXERCISES (pages 372–375)

1. $\frac{3}{4}$ **2.** $\frac{4}{1}$ **3.** great white shark to whale shark; whale shark to blue whale **4.** $\frac{2}{1}$ **5.** $\frac{2}{3}$ **6.** $\frac{5}{2}$ **7.** $\frac{1}{6}$ **8.** $\frac{3}{1}$ **9.** $\frac{3}{8}$ **10.** $\frac{4}{3}$ **11.** $\frac{1}{9}$ **12.** $\frac{10}{7}$ **13.** $\frac{7}{5}$ **14.** $\frac{5}{6}$ **15.** $\frac{\$11}{1 \text{ dozen}}$ **16.** $\frac{12 \text{ children}}{5 \text{ families}}$

17. 0.2 page/minute or $\frac{1}{5}$ page/minute; 5 minutes/page

18. \$20/hour; 0.05 hour/dollar or $\frac{1}{20}$ hour/dollar

19. 8 ounces for \$4.98, about \$0.623/ounce

20. 17.6 pounds for \$18.69 − \$1 coupon, about \$1.005/pound

21. $\frac{3}{5} = \frac{3}{5}$ or 90 = 90; true **22.** $\frac{1}{8} \neq \frac{1}{4}$ or 432 ≠ 216; false

23. $\frac{47}{10} \neq \frac{49}{10}$ or 980 ≠ 940; false **24.** $\frac{16}{9} = \frac{16}{9}$ or 3456 = 3456; true

25. 4.8 = 4.8; true **26.** 14 = 14; true **27.** x = 1575 **28.** x = 20 **29.** x = 400 **30.** x = 12.5 **31.** x ≈ 14.67 (rounded) **32.** x ≈ 8.17 (rounded) **33.** x = 50.4 **34.** x ≈ 0.57 (rounded) **35.** x ≈ 2.47 (rounded) **36.** 27 cats **37.** 46 hits **38.** \$15.63 (rounded) **39.** 3299 students (rounded) **40.** 68 feet **41.** $27\frac{1}{2}$ hours or 27.5 hours **42.** 511 calories (rounded) **43.** 14.7 milligrams **44.** x = 105 **45.** x = 0 **46.** x = 128 **47.** x ≈ 23.08 (rounded) **48.** x = 6.5 **49.** x ≈ 117.36 (rounded) **50.** 1440 ≠ 1485; False

51. 10.8 ≠ 10.864; False **52.** $2\frac{1}{3} = 2\frac{1}{3}$; True **53.** $\frac{8}{5}$ **54.** $\frac{33}{80}$

55. $\frac{15}{4}$ **56.** $\frac{4}{1}$ **57.** $\frac{4}{5}$ **58.** $\frac{37}{7}$ **59.** $\frac{3}{8}$ **60.** $\frac{1}{12}$ **61.** $\frac{45}{13}$

62. 24,900 fans (rounded) **63.** $\frac{8}{3}$ **64.** 75 ft for \$1.99 − \$0.50 coupon, about \$0.020/ft **65.** 21 ft long; 15 ft wide **66.** 7.5 hours or $7\frac{1}{2}$ hours **67.** $\frac{1}{2}$ teaspoon or 0.5 teaspoon **68.** 21 points (rounded)

69. Set up the proportion to compare teaspoons to pounds on both sides.

$$\frac{1.5 \text{ teaspoons}}{24 \text{ pounds}} = \frac{x \text{ teaspoons}}{8 \text{ pounds}}$$

Show that cross products are equal. $(24)(x) = (1.5)(8)$

Divide both sides by 24. $\frac{\overset{1}{\cancel{24}}(x)}{\underset{1}{\cancel{24}}} = \frac{12}{24}$ so $x = \frac{1}{2}$ teaspoon or 0.5 teaspoon

70. (a) 1400 milligrams **(b)** 100 milligrams

Chapter 5 TEST (pages 376–377)

1. $\frac{4}{5}$ **2.** $\frac{20 \text{ miles}}{1 \text{ gallon}}$ **3.** $\frac{\$1}{5 \text{ minutes}}$ **4.** $\frac{9}{2}$ **5.** $\frac{15}{4}$ **6.** Best buy is 8 in. sub, about \$0.861/inch **7.** 16 ounces for \$1.89 − \$0.50 coupon, about \$0.087/ounce **8.** You earned less this year. An example is: $\frac{\text{Last year} \rightarrow \$30{,}000}{\text{This year} \rightarrow \$20{,}000} = \frac{3}{2}$ **9.** $\frac{3}{7} \neq \frac{2}{5}$ or $252 \neq 270$; false **10.** 5.88 = 5.88; true **11.** $x = 25$ **12.** $x \approx 2.67$ (rounded) **13.** $x = 325$ **14.** $x = 10\frac{1}{2}$ **15.** 24 orders **16.** 3.6 ounces **17.** 87 students (rounded) **18.** No, 4875 cannot be correct because there are only 650 students in the whole school. **19.** 23.8 grams (rounded) **20.** 60 ft

Chapters 1–5 CUMULATIVE REVIEW EXERCISES (pages 378–379)

1. 9900 **2.** 617.1 **3.** \$100 **4.** \$3.06 **5.** 29.34 **6.** 610 R27 **7.** 0.0076 **8.** 2312 **9.** 68.381 **10.** 55.6 **11.** 39 **12.** 18 **13.** 64 **14.** 0.95 **15.** one hundred five ten-thousandths **16.** 60.071 **17.** $\frac{1}{5}$ **18.** $\frac{4}{1}$ **19.** $1\frac{1}{8}$, 1.25, $1\frac{3}{8}$, 1.5, $1\frac{9}{16}$ (all inches) **20.** $\frac{1}{16}$ inch or 0.063 inch (rounded) **21.** $x = 21$ **22.** $x \approx 17.14$ (rounded) **23.** $x \approx 0.98$ (rounded) **24.** 250 pounds **25.** 26.7 centimeters (rounded) **26.** $7\frac{3}{20}$ miles **27.** 34 servings for \$3.35 − \$0.50 coupon, about \$0.084/serving **28.** 140 residents **29.** $1\frac{1}{4}$ teaspoons **30.** \$0.018, \$0.02, \$0.084, \$0.099, \$0.10, \$0.106, \$0.134 **31.** 202 minutes (rounded) **32.** 189 minutes (rounded) per \$20 card, so buy 2 cards to cover 360 minutes (6 hours). **33.** $\frac{1000}{200} = \frac{5}{1}$ **34.** 89.7 min rounds to 90, but the call would be cut off after 89 min.

CHAPTER 6 Percent

SECTION 6.1 (pages 387–392)

1. %; 100 **2.** decimal point; left **3.** 0.12 **5.** 0.70 or 0.7 **7.** 0.25 **9.** 1.40 or 1.4 **11.** 0.055 **13.** 1.00 or 1 **15.** 0.005 **17.** 0.0035 **19.** 100; % **20.** decimal point; right **21.** 60% **23.** 1% **25.** 37.5% **27.** 200% **29.** 370% **31.** 3.12% **33.** 416.2% **35.** 0.28% **37.** Answers will vary. Some possibilities are: No common denominators are needed with percents. The denominator is always 100 with percent, which makes comparisons easier to understand. **38.** Answers will vary. Some answers might be: when using discounts on purchases, calculating sales tax, figuring interest on loans, examining investments, finding tips in restaurants, calculating interest on savings, and doing math problems in this book. **39.** 0.13 **41.** 21.8% **43.** 0.151 **45.** 17.7% **47.** 0.146 **49.** 8% **51.** 0.30 or 0.3 **53.** 12 children **54.** 500 adults **55.** 420 employees **57.** 270 chairs **59.** \$377.50 **61.** 820 commuters **63.** 26 plants **65. (a)** Since 100% means 100 parts out of 100 parts, 100% is all of the number. **(b)** Answers will vary. For example, 100% of \$72 is \$72. **66. (a)** 50% means 50 parts out of 100 parts. That's half of the number. A shortcut for finding 50% of a number is to divide the number by 2. **(b)** Answers will vary. For example, 50% of \$14 is \$14 ÷ 2 = \$7. **67. (a)** Since 200% is two times a number, find 200% of the number by multiplying the number by 2 (double it). **(b)** Answers will vary. For example, 200% of \$20 is 2 • \$20 = \$40. **68. (a)** Since 300% is three times a number, find 300% of the number by multiplying the number by 3 (triple it). **(b)** Answers will vary. For example, 300% of \$10 is 3 • \$10 = \$30. **69. (a)** Since 10% means 10 parts out of 100 parts or $\frac{1}{10}$, the shortcut for finding 10% of a number is to move the decimal point in the number one place to the left. **(b)** Answers will vary. For example, 10% of \$90 is \$9. **70. (a)** Since 1% means 1 part out of 100 parts or $\frac{1}{100}$, the shortcut for finding 1% of a number is to move the decimal point in the number two places to the left. **(b)** Answers will vary. For example, 1% of \$500 is \$5. **71.** 12%; 0.12 **73. (a)** witch costume **(b)** 4.5%; 0.045 **75.** 9%; 0.09 **77. (a)** speeding **(b)** 5%; 0.05 **79.** 21%; 0.21 **81. (a)** candy **(b)** 10%; 0.10 **83.** 95% shaded; 5% unshaded **85.** 30% shaded; 70% unshaded **87.** 55% shaded; 45% unshaded **89.** 64% shaded; 36% unshaded

SECTION 6.2 (pages 398–403)

1. false **2.** true **3.** true **4.** false **5.** $\frac{17}{20}$ **7.** $\frac{5}{8}$ **9.** $\frac{1}{16}$ **11.** $\frac{1}{6}$ **13.** $\frac{1}{15}$ **15.** $\frac{1}{200}$ **17.** $1\frac{4}{5}$ **19.** $3\frac{3}{4}$ **21.** true **22.** false **23.** false **24.** true **25.** 70% **27.** 37% **29.** 62.5% **31.** 87.5% **33.** 48% **35.** 46% **37.** 35% **39.** 83.3% (rounded) **41.** 55.6% (rounded) **43.** 14.3% (rounded) **45.** (b) **46.** (d) **47.** $\frac{1}{2}$; 50% **49.** $\frac{7}{8}$; 0.875 **51.** 0.167 (rounded); 16.7% (rounded) **53.** $\frac{7}{10}$; 70% **55.** $\frac{1}{8}$; 0.125 **57.** 0.667 (rounded); 66.7% (rounded) **59.** 0.06; 6% **61.** 0.08; 8% **63.** 0.005; 0.5% **65.** $2\frac{1}{2}$; 250% **67.** 3.25; 325% **69.** There are many possible answers. Examples 2 and 3 show the steps that students should include in their answers. **70.** There are many correct answers. The table of percent equivalents shows some of the possibilities.

71. $\frac{9}{50}$; 0.18; 18% **73.** $\frac{13}{100}$; 0.13; 13% **75.** $\frac{1}{5}$; 0.2; 20% **77. (a)** $\frac{4}{5}$; 0.80; 80% **(b)** $\frac{1}{5}$; 0.20; 20% **79.** $\frac{1}{10}$; 0.1; 10% **81.** $\frac{2}{25}$; 0.08; 8% **83.** $\frac{13}{50}$; 0.26; 26% **85.** 100; 100 **86. (a)** 765 workers **(b)** 96 letters **(c)** 21 DVDs **87.** 50; 100; half or $\frac{1}{2}$ **88.** 10; 100; 1; left **89.** 1; 100; 2; left **90. (a)** 525 homes **(b)** 37 printers **(c)** \$0.08

91. Find 10% of \$160, then add $\frac{1}{2}$ of 10%.

$$\begin{array}{ccccc} 10\% & + & 5\% & = & 15\% \\ \downarrow & & \downarrow & & \downarrow \\ \$16 & + & \$8 & = & \$24 \end{array}$$

92. Find 100% of \$160, then add 50% of 160.

$$\begin{array}{ccccc} 100\% & + & 50\% & = & 150\% \\ \downarrow & & \downarrow & & \downarrow \\ \$160 & + & \$80 & = & \$240 \end{array}$$

93. From 100% of \$450, subtract 10% of \$450.

$$\begin{array}{ccccc} 100\% & - & 10\% & = & 90\% \\ \downarrow & & \downarrow & & \downarrow \\ \$450 & - & \$45 & = & \$405 \end{array}$$

94. To 100% of \$800, add 100% of \$800, and then add 10% of \$800.

$$\begin{array}{ccccccc} 100\% & + & \$100\% & + & 10\% & = & 210\% \\ \downarrow & & \downarrow & & \downarrow & & \downarrow \\ \$800 & + & \$800 & + & \$80 & = & \$1680 \end{array}$$

SECTION 6.3 (pages 408–411)

1. whole; $\frac{5}{\text{unknown}} = \frac{10}{100}$ **2.** percent; $\frac{1.5}{4.5} = \frac{\text{unknown}}{100}$

3. percent; $\frac{36}{24} = \frac{\text{unknown}}{100}$ **4.** part; $\frac{\text{unknown}}{72} = \frac{30}{100}$

5. part; $\frac{\text{unknown}}{160} = \frac{35}{100}$ **6.** whole; $\frac{20}{\text{unknown}} = \frac{25}{100}$

7. whole = 150 **9.** whole = 70 **11.** percent = 25% **13.** percent = 33.3% (rounded) **15.** part = 26 **17.** percent = 26.5% (rounded) **19.** 115 **21.** 0.4% **23.** 2.5% **25.** 0.3% **27.** percent; % **28.** entire or total; of **29.** whole **30.** cross multiply

31. $\begin{array}{l} \text{Part} \rightarrow \\ \text{Whole} \rightarrow \end{array} \frac{60}{\text{unknown}} = \frac{10}{100} \begin{array}{l} \leftarrow \text{Percent} \\ \leftarrow \text{Always } 100 \end{array}$

33. $\frac{600}{800} = \frac{75}{100}$ **35.** $\frac{\text{unknown}}{970} = \frac{25}{100}$ **37.** $\frac{12}{\text{unknown}} = \frac{20}{100}$

39. $\frac{54.34}{\text{unknown}} = \frac{3.25}{100}$ **41.** $\frac{\text{unknown}}{487} = \frac{0.68}{100}$

43. Percent—the ratio of the part to the whole. It appears with the word *percent* or "%" after it. Whole—the entire quantity. Often appears after the word *of.* Part—the part being compared with the whole.

44. A possible sentence is: Of the 580 cars entering the parking lot, 464 cars, or 80%, had parking stickers on their windshields.
percent = 80; whole = 580; part = 464

45. $\frac{730}{1262} = \frac{\text{unknown}}{100}$ **47.** $\frac{86}{142} = \frac{\text{unknown}}{100}$ **49.** $\frac{\text{unknown}}{610} = \frac{23}{100}$

51. $\frac{4300}{8600} = \frac{\text{unknown}}{100}$ **53.** $\frac{\text{unknown}}{480} = \frac{55}{100}$ **55.** $\frac{\text{unknown}}{822} = \frac{49.5}{100}$

57. $\frac{168}{\text{unknown}} = \frac{12}{100}$ **59.** $\frac{\text{unknown}}{680} = \frac{45}{100}$

SECTION 6.4 (pages 419–424)

1. part = percent • whole **2.** 76 **3.** 42 test tubes **5.** 1836 military personnel **7.** 4.8 ft **9.** 315 files **11.** 819 trucks **13.** \$3.28 **15.** 1530 tables **17.** 182 cell phones **19.** \$21.60

21. $\frac{80}{\text{unknown}} = \frac{25}{100}$ **22.** $\frac{32}{\text{unknown}} = \frac{5}{100}$ **23.** 160 hay bales **25.** 550 students **27.** 330 mountain bikes **29.** 2800

31. $\frac{18}{36} = \frac{\text{unknown}}{100}$ **32.** $\frac{62}{248} = \frac{\text{unknown}}{100}$ **33.** 52% **35.** 1.5; 1.5%

37. 18.6% (rounded) **39.** 9.2% **41.** 150% of \$30 cannot be less than \$30 because 150% is greater than 1 (100%). The answer must be greater than \$30. 25% of \$16 cannot be greater than \$16 because 25% is less than 1 (100%). The answer must be less than \$16. **42.** Answers will vary. One example is: There are 600 vehicles in the parking lot and 45 of them are pickup trucks. What percent are pickup trucks? Total vehicles, 600, is the whole, and the number of pickup trucks, 45, is the part.

$$\frac{45}{600} = \frac{x}{100}$$

$$600 \cdot x = 4500$$

$$\frac{\overset{1}{\cancel{600}} \cdot x}{\underset{1}{\cancel{600}}} = \frac{4500}{600}$$

$$x = 7.5 \text{ or } 7.5\%$$

43. **(a)** \$52.80 **(b)** \$187.20 **45.** 1963 people (rounded) **47.** 2144 people (rounded) **49.** **(a)** 0.96 million or 960,000 trips **(b)** 47.04 million or 47,040,000 trips **51.** 33% **53.** 2074 children **55.** 2% **57.** \$3200 monthly earnings; \$38,400 yearly earnings **59.** Breyers **61.** 76 people (rounded) **63.** 84.7% **65.** 1072 drivers **67.** 2156 products **69.** 231 customers **71.** whole; 100 **72.** whole **73.** 108 calories **74.** 300 grams **75.** 67 grams (rounded) **76.** 20 grams **77.** Yes, since they would eat 28% × 4 servings = 112% of the daily value. **78.** 17 packages (rounded). It may be possible but would result in a diet that is high in total fat, saturated fat, sodium, and total carbohydrates.

SECTION 6.5 (pages 429–432)

1. 46 percent; 780 whole **2.** 1800 whole; 14 percent **3.** 270 donors **5.** 1350 bath towels **7.** 83.2 quarts **9.** 3500 airbags **11.** 1029.2 meters **13.** \$4.16 **15.** 70 percent; 476 part **16.** 270 part; 45 percent **17.** 160 patients **19.** 325 salads **21.** 1080 people **23.** 300 gallons **25.** percent; two; right; % **26.** decimal; two; left; % **27.** 76% **29.** 1.5% **31.** 250% **33.** You must first change the fraction in the percent to a decimal, then divide the percent by 100 to change it to a decimal.

$$\underbrace{2\tfrac{1}{2}\%}_{\text{Write } 2\frac{1}{2} \text{ as } 2.5.} = 2.5\% = 0.025 \leftarrow 2.5\% \text{ as a decimal}$$

34. The correct answer is \$6.50. The error is in changing $\frac{1}{2}\%$ to a decimal.

$\frac{1}{2}\% = 0.5\% = 0.005;$
$(0.005)(\$1300) = \$6.50$ Correct

Here are the incorrect answers and how your classmates got them.

$$\left.\begin{array}{ll} \frac{1}{2}\% = 0.0005; & (0.0005)(\$1300) = \$0.65 \\ \frac{1}{2}\% = 0.05; & (0.05)(\$1300) = \$65 \\ \frac{1}{2}\% = 0.5; & (0.5)(\$1300) = \$650 \end{array}\right\} \text{Incorrect}$$

35. 3.78 million or 3,780,000 office workers **37. (a)** 9625 people **(b)** 1875 people **(c)** 1000 people **39. (a)** 17,640 have a refrigerator **(b)** 360 do not have a refrigerator **41. (a)** 4% **(b)** 96% **43.** 36.9% (rounded) **45. (a)** relatives **(b)** 1950 families **47.** 1326 families **49.** $24,009 **51.** 478,175 Mustangs (rounded) **53.** $510,390 **55.** $439.61

SUMMARY EXERCISES Using Percent Proportion and Percent Equation (pages 433–434)

1. 0.0625 **2.** 3.80 or 3.8 **3.** 37.5% **4.** 0.6% **5.** $\frac{7}{8}$ **6.** $1\frac{3}{5}$ **7.** 62.5% **8.** 0.8% **9.** $\frac{13}{50}$; 0.26; 26% **10.** $\frac{17}{100}$; 0.17; 17% **11.** $\frac{3}{50}$; 0.06; 6% **12.** $\frac{9}{100}$; 0.09; 9% **13.** $46.80 **14.** 980 customers **15.** 55% **16.** 28 screening exams **17.** 363 acres **18.** 2.7% (rounded) **19.** 168 people (rounded) **20.** $429.92 **21.** 243,000 loans (rounded) **22.** 1749 owners **23.** 40% **24.** 3.7% (rounded) **25.** 142.5 million or 142,500,000 returns (rounded) **26.** 3419 ski-lift tickets (rounded)

SECTION 6.6 (pages 441–444)

1. whole; percent; part **2.** rate; cost of the item; sales **3.** $0.24; $6.24 **5.** 3%; $437.75 **7.** $672.60 (rounded); $12,901.60 (rounded) **9.** sales amount; commission rate; commission **10.** commission; sales amount **11.** $22.40 **13.** 20% **15.** $185.51 (rounded) **17.** whole; percent; part **18.** discount; original **19.** $20 (rounded); $179.99 (rounded) **21.** 30%; $126 **23.** $8.76; $49.64 **25.** On the basis of commission alone you would choose Company A. Other considerations might be: reputation of the company; expense allowances; other fringe benefits; travel; promotion and training, to name a few. **26.** Some answers might be: calculating percent pay increases or decreases; changes in the cost of utilities, groceries, gasoline, and insurance; changes in the value of investments; the economy (inflation or deflation)—to name a few. **27.** $203.93 (rounded) **29.** $90.62 (rounded) **31.** 5% **33.** 35.9% (rounded) **35.** 25% **37.** $134 **39.** $568.80 **41.** 4% **43.** $85.80; $304.20 **45.** 22% **47.** $18.43 (rounded) **49.** $6087.20 **51.** $14,032.12 (rounded) **53.** percent; whole **54.** rate (percent) of tax; cost of item **55.** $110.68 (rounded) **56.** Yes, the same; (10% • $95) + (6.5% • $95) = $15.68 (rounded) and (10% + 6.5%) • ($95) = $15.68 **57.** $1360.12 **58.** No. In Exercise 57 the excise tax of $15.40 cannot be added to the sales tax rate of $7\frac{3}{4}$% because the excise tax is an amount, not a percent.

SECTION 6.7 (pages 447–450)

1. 0.035 **2.** 0.075 **3.** 2.25 years **4.** 5.75 years **5.** $6 **7.** $105 **9.** $258.75 **11.** $2227.50 **13.** $\frac{3}{12}$ or $\frac{1}{4}$; 0.25 **14.** $\frac{6}{12}$ or $\frac{1}{2}$; 0.5 **15.** $\frac{9}{12}$ or $\frac{3}{4}$; 0.75 **16.** $\frac{15}{12}$ or $\frac{5}{4}$; 1.25 **17.** $6 **19.** $42.30 **21.** $16.84 (rounded) **23.** $647.06 (rounded) **25.** $210 **27.** $773.30 **29.** $2043 **31.** $3306.88 (rounded) **33.** $17,797.81 (rounded) **35.** The answer should include: Amount of principal—This is the amount of money borrowed or loaned. Interest rate—This is the percent used to calculate the interest. Time of loan—The length of time that money is loaned or borrowed is an important factor in determining interest. **36.** When time is given in months, the number of months are placed over 12 because there are 12 months in a year. This becomes a fraction of a year. Here is an example.

$$6 \text{ months} = \frac{6}{12}\text{ year} = \frac{1}{2}\text{ year} = 0.5 \text{ year}$$

37. $345.68 **39.** $26,250 **41.** $4746.88 (rounded) **43.** $277.50 **45.** $159.50 **47. (a)** $222.75 **(b)** $5622.75 **49.** $3258.50

SECTION 6.8 (pages 455–458)

1. false **2.** interest; principal **3.** $540.80 **5.** $1966.91 (rounded) **7.** $4587.79 (rounded) **9.** 1.02; 1.02; 1.02 **10.** 1.05; 1.05 **11.** $1102.50 **13.** $1873.52 (rounded) **15.** $2027.46 (rounded) **17.** $13,842.59 (rounded) **19.** 1.1941 **20.** 1.5530 **21.** 1.1717 **22.** 1.1742 **23.** $216.70 **25.** $9752; $1752 **27.** $10,976.01 (rounded); $2547.84 (rounded) **29.** Interest paid on past interest as well as on the principal. Many people describe compound interest as "interest on interest." **30.** Compound amount is the total amount, original deposit plus interest on deposit, at the end of the compound interest period. Compound interest is found by subtracting the original deposit from the compound amount. **31.** $60,835 **33. (a)** $128,402 **(b)** $52,402 **35. (a)** $87,786.23 **(b)** $17,786.23 **37.** principal; rate; time; *p*; *r*; *t* **38.** principal; interest **39.** principal; interest **40.** principal; interest **41. (a)** $254.48 **(b)** No; it has more than doubled (about five times). **(c)** Examples will vary. **42. (a)** the compound interest account; $474.55 more **(b)** Greater amounts of interest are earned with compound interest than simple interest because interest is earned on both principal and past interest.

Chapter 6 REVIEW EXERCISES (pages 466–471)

1. 0.35 **2.** 1.5 **3.** 0.9944 **4.** 0.00085 **5.** 315% **6.** 2% **7.** 87.5% **8.** 0.2% **9.** $\frac{3}{20}$ **10.** $\frac{3}{8}$ **11.** $1\frac{3}{4}$ **12.** $\frac{1}{400}$ **13.** 75% **14.** 62.5% or $62\frac{1}{2}$% **15.** 325% **16.** 0.5% **17.** 0.125 **18.** 12.5% **19.** $\frac{1}{4}$ **20.** 25% **21.** $1\frac{4}{5}$ **22.** 1.8 **23.** whole = 250 **24.** part = 24

25. $\text{Part} \rightarrow \frac{287}{820} = \frac{35}{100} \leftarrow \text{Percent}$, $\text{Whole} \rightarrow 820$, $100 \leftarrow \text{Always } 100$ **26.** $\frac{73}{90} = \frac{\text{unknown}}{100}$

27. $\frac{\text{unknown}}{160} = \frac{14}{100}$ **28.** $\frac{418}{\text{unknown}} = \frac{16}{100}$ **29.** $\frac{3}{8} = \frac{\text{unknown}}{100}$

30. $\frac{\text{unknown}}{1280} = \frac{88}{100}$ **31.** 171 programs **32.** 870 reference books

33. 31.2 acres **34.** 2.8 kilograms **35.** 750 crates **36.** 2320 test tubes **37.** 484 miles **38.** 17,000 cases **39.** 55% **40.** 5.2% (rounded) **41.** 9.5% (rounded) **42.** 30.8% (rounded) **43.** 79 shark attacks (rounded) **44.** 28% **45.** $145.28 **46.** 186 trucks **47.** 0.4% **48.** 175% **49.** 120 miles **50.** $575 **51.** $31.50; $661.50 **52.** $7\frac{1}{2}$%; $838.50 **53.** $276 **54.** 5% **55.** $33.75; $78.75 **56.** 25%; $189 **57.** $8 **58.** $67.50 **59.** $3.50 **60.** $152.10 **61.** $832.50 **62.** $1595.10 **63.** $5375.60; $1375.60 **64.** $2187.71 (rounded); $317.71 (rounded) **65.** $4108.32; $508.32 **66.** $16,337.50; $3837.50 **67.** part = 12 **68.** whole = 1640 **69.** 23.28 meters **70.** 150% **71.** $0.51 **72.** 1980 employees **73.** 40% **74.** 498.8 liters **75.** 0.55 **76.** 3

ANSWERS

77. 500% **78.** 471% **79.** 0.086 **80.** 62.1% **81.** 0.00375

82. 0.06% **83.** 75% **84.** $\frac{21}{50}$ **85.** $\frac{7}{8}$ **86.** 37.5% or $37\frac{1}{2}\%$

87. $\frac{13}{40}$ **88.** 60% **89.** $\frac{1}{400}$ **90.** 375% **91.** \$3331.25 **92.** \$16,520

93. (a) 96.0% (rounded) **(b)** 45.9% (rounded) **94. (a)** \$179,894.93 (rounded) **(b)** \$31,894.93 (rounded) **95.** \$6225 **96.** 18.2% (rounded)
97. \$848.40 (rounded) **98.** 1100 boaters **99.** \$13,005 **100.** 33.4% (rounded)

Chapter 6 TEST (pages 472–473)

1. 0.65 **2.** 80% **3.** 175% **4.** 87.5% or $87\frac{1}{2}\%$ **5.** 3.00 or 3 **6.** 0.02
7. $\frac{1}{8}$ **8.** $\frac{1}{400}$ **9.** 60% **10.** 62.5% or $62\frac{1}{2}\%$ **11.** 250%
12. 800 sacks **13.** 20% **14.** 125,000,000 households **15.** \$3450.60

16. 8% **17.** 25% **18.** A possible answer is: Part is the increase in salary. Whole is last year's salary. Percent of increase is unknown.

$$\frac{\text{amount of increase}}{\text{last year's salary}} = \frac{p}{100}$$

19. The interest formula is $I = p \cdot r \cdot t$. If time is in months, it is expressed as a fraction with 12 as the denominator. If time is expressed in years, it is placed over 1 or shown as a decimal number.

Problems will vary.

Some possibilities are:

$$I = (1000)\,(0.05)\left(\frac{9}{12}\right) = \$37.50$$

$$I = (1000)\,(0.05)\,(2.5) = \$125$$

20. \$11.52; \$84.48 **21.** \$91; \$189 **22.** \$378 **23.** \$192 **24.** \$20,880
25. (a) \$10,668 (rounded) **(b)** \$1668

Chapters 1–6 CUMULATIVE REVIEW EXERCISES (pages 474–475)

1. *Estimate:* 80,000 − 50,000 = 30,000; *Exact:* 28,988
2. *Estimate:* 8 − 4 = 4; *Exact:* 4.2971
3. *Estimate:* 7000 × 700 = 4,900,000; *Exact:* 4,628,904
4. *Estimate:* 70 × 9 = 630; *Exact:* 568.11
5. *Estimate:* 40,000 ÷ 40 = 1000; *Exact:* 902
6. *Estimate:* 7 ÷ 1 = 7; *Exact:* 8.45

7. 18 **8.** 19 **9.** 39 **10.** 7,600,000 **11.** \$513 **12.** \$362.74

13. $13\frac{3}{8}$ **14.** $3\frac{7}{8}$ **15.** $28\frac{4}{5}$ **16.** 16 **17.** $12\frac{1}{2}$ ft^2 **18.** $22\frac{1}{2}$ hours

19. $<$ **20.** $<$ **21.** $>$ **22.** $\frac{1}{4}$ **23.** 1 **24.** $\frac{1}{4}$ **25.** 0.75 **26.** 0.375

27. 0.583 (rounded) **28.** 0.55 **29.** $x = 6$ **30.** $x = 4$ **31.** $x = 16$

32. $x = 30$ **33.** 0.03 **34.** 2.00 or 2 **35.** 87% **36.** 380% **37.** $\frac{2}{25}$

38. $\frac{5}{8}$ **39.** $1\frac{3}{4}$ **40.** 87.5% or $87\frac{1}{2}\%$ **41.** 420% **42.** 2132 DVDs

43. 1700 miles **44.** 140% **45.** 6.5% or $6\frac{1}{2}\%$ **46.** \$117.18

47. \$205.20; \$250.80 **48.** \$47,341.75 **49.** 45 children
50. 255 ounces **51.** 12.5% **52.** 2% **53.** 5.5% (rounded)
54. 8.1% (rounded)

Appendix

(pages A-5–A-6)

1. 37; add 7 **3.** 26; add 10, subtract 2 **5.** 16; multiply by 2
7. 243; multiply by 3 **9.** 36; add 3, add 5, add 7, etc.; or $1^2, 2^2, 3^2$, etc.

11. **13.**

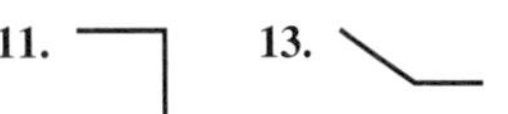

15. Conclusion follows. **17.** Conclusion does not follow.

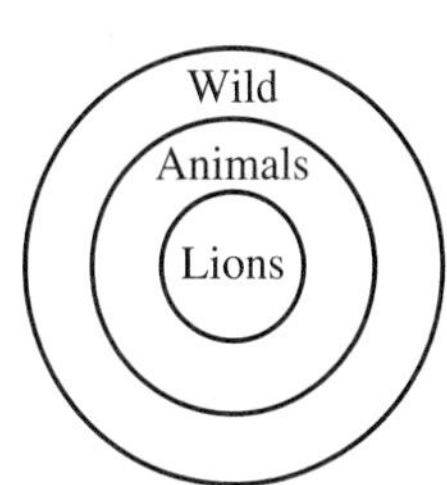

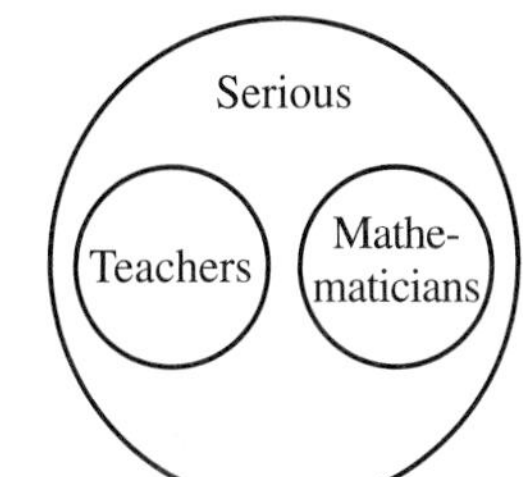

19.

Together
2 18 7
Husband
Wife

2 + 18 + 7 = 27
30 − 27 = 3

Neither watched TV on 3 days.

21. Dick is the computer operator.

ANSWERS

Solutions to Selected Exercises

CHAPTER 1 Whole Numbers

SECTION 1.1 (pages 8–9)

9. 3,561,435; millions: 3; thousands: 561; ones: 435

15. 346,009 is three hundred forty-six thousand, nine.

31. The least-used method of transportation is public transportation. 6,069,589 in words is six million, sixty-nine thousand, five hundred eighty-nine.

SECTION 1.2 (pages 18–21)

11. Line up the numbers in columns. Then start at the right and add the ones digits. Add the ten digits next, and, finally, the hundreds digits.

$$\begin{array}{r} 932 \\ 44 \\ +\ 613 \\ \hline 1589 \end{array}$$

49. *Step 1*

Add the digits in the ones column.

Step 2

Add the digits in the tens column, including the regrouped 2.

Step 3

Add the hundreds column, including the regrouped 1.

Step 4

Add the thousands column, including the regrouped 1.

$$\begin{array}{r} {\scriptstyle 1\,12\ \ } \\ 18 \\ 708 \\ 9\,286 \\ +\ 636 \\ \hline 10{,}648 \end{array}$$

63. Add up to check addition.

$$\begin{array}{rl} 769 & \\ \hline 179 & \\ 214 & \\ +\ 376 & \\ \hline 759 & \text{incorrect; should be 769} \end{array}$$

77. To find the total amount raised, add the amounts raised at each event.

\$3 482	flea market
+ 12,860	annual auction
\$16,342	total amount raised

SECTION 1.3 (pages 28–31)

33. In the ones column, 5 is less than 7, so in order to subtract, regroup 1 ten as 10 ones. Then subtract 7 ones from 15 ones in the ones column. Finally, subtract 3 tens from 6 tens in the tens column.

$$\begin{array}{r} {\scriptstyle 6\,15} \\ \not{7}\not{5} \\ -37 \\ \hline 38 \end{array}$$

65.

$$\begin{array}{r} {\scriptstyle 9\ \ 9\ \ \ } \\ {\scriptstyle 5\ \ \not{10}\ \not{10}\ 10} \\ 6\not{6},\not{0}\,\not{0}\,\not{0} \\ -34,4\,4\,4 \\ \hline 31,5\,5\,6 \end{array}$$

81. To find how many fewer calories a woman burns, subtract the number of calories a woman burns from the number of calories a man burns.

187	calories man burns
−140	calories woman burns
47	fewer calories for a woman

The woman burned 47 fewer calories.

SECTION 1.4 (pages 38–41)

11. $\underbrace{(4)(5)}_{20(2)\,=\,40}(2)$ or $(4)\underbrace{(5)(2)}_{(4)(10)\,=\,40}$

37. $\begin{array}{r} 600 \\ \times\ 6 \\ \hline \end{array}$ $\begin{array}{r} 6 \\ \times 6 \\ \hline 36 \end{array}$ $\begin{array}{r} 600 \\ \times\ 6 \\ \hline 3600 \end{array}$ Attach 00.

73. First multiply 9352 by 4. Then multiply 9352 by 6, making sure to line up the tens. Then multiply 9352 by 2, making sure to line up the hundreds. Then add the partial products.

$$\begin{array}{rl} 9\,3\,5\,2 & \\ \times\quad 2\,6\,4 & \\ \hline 3\,7\,4\,0\,8 & \leftarrow 4 \times 9352 \\ 5\,6\,1\,1\,2\ \ & \leftarrow 6 \times 9352 \\ 1\,8\,7\,0\,4\ \ \ \ & \leftarrow 2 \times 9352 \\ \hline 2{,}4\,6\,8{,}9\,2\,8 & \end{array}$$

85. *Approach* To find the number of balls purchased, multiply the number of cartons (300) by the number of balls per carton (10).

Solution

300	cartons
× 10	balls per carton
3000	balls

3000 balls were purchased.

103. Multiply the number of laptop computers purchased (12) by the price per laptop (\$970). Then multiply the number of printers purchased (8) by the price per printer (\$315).

$$\begin{array}{r} 970 \\ \times\ \ 12 \\ \hline 11{,}640 \end{array} \qquad \begin{array}{r} 315 \\ \times\ \ 8 \\ \hline 2520 \end{array}$$

The total cost is the sum of these values: \$11,640 + \$2520 + \$14,160

SECTION 1.5 (pages 51–54)

3. $24 \div 4 = 6$: $4\overline{)24}$ with quotient 6, or $\dfrac{24}{4} = 6$

35. $\begin{array}{r} 1\ 5\ 2\ 2\ \mathbf{R}5 \\ 6\overline{)9\,{}^{3}1\,{}^{1}3\,{}^{1}7} \end{array}$

CHECK $(6 \times 1522) + 5 = 9132 + 5$
$= 9137$

51. $\begin{array}{r} 1\,0,\ 2\ 5\ 3\ \mathbf{R}5 \\ 7\overline{)7\,1,{}^{1}7\,{}^{3}7\,{}^{2}6} \end{array}$

CHECK $(7 \times 10{,}253) + 5 = 71{,}771 + 5$
$= 71{,}776$

67. $\begin{array}{r} 9\ 6\,2\,8\ \mathbf{R}7 \\ 9\overline{)8\,6,6\,5\,5} \end{array}$

CHECK $(9 \times 9628) + 7 = 86{,}652 + 7$
$= 86{,}659$

Does not match dividend, incorrect

Rework:

$\begin{array}{r} 9\ 6\ 2\ 8\ \mathbf{R}3 \\ 9\overline{)8\,6,{}^{5}6\,{}^{2}5\,{}^{7}5} \end{array}$

CHECK $(9 \times 9628) + 3 = 86{,}652 + 3$
$= 86{,}655$

Matches dividend, correct

73. To find the number of tables that can be set, divide the number of napkins (2624) by the number of napkins it takes to set each table (8).

$\begin{array}{r} 3\ 2\ 8 \\ 8\overline{)2\,6\,{}^{2}2\,{}^{6}4} \end{array}$

328 tables can be set.

SECTION 1.6 (pages 60–62)

13. Use 2 as a trial divisor, since 18 is closer to 20 than to 10.

$\dfrac{131}{2} = 6$ with 1 left over

Because $6 \times 18 = 108$ and $131 - 108 = 23$, which is greater than the divisor, use 7 instead. To find the next digit in the quotient, use 2 as a trial divisor again.

$\dfrac{5}{2} = 2$ with 1 left over

(Continued)

SOLUTIONS

Because $2 \times 18 = 36$ and $59 - 36 = 23$, which is greater than the divisor, use 3 instead.

```
     7 3 R5      CHECK     7 3
18 )1 3 1 9              × 1 8
    1 2 6                5 8 4
      5 9                7 3
      5 4              1 3 1 4
        5            +     5
                       1 3 1 9 ✓ matches
```

25.

```
     1 0 1 R4    CHECK     1 0 1
35 )3 5 4 9              ×   3 5
                           5 0 5
                         3 0 3
                         3 5 3 5
                       +       4
                         3 5 3 9   incorrect
```

Rework:

```
     1 0 1 R14   CHECK     1 0 1
35 )3 5 4 9              ×   3 5
    3 5                    5 0 5
      0 4 9              3 0 3
        3 5              3 5 3 5
        1 4            +     1 4
                         3 5 4 9 ✓ matches
```

The correct answer is 101 **R**14.

39. **(a)** Divide the number of Big Macs eaten by 39 years

```
        6 4 8
39 )2 5, 2 7 2
    2 3 4
      1 8 7
      1 5 6
        3 1 2
        3 1 2
            0
```

The average number of Big Macs eaten each year was 648.

(b) 2 Big Macs eaten each day would be $2 \times 365 = 730$. He has eaten less than 2 Big Macs each day.

SECTION 1.7 (pages 72–75)

9. 855 rounded to the nearest ten: 860

8$\underline{5}$5 Underline the 5 in the tens place. Next digit is 5 or more, so add one to the 5 in the tens place. All digits to the right of the underlined place change to zero.

25. **To the nearest ten:** 44$\underline{7}$6

Underline the 7 in the tens place. Next digit is 5 or more, so add 1 to the 7. All digits to the right of the underlined place are changed to zero.

4480

To the nearest hundred: 4$\underline{4}$76

Underline the 4 in the hundreds place. Next digit is 5 or more, so add 1 to the 4. All digits to the right of the underlined place are changed to zero.

4500

To the nearest thousand: $\underline{4}$476

Underline the 4 in the thousands place. Next digit is 4 or less, so leave 4 as 4 in the thousands place. All digits to the right of the underlined place are changed to zero.

4000

41.

Estimate:	*Exact:*
30	25
60	63
50	47
+ 80	+ 84
220	219

47.

Estimate:	*Exact:*
900	863
700	735
400	438
+ 800	+ 792
2800	2828

53.

Estimate:	*Exact:*
8000	8 215
60	56
700	729
+ 4000	+ 3 605
12,760	12,605

61. $\underline{7}$6,000,000 Next digit is 5 or more, so add 1 to 7. Change all digits to the right of the underlined place to zero. Add 1 to 7.

80 million people

3$\underline{1}$1,000,000 Next digit is 4 or less, so leave 1 as 1. Change all digits to the right of the underlined place to zero.

310 million people

SECTION 1.8 (pages 78–81)

5. 8^2: exponent is 2, base is 8.

$8^2 = 8 \cdot 8 = 64$

21. $6^2 = 36$, so $\sqrt{36} = 6$.

37.

$5 \cdot 3^2 + \frac{0}{8}$	Exponent
$5 \cdot 9 + \frac{0}{8}$	Multiply.
$45 + \frac{0}{8}$	Divide.
$45 + 0 = 45$	Add.

67.

$8 \cdot \sqrt{49} - 6(9 - 4)$	Parentheses
$8 \cdot \sqrt{49} - 6(5)$	Square root
$8 \cdot 7 - 6(5)$	Multiply.
$56 - 6(5)$	Multiply.
$56 - 30 = 26$	Subtract.

85.

$8 \cdot 9 \div \sqrt{36} - 4 \div 2 + (14 - 8)$	Parentheses
$8 \cdot 9 \div \sqrt{36} - 4 \div 2 + 6$	Square root
$8 \cdot 9 \div 6 - 4 \div 2 + 6$	Multiply.
$72 \div 6 - 4 \div 2 + 6$	Divide.
$12 - 4 \div 2 + 6$	Divide.
$12 - 2 + 6$	Subtract.
$10 + 6 = 16$	Add.

SECTION 1.9 (pages 86–89)

7. According to the pictograph, Family Dollar has 4500 stores. Subtract 4500 from 5750 to find the difference.

$5750 - 4500 = 1250$ fewer stores

15. According to the bar graph, $18 - 9 = 9$ more people out of 100 found their career as a result of "Studied in school" than "Luck or chance."

19. According to the line graph the increase in the number of installations from 2113 to 2014 is $7000 - 2500 = 4500$.

23. We want to insert parentheses in $7 - 2 \cdot 3 - 6$ so that it simplifies to 9. There isn't a method to use, so just use trial and error.

$(7 - 2) \cdot 3 - 6 = 5 \cdot 3 - 6$
$= 15 - 6 = 9$

SECTION 1.10 (pages 94–97)

11. ***Step 1***

Find the number of toys each child will receive.

Step 2

"Same number of toys to each" indicates division should be used.

Step 3

3000 divided by 700 gives an estimate of about 4 toys.

Step 4

```
          4
657 )2 6 2 8
     2 6 2 8
           0
```

Step 5

Each child will receive 4 toys.

Step 6
Exact answer is close to estimate. Also, $4 \times 657 = 2628$ which matches the number given in the problem.

21. *Step 1*
Find her monthly savings.

Step 2
Her monthly take home pay and expenses are given. "Remainder" indicates subtraction may be used.

Step 3
Estimate:
$2000 − $700 − $300 − $400 − $200 − $200 = $200

Step 4

$695	$2240
340	− 1890
435	$350
240	
+ 180	
1890	

Step 5
Her monthly savings are $350.

Step 6
The exact answer seems reasonable compared to the estimate.

Check: $350 + $1890 = $2240. Re-add the expenses to check the sum of $1890.

29. *Step 1*
The number and cost of wheelchairs and recorder-players are given, and the total cost of all items must be found.

Step 2
Find the cost of all wheelchairs and the cost of all recorder-players. Then add these costs to get the total cost.

Step 3
The cost of the wheelchairs is about $1000 × 6 = $6000.
The cost of the recorder-players is about $900 × 20 = $18,000.
Estimate: $6000 + $18,000 = $24,000

Step 4

cost of wheelchairs: $1256 × 6 = $7536

cost of recorder-players:

$8 9 5
× 1 5
4 4 7 5
8 9 5
$1 3,4 2 5

Step 5
The total cost is $13,425 + $7536 = $20,961.

Step 6
The exact answer is reasonably close to the estimate. Check by repeating Step 4.

41. *Step 1*
The total seating is given along with the information to find the number of seats on the main floor. The number of rows of seats in the balcony is given, and the number of seats in each row of the balcony must be found.

Step 2
First, the number of seats on the main floor must be found. Next, the number of seats on the main floor must be subtracted from the total number of seats to find the number of seats in the balcony. Finally, the number of seats in the balcony must be divided by the number of rows of seats in the balcony.

Step 3
Seats on main floor: $30 \times 25 = 750$
Seats in balcony: $1250 - 750 = 500$
Seats in each row in balcony:
$500 \div 25 = 20$
Since we didn't round, our estimate matches our exact answer.

Step 4
See the calculations in Step 3.

Step 5
The number of seats in each row of the balcony is 20.

Step 6
The answer matches the estimate, as expected.

Check:
Seats in balcony: $20 \times 25 = 500$
Seats on main floor: $30 \times 25 = 750$
Total seats: $500 + 750 = 1250$, which matches the number given in the problem.

CHAPTER 2 Multiplying and Dividing Fractions

SECTION 2.1 (pages 116–118)

13. Each of the two figures is divided into 5 parts and 7 are shaded: $\frac{7}{5}$

Three are unshaded: $\frac{3}{5}$

21. Proper fractions: numerator *less* than denominator.

$$\frac{1}{3}, \frac{5}{8}, \frac{7}{16}$$

Improper fractions: numerator *greater than or equal to* denominator.

$$\frac{8}{5}, \frac{6}{6}, \frac{12}{2}$$

SECTION 2.2 (pages 123–125)

17. Write $5\frac{4}{5}$ as an improper fraction.

$5 \cdot 5 = 25$ Multiply 5 and 5.
$25 + 4 = 29$ Add 4. The numerator is 29.
$5\frac{4}{5} = \frac{29}{5}$ Use the same denominator.

43. $\frac{63}{4}$

Divide 63 by 4.

$$\begin{array}{r} 15 \\ 4\overline{)63} \\ \underline{4} \\ 23 \\ \underline{20} \\ 3 \end{array}$$

15 ← Whole number part; 3 ← Remainder

The quotient 15 is the whole number part of the mixed number. The remainder 3 is the numerator of the fraction, and the denominator remains as 4.

$$\frac{63}{4} = 15\frac{3}{4}$$

61. Write $522\frac{3}{8}$ as an improper fraction.

$522 \cdot 8 = 4176$
$4176 + 3 = 4179$ Add 3. The numerator is 4179.
$522\frac{3}{8} = \frac{4179}{8}$ Use the same denominator.

SECTION 2.3 (pages 130–131)

5. Factorizations of 48:
$1 \cdot 48 = 48$ $2 \cdot 24 = 48$ $3 \cdot 16 = 48$
$4 \cdot 12 = 48$ $6 \cdot 8 = 48$
The factors of 48 are 1, 2, 3, 4, 6, 8, 12, 16, 24, and 48.

41. 44

$11\overline{)11}$ = 1 Divide 11 by 11.
$2\overline{)22}$ = 11 Divide 22 by 2.
$2\overline{)44}$ = 22 Divide 44 by 2.

(Continued)

Quotient is 1.

Because all factors (divisors) are prime, the prime factorization of 44 is $2 \cdot 2 \cdot 11 = 2^2 \cdot 11$

SECTION 2.4 (pages 136–137)

21. Because the greatest common factor of 56 and 64 is 8, divide both numerator and denominator by 8.

$$\frac{56}{64} = \frac{56 \div 8}{64 \div 8} = \frac{7}{8}$$

33. Write the prime factorization of both numerator and denominator. Then divide both numerator and denominator by any common factors, and write a 1 by each factor that has been divided. Finally, multiply the remaining factors in both numerator and denominator.

$$\frac{18}{24} = \frac{\overset{1}{\cancel{2}} \cdot \overset{1}{\cancel{3}} \cdot 3}{\underset{1}{\cancel{2}} \cdot 2 \cdot 2 \cdot \underset{1}{\cancel{3}}} = \frac{1 \cdot 1 \cdot 3}{1 \cdot 2 \cdot 2 \cdot 1} = \frac{3}{4}$$

43. $\frac{3}{6}$ and $\frac{18}{36}$ Write each fraction in lowest terms.

$$\frac{3}{6} = \frac{1 \cdot \overset{1}{\cancel{3}}}{2 \cdot \underset{1}{\cancel{3}}} = \frac{1}{2}$$

$$\frac{18}{36} = \frac{\overset{1}{\cancel{2}} \cdot \overset{1}{\cancel{3}} \cdot \overset{1}{\cancel{3}}}{\underset{1}{\cancel{2}} \cdot 2 \cdot \underset{1}{\cancel{3}} \cdot \underset{1}{\cancel{3}}} = \frac{1}{2}$$

The fractions are *equivalent* $\left(\frac{1}{2} = \frac{1}{2}\right)$.

57. Write the prime factorization of both numerator and denominator. Then divide both numerator and denominator by any common factors, and write a 1 by each factor that has been divided. Finally, multiply the remaining factors in both numerator and denominator.

$$\frac{160}{256} = \frac{\overset{1}{\cancel{2}} \cdot \overset{1}{\cancel{2}} \cdot \overset{1}{\cancel{2}} \cdot \overset{1}{\cancel{2}} \cdot \overset{1}{\cancel{2}} \cdot 5}{\underset{1}{\cancel{2}} \cdot \underset{1}{\cancel{2}} \cdot \underset{1}{\cancel{2}} \cdot \underset{1}{\cancel{2}} \cdot \underset{1}{\cancel{2}} \cdot 2 \cdot 2 \cdot 2} = \frac{5}{8}$$

SECTION 2.5 (pages 148–151)

11. Divide a numerator and denominator by the same number to use the multiplication shortcut.

$$\frac{2}{3} \cdot \frac{7}{12} \cdot \frac{9}{14} = \frac{\overset{1}{\cancel{2}}}{\underset{1}{\cancel{3}}} \cdot \frac{\overset{1}{\cancel{7}}}{\underset{\underset{2}{\cancel{6}}}{\cancel{12}}} \cdot \frac{\overset{\overset{1}{\cancel{3}}}{\cancel{9}}}{\underset{2}{\cancel{14}}}$$

$$= \frac{1 \cdot 1 \cdot 1}{1 \cdot 2 \cdot 2} = \frac{1}{4}$$

25. Write 36 as $\frac{36}{1}$ and multiply.

$$36 \cdot \frac{5}{8} \cdot \frac{9}{15} = \frac{\overset{9}{\cancel{36}}}{1} \cdot \frac{\overset{1}{\cancel{5}}}{\underset{2}{\cancel{8}}} \cdot \frac{\overset{3}{\cancel{9}}}{\underset{\underset{1}{\cancel{3}}}{\cancel{15}}} = \frac{27}{2} = 13\frac{1}{2}$$

41. Area = length • width

$$\text{Area} = \frac{5}{6} \cdot \frac{3}{10}$$

$$= \frac{\overset{1}{\cancel{5}}}{\underset{2}{\cancel{6}}} \cdot \frac{\overset{1}{\cancel{3}}}{\underset{2}{\cancel{10}}} \quad \begin{array}{l}\text{Divide 5 and 10 by 5.}\\ \text{Divide 3 and 6 by 3.}\end{array}$$

$$= \frac{1}{4} \text{ square mile}$$

45. Area = length • width

$$\text{Area} = 2 \cdot \frac{3}{4}$$

$$= \frac{\overset{1}{\cancel{2}}}{1} \cdot \frac{3}{\underset{2}{\cancel{4}}} \quad \text{Divide numerator and denominator by 2.}$$

$$= \frac{3}{2} \quad \text{Write as a mixed number.}$$

$$= 1\frac{1}{2} \quad \text{Square yards}$$

SECTION 2.6 (pages 155–158)

5. ***Step 1***

The problem asks for the area of the front surface of a digital photo frame.

Step 2

To find area, multiply the length of $\frac{3}{4}$ foot by the width of $\frac{2}{3}$ foot.

Step 3

An estimate is $\frac{6}{12}$ square foot.

$$\frac{3}{4} \cdot \frac{2}{3} = \frac{\overset{1}{\cancel{3}} \cdot \overset{1}{\cancel{2}}}{\underset{2}{\cancel{4}} \cdot \underset{1}{\cancel{3}}} = \frac{1}{2}$$

Step 4

The exact value is $\frac{1}{2}$, which is the same as the estimate since we didn't round.

Step 5

The area of the digital photo frame is $\frac{1}{2}$ square foot.

Step 6

The answer, $\frac{1}{2}$ square foot, matches our estimate.

15. ***Step 1***

The problem asks for the number of runners who are women.

Step 2

$\frac{7}{12}$ of the 1560 runners are women, so multiply $\frac{7}{12}$ by 1560 to find the number of runners who are women.

Step 3

Round the number of runners to 1600; then, $\frac{1}{2}$ of 1600 is 800. Since $\frac{7}{12}$ is more than $\frac{1}{2}$, our estimate is that "more than 800 runners" are women.

Step 4

$$\frac{7}{12} \cdot 1560 = \frac{7 \cdot \overset{130}{\cancel{1560}}}{\underset{1}{\cancel{12}} \cdot 1} = 910$$

Step 5

(a) 910 runners are women.

(b) 1560 total runners − 910 women = 650 men

Step 6

The exact answer, 910, fits our estimate of "more than 800."

25. From Exercise 23 the total income is \$76,000. The circle graph shows that $\frac{1}{5}$ of the income is for rent.

$$\frac{1}{5} \cdot 76{,}000 = \frac{1}{\underset{1}{\cancel{5}}} \cdot \frac{\overset{15{,}200}{\cancel{76{,}000}}}{1} = 15{,}200$$

The amount of their rent is \$15,200.

37. To find the remaining amount of the estate, subtract $\frac{7}{8}$ from 1.

$$1 - \frac{7}{8} = \frac{8}{8} - \frac{7}{8} = \frac{1}{8}$$

Multiply the remaining $\frac{1}{8}$ of the estate by the fraction going to the American Cancer Society.

$$\frac{1}{4} \cdot \frac{1}{8} = \frac{1}{32}$$

$\frac{1}{32}$ of the estate goes to the American Cancer Society.

SECTION 2.7 (pages 164–167)

5. The reciprocal of $\frac{3}{8}$ is $\frac{8}{3}$ because

$$\frac{3}{8} \cdot \frac{8}{3} = \frac{24}{24} = 1.$$

SOLUTIONS

29. $\dfrac{18}{\frac{3}{4}} = 18 \div \dfrac{3}{4}$ Rewrite using the ÷ symbol for division.

$= \dfrac{18}{1} \div \dfrac{3}{4}$ Write 18 as $\frac{18}{1}$.

$= \dfrac{\overset{6}{\cancel{18}}}{1} \cdot \dfrac{4}{\underset{1}{\cancel{3}}}$ The reciprocal of $\frac{3}{4}$ is $\frac{4}{3}$. Change "÷" to "•". Divide the numerator and denominator by 3.

$= \dfrac{6 \cdot 4}{1 \cdot 1}$ Multiply.

$= \dfrac{24}{1} = 24$

35. ***Step 1***
The problem asks for the number of times a measuring cup needs to be filled.
Step 2
Solve the problem by dividing the total number of cups (5) by the size of the measuring cup $\left(\frac{1}{3}\right)$.
Step 3
Because there are three $\frac{1}{3}$-cups in one cup, multiply 3 by 5 to get 15. So, our estimate is 15.
Step 4
$5 \div \dfrac{1}{3} = \dfrac{5}{1} \div \dfrac{1}{3} = \dfrac{5}{1} \cdot \dfrac{3}{1} = 15$
Step 5
They need to fill the measuring cup 15 times.
Step 6
The exact answer, 15, is the same as our estimate.

49. ***Step 1***
The problem asks for the number of towels that can be made.
Step 2
Solve the problem by dividing the 912 yards of fabric by the fraction of a yard $\left(\frac{3}{8}\right)$ needed for each dish towel.
Step 3
Round 912 yards to 900 yards. Round $\frac{3}{8}$ to $\frac{1}{2}$.
Multiply 900 by 2, the reciprocal of $\frac{1}{2}$, to get the estimate of 1800 towels.
Step 4
$912 \div \dfrac{3}{8} = \dfrac{912}{1} \cdot \dfrac{8}{3} = \dfrac{\overset{304}{\cancel{912}}}{1} \cdot \dfrac{8}{\underset{1}{\cancel{3}}} = 2432$
Step 5
2432 towels can be made.
Step 6
The exact answer, 2432, is close to our estimate, 1800.

SECTION 2.8 (pages 175–179)

5. $4\frac{1}{2} \cdot 1\frac{3}{4}$

Estimate: $5 \cdot 2 = 10$

To find the exact answer, change each mixed number to an improper fraction and then multiply.

Exact: $4\frac{1}{2} \cdot 1\frac{3}{4} = \frac{9}{2} \cdot \frac{7}{4} = \frac{63}{8} = 7\frac{7}{8}$

21. $1\frac{1}{4} \div 3\frac{3}{4}$

Estimate: $1 \div 4 = \frac{1}{4}$

To find the exact answer, first change each mixed number to an improper fraction. Then, use the reciprocal of the divisor (the second fraction) and multiply.

Exact:

$1\frac{1}{4} \div 3\frac{3}{4} = \frac{5}{4} \div \frac{15}{4} = \frac{\overset{1}{\cancel{5}}}{\underset{1}{\cancel{4}}} \cdot \frac{\overset{1}{\cancel{4}}}{\underset{3}{\cancel{15}}} = \frac{1}{3}$

CHAPTER 3 Adding and Subtracting Fractions

SECTION 3.1 (pages 201–202)

15. Add the numerators. Keep the denominator.
$\frac{3}{8} + \frac{7}{8} + \frac{2}{8} = \frac{3+7+2}{8} = \frac{12}{8} = 1\frac{1}{2}$

29. First, subtract the numerators and keep the denominator. Then write the fraction in lowest terms.
$\frac{47}{36} - \frac{5}{36} = \frac{47-5}{36} = \frac{42}{36}$
$= \frac{42 \div 6}{36 \div 6} = \frac{7}{6} = 1\frac{1}{6}$

39. First, add the two fractions of land that were purchased.
$\frac{9}{10} + \frac{3}{10} = \frac{9+3}{10} = \frac{12}{10}$
Next, subtract the fraction of land that was planted in carrots.
$\frac{12}{10} - \frac{7}{10} = \frac{12-7}{10} = \frac{5}{10} = \frac{1}{2}$
$\frac{1}{2}$ acre is planted in squash.

SECTION 3.2 (pages 211–214)

13. 20 and 50
Multiples of 50:
50, 100, 150, 200, 250, . . .
100 is the first multiple of 50 that is divisible by 20. (100 ÷ 20 = 5)
The least common multiple of 20 and 50 is 100.

21. Find the prime factorization for each number.
4, 6, 8, 10
4 = 2 • 2
6 = 2 • ③
8 = ②•②•②
10 = 2 •⑤
LCM = 2 • 2 • 2 • 3 • 5 = 120
The LCM of 4, 6, 8, and 10 is 120.

37. $\frac{2}{3} = \frac{?}{9}$ Divide 9 by 3, to get 3. Now multiply both numerator and denominator of $\frac{2}{3}$ by 3.
$\frac{2}{3} = \frac{2 \cdot 3}{3 \cdot 3} = \frac{6}{9}$

41. $\frac{3}{16} = \frac{?}{64}$ Divide 64 by 16, to get 4. Now multiply both numerator and denominator of $\frac{3}{16}$ by 4.
$\frac{3}{16} = \frac{3 \cdot 4}{16 \cdot 4} = \frac{12}{64}$

55. $\frac{109}{1512}, \frac{23}{392}$

2	1512	392
2	756	196
2	378	98
3	189	~~49~~
3	63	~~49~~
3	21	~~49~~
7	7	49
7	~~7~~	7
	1	1

The LCM of the denominators is
2 • 2 • 2 • 3 • 3 • 3 • 7 • 7 = 10,584.

61. Find the prime factorization for each number.
5, 7, 14, 10
5 = 5
7 = ⑦
14 = ②• 7
10 = ②• 5
LCM = 2 • 5 • 7 = 70
The LCM of 5, 7, 14, and 10 is 70.

SECTION 3.3 (pages 219–222)

3. The least common multiple of 4 and 8 is 8. Rewrite both fractions as fractions with a least common denominator of 8. Then add the numerators.

$$\frac{3}{4}+\frac{1}{8}=\frac{6}{8}+\frac{1}{8}=\frac{6+1}{8}=\frac{7}{8}$$

29. The least common multiple of 12 and 4 is 12. Rewrite both fractions as fractions with a least common denominator of 12. Then subtract the numerators and write the resulting fraction in lowest terms.

$$\frac{5}{12}-\frac{1}{4}=\frac{5}{12}-\frac{3}{12}=\frac{5-3}{12}=\frac{2}{12}=\frac{1}{6}$$

41. Add the fractions to find the total length of the screw. The least common denominator of the fractions is 40. Rewrite all three fractions with a denominator of 40. Then add the numerators.

$$\frac{1}{8}+\frac{1}{4}+\frac{2}{5}=\frac{5}{40}+\frac{10}{40}+\frac{16}{40}=\frac{5+10+16}{40}=\frac{31}{40}$$

The total length of the screw is $\frac{31}{40}$ inch.

49. One way to compare fractions accurately is to rewrite each fraction with a common denominator.

$$\frac{1}{4}=\frac{15}{60};\ \frac{1}{3}=\frac{20}{60};\ \frac{1}{5}=\frac{12}{60};\ \frac{13}{60}$$

The fraction with the largest numerator is the largest fraction. This is $\frac{20}{60}$ or $\frac{1}{3}$, which is "totally honest."
To find the number of hours, multiply.

$$\frac{1}{3}\cdot 1200=\frac{1}{\cancel{3}_1}\cdot\frac{\cancel{1200}^{400}}{1}=\frac{400}{1}=400$$

400 people responded "totally honest."
To find what fraction of the users responding to "totally honest" and "fib a little" add the fractions $\frac{1}{3}$ and $\frac{1}{4}$. The least common denominator is 12. Rewrite $\frac{1}{3}$ and $\frac{1}{4}$ with denominators of 12. Then add the numerators and write in lowest terms.

$$\frac{1}{3}+\frac{1}{4}=\frac{4}{12}+\frac{3}{12}=\frac{4+3}{12}=\frac{7}{12}$$

SECTION 3.4 (pages 227–234)

19. *Estimate:* *Exact:*

$$\begin{array}{rcl} 23 & \xleftarrow{\text{Rounds to}} & 22\frac{3}{4}=22\frac{21}{28} \\ +15 & \xleftarrow{\text{Rounds to}} & +15\frac{3}{7}=15\frac{12}{28} \\ \hline 38 & & 37\frac{33}{28} \end{array}$$

$$37\frac{33}{28}=37+1\frac{5}{28}=38\frac{5}{28}$$

29. *Estimate:* *Exact:*

$$\begin{array}{rcl} 17 & \xleftarrow{\text{Rounds to}} & 17 \\ -\ 7 & \xleftarrow{\text{Rounds to}} & -\ 6\frac{5}{8} \\ \hline 10 & & \end{array}$$

To subtract $\frac{5}{8}$, first regroup the whole number 17 into 16 + 1.

$$17=16+1=16+\frac{8}{8}$$

Now you can subtract.

$$\begin{array}{r} 16\frac{8}{8} \\ -\ 6\frac{5}{8} \\ \hline 10\frac{3}{8} \end{array}$$

55.

$$\begin{array}{rl} 8\frac{3}{4}=\frac{35}{4}= & \frac{70}{8} \\ -\ 5\frac{7}{8}=\frac{47}{8}= & \frac{47}{8} \\ \hline & \frac{23}{8}=2\frac{7}{8} \end{array}$$

71. Subtract the size of the smallest hose clamp from the size of the largest hose clamp.
Estimate: 3 − 1 = 2 in.

Exact:

$$\begin{array}{rl} 2\frac{3}{4}=\frac{11}{4}= & \frac{44}{16} \\ -\ \frac{9}{16}=\frac{9}{16}= & \frac{9}{16} \\ \hline & \frac{35}{16}=2\frac{3}{16} \end{array}$$

The difference in size is $2\frac{3}{16}$ inches.

81. Add the weights.

Estimate:

59 + 24 + 17 + 29 + 58 = 187 tons

Exact:

$$\begin{array}{rl} 58\frac{1}{2}= & 58\frac{12}{24} \\ 23\frac{5}{8}= & 23\frac{15}{24} \\ 16\frac{5}{6}= & 16\frac{20}{24} \\ 29\frac{1}{4}= & 29\frac{6}{24} \\ +\ 58\frac{1}{3}= & 58\frac{8}{24} \\ \hline & 184\frac{61}{24}=184+2\frac{13}{24}=186\frac{13}{24} \end{array}$$

The total weight is $186\frac{13}{24}$ tons.

SECTION 3.5 (pages 243–246)

7. Number line from 0 to 4 with a dot at $2\frac{1}{6}$.

33. $\left(\frac{3}{2}\right)^4=\frac{3}{2}\cdot\frac{3}{2}\cdot\frac{3}{2}\cdot\frac{3}{2}=\frac{81}{16}=5\frac{1}{16}$

49. $6\left(\frac{2}{3}\right)^2\left(\frac{1}{2}\right)^3=6\left(\frac{2}{3}\cdot\frac{2}{3}\right)\left(\frac{1}{2}\cdot\frac{1}{2}\cdot\frac{1}{2}\right)$

Simplify the expressions with exponents.

$$=\frac{\cancel{6}^{2}\cdot\cancel{2}^{1}\cdot\cancel{2}^{1}\cdot 1\cdot 1\cdot 1}{\cancel{3}_{1}\cdot 3\cdot\cancel{2}_{1}\cdot\cancel{2}_{1}\cdot\cancel{2}_{1}}$$

$$=\frac{1}{3}$$

63. $\left(\frac{3}{4}\right)^2-\left(\frac{1}{2}-\frac{1}{6}\right)\div\frac{4}{3}$

$$=\left(\frac{3}{4}\right)^2-\left(\frac{3}{6}-\frac{1}{6}\right)\div\frac{4}{3}$$

Work inside parentheses first.

$$=\left(\frac{3}{4}\right)^2-\left(\frac{2}{6}\right)\cdot\frac{3}{4}$$

Simplify the expression with the exponent.

$$=\frac{3}{4}\cdot\frac{3}{4}-\left(\frac{2}{6}\right)\cdot\frac{3}{4}$$ Multiply.

$$=\frac{9}{16}-\frac{\cancel{2}^{1}}{\cancel{6}_{2}}\cdot\frac{\cancel{3}^{1}}{\cancel{4}_{2}}$$

$$=\frac{9}{16}-\frac{1}{4}$$ The LCD is 16. Rewrite $\frac{1}{4}$ as a fraction with a denominator of 16.

$$=\frac{9}{16}-\frac{4}{16}$$ Subtract.

$$=\frac{5}{16}$$

SOLUTIONS

CHAPTER 4 Decimals

SECTION 4.1 (pages 269–273)

5. 7 0 . 4 8 9 (tens, ones, tenths)

17. Reorder as 5 thousands, 4 hundreds, 0 tens, 6 ones, 0 tenths, 4 hundredths, 5 thousandths: 5406.045

31. $0.205 = \frac{205}{1000} = \frac{205 \div 5}{1000 \div 5} = \frac{41}{200}$

45. Three decimal places is *thousandths,* so 1.075 is one and seventy-five *thousandths.*

53. Seven hundred three ten-thousandths: $\frac{703}{10{,}000} = 0.0703$

59. 8-pound test line has a diameter of 0.010 inch. 0.010 inch is read "ten thousandths inch."

$0.010 = \frac{10}{1000} = \frac{10 \div 10}{1000 \div 10} = \frac{1}{100}$ inch

SECTION 4.2 (pages 279–280)

13. Draw a cut-off line after the tenths place: 793.9|88

The first digit cut is 8, which is *5 or more,* so round up the tenths place.

$$\begin{array}{r} 793.9 \\ +\ \ 0.1 \\ \hline 794.0 \end{array}$$

Answer: 793.988 ≈ 794.0

17. Round 9.0906 to the nearest hundredth.

Draw a cut-off line: 9.09|06

The first digit cut is 0, which is *4 or less*. The part you keep stays the same.

9.0906 ≈ 9.09

35. Round $999.73 to the nearest dollar.

Draw a cut-off line: $999|.73

The first digit cut is 7, which is *5 or more,* so round up.

$$\begin{array}{r} \$999 \\ +\ \ \ \ 1 \\ \hline \$1000 \end{array}$$

Answer: $999.73 ≈ $1000

SECTION 4.3 (pages 285–288)

13. Subtract 0.291 from 0.4

Line up decimal points.

$$\begin{array}{r} 0.400 \\ -\,0.291 \\ \hline 0.109 \end{array}$$ ← Write two zeros.

25. *Estimate:* / *Exact:*

Estimate	Exact
400	392.700
1	0.865
+ 20	+ 21.080
421	414.645

43. First add the three players' heights.

Estimate	Exact
2	2.13
2	1.98
+ 2	+ 2.08
6 meters	6.19 meters

Now subtract the players' combined height of 6.19 meters from the rhino's height of 6.4 meters.

6.4 − 6.19 = 0.21

So, the NBA stars' combined height is 0.21 meter less than the rhino's height.

49. Subtract the price of the regular fishing line, $4.84, from the price of the fluorescent fishing line, $5.14.

Estimate	Exact
$5	$5.14
− 5	− 4.84
$0	$0.30

The fluorescent fishing line costs $0.30 more than the regular fishing line.

SECTION 4.4 (pages 291–296)

15. (7.2) (0.06)

72	7.2 ← 1 decimal place
× 6	× 0.06 ← 2 decimal places
432	0.432 ← 3 decimal places

19. (0.006) (0.0052)

0.0052 ← 4 decimal places
× 0.006 ← 3 decimal places
0.0000312 ← 7 decimal places

43. Multiply the number of gallons that she pumped into her pickup truck by the price per gallon. Use a calculator.

20.510 × 3.759 = 77.09709

Round 77.09709 to the nearest cent.

Michelle paid $77.10 for the gas.

55. (a) Find the cost for three short-sleeved, solid-color shirts

$$\begin{array}{r} \$14.75 \\ \times\ \ \ 3 \\ \hline \$44.25 \end{array}$$

Based on this subtotal, shipping is $7.95.

Next, find the cost of the three monograms.

$$\begin{array}{r} \$4.95 \\ \times\ \ \ 3 \\ \hline \$14.85 \end{array}$$

Add these amounts, plus $5.00 for a gift box.

$44.25 ← Shirts
7.95 ← Shipping
14.85 ← Monograms
+ 5.00 ← Gift box
$72.05

The total cost is $72.05.

(b) Subtract the cost of the shirts to find the difference.

$$\begin{array}{r} \$72.05 \\ -\ 44.25 \\ \hline \$27.80 \end{array}$$

The monograms, gift box, and shipping added $27.80 to the cost of the gift.

SECTION 4.5 (pages 303–306)

11. Move decimal point in divisor and dividend 1 place; write 0 in the dividend.

$$\begin{array}{r} 36. \\ 1.5\overline{)54.0} \\ 45 \\ \hline 90 \\ 90 \\ \hline 0 \end{array}$$

13. Given: 108 ÷ 18 = 6

Find: 0.108 ÷ 1.8 by moving the decimal points.

$$\begin{array}{r} .06 \\ 1.8\overline{)0.1\,08} \end{array}$$

So, 0.108 ÷ 1.8 = 0.06

19. $\frac{3.1}{0.006}$

$$\begin{array}{r} 516.666 \\ 0.006\overline{)3.100\,000} \\ 30 \\ \hline 10 \\ 6 \\ \hline 40 \\ 36 \\ \hline 40 \\ 36 \\ \hline 40 \\ 36 \\ \hline 40 \\ 36 \\ \hline 4 \end{array}$$

Line up decimal points. Move decimal point in divisor and dividend three places. Write 000 in dividend.

Write 0 in dividend.

Write 0 in dividend.

Write 0 in dividend.

Stop and round answer to the nearest hundredth.

The quotient is 516.67 (rounded).

29. $307.02 \div 5.1 = 6.2$

Estimate: $300 \div 5 = 60$

The answer 6.2 is *unreasonable* because it is so much less than 60.

$$\begin{array}{r} 60.2 \\ 5.1\overline{)307.02} \\ \underline{306} \\ 1\,0 \\ \underline{0} \\ 1\,0\,2 \\ \underline{1\,0\,2} \\ 0 \end{array}$$

The correct answer is 60.2, which is close to the estimate of 60.

55. $33 - 3.2(0.68 + 9) - 1.3^2$ — Parentheses, Exponent

$33 - 3.2(9.68) - 1.69$ — Multiply

$33 - 30.976 - 1.69$ — Subtract

$2.024 - 1.69 = 0.334$ — Subtract

61. Divide 100,000 box tops (the answer from Exercise 59) by 38 weeks.

$$\begin{array}{r} 2631.5 \\ 38\overline{)100{,}000.0} \\ \underline{76\phantom{0{,}000.0}} \\ 24\,0 \\ \underline{22\,8} \\ 1\,20 \\ \underline{1\,14} \\ 60 \\ \underline{38} \\ 22\,0 \\ \underline{19\,0} \\ 3\,0 \end{array}$$

2631.5 rounds to 2632.

The school needs to collect 2632 box tops (rounded) during each of the 38 weeks.

SECTION 4.6 (pages 311–314)

23. $3\frac{5}{8} = \frac{29}{8} = 3.625$

$$\begin{array}{r} 3.625 \\ 8\overline{)29.000} \\ \underline{24} \\ 5\,0 \\ \underline{4\,8} \\ 20 \\ \underline{16} \\ 40 \\ \underline{40} \\ 0 \end{array}$$

35. $0.35 = \frac{35}{100} = \frac{35 \div 5}{100 \div 5} = \frac{7}{20}$

51. Write zeros so that all the numbers have four decimal places. The acceptable lengths must be greater than 0.9980 cm and less than 1.0020 cm.

$1.0100 > 1.0020$ *unacceptable*

$0.9991 > 0.9980$ and

$0.9991 < 1.0020$ *acceptable*

$1.0007 > 0.9980$ and

$1.0007 < 1.0020$ *acceptable*

$0.9900 < 0.9980$ *unacceptable*

The lengths of 0.9991 cm and 1.0007 cm are acceptable.

63. $\frac{3}{8}, \frac{2}{5}, 0.37, 0.4001$

$\frac{3}{8} = 0.3750$

$\frac{2}{5} = 0.4000$

$0.37 = 0.3700 \leftarrow$ least

$0.4001 = 0.4001 \leftarrow$ greatest

From least to greatest: $0.37, \frac{3}{8}, \frac{2}{5}, 0.4001$

CHAPTER 5 Ratio and Proportion

SECTION 5.1 (pages 335–338)

7. \$100 to \$50

$$\frac{\$100}{\$50} = \frac{100}{50} = \frac{100 \div 50}{50 \div 50} = \frac{2}{1}$$

19. $1\frac{1}{4}$ to $1\frac{1}{2}$

$$\frac{1\frac{1}{4}}{1\frac{1}{2}} = \frac{\frac{5}{4}}{\frac{3}{2}} = \frac{5}{4} \div \frac{3}{2} = \frac{5}{\cancel{4}_2} \cdot \frac{\cancel{2}^1}{3} = \frac{5}{6}$$

47. The increase in price is $\$12.50 - \$10 = \$2.50$

$$\frac{\$2.50}{\$10} = \frac{2.50}{10} = \frac{2.50 \cdot 10}{10 \cdot 10} = \frac{25}{100}$$

$$= \frac{25 \div 25}{100 \div 25} = \frac{1}{4}$$

The ratio of the increase in price to the original price is $\frac{1}{4}$.

SECTION 5.2 (pages 343–346)

15. Miles traveled:

$28{,}396.7 - 28{,}058.1 = 338.6$

Miles per gallon:

$$\frac{338.6 \text{ miles} \div 16.2}{16.2 \text{ gallons} \div 16.2} \approx 20.90 \text{ or } 20.9$$

19.

Size	*Cost per Unit*
12 ounces	$\frac{\$2.49}{12 \text{ ounces}} \approx \$0.208$
14 ounces	$\frac{\$2.89}{14 \text{ ounces}} \approx \$0.206$
18 ounces	$\frac{\$3.96}{18 \text{ ounces}} = \$0.22$

Because \$0.206 is the lowest cost per ounce, the best buy is 14 ounces for \$2.89.

33. One battery for \$1.79 is like getting 3 batteries, so $\$1.79 \div 3 \approx \$0.597$ per battery. An eight-pack of AA batteries for \$4.99 is $\$4.99 \div 8 \approx \$0.624$ per battery. Because \$0.597 is the lower cost per battery, the better buy is the package with one battery.

SECTION 5.3 (pages 350–351)

11. $\frac{150}{200} = \frac{200}{300}$

$\frac{150 \div 50}{200 \div 50} = \frac{3}{4}$ and $\frac{200 \div 100}{300 \div 100} = \frac{2}{3}$

Because $\frac{3}{4}$ is *not* equal to $\frac{2}{3}$, the proportion is *false*.

33. $\frac{2\frac{5}{8}}{3\frac{1}{4}} = \frac{21}{26}$

Cross products:

$2\frac{5}{8} \cdot 26 = \frac{21}{\cancel{8}_4} \cdot \frac{\cancel{26}^{13}}{1} = \frac{273}{4} = 68\frac{1}{4}$

$3\frac{1}{4} \cdot 21 = \frac{13}{4} \cdot \frac{21}{1} = \frac{273}{4} = 68\frac{1}{4}$

The cross products are *equal,* so the proportion is *true*.

37. $\frac{2\frac{3}{10}}{8.05} = \frac{\frac{1}{4}}{0.9}$

Cross products:

$2\frac{3}{10}(0.9) = \frac{23}{10} \cdot \frac{9}{10} = \frac{207}{100} = 2\frac{7}{100}$

or $(2.3)(0.9) = 2.07$

$(8.05)\frac{1}{4} = (8.05)(0.25) = 2.0125$

or $8\frac{5}{100} \cdot \frac{1}{4} = \frac{805}{100} \cdot \frac{1}{4} = \frac{805}{400} = 2\frac{1}{80}$

The cross products are *not* equal so the proportion is *false*.

$\left(2\frac{7}{100} \neq 2\frac{1}{80} \text{ or } 2.07 \neq 2.0125\right)$

SECTION 5.4 (pages 356–357)

9. $\frac{42}{x} = \frac{18}{39}$

$x \cdot 18 = 42 \cdot 39$ Cross products are equal.

$\frac{x \cdot \cancel{18}}{\cancel{18}} = \frac{1638}{18}$ Divide both sides by 18.

$x = 91$

CHECK

$42 \cdot 39 = 1638$

$91 \cdot 18 = 1638$

15. $\frac{99}{55} = \frac{44}{x}$ **OR** $\frac{9}{5} = \frac{44}{x}$

$9 \cdot x = 5 \cdot 44$ Cross products are equal.

$\frac{\cancel{9} \cdot x}{\cancel{9}} = \frac{220}{9}$ Divide both sides by 9.

$x = \frac{220}{9}$

$x \approx 24.44$ (rounded)

CHECK

$55 \cdot 44 = 2420$

$99 \cdot 24.44 = 2419.56$

Slightly different because of rounding.

23. $\frac{2\frac{1}{3}}{1\frac{1}{2}} = \frac{x}{2\frac{1}{4}}$

$1\frac{1}{2} \cdot x = 2\frac{1}{3} \cdot 2\frac{1}{4}$

$\frac{3}{2} \cdot x = \frac{7}{\cancel{3}} \cdot \frac{\cancel{9}}{4}$

$\frac{3}{2} \cdot x = \frac{21}{4}$

$\frac{\frac{3}{2} \cdot x}{\frac{3}{2}} = \frac{\frac{21}{4}}{\frac{3}{2}}$

$x = \frac{21}{4} \div \frac{3}{2} = \frac{\cancel{21}}{\cancel{4}} \cdot \frac{\cancel{2}}{\cancel{3}} = \frac{7}{2} = 3\frac{1}{2}$

27. $\frac{x}{\frac{3}{50}} = \frac{0.15}{1\frac{4}{5}}$

Change to decimals:

$\frac{3}{50} = 3 \div 50 = 0.06$

$1\frac{4}{5} = \frac{9}{5}$ and $9 \div 5 = 1.8$

$\frac{x}{0.06} = \frac{0.15}{1.8}$

$x \cdot 1.8 = (0.06)(0.15)$

$\frac{x \cdot \cancel{1.8}}{\cancel{1.8}} = \frac{0.009}{1.8}$

$x = 0.005$

Change to fractions:

$0.15 = \frac{15 \div 5}{100 \div 5} = \frac{3}{20}$

$\frac{x}{\frac{3}{50}} = \frac{\frac{3}{20}}{1\frac{4}{5}}$

$1\frac{4}{5} \cdot x = \frac{3}{50} \cdot \frac{3}{20}$

$\frac{\cancel{1\frac{4}{5}} \cdot x}{\cancel{1\frac{4}{5}}} = \frac{\frac{9}{1000}}{1\frac{4}{5}}$

$x = \frac{9}{1000} \div 1\frac{4}{5} = \frac{\cancel{9}}{\cancel{1000}} \cdot \frac{\cancel{5}}{\cancel{9}}$ (1000 reduces to 200)

$= \frac{1}{200}$

Compare answers.

$0.005 = \frac{5 \div 5}{1000 \div 5} = \frac{1}{200}$ ← Matches

SECTION 5.5 (pages 363–366)

11. $\frac{6 \text{ ounces}}{7 \text{ servings}} = \frac{x \text{ ounces}}{12 \text{ servings}}$

$7 \cdot x = 6 \cdot 12$

$\frac{\cancel{7} \cdot x}{\cancel{7}} = \frac{72}{7}$

$x \approx 10.3 \approx 10$ (rounded)

You need about 10 ounces for 12 servings.

17. The length of the dining area is the same as the length of the kitchen, which is 14 feet (from Exercise 15).

Find the width of the dining area.

4.5 inches − 2.5 inches = 2 inches on the floor plan.

$\frac{1 \text{ inch}}{4 \text{ feet}} = \frac{2 \text{ inches}}{x \text{ feet}}$

$1 \cdot x = 4 \cdot 2$

$x = 8$

The dining area is 8 feet wide.

23. $\frac{7 \text{ refresher}}{10 \text{ entering}} = \frac{x \text{ refresher}}{2950 \text{ entering}}$

$10 \cdot x = 7 \cdot 2950$

$\frac{\cancel{10} \cdot x \cdot}{\cancel{10}} = \frac{20{,}650}{10}$

$x = 2065$

2065 students will probably need a refresher course. This is a reasonable answer because it's more than half the students, but not all the students.

Incorrect set-up

$\frac{10 \text{ entering}}{7 \text{ refresher}} = \frac{x \text{ refresher}}{2950 \text{ entering}}$

$7 \cdot x = 10 \cdot 2950$

$\frac{\cancel{7} \cdot x}{\cancel{7}} = \frac{29{,}500}{7}$

$x \approx 4214$ (rounded)

The *incorrect* set-up gives an *unreasonable* estimate of 4214 entering students; there are only 2950 entering students.

33. Coretta $\left\{\frac{1.05 \text{ meters}}{1.68 \text{ meters}} = \frac{6.58 \text{ meters}}{x \text{ meters}}\right\}$ tree

$1.05 \cdot x = 1.68(6.58)$

$\frac{\cancel{1.05} \cdot x}{\cancel{1.05}} = \frac{11.0544}{1.05}$

$x = 10.528 \approx 10.53$

The height of the tree is about 10.53 meters.

39. First find the number of calories in a $\frac{1}{2}$-cup serving of bran cereal.

$\frac{\frac{1}{3} \text{ cup}}{80 \text{ calories}} = \frac{\frac{1}{2} \text{ cup}}{x \text{ calories}}$

$\frac{1}{3} \cdot x = 80 \cdot \frac{1}{2}$

$\frac{\cancel{\frac{1}{3}} \cdot x}{\cancel{\frac{1}{3}}} = \frac{40}{\frac{1}{3}}$

$x = \frac{40}{1} \cdot \frac{3}{1}$

$x = 120$ calories

Then find the number of grams of fiber in a $\frac{1}{2}$-cup serving of bran cereal.

$\frac{\frac{1}{3} \text{ cup}}{8 \text{ grams}} = \frac{\frac{1}{2} \text{ cup}}{x \text{ grams}}$

$\frac{1}{3} \cdot x = 8 \cdot \frac{1}{2}$

$\frac{\cancel{\frac{1}{3}} \cdot x}{\cancel{\frac{1}{3}}} = \frac{4}{\frac{1}{3}}$

$x = \frac{4}{1} \cdot \frac{3}{1}$

$x = 12$ grams

A $\frac{1}{2}$-cup serving of bran cereal provides 120 calories and 12 grams of fiber.

CHAPTER 6 Percent

SECTION 6.1 (pages 387–392)

15. $0.5\% = 0.005$

Drop the percent symbol. Attach two zeros so the decimal point can be moved two places to the left.

27. $2 = 200\%$

Two zeros are attached so the decimal point can be moved two places to the right. Attach a percent symbol.

53. 100% is all of the children.

So, 100% of 12 children is 12 children.

12 children are present.

79. 21% of the children eat french fries. To write 21% as a decimal, drop the percent symbol and move the decimal point two places to the left, resulting in 0.21. So, 0.21 of the children eat french fries.

SECTION 6.2 (pages 398–403)

9. First write 6.25 over 100. Then to get a whole number in the numerator, multiply the numerator and denominator by 100. Finally, write the fraction in lowest terms.

$$6.25\% = \frac{6.25}{100} = \frac{6.25(100)}{100(100)}$$
$$= \frac{625 \div 625}{10{,}000 \div 625} = \frac{1}{16}$$

41. $\frac{5}{9} = \frac{p}{100}$

$9 \cdot p = 5 \cdot 100$ Find cross products.

$\frac{\not{9} \cdot p}{\not{9}} = \frac{500}{9}$ Divide both sides by 9.

$p = \frac{500}{9}$

$p \approx 55.5\overline{5}$

Thus, $\frac{5}{9} = 55.6\%$ (rounded).

49. $87.5\% = 0.875$ decimal

$= \frac{875 \div 125}{1000 \div 125} = \frac{7}{8}$ fraction

63. $\frac{1}{200} = 1 \div 200 = 0.005$ decimal

$= 0.5\%$ percent

67. $3\frac{1}{4} = \frac{13}{4} = 13 \div 4 = 3.25$ decimal

$= 325\%$ percent

77. 64 out of 80 employees have cell phones.

(a) $\frac{64}{80} = \frac{64 \div 16}{80 \div 16} = \frac{4}{5}$ fraction

$\frac{4}{5} = 4 \div 5 = 0.8$ decimal

$0.8 = 0.80 = 80\%$ percent

(b) If $\frac{4}{5}$ (have cell phones)

$1 - \frac{4}{5} = \frac{1}{5}$ do not have cell phones.

$\frac{1}{5} = 1 \div 5 = 0.2$ decimal

$0.2 = 0.20 = 20\%$ percent

SECTION 6.3 (pages 408–411)

11. part = 15, whole = 60

$\frac{15}{60} = \frac{x}{100}$

$\frac{1}{4} = \frac{x}{100}$ $\frac{15}{60}$ is $\frac{1}{4}$ in lowest terms.

$4 \cdot x = 1 \cdot 100$ Find cross products.

$\frac{\not{4} \cdot x}{\not{4}} = \frac{100}{4}$ Divide both sides by 4.

$x = 25$

The percent is 25, written as 25%.

21. whole = 5000, part = 20

$\frac{20}{5000} = \frac{x}{100}$ Percent proportion

$5000 \cdot x = 20 \cdot 100$ Find cross products.

$\frac{\not{5000} \cdot x}{\not{5000}} = \frac{2000}{5000}$ Divide both sides by 5000.

$x = 0.4$

The percent is 0.4, written as 0.4%.

37. 12 injections is 20% of what number of injections?

12 injections is → part; 20% → percent; of what number of injections? → whole (unknown)

$\frac{12}{\text{unknown}} = \frac{20}{100}$

47. 86 of 142 people is what percent?

86 → part; of 142 people → whole; is what percent? → percent (unknown)

$\frac{86}{142} = \frac{\text{unknown}}{100}$

SECTION 6.4 (pages 419–424)

15. Write 225% as a decimal, 2.25.

225% of 680 tables

$(2.25)(680) = 1530$

part = 1530 tables

29. part is 350; percent is 12.5

$\frac{\text{part}}{\text{whole}} = \frac{\text{percent}}{100}$

so $\frac{350}{x} = \frac{12.5}{100}$

$12.5 \cdot x = 35{,}000$ Cross products

$\frac{\not{12.5} \cdot x}{\not{12.5}} = \frac{35{,}000}{12.5}$ Divide both sides by 12.5.

$x = 2800$

$12\frac{1}{2}\%$ of 2800 is 350.

37. part is 64; whole is 344

$\frac{\text{part}}{\text{whole}} = \frac{\text{percent}}{100}$

$\frac{64}{344} = \frac{x}{100}$ **OR** $\frac{8}{43} = \frac{x}{100}$

Cross products $43 \cdot x = 8 \cdot 100$

Divide both sides by 43. $\frac{\not{43} \cdot x}{\not{43}} = \frac{800}{43}$

$x \approx 18.6$

\$64 is 18.6% (rounded) of \$344.

43. (a) part is unknown; whole is 240; percent is 22

$\frac{\text{part}}{\text{whole}} = \frac{\text{percent}}{100}$

$\frac{x}{240} = \frac{22}{100}$ **OR** $\frac{x}{240} = \frac{11}{50}$

Cross products $x \cdot 50 = 240 \cdot 11$

Divide both sides by 50. $\frac{x \cdot \not{50}}{\not{50}} = \frac{2640}{50}$

$x = 52.8$

The amount withheld is \$52.80.

(b) \$240 earnings –\$52.80 withheld = \$187.20 amount remaining. The amount remaining is \$187.20.

55. part is 960; whole is 48,000; percent is unknown

$\frac{\text{part}}{\text{whole}} = \frac{\text{percent}}{100}$

$\frac{960}{48{,}000} = \frac{x}{100}$ **OR** $\frac{1}{50} = \frac{x}{100}$

Cross products $50 \cdot x = 1 \cdot 100$

Divide both sides by 50. $\frac{\not{50} \cdot x}{\not{50}} = \frac{100}{50}$

$x = 2$

There are 2% of these jobs filled by women.

SOLUTIONS

SECTION 6.5 (pages 429–432)

3. Write 25% as the decimal 0.25. The whole is 1080. Let x represent the unknown part.

part = percent • whole

$x = (0.25)(1080)$

$x = 270$

25% of 1080 blood donors is 270 blood donors.

23. The part is 3.75 and the percent is $1\frac{1}{4}\% = 1.25\%$ or 0.0125 as a decimal.

The whole is unknown.

part = percent • whole

$3.75 = (0.0125)(x)$

$\frac{3.75}{0.0125} = \frac{\overset{1}{\cancel{(0.0125)}}(x)}{\underset{1}{\cancel{0.0125}}}$

$300 = x$

$1\frac{1}{4}\%$ of 300 gallons is 3.75 gallons.

29. Because 160 follows *of*, the whole is 160. The part is 2.4, and the percent is unknown.

part = percent • whole

$2.4 = x \cdot 160$

$\frac{2.4}{160} = \frac{x \cdot \overset{1}{\cancel{160}}}{\underset{1}{\cancel{160}}}$

$0.015 = x$

0.015 is 1.5%

1.5% of 160 liters is 2.4 liters.

43. Because 1250 follows *of*, the whole is 1250. The part is 461, and the percent is unknown.

part = percent • whole

$461 = x \cdot 1250$

$\frac{461}{1250} = \frac{x \cdot \overset{1}{\cancel{1250}}}{\underset{1}{\cancel{1250}}}$

$0.369 \approx x$

0.369 is 36.9%

36.9% (rounded) of these Americans rate their health as excellent.

55. Find the sales tax on the new Polaris unit. The whole is 524. The percent is $7\frac{3}{4}\% = 7.75\%$, which is 0.0775 as a decimal.

part = percent • whole

$x = (0.0775)(524)$

$x = 40.61$

Add the sales tax to the purchase price.

$\$524 + \$40.61 = \$564.61$

Subtract the trade-in.

$\$564.61 - \$125 = \$439.61$

The total cost to the customer is $439.61.

SECTION 6.6 (pages 441–444)

5. Find the sales tax rate using the sales tax formula. The sales tax is $12.75, and the cost of the item is $425.

sales tax = *rate of tax* • *cost of item*

$\$12.75 = r \cdot \$425$

$\frac{12.75}{425} = \frac{r \cdot \overset{1}{\cancel{425}}}{\underset{1}{\cancel{425}}}$ Divide both sides by 425.

$0.03 = r$

0.03 is 3% Write the decimal as a percent.

The tax rate is 3% and the total cost is $\$425 + \$12.75 = \$437.75$.

11. The problem asks for the amount of commission. Use the commission formula. The rate of commission is 8%, and the sales are $280.

commission = *rate of commission* • *sales*

= (8%)($280)

= (0.08)($280)

= $22.40

The amount of commission is $22.40.

23. The problem asks for the amount of the discount and the sale price. First, find the amount of discount using the discount formula. The rate of discount is 15%, and the original price is $58.40.

amount of discount = *rate of discount* • *original price*

= (15%)($58.40)

= (0.15)($58.40) Write 15% as a decimal.

= $8.76 Amount of discount

Now find the sale price by subtracting the amount of the discount ($8.76) from the original price: $\$58.40 - \$8.76 = \$49.64$. The amount of discount is $8.76 and the sale price is $49.64.

31. The problem asks for the rate of sales tax. Use the sales tax formula. The sales tax is $99. The cost of the door is $1980. Use r to represent the unknown rate of tax.

sales tax = *rate of tax* • *cost of item*

$\$99 = r \cdot \$1980$

$\frac{99}{1980} = \frac{r \cdot \overset{1}{\cancel{1980}}}{\underset{1}{\cancel{1980}}}$ Divide both sides by 1980.

$0.05 = r$

0.05 is 5% Write the decimal as a percent.

The rate of sales tax is 5%.

47. The problem asks for the cost of the dictionary. First find the amount of discount using the discount formula. The rate of discount is 6%, and the original price is $18.50.

discount = *rate of discount* • *original price*

= (6%)($18.50)

= (0.06)($18.50) Write 6% as a decimal.

= $1.11 Amount of discount

To find the sale price, subtract the amount of discount ($1.11) from the original price.

Sale price

= original price − amount of discount

= $18.50 − $1.11 = $17.39 Sale price

To find the sales tax, use the sales tax formula. The rate of tax is 6% and the cost of the dictionary is $17.39.

sales tax = *rate of tax* • *cost of item*

= (6%)($17.39)

= (0.06)($17.39) Write 6% as a decimal.

≈ $1.04 Sales tax

The total cost of the dictionary is $\$17.39 + \$1.04 = \$18.43$.

57. To find the sales tax, use the sales tax formula.

sales tax = *rate of tax* • *cost of item*

$= \left(7\frac{3}{4}\%\right)(\$1248)$

= (0.0775)($1248) Write $7\frac{3}{4}\%$ as a decimal.

= $96.72 Sales tax

The total cost is the sum of the ticket, the sales tax, and the excise tax.

$\$1248 + \$96.72 + \$15.40 = \$1360.12$

SECTION 6.7 (pages 447–450)

9. $2300 at $4\frac{1}{2}\%$ for $2\frac{1}{2}$ years

The principal (p) is $2300. The rate ($r$) is $4\frac{1}{2}\%$, or 0.045 as a decimal, and the time (t) is $2\frac{1}{2}$ or 2.5 years.

$I = p \cdot r \cdot t$

$= (2300)(0.045)(2.5)$

$= 258.75$

The interest is $258.75.

19. \$940 at 3% for 18 months

The principal is \$940. The rate is 3% or 0.03, and the time is $\frac{18}{12}$ of a year.

$$I = p \cdot r \cdot t$$
$$= (940)(0.03)\left(\frac{18}{12}\right) \quad \text{18 months} = \tfrac{18}{12} \text{ of a year.}$$
$$= (28.2)(1.5) \quad \tfrac{18}{12} = (1.5)$$
$$= 42.3$$

The interest is \$42.30.

33. \$16,850 at $7\frac{1}{2}$% for 9 months

First find the interest. The principal is \$16,850. The rate is $7\frac{1}{2}$% or 0.075, and the time is $\frac{9}{12}$ of a year.

$$I = p \cdot r \cdot t$$
$$= (16{,}850)(0.075)\left(\frac{9}{12}\right)$$
$$= (1263.75)\left(\frac{3}{4}\right) \quad \text{Write } \tfrac{9}{12} \text{ in lowest terms as } \tfrac{3}{4}.$$
$$= 947.81 \text{ (rounded)}$$

The interest is \$947.81.

To find the total amount due, add the principal and the interest.

amount due = *principal* + *interest*
= \$16,850 + \$947.81
= \$17,797.81

The total amount due is \$17,797.81.

43. \$14,800 at $2\frac{1}{4}$% for 10 months

The principal is \$14,800. The rate is $2\frac{1}{4}$% or 0.0225, and the time is $\frac{10}{12}$ of a year.

$$I = p \cdot r \cdot t$$
$$= (14{,}800)(0.0225)\left(\frac{10}{12}\right)$$
$$= (333)\left(\frac{5}{6}\right) \quad \text{Write } \tfrac{10}{12} \text{ in lowest terms as } \tfrac{5}{6}.$$
$$= 277.50$$

She will earn \$277.50 in interest.

SECTION 6.8 (pages 455–458)

3. \$500 at 4% for 2 years

Year	*Interest*	*Compound Amount*
1	(\$500)(0.04)(1) = \$20	\$500 + \$20 = \$520
2	(\$520)(0.04)(1) = \$20.80	\$520 + \$20.80 = \$540.80

The compound amount is \$540.80.

13. \$1400 at 6% for 5 years

Year 1 Year 2 Year 3 Year 4 Year 5

(\$1400) (1.06) (1.06)(1.06)(1.06)(1.06) ≈ \$1873.52

Original deposit — 100% + 6% = 106% = 1.06 — Compound amount

The compound amount is \$1873.52 (rounded).

23. \$1000 at 4% for 5 years

Look down the column headed 4%, and across to row 5 (because 5 years = 5 time periods). At the intersection of the column and the row, read the compound amount, 1.2167.

compound amount = (\$1000)(1.2167)
= \$1216.70

Find the interest by subtracting the principal (\$1000) from the compound amount.

interest = \$1216.70 − \$1000
= \$216.70

33. **(a)** 6% column, row 9 from the table is 1.6895.

compound amount = (\$76,000)(1.6895)
= \$128,402

The total amount that should be repaid is \$128,402.

(b) To find the amount of interest earned, subtract the principal (\$76,000) from the compound amount.

The amount of interest earned is \$128,402 − \$76,000 = \$52,402

Index

Photo Credits

1 Marcito/Fotolia; **4** Yuri Arcurs/Fotolia; **9** (top left) Nicola Sutton/Life File/Photodisc/Getty Images, (top right) DKatz/Fotolia, (bottom left) Alexander Crispin/Getty Images, (bottom right) Monart Design/Fotolia; **15** Bew111/Dreamstime; **20** (left) Andres Rodriguez/Fotolia, (right) Emiliau/Fotolia; **30** (top left) Victor Zastol'skiy/Fotolia, (top right) Robhainer/Fotolia, (bottom left) Kenneth Graff/Shutterstock, (bottom right) Stan Salzman; **40** (top) Jonathan Alcorn/ZUMA Press/Newscom, (bottom) John A. Rizzo/Photodisc/Getty Images; **53** (top left) Anette Linnea Rasmussen/Shutterstock, (top right) Beth Anderson/Pearson Education, Inc., (bottom left) Sueharper/Dreamstime, (bottom right) Stan Salzman; **61** (left) Edwin Remsberg/Alamy, (right) Stan Salzman; **62** (left) Spflaum/Dreamstime, (right) Todd Taulman/Fotolia; **65** Stan Salzman; **84** A.F. Archive/Alamy; **87** Stan Salzman; **88** Elenathewise/Fotolia; **92** Stan Salzman; **95** Lightpainter/Dreamstime; **96** Svlumagraphica/Dreamstime; **97** (left) Beth Anderson/Pearson Education, Inc. SUBWAY® is a registered trademark of Doctor's Associates Inc. © Doctor's Associates Inc. All Rights Reserved, (right) Vacclav/Fotolia; **108** (left) Sascha Burkard/Fotolia, (right) Vladimir Caplinskij/Shutterstock; **109** (left) Jason Stitt/Shutterstock, (right) Lily/Fotolia; **110** Karl Weatherly/Photodisc/Getty Images; **113** (both) Jelly Belly Candy Company. Used with permission; **117** (left) Photodisc/Getty Images, (right) U.S. Mint; **150** (left) Sascha Burkard/Shutterstock, (top right) Pakhnyushchyy/Fotolia, (bottom right) Kotafoty/Fotolia; **151** Iofoto/Fotolia; **155** (left) Antares614/Dreamstime, (right) Mark Breck/Shutterstock; **156** (left) Lynn Watson/Shutterstock, (right) Brodogg1313/Dreamstime; **157** Stan Salzman; **158** (left) Slocummedia/Fotolia, (right) Arim44/Dreamstime; **165** (left) V. J. Matthew/Shutterstock, (right) Ulrich Willmünder/Shutterstock; **166** (left) Stephen Coburn/Fotolia, (right) Oleg Zabielin/Fotolia; **167** Collection of Stan Salzman; **177** Jelly Belly Candy Company. Used with permission; **188** Giuseppe Porzani/Fotolia; **191** (left) Sylvana Rega/Fotolia, (right) Kropic/Fotolia; **196** Olga Nayashkova/Fotolia; **197** ZUMA Press/Newscom; **202** (both) Stan Salzman; **221** (left) Chris Hill/Shutterstock, (right) Stan Salzman; **222** Elenathewise/Fotolia; **230** (left) ZUMA Press/Newscom, (right) Imageegami/Fotolia; **232** (left) Pavel Losevsky/Fotolia, (right) Stan Salzman; **233** (left) Gudellaphoto/Fotolia, (right) ZUMA Press/Newscom; **245** (left) Konstantin Sutyagin/Shutterstock, (right) Stan Salzman; **253** (left) Kim Steele/Digital Vision/Getty Images, (right) Stan Salzman; **254** (left) Amy Myers/Shutterstock, (right) Stan Salzman; **256** (both) Stan Salzman; **258** (left) Alisonhancock/Fotolia, (right) Lise F. Young/Shutterstock; **262** (left) Stan Salzman, (right) Cosmin Manci/Fotolia; **263** Pressmaster/Fotolia; **272** Pressmaster/Fotolia; **273** Carlosseller/Fotolia; **279** (both) Beth Anderson/Pearson Education, Inc.; **280** (top) Perry Mastrovito/Corbis, (bottom) Charles Smith/Corbis; **296** (top) Stan Salzman, (bottom) JGI/Blend Images/Getty Images; **305** U.K. History/Alamy; **306** (left) Feng Yu/Fotolia, (top right) National Motor Museum/Motoring Picture Library/Alamy, (bottom right) Used with permission of General Mills Marketing, Inc. (GMMI); **312** (left) Diane Macdonald/Stockbyte/Getty Images, (right) Flying Colours Ltd/Photodisc/Getty Images; **313** Neo Edmund/Fotolia; **321** (left) Stockbyte/Getty Images, (right) Beth Anderson/Pearson Education, Inc.; **323** Beth Anderson/Pearson Education, Inc.; **327** Stockbroker/Valueline/PunchStock; **328** Twixx/Fotolia; **329** Edbockstock/Fotolia; **336** Picturequest/JupiterImages; **338** Alan Pappe/Photodisc/Getty Images; **345** Graca Victoria/Shutterstock; **346** Edbockstock/Fotolia; **351** Sly/Fotolia; **358** AP Images; **364** Ryan McVay/Photodisc/Getty Images; **366** Jjava/Fotolia; **374** (left) Doug Menuez/Photodisc/Getty Images, (right) Jack Hollingsworth/Photodisc/Getty Images; **375** (left) Comstock/Thinkstock, (right) David Buffington/Photodisc/Getty Images; **378** Tony Campbell/Fotolia; **380** Steve Byland/Fotolia; **381** Pavel Losevsky/Fotolia; **388** (left) Stan Salzman, (right) Digital Vision/Getty Images; **389** (left) Getty Images/Digital Vision, (right) Yuri Arcurs/Fotolia; **390** Annette Shaff/Shutterstock; **401** (left) Stockbyte/Getty Images, (right) Beth Anderson/Pearson Education, Inc.; **402** (left) Regien Paassen/Shutterstock, (right) Siri Stafford/Digital Vision/Getty Images; **410** (top left) Yuri Arcurs/Fotolia, (top right, bottom left) Beth Anderson/Pearson Education, Inc., (bottom right) Lisa F. Young/Fotolia; **411** (left) Beth Anderson/Pearson Education, Inc., (right) Actionpics/Fotolia; **414** Pete Saloutos/Fotolia; **422** (left) Alexander/Fotolia, (right) Stan Salzman; **423** Okea/Fotolia; **424** Givaga/Fotolia; **431** Stan Salzman; **432** (left) Vac/Fotolia, (right) Sean Locke/Photodisc/Getty Images; **433** Evgenyb/Fotolia; **434** (top left) Monkey Business/Fotolia, (top right) Kameel4u/Shutterstock, (bottom left) U.S. Internal Revenue Service, (bottom right) Valueline/PunchStock/Getty Images; **443** (left) PA Saez/Alamy, (top right) Michael Shake/Fotolia, (bottom right) Marc Romanelli/Photodisc/Getty Images; **450** (left) Stan Salzman, (right) Andresr/Shutterstock; **457** (left) Roza/Fotolia, (right) Gabriel Blaj/Fotolia; **467** (left) Peternile/Fotolia, (right) Kratuanoiy/Fotolia; **471** (top left) Jj2theb/Fotolia, (top right) Stan Salzman, (bottom left) Halfdark/PhotoAlto/Getty Images, (bottom right) Jack Hollingsworth/Digital Vision/Getty Images; **476** Michael Flippo/Fotolia